# Seismic Performance of Corroded Reinforced Concrete Structures Retrofitted with FRP

Dejian Shen

# Seismic Performance of Corroded Reinforced Concrete Structures Retrofitted with FRP

Science Press
Beijing

Springer

Dejian Shen
College of Civil and Transportation
Engineering
Hohai University
Nanjing, China

ISBN 978-981-99-7983-7 ISBN 978-981-99-7984-4 (eBook)
https://doi.org/10.1007/978-981-99-7984-4

Jointly published with Science Press
The print edition is not for sale in China (Mainland). Customers from China (Mainland) please order the print book from: Science Press.



This Springer imprint is published by the registered company Springer Nature Singapore Pte Ltd.
The registered company address is: 152 Beach Road, #21-01/04 Gateway East, Singapore 189721, Singapore

Paper in this product is recyclable.

# Foreword

Reinforced concrete (RC) structures have been widely utilized in civil engineering and hydraulic engineering. However, the carbonation of the concrete cover or excessive chloride penetration of RC structures will lead to reinforcement corrosion. Reinforcement corrosion will deteriorate the safety and durability of RC structures and members such as beams, columns, beam–column joints, and shear walls. Therefore, reliable and effective methods should be proposed to retrofit these corroded RC structures and members. Recently, fiber-reinforced polymer (FRP) has been commonly utilized for strengthening or retrofitting RC members due to the advantages of lightweight, high tensile strength, and corrosion resistance.

In order to explore the influence of reinforcement corrosion on the performance of RC structures and propose practical retrofitting techniques, Prof. Dejian Shen and his team made several innovative and in-depth achievements by combining theoretical research with engineering practice to evaluate the performance and retrofit corroded RC structures with basalt FRP (BFRP) sheets. This monograph introduces the evaluation technique for the performance of corroded RC members, develops the effective anchorage techniques for retrofitting the corroded RC members with BFRP sheets, proposes the models for the load-bearing capacities of corroded RC members and corresponding members retrofitted with BFRP sheets, and explores the bond behavior between BFRP sheets or bars and concrete under static and dynamic loadings. The achievements have been successfully applied in some projects to increase the safety and durability of RC structures, and promote the development of the retrofitting technique of corroded RC structures.

This monograph is a systematic summary of the research of the author and his team on seismic assessment and retrofitting of corroded RC structures during the recent two decades. As a high-level professional monograph, it provides valuable knowledge, useful methods, and practical experience that can be utilized for the retrofitting of RC structures. The research achievements that make up this monograph are well presented and the evaluating and retrofitting techniques for different corroded RC members are clearly clarified. This monograph can not only be used as a reference for teachers and students in university, but also guide the majority of engineering and technical personnel to promote the development of the retrofitting technique of corroded RC structures with FRPs in practice.

October 2023

Xilin Lu
Academician
Chinese Academy of Engineering
Shanghai, China

# Preface

Reinforced concrete (RC) structure is the main structural form and has been widely used in the building engineering, marine engineering, transportation engineering, etc. RC structure permanently exposed to the atmosphere or other situations such as the deicing salt, acid environment, and environmental pollution, or chloride penetration in the marine environment, resulting in the reinforcement corrosion and concrete carbonation. The effective areas and nominal yielding strength of bars decreased since the reinforcing bars corroded; subsequently, the bond behavior between the reinforcing bars and concrete deteriorated. Then, the seismic performance of RC members such as beams, columns, beam-column joints, and shear walls decreased after the reinforcing bars corroded, affecting the durability and service performance of structures. Besides, several methods had been developed to repair or retrofit the corroded RC members, such as the increasing section area, concrete replacement, and bonded steel method. Nowadays, the fibers such as steel fiber or fiber-reinforced polymer (FRP) have been more available for retrofitting or strengthening the RC structures. Due to the benefits of lightweight, high tensile strength, and outstanding corrosion resistance, FRP has attracted the concentrations of scholars. Ordinary FRP composites such as AFRP, GFRP, CFRP, and HFRP have been commonly utilized to retrofit or strengthen the RC beams, columns, beam-column joints, and shear walls. However, the expenses of GFRP and CFRP are extremely high for the engineering. The poor chemical stability, heat resistance, and alkali resistance of GFRP can be found in literatures. Electric conduction of CFRP sheets can be found in literatures, resulting in potential dangers for the practical application in engineering. Furthermore, BFRP considered as a preferable high-performance fiber material has been applied for the sake of high resistance of acid, high tensile stress, low expense, and excellent ductility. Despite the fact that the elastic modulus and failure stress of BFRP are lower than those of CFRP, the tensile strength of BFRP is higher than those of AFRP or GFRP, and better performance found in acidic environments than GFRP. Besides, better bond behavior between BFRP sheet and concrete than that between CFRP sheet and concrete can also be found in literatures. In particular, the RC members strengthened with BFRP exhibit higher ductility than those strengthened with CFRP, which may attribute to the larger ultimate strain of BFRP.

Therefore, BFRP exhibited excellent advantages in the application of engineering. However, systematic investigations on the seismic performance of corroded RC structures retrofitted with BFRP sheets remain lacking. Therefore, this monograph focus on the investigations on the applications of BFRP sheets in retrofitting the corrosion-damaged RC structures.

The financial supports from the National Natural Science Foundation of China (51008113 and 51578215), Bingtuan Science and Technology Program (2023AB016-01), Fundamental Research Funds for the Central Universities (B230201060), Science and Technology Project of Ministry of Housing and Urban Rural Development of the People's Republic of China (2014-K3 -011, 2018-K5 -015, and 2018-K8 -010), Transportation Science and Technology Project of Jiangsu Province (2015T55-1), Project of Natural Science Foundation of Jiangsu Province (BK2010512), and Special Fund for Water Conservation Research in the Public Interest (200701014, 201101014) are gratefully acknowledged.

Dr. Ming Li prepared the draft for Chaps. 1, 5, and 6. Dr. Chuyuan Wen prepared the draft for Chaps. 1, 2, and 7. Dr. Jiaojiao Yuan prepared the draft for Chaps. 1, 2, and 4. Dr. Da Zong prepared the draft for Chaps. 1 and 3. Team members in the author's research group made a great contribution to the preparation of this monograph, including Mr. Zhihao Wang, Mr. Yifan Wei, Mr. Haokang Wang, Mr. Jingwang Yao, Mr. Hanbin Xu. Graduate students in the author's research group contributed a great deal to the research work and completion of this monograph, including Dr. Chengcai Li, Dr. Yang Jiao, Mr. Xingzuo Liu, Mr. Xuan Zeng, Mr. Huafeng Shi, Mr. Xiang Shi, Mr. Yong Ji, Mr. Shucheng Deng, Mr. Qun Yang, Mr. Zhenghua Cui, Mr. Ming Li, Mr. Yunshang Qi, Ms. Tian Tian, Mr. Wei Wang, Mr. Peng Du, Mr. Haifeng Su, Mr. Bin Wang, Mr. Jun Du, Mr. Binod Ojha, Ms. Fenfang Yin, Ms. Qianyan Jiang, Ms. Jiaojiao Yuan, Mr. Congbin Huang, Ms. Jie Yang, Ms. Ling Song, Mr. Sen Qiu, Mr. Jinkun Li, and Mr. Cong Li. Their hard-working, dedication, and intelligence have contributed considerably to the work presented in this monograph.

Nanjing, China
October 2023

Dejian Shen

# Contents

# Abbreviations

| | |
|---|---|
| AFRP | Aramid fiber-reinforced polymer |
| BFRP | Basalt fiber-reinforced polymer |
| CFRP | Carbon fiber-reinforced polymer |
| FE | Finite element |
| FRP | Fiber-reinforced polymer |
| GFRP | Glass fiber-reinforced polymer |
| HFRP | Hybrid fiber-reinforced polymer |
| HPB | Hot-rolled plain steel bar |
| HRB | Hot-rolled ribbed steel bar |
| IAE | Integral absolute error |
| LVDT | Linear variable differential transducer |
| RC | Reinforced concrete |

# Chapter 1
# Introduction

## 1.1 Corrosion and Retrofitting of Structures

### *1.1.1 Seismic Performance of Corrosion-Damaged Structures*

RC structure is the main structural form and has been widely used in the building engineering, marine engineering, transportation engineering, etc. The durability and seismic performances of RC structure have always been a hot topic [1–5]. However, the reinforcement corrosion is one of the main reasons for the deterioration of the RC structure. According to relevant reports, nearly 50% of China's existing houses entered the aging stage, and nearly 250 billion square meters of buildings had to face durability issues [6]. According to the NIST report [7], the United States lost nearly $300 billion annually due to corrosion of reinforcing bars. In Japan, the annual cost of only maintaining housing structures has reached 40 billion yen, and about 21.4% of the damage to RC structures is caused by corrosion of reinforcing bars. In China, the direct economic loss caused by the corrosion is 500 billion yuan per year, and the loss of concrete structure damage caused by the reinforcement corrosion is 100 billion yuan per year [8]. Therefore, the reinforcement corrosion has aroused relevant attentions.

Reinforcement corrosion is mainly caused by the exposure to atmosphere or other situations such as the deicing salt, acid environment, and environmental pollution, or chloride penetration in the marine environment [9]. The effective areas and the nominal yielding strength of reinforcing bars decreased since the reinforcing bars corroded. Results reported in [10] indicate that the main reason for the degradation of bearing capacity of RC member is that the corrosion of reinforcing bars contributes to the biaxial tensile and compressive stresses in the concrete in the compression zone. Corrosion of reinforcing bars not only leads to loss of cross-sectional area, but also leads to degradation of mechanical properties of reinforcing bars. Results reported in [11] find that when the section loss rate of reinforcing bars is less than 5%, the mechanical properties of corroded reinforcing bars are the same as those



D. Shen, *Seismic Performance of Corroded Reinforced Concrete Structures Retrofitted with FRP*, https://doi.org/10.1007/978-981-99-7984-4_1

of the base metal, indicating that small corrosion rates have no impact on structural performance. When the loss rate of reinforcing bars is between 5% and 10%, the yielding strength, tensile strength, and elongation decrease. When the slope loss rate of the reinforcing bars is between 10% and 60% and the reinforcing bar is severely pitted, the yielding point is no longer obvious, and the actual strength of reinforcing bars decreases significantly. The non-uniformity and dispersion, as well as the losing strength of reinforcing bars increase with the increasing of corrosion rates. Results reported in [12] indicate that the ductility of reinforcing bars decreases significantly with the increasing of the corrosion rate, and brittle failure of reinforcing bars occurs when the corrosion rate is greater than 12%. Results reported in [13] indicate the mechanical properties of 123 corroded reinforcing bars in atmospheric condition, indicating that the significant linear relationship between yielding load, ultimate load, and ultimate elongation of corroded reinforcing bars and corrosion rate can be found. Besides, a three-dimensional ellipsoidal model of corroded reinforcing bars and a method for measuring the corrosion rate of reinforcing bars are proposed.

Besides, corrosion of reinforcing bars not only reduced the mechanical properties and effective cross-sectional area of the reinforcing bars, but also changed the bond behavior between the reinforcing bars and concrete. Results reported in [14] conduct the experimental tests on the performance of corroded specimens by the pull-out tests, indicating that the bond strength between the reinforcing bars and concrete slightly increases when the corrosion rate of reinforcement is less than 1%. As the corrosion rate of reinforcing bars increases, the bond strength between reinforcing bars and concrete begins to decrease. Results reported in [15] find that the average bond stress decreases rapidly by 9% and 92% when the corrosion rates are 4% and 17.5%, respectively, indicating that transverse ribs of reinforcing bars are severely rusted, and the bond strength between reinforcing bars and concrete decreases when the reinforcing bars significantly corroded. Results reported in [16] indicate that the bond behavior between reinforcing bars and concrete significantly degrades after the reinforcing bars corroded, and the degradation is mainly related to the corrosion rate, the diameter, and the thickness of concrete protective layer.

Furthermore, since the reinforcing bars corroded, the load-bearing capacity and safeness of RC members would also decrease significantly. Results reported in [17] find that the ununiformity of corrosion and rust pits can be observed on the corroded RC beams. When there is a coupling effect of load and corrosion on the beam, the deflection of the beam will increase with the increase of corrosion rate, and with the increase of corrosion rate, the failure mode of RC beams will change to bending failure accompanied by bond splitting failure. For beams with cracks, the bearing capacity decreases significantly when the corrosion rate is low, indicating that the reduction in bearing capacity is related to the stress concentration caused by uneven corrosion of the reinforcement. Results reported in [18] find that the shear resistance of beams decrease after the stirrups corroded, contributing to the brittle failure of beams. The degradation of the bond behavior between corroded stirrups and concrete results in a significant nonlinear relationship between the bending moment and deflection in the beam span. Moreover, the section ductility and stiffness of severely corroded beams will decrease as the corrosion rate increases, and a method for predicting the bearing capacity of corroded beams is proposed. Results reported in [19] indicate that the energy dissipation and load-bearing capacities of corroded beam-column joints apparently decrease after reinforcing bars corroded, which results in the brittle failure mode.

Furthermore, the mechanical property of the reinforcing bars, bond behavior between reinforcing bars and concrete, as well as the performance of RC beams, columns, beam-column joints, and shear walls significantly decreased after the reinforcing bars corroded. Therefore, investigations on the seismic performance of corroded RC members retrofitted with certain methods are necessary.

### *1.1.2 Retrofitting of Structures*

Buildings are one of the important disaster bearers of earthquake disasters. The increasing earthquake casualties and economic losses are mainly attributed by damage, failure, destruction, and collapse of the disaster bearers caused by strong earthquake action. Therefore, the seismic resistance of building structures is the key to ensure structural safety and use functions [20]. As an important component of the frame, RC beam-column joints are also weak links in the structure, and their seismic performance is the top priority in design. Due to the frequent occurrence of earthquake disasters, changes in temperature environment, construction factors, design changes, changes in use functions, and other reasons in recent years, the seismic performance of existing RC members cannot meet the seismic fortification standards, and it is necessary to repair and strengthen them [21].

Presently, the traditional retrofitting methods of concrete structures mainly include the followings [22]: increasing section method, replacing concrete method, external prestressing method, external bonding steel method, bonding steel plate method, bonding fiber-reinforced composite material method, and so on. Therefore, scholars have conducted extensive researches on repairing and retrofitting techniques.

Recently, FRP has been commonly utilized for strengthening or retrofitting the RC members owing to the advantages of light weight, high tensile strength, corrosion resistance, and easy installation [23–26]. Although conventional FRP composites, such as the CFRP [27–29], GFRP [30], AFRP [31], and HFRP [32], have been applied to strengthen the RC members, a few deficiencies cannot be avoided. As the most expensive composite among those mentioned above, the CFRP exhibited the highest strength and elastic modulus; however, the ultimate tensile strength could not be obtained attributed to the debonding instead of rupturing. The GFRP exhibited relatively weak performance in the properties of thermal stability, heat resistance, and alkali resistance [33, 34]. In addition, as a new type of inorganic eco-friendly high-performance fiber material, the BFRP with the advantages of high tensile stress, acid resistance, and excellent ductility is also commonly applied [35–37]. A higher ductility than that of CFRP makes the BFRP exhibit a larger ultimate strain, which is beneficial for RC members to maintain the ductility under the earthquake [38, 39]. Presently, investigations on the bond behavior of concrete members reinforced by BFRP bars under the static load [36, 37], dynamic load [24], cyclic load [40], the bond stress-slip relationship between BFRP sheets and concrete [41, 42], as well as the effective bond length of the BFRP-concrete interface [43] have been conducted. Besides, utilizing the BFRP to strengthen the RC beams and shear walls was preliminarily studied as well. Results reported in [44] indicate that the BFRP sheets can bear moment prior to the steel plate, which is also utilized to strengthen the pre-cracked RC beams, and completely exhibit the excellent tensile performance. Results reported in [33] indicate that the load-bearing capacity and ductility of RC shear walls are significantly improved after strengthening with BFRP strips. Furthermore, results reported in [34] also indicate that the failure mode of unreinforced masonry walls is altered from the shear failure to flexural failure, and the shear capacities are significantly improved with the application of BFRP. Although investigations on the seismic performance of RC members strengthened with BFRP have been studied, investigations on the comparisons of RC members strengthened with different methods are relatively limited. Therefore, further investigations on the RC members strengthened or retrofitted with BFRP for better understanding the efficiency of BFRP are needed.

## 1.2 State-of-the-Art of Retrofitting of Structures

### *1.2.1 Bond Stress-Slip Relationship Between FRP Sheets and Concrete*

Factors such as environmental attack, fatigue damage, and lacking of appropriate maintenance have a negative influence on the service life of civil infrastructures [33]. The strengthening and rehabilitation of these structures proved to be costly. Accordingly, FRP sheets, which are less demolition of damaged RC with lower related costs, have been widely used to strengthen damaged RC [45, 46]. Determining the bond behavior between the FRP sheets and concrete is an important issue because the typical failure mode between the FRP sheets and concrete is debonding. When cracks are developed in concrete, FRP debonding is triggered by the stress concentration [47], which is considered an early failure mode of rehabilitated RC structures [48]. The bond behavior between FRP and concrete is mainly affected by the stiffness of FRP [49–52], the mechanical and physical properties of concrete [53–57], the thickness of the adhesive [35], and bond length [39]. In addition, numerous methods have been suggested to predict the effective bond length [58–61], the failure load [62–64], and the local shear stress-slip relationship between FRP sheets and concrete [65, 66].

Although BFRP sheets are available to be used in damaged RC, the bond behavior between BFRP sheets and concrete is still lacking. Three methods can be used for evaluating the bond behavior between FRP sheets and concrete: (1) FE analyses

[67]; (2) shear tests with strain gauges attached to the FRP sheet [68]; and (3) local bond-slip law [69, 70]. The second method was used in this book.

The force induced in the FRP sheets is transmitted to concrete mainly through bond stress in the adhesive. However, the transfer length of bond stress is limited, and the effective bond length is defined as the bond length beyond which no further increase of failure load can be achieved [59]. Then, extending the bond length to raise the failure load cannot be proved beneficial when it is beyond the effective bond length. Three methods can be used to assess the effective bond length: (1) the failure load [71, 72]; (2) the bond length obtained from the distance between 10% of the maximum bond stress [35, 57]; and (3) length over which strain decreases from the maximum value to 0 [39, 51]. In the first method, the range of the effective bonding length could be achieved. In the second or third method, the accurate value of the effective bonding length could be achieved.

RC structures in civil engineering are exposed to dynamic loadings. Moreover, concrete has a higher dynamic strength and fracture toughness than its corresponding static values [73, 74]. Numerous investigations on RC structures retrofitted with FRP under dynamic loadings have been conducted. Through experiments and models, the blast resistance of structures externally reinforced with FRP has also been studied [75, 76]. By regression analysis on bond data, it was discovered that the bond between FRP and concrete grows logarithmically with strain rate [77]. The effective length of the interface between FRP sheet and concrete will shorten as the strain rate increases and the interfacial fracture energy of FRP and concrete increases with strain rate as a logarithmic function. To date, the models for estimating the dynamic effective bond length of the interface between FRP and concrete under dynamic loadings at various strain rates are lacking.

### *1.2.2 Behavior of Beams Retrofitted with FRP*

Cracking is normal in RC structures [78–81], and local cracks that result from overloading, the flaws of the construction technology, and other causes inevitably occur during the design life of RC structures. The load-bearing capacity of RC structures decreases after years of service because of the cracking of concrete and the reinforcement corrosion [82]. The occurrence of cracks in a structure affects its dynamic characteristics, such as natural frequency, stiffness, damping properties, and mode shape [83], and even adversely affects the durability and security of the structure [84]. The natural frequencies of cracked beams can provide insight into the extent of damage, and the research has been conducted to develop vibration-based inspection techniques for a wide range of applications [84]. Damage detection and structural performance monitoring, especially for the case of crack detection in RC structures, have been conducted on the basis of this effect [85]. Data on the performance of old RC beams with cracks is required to offer a reference for detection and repair [86].

The natural frequency of cracked beams has raised considerable research interest in recent years [87]. Natural frequency is a global property of structures, in which any change signifies a change in characteristics [78]. Monitoring the performance of RC beams on the basis of the vibration response is reliable and economical [88]. Many experimental modal tests have been conducted to detect damage. The frequency response obtained with an appropriate signal processing technique such as Hilbert-Huang transform is used to infer the performance of monitored structures [78]. The damage detection schemes used in [85] depended on the measured changes in the first three natural frequencies and the corresponding amplitudes of the measured acceleration frequency response functions. Various methods based on the measurement of natural frequencies have been proposed to detect the location and size of a crack in RC beams. The point of intersection of the three curves gives the crack location, and the crack size is then computed using the standard relation between stiffness and crack size [89]. A method is presented which uses the modal parameters of the lower modes for the non-destructive detection and sizing of cracks in beams [90, 91]. A simple and easy non-destructive evaluation procedure is presented for identifying a crack in a beam type structure using the lowest four natural frequencies test data [92]. The results of the forced responses of the free end point for a cracked cantilever beam are also shown to present the crack effects for crack extents and crack locations [93]. Several methods consider RC beams as Euler-Bernoulli beams [94–96], while the effects of shear deformation and rotational inertia are considered through Timoshenko beam theory in other methods [97–99]. Meanwhile, several techniques for the calculation of natural frequencies of RC beams with a number of open cracks have been proposed. The natural frequencies of a cracked beam are evaluated by representing cracks as massless springs and using a continuous mathematical model of the beam in transverse vibration [87]. The natural frequencies and mode shapes of a cracked beam are obtained using the FE method [100]. The FE and component mode synthesis methods are used together to analyze the free vibration [101]. The forced response is determined by employing modal series expansion technique [102]. Although the model for the frequency of RC beams with one or two flexural cracks has been proposed [88], the method for calculation of the frequency of RC beams with numerous cracks remains lacking.

Deflection is often utilized to evaluate beam performance, which is affected by stiffness. Stiffness reduction is an important indication of crack development. The full-scale loading test is effective in evaluating the performance of an old beam [86]. Destructive tests on 40-year-old girders are conducted to determine their ultimate load-bearing capacities and deflections [103, 104]. Full-scale tests on 30-year-old girders at varying shear span-to-depth ratios are conducted to evaluate their behavior, failure mode, and capacity [105]. Tests on the load-bearing capacity and deflection of a severely distressed and deteriorated 50-year-old RC box beam of bridge are conducted [106]. Full-scale tests on old RC box beams are rare because of the high cost of performing the tests and the limited availability of existing deteriorated beams for testing [86]. The test on the deflection of RC beams with damage has also been rarely investigated. Severe damages on the beam indicates high loss of stiffness [107], and stiffness change affects the deformability of RC beams [108]. The effective moment of inertia for the short-term deflection of cracked non-prestressed and prestressed concrete members is proposed [109], which is called Branson's method and is recommended by several standards [110, 111]. The deflection of cracked RC beams under long-term has also been investigated, and test results show that the overall stiffness of beams is substantially reduced after six months under sustained loading [112].

Stiffness is a significant parameter of the structural characteristics [113]. Dynamic tests have validated that the natural frequency is useful for detecting damage and stiffness change in RC beams [114]. However, the dynamic behavior of beams with initial flexural cracks strengthened with FRP sheets has not been investigated thoroughly [115]. Results reported in [116, 117] investigated the effectiveness of modal testing in assessing the variations of overall stiffness of RC beams extensively cracked and subsequently repaired with FRP sheets. Zanardo et al. conduct vibration tests and modal analysis on a bridge structure before and after retrofitting with FRP and determine that the upgrading works resulted in an increase of 26% to 32% in the experimental flexural stiffness, respectively [113].

Favorable failure mode is the flexural failure due to either concrete crushing or rupture of tensile reinforcement. Adequate precautions should be taken to postpone the other two failure modes: shear failure and FRP debonding [118]. The most available methods in calculating load-bearing capacity of strengthening RC beams have not considered the influence of end anchorage. Common strengthening method without end anchorage may lead to premature failure. Some studies are conducted to investigate the premature failure of FRP systems. Results in [60] show that interfacial stresses between concrete and reinforcing FPR may produce a sudden and premature failure and a simplified procedure for verifying the interfacial stress state. A cohesive zone model is adopted in a FE code for simulating the debonding failure in composite structures [119]. Results in [120] show that surface preparation prior to bonding of FRP sheets increases ultimate rupture strength and can postpone the premature failure. End anchorage is an important method that can reduce the probability of FRP debonding and has been recently suggested for prevention of the premature FRP debonding [121]. End anchorage improves the performance of the strengthened beams, particularly increasing the load-bearing capacity and cracking characteristics, and also changes the failure modes [122]. There are different methods for end anchorage, the most common being the use of FRP U-strips, FRP fan anchors, steel plates with bolts [123], anchors bolts or fasteners [124], and steel clamps [125]. In addition, there are no design codes or guides for the calculation for BFRP-strengthened beams anchored with new methods [126]. In most cases, orientation of U-strips transverse FRP composites perpendicular to the beam longitudinal axis can meet the general requirements for shear strengthening [127]. Results in [128] show that the RC beam shear-strengthened with FRP strips can significantly increase the shear capacity. Placement of U-strips nearby the end of beam also prevents the concrete cover delamination, which in turn postpones the FRP debonding. However, in previous studies, the debonding failure is still appeared to be almost inevitable in the presence of U-strips [129, 130]. Grooving is widely utilized in near-surface mounted method. The steel reinforcement is embedded in grooves cut onto the surface of the beam to be strengthened and filled with an appropriate binding agent such as epoxy paste or cement grout [131]. Results in [132] show that grooving as near-surface mounted method can increase the load-bearing capacities of RC beams. Results in [120] show that grooving as an alternative method of surface preparation is able to postponed FRP debonding in RC beams.

The behavior of RC beams strengthened with FRP strips is significantly influenced by cracks. For RC beams strengthened with FRP strips in tension face, intermediate crack-induced debonding may arise as a major flexural crack or flexural-shear crack [133]. Externally bonded CFRP strips can achieve adequate repair on corroded and cracked RC beams [134]. CFRP or GFRP sheets placed perpendicular to the shear crack significantly strengthen RC beams [135, 136]. Standards corresponding to externally bonded FRP systems to strengthen concrete structures have been published [137, 138]. Most investigations have been conducted to focus on undamaged or uncracked beams [139]. Seven RC beams were poured in advance in [140] to study the failure modes of RC beams strengthened in shear with CFRP strips. Six RC beams were cast in place in [141] to analyze the RC beams strengthened in flexure using CFRP laminates under sustaining load. The stress-strain relationship of FRP-strengthened beams without initial cracks is studied in [142], and the results show that the load history can be used to define a design criterion for predicting the safe load levels.

The flexural failure of structures is a favorable failure mode, which may be caused by the crush of concrete or the rupture of tensile bar [143]. FRP debonding and shear failure should be postponed through adequate precautions as other two failure modes [144]. The application of end anchorages improves the load-bearing capacity and cracking characteristics of strengthened beams, and the failure modes may be changed [120]. Various methods have been proposed to meet the requirement of the anchorage, among which the FRP U-strips are the most common ones [144]. However, the effect of anchorage is not considered in most existing calculating methods when evaluating the load-bearing capacity of FRP-strengthened beams [145]. This condition may lead to premature failure of the beams, which is a hot issue and has been investigated in recent years [60]. The results demonstrated in [128] show that the concrete cover delamination will be prevented when the FRP U-strips are placed nearby the ends of FRP-strengthened RC beams, and the FRP debonding will be postponed under this condition. However, several investigations show that the debonding failure still appears when the FRP U-strips are utilized in the strengthened RC beams [146]. Steel plates have been utilized widely to strengthen the RC beams with FRP composites in recent years and proved to be effective and valuable [147]. The steel plate anchorage refers to the application of steel plates with anchor bolts to fix the FRP composites on the concrete structures and members [42]. The steel plates can be placed along the entire length of the FRP strips, which is called full-length mechanical anchoring or only at both ends of the FRP sheet, which is called end mechanical anchoring. The results demonstrated in [148] show that the application of steel plates as well as externally bonded FRP can prevent the premature FRP debonding.

### 1.2.3 Seismic Performance of Columns Retrofitted with FRP

Many coastal zones are vulnerable to encountering earthquakes. RC columns in these areas are vulnerable to the combined effects of tremors and corrosion. Corroded RC columns are anticipated to function seismically much worse than their intended performance [149, 150]. Characterizing the seismic performance of corroded RC columns has been the subject of studies over the past 20 years [151–154]. Higher corrosion levels and axial loads produce less stable hysteretic loops with more extreme stiffness and strength degradations as well as worse ductility. Six rectangular cast-in-place RC columns of varying corrosion levels are studied in [155], which identifies two main effects of structural performance degradation: concrete cover crack and diminished mechanical properties of corroded reinforcing bars. Four columns with varying degrees of corrosion are subjected to cyclic pseudo-static tests in [156], which concludes that corrosion has a substantial impact on load-bearing capacity.

In order to solve the corrosion of reinforcing bars, the traditional method is to repour the concrete cover. Damaged concrete covers and rust from corroded reinforcing bars are normally removed from the degraded columns to restore them. Next, cleaning reinforcing bar is covered with a corrosion-resistant coating layer. And finally, a fresh layer of concrete is added to the columns. This conventional procedure is time-consuming, needs highly trained labor, and is particularly less effective than employing FRP strengthening techniques [157–162]. FRP has great corrosion resistance, high tensile strength, and lightweight and is easy to fabricate. Much research on the structural behavior (capacity, stiffness, and deformation) of FRP retrofitting RC columns with virgin or corroded reinforcing bars has been systematically studied since the 1990s [163–169]. The ductility of the columns will reduce even at low levels of corrosion, and the retrofitting measure with CFRP can increase the ductility but will not increase the lateral strength in [170]. Above all, it is obvious that corrosion attributed by chloride could result in a deterioration in the seismic performance of columns. These studies have demonstrated that using FRP sheets is very efficient in increasing the load-bearing capacity of columns and preventing harmful attacks from outside sources, such as water, chloride ions, and sulfate ions. This reduces the corrosion of bar and increases the service life of RC structures. Although seismic performance of corroded concrete columns and corroded RC columns retrofitted with FRP has been studied, the seismic performance of corroded RC columns retrofitted with BFRP is still lacking.

### 1.2.4 Seismic Performance of Beam-Column Joints Retrofitted with FRP

The beam-column joint designed as the critical region of RC structures suffers high shear stresses and large shear deformations during earthquake, which may result in the brittle failure after the reinforcing bars corroded [19]. Results reported in

[171] indicate that the damaged characteristics of corroded beam-column joints can be evaluated by the acoustic emission technique, and the initial diagonal cracks of corroded specimens occur earlier than those of reference specimen. Tendency of b-values obtained from the acoustic emission of corroded specimens is more stable than that of reference specimen, which indicates that more microcracks and the reappearance of old cracks occur with the decrease of the load-bearing capacity. Indeed, the load-bearing capacity of corroded RC beams [145], columns [172], and shear walls [46] can be increased by repairing with BFRP and together with the seismic behavior of pre-damaged beam-column joints [173] can also be improved.

The seismic behavior of RC members also decreased after the earthquake damage, especially the beam-column joints, while investigations on the damaged joints retrofitted with BFRP were still limited. In recent times, frequent earthquakes have provoked considerable damages to RC structures or buildings [174]. These buildings, built in the 1960s and 1970s, are insufficiently detailed and have some deficiencies, such as deficient structural system or seismic design, low concrete strength, and unsatisfiable meeting requirements, which will cause those buildings to vulnerably experience considerable damages under the earthquake [175]. As mentioned, the beam-column joints are the momentous parts and suffer severely inelastic deformations in the earthquake [176–178]. Phenomena from the 1999 Kocaeli and Chi-Chi earthquakes exhibit brittle failures in the joints, which may lead to the collapse of whole buildings [179]. Cracks are significantly found at the surfaces of the column with the yielded longitudinal reinforcements of the beams penetrating the core area, which contributes to the degradation of the strength and ductility of joints [180]. Although BFRP and other FRPs sheets have been applied in retrofitting the seismic-damaged beams [181, 182] and columns [183, 184], assessments on the seismic-damaged beam-column joints retrofitted with BFRP are still limited.

### *1.2.5 Seismic Performance of Shear Walls Retrofitted with FRP*

Investigations on the behavior of RC members with externally bonded FRP sheets have mainly concentrated on either CFRP or GFRP in last decades. Two key issues had a considerable influence on seismic performance of strengthened walls, involving the configuration of FRP strips and anchor system between FRP strips and concrete surface. X-shaped, U-shaped, lateral and parallel or combination of them are widely adopted in [185–189]. Results reported in [189] show that the best performance for the improvement of lateral displacement resistance capacity and load-bearing capacity of RC shear walls has been obtained from the strengthening with lateral strips. At present, epoxy was used to bond FRP materials to the exterior surface of concrete; however, it was commonly reported that the FRP strips debond from the surface prior to fracture under cyclic loads. Despite FRPs' high ultimate strength, structures strengthened with FRPs could be failed in a brittle manner potentially below its mechanical properties due to a bonding problem at the surface between FRP sheets and concrete. In this case, the FRP materials did not play the most important role in strengthening RC structures and it was a waste of material to some extent. To enhance the bond behavior between FRP materials and concrete effectively, anchor systems including steel plate anchor [186], U-shaped FRP and bonded metal anchor [187, 188], fan type anchor [189], and improved hybrid bonded FRP anchor [190] are widely used. Experimental results show that as results of prevention of debonding of FRP strips totally by the anchors, the tensile forces of the strips are provided the load-bearing capacity to continue till reaching considerable lateral displacement [189]. The flexibility of FRP materials has been fully studied.

Investigations on BFRP in retrofitting RC beams and columns have been conducted; however, investigations on seismic performance of corroded RC shear walls after repairing with BFRP sheets are comparatively limited.

### *1.2.6 Bond Behavior Between FRP Bars and Concrete*

The bond strength between FRP bars and concrete is influenced by bar diameter [191, 192]. The bond strength between concrete and GFRP bars with diameter of 12.7 and 19.1 mm is experimentally investigated using a pull-out test [191], which shows that the bond strength decreases with the increase of bar diameter. A similar test on the bond strength between CFRP, GFRP, AFRP bars with different diameters and concrete is conducted in [192]. Although the effect of bar diameter on the bond strength between CFRP, GFRP, and AFRP bars and concrete has been investigated, the effect of bar diameter on the bond strength between BFRP bars and concrete remains lacking. Thus, whether and how the bar diameter influences the bond strength between BFRP bars and concrete need to be studied for better understanding the bond performance. The bond strength between FRP bars and concrete is also influenced by concrete strength [193]. Experimental results in [194] show that the bond strength between FRP bars and concrete is directly proportional to the concrete strength up to 20 MPa because bond failure is entirely due to the concrete crushing for low concrete strength. However, the effect of concrete strength on the bond strength and bond failure between BFRP bars and concrete remains lacking. Thus, the investigation on the effect of concrete strength on the bond strength and bond failure between BFRP bars and concrete needs to be studied further. The relationship between FRP bars and concrete is necessary to evaluate the load-bearing capacity of concrete components or structures reinforced with FRP bars. Experimental and theoretical investigations on the bond performance between FRP bars and concrete have been conducted in [192, 195], which suggest that the geometrical and mechanical parameters of FRP bars influence the bond behavior, and sometimes various parameters cause the complexity in determining the actual bond stress-slip relationship between FRP bars and concrete. The bond stress-slip relationships between FRP bars and concrete are presented based on range of slip [196]. Another bond stress-slip relationship for recycle aggregate concrete and bars was proposed in consideration of maximum bond stress and corresponding slip [197]. Although some models for bond stress-slip relationship between some FRP bars or reinforcing bars and concrete are proposed, the model for BFRP bars remains lacking. Thus, the model for bond stress-slip relationship between BFRP bars and concrete needs to be studied.

Dynamic loading, such as earthquake, blast, traffic loading, wind loading, and machinery, is inevitable for concrete structures during their service life [198–200]. The bond behavior of concrete members reinforced with FRP bars under dynamic loading is essential for evaluating the load-bearing capacity [201]. The strain rate is a factor that influences the bond strength of concrete members reinforced with FRP bars. Results in [202] show that the stiffness of stress-strain relationship curves of concrete in direct compression and tension increases as the strain rates increase. Results in [203] show that the strain rate has a pronounced influence on the bond strength of concrete members reinforced with deformed bars, and the higher the strain rate is, the greater the bond strength and the bond stiffness are. Result in [204] shows that the bond strength of GFRP sheet-concrete interface under dynamic loading is larger than the corresponding static value. However, investigations on the influence of strain rate on the bond strength of concrete members reinforced with BFRP bars are still lacking. The bond behavior of RC members is also determined by the slip corresponding to bond strength. A constant value is utilized for evaluating the slip corresponding to bond strength [205], which compares reasonably well with the results reported in [206]. However, a variable value 0.04 times bar diameter is utilized as the slip corresponding to bond strength [207]. Results in [208, 209] show that the value of the slip corresponding to bond strength is related to the clear distance between the lugs of reinforcing bars. However, investigation on the influence of strain rate on the slip corresponding to bond strength of concrete members reinforced with BFRP bars is still lacking. The bond stress-slip relationship model of concrete members reinforced with BFRP bars considering strain rate is of fundamental importance to better understand the bond behavior of concrete members reinforced with BFRP bars under dynamic loading. Thus, investigations on the bond stress-slip relationship of concrete members reinforced with BFRP bars under different strain rates need to be conducted.

Under severe seismic excitation, cyclic loading can cause bond deterioration between reinforcing bars and concrete [210] while the hysteretic behavior of concrete structures is highly dependent on the bond behavior between reinforcing bars and concrete. Investigations on the bond behavior between concrete and steel bars [205, 211–213], GFRP bars [214, 215], CFRP bars [215, 216], and FRP sheets [185] have been conducted under cyclic loading. The results in [217] show that the bond stress-slip relationship under cyclic loading between reinforcing bars and concrete is similar to that under monotonic loading while the slip corresponding to bond strength under cyclic loading is larger than that under monotonic loading. The results in [216] show that the bond strength of CFRP bars under cyclic loading is lower than that under monotonic loading. The results in [211] show that the slip at the loaded end and residual slip increase with the increase in the number of cycles, and the bond strength is not affected by repeated loading, providing that bond failure does not occur. The results in [212] show that bond behavior between reinforcing bars and concrete deteriorates gradually with the increase of number of cycles, and the deteriorative speed slows down with the increase of number of cycles. Although the bond behavior between concrete and reinforcing bars, GFRP bars, CFRP bars, and FRP sheets under cyclic loading has been investigated, the bond behavior between BFRP bars and concrete under cyclic loading remains lacking. Therefore, the effect of cyclic loading on the bond strength, the slip corresponding to bond strength, and the hysteretic curve area between BFRP bars and concrete need to be investigated to further understand the bond behavior.

## 1.3 Objectives and Scope

The authors presented systematically their experimental and theoretical researches on the seismic performance of corroded RC structures retrofitted with FRP in recent years in this monograph. Seven chapters were organized in this monograph, and the main contents were introduced briefly as follows.

Chapter 2 investigated the bond behavior between BFRP sheets and concrete under static, dynamic, and initial static loading. The dynamic effective bond length, dynamic bond stress, dynamic ultimate load, and dynamic local bond-slip relationship between BFRP sheets and concrete under different strain rates or initial static loading proportion were studied. Besides, the models for the dynamic ultimate load, bond stress, slip, effective bond length, and the bond stress-slip relationship between BFRP sheets and concrete were proposed in consideration of the strain rate or initial static loading.

Chapter 3 investigated the fundamental natural frequency of deteriorated RC box beam with initial flexural cracks and performance of deteriorated RC box beams retrofitted with BFRP sheets by three types of anchorage, such as U-strips anchorage, end anchorage with grooving, and steel plate anchorage. The failure mode, ductility, stiffness, load-bearing capacity, and fundamental natural frequency of deteriorated RC box beams without retrofitting or retrofitted with BFRP sheets were analyzed.

Chapter 4 investigated the seismic performance of corroded RC columns without retrofitting or retrofitted with BFRP sheets. The failure mode, stiffness, displacement ductility ratios, and cumulative energy dissipation of corroded RC columns without retrofitting or retrofitted with BFRP sheets were discussed. Besides, the corresponding results were also compared with those of corroded RC columns.

Chapter 5 investigated the seismic performance of corroded RC beam-column joints, corroded joints retrofitted with BFRPs, earthquake-damaged and corrosion-damaged joints retrofitted with BFRPs, uncorroded joints strengthened with BFRPs in five types. The failure mode, load-bearing capacity, stiffness, ultimate rotations, displacement ductility ratios, and cumulative energy dissipation were analyzed. The models for shear strength of core area of joints in consideration of corrosion rate, earthquake damage, and FRP confinement were proposed.

Chapter 6 investigated the seismic performance of corroded RC shear walls, corroded walls retrofitted with BFRPs, and uncorroded walls strengthened with

BFRPs in fan anchor technique. The failure mode, yielding strength, ultimate shear strength, and cumulative energy dissipation were discussed. The models for the ultimate shear strength of corroded RC shear walls without retrofitting or retrofitted with BFRP sheets based on Strut-and-Tie model were proposed in consideration of corrosion rate as well as FRP confinement.

Chapter 7 investigated the bond behavior between BFRP bars and concrete under static, dynamic, and cyclic loadings. Firstly, the bond stress-slip relationship between BFRP bars and concrete under static loading was investigated considering the ratio of concrete cover depth to diameter of BFRP bars and concrete cubic compressive strength. The bond strength and the slip corresponding to the bond strength were investigated considering the BFRP bar diameter, and the mechanisms of the bond failure between BFRP bars and concrete were revealed. Secondly, the bond behavior between concrete and BFRP bars under dynamic loadings was investigated. The models for the bond stress-slip relationship, bond strength, and the slip corresponding to the bond strength were proposed in consideration of the strain rate. Thirdly, the bond behavior between BFRP bars and concrete under cyclic loading was investigated, and the models for the bond strength, slip corresponding to bond strength, and hysteretic curve area were proposed considering the number of cycles.

## References

1. Lu XL, Wu DY, Zhou Y (2019) State-of-the-art of earthquake resilient structures. J Build Struct 40(02):1–15 (in Chinese)
2. Lu XL, Zhu QY (2019) Shaking table test of earthquake-damaged recycled aggregate concrete frame retrofitted with steel dampers. J Tongji Univ (Nat Sci) 47(07):914–924 (in Chinese)
3. Lu XL, Wu XH (2000) Study on a new shear wall system with shaking table test and finite element analysis. Earthquake Eng Struct Dynam 29(10):1425–1440
4. Lu XL, Jiang C, Yang B et al (2019) Seismic design methodology for self-centering reinforced concrete frames. Soil Dyn Earthq Eng 119:358–374
5. Lu XL (2014) Precast concrete structures in the future. Struct Concr 15(1):1–2
6. Shen JM, Zhou XY, Gao XW et al (2000) Aseismic engineering. China Architecture and Building Press, Beijing (in Chinese)
7. Behshid BNA (2001) A comparative study of deterioration of bond due to corrosion in different concretes. McGill University, Canada
8. Sun W (2006) Durability and service life of structure concrete under load and environment coupling effects. J SE Univ (Nat Sci Ed) S2:7–14 (in Chinese)
9. Rajput AS, Sharma UK, Engineer K (2018) Seismic retrofitting of corroded RC columns using advanced composite materials. Eng Struct 181:35–46
10. Capozucca R, Cerri MN (2000) Identification of damage in reinforced concrete beams subjected to corrosion. Struct J 97(6):902–909
11. Hui YL, Lin ZS, Li R (1997) Experimental study and analysis on the property of corroded rebar. Ind Constr 27(6):10–13 (in Chinese)
12. Almusallam AA (2001) Effect of degree of corrosion on the properties of reinforcing steel bars. Constr Build Mater 15(8):361–368
13. Shen DJ, Wu SX (2004) Experimental study on the performance and corrosion model of corroded bars in concrete under the atmospheric environment. China Concr Cem Prod 03:46–50 (in Chinese)

14. Al-Sulaimani G, Kaleemullah M, Basunbul I (1990) Influence of corrosion and cracking on bond behavior and strength of reinforced concrete members. Struct J 87(2):220–231
15. Amleh L, Mirza S (1999) Corrosion influence on bond between steel and concrete. Struct J 96(3):415–423
16. Yuan YS, Yu S, Jia FP (1999) Deterioration of bond behavior of corroded reinforced concrete. Ind Constr 11:47–50 (in Chinese)
17. He SQ, Wang HC, Gong JX (2007) Study on flexural experiment of reinforced concrete beams under simultaneous service loading and corrosion. J Hydroelectric Eng 6:46–51 (in Chinese)
18. Shen DJ, Wu SX (2007) Experimental study of mechanical properties of severely corroded reinforced concrete beams in atmospheric environment. Bridge Constr 01:28–31 (in Chinese)
19. Kanchanadevi A, Ramanjaneyulu K (2018) Effect of corrosion damage on seismic behaviour of existing reinforced concrete beam-column sub-assemblages. Eng Struct 174:601–617
20. Budelmann H, Holst A, Wachsmann A (2012) Durability related life-cycle assessment of concrete structures: mechanisms, models, implementation. In: Life-cycle and sustainability of civil infrastructure systems—proceedings of the 3rd international symposium on life-cycle civil engineering, 2012, pp 75–86
21. Lu XL (2010) Retrofitting design of building structures. CRC Press
22. Bu LT, Zhou XQ (2009) Reliability appraisal and retrofitting of engineering structures. China Architecture and Building Press (in Chinese)
23. Belarbi A, Acun B (2013) FRP systems in shear strengthening of reinforced concrete structures. Procedia Eng 57:2–8
24. Shen DJ, Li CC, Feng ZZ et al (2019) Influence of strain rate on bond behavior of concrete members reinforced with basalt fiber-reinforced polymer rebars. Constr Build Mater 228:116755
25. Hadhood A, Agamy MH, Abdelsalam MM et al (2019) Shear strengthening of hybrid externally-bonded mechanically-fastened concrete beams using short CFRP strips: experiments and theoretical evaluation. Eng Struct 201:109795
26. Shi JZ, Wang X, Wu ZS et al (2017) Fatigue behavior of basalt fiber-reinforced polymer tendons under a marine environment. Constr Build Mater 137:46–54
27. Alhassan MA, Al-Rousan RZ, Abu-Elhija AM (2020) Anchoring holes configured to enhance the bond-slip behavior between CFRP composites and concrete. Constr Build Mater 250:118905
28. Hsu WL, Liu CC, Shiau YC et al (2019) Discussion on the reinforcement of reinforced concrete slab structures. Sustainability 11(6):1756
29. Jiang C, Wu YF, Wu G (2014) Plastic hinge length of FRP-confined square RC columns. J Compos Constr 18(4):04014003
30. Khaloo A, Moradi H, Kazemian A et al (2020) Experimental investigation on the behavior of RC arches strengthened by GFRP composites. Constr Build Mater 235:117519
31. Zhang S, Wu B (2019) Effects of salt solution on the mechanical behavior of concrete beams externally strengthened with AFRP. Constr Build Mater 229:117044
32. Ibrahim M, Wakjira T, Ebead U (2020) Shear strengthening of reinforced concrete deep beams using near-surface mounted hybrid carbon/glass fibre reinforced polymer strips. Eng Struct 210:110412
33. Shen DJ, Yang Q, Jiao Y et al (2017) Experimental investigations on reinforced concrete shear walls strengthened with basalt fiber-reinforced polymers under cyclic load. Constr Build Mater 136:217–229
34. Wang ZK, Zhao XL, Xian GJ et al (2017) Long-term durability of basalt- and glass-fibre reinforced polymer (BFRP/GFRP) bars in seawater and sea sand concrete environment. Constr Build Mater 139:467–489
35. Shen DJ, Ojha B, Shi X et al (2016) Bond stress-slip relationship between basalt fiber-reinforced polymer bars and concrete using a pull-out test. J Reinf Plast Compos 35(9):747–763
36. Wei B, Cao HL, Song SH (2010) Tensile behavior contrast of basalt and glass fibers after chemical treatment. Mater Des 31(9):4244–4250

37. Sim J, Park C, Moon DY (2005) Characteristics of basalt fiber as a strengthening material for concrete structures. Compos B Eng 36(6):504–512
38. Shen DJ, Shi X, Ji Y et al (2015) Strain rate effect on bond stress-slip relationship between basalt fiber-reinforced polymer sheet and concrete. J Reinf Plast Compos 34(7):547–563
39. Shen DJ, Ji Y, Yin FF et al (2015) Dynamic bond stress-slip relationship between basalt FRP sheet and concrete under initial static loading. J Compos Constr 19(6):04015012
40. Shi JW, Zhu H, Wu ZS et al (2013) Bond behavior between basalt fiber-reinforced polymer sheet and concrete substrate under the coupled effects of freeze-thaw cycling and sustained load. J Compos Constr 17(4):530–542
41. Lin XS, Zhang YX (2013) Bond-slip behaviour of FRP-reinforced concrete beams. Constr Build Mater 44:110–117
42. Qin ZP, Tian Y, Li G et al (2019) Study on bending behaviors of severely pre-cracked RC beams strengthened by BFRP sheets and steel plates. Constr Build Mater 219:131–143
43. Zhou DY, Lei Z, Wang JB (2013) In-plane behavior of seismically damaged masonry walls repaired with external BFRP. Compos Struct 102:9–19
44. Mostofinejad D, Hajrasouliha M (2019) 3D beam-column corner joints retrofitted with X-shaped FRP sheets attached via the EBROG technique. Eng Struct 183:987–998
45. Grazide C, Ferrier E, Michel L (2020) Rehabilitation of reinforced concrete structures using FRP and wood. Constr Build Mater 234:117716
46. Shen DJ, Yang Q, Huang C et al (2019) Tests on seismic performance of corroded reinforced concrete shear walls repaired with basalt fiber-reinforced polymers. Constr Build Mater 209:508–521
47. Franco A, Royer-Carfagni G (2014) Effective bond length of FRP stiffeners. Int J Non-Linear Mech 60:46–57
48. Lu XZ, Teng JG, Ye LP et al (2005) Bond–slip models for FRP sheets/plates bonded to concrete. Eng Struct 27(6):920–937
49. Ferracuti B, Savoia M, Mazzotti C (2006) A numerical model for FRP–concrete delamination. Compos B Eng 37(4–5):356–364
50. Ceroni F, Pecce M, Bilotta A et al (2012) Bond behavior of FRP NSM systems in concrete elements. Compos B Eng 43(2):99–109
51. Kabir MI, Shrestha R, Samali B (2016) Effects of applied environmental conditions on the pull-out strengths of CFRP-concrete bond. Constr Build Mater 114:817–830
52. Shi JW, Cao WH, Wu ZS (2019) Effect of adhesive properties on the bond behaviour of externally bonded FRP-to-concrete joints. Compos B Eng 177:107365
53. Freddi F, Savoia M (2008) Analysis of FRP-concrete debonding via boundary integral equations. Eng Fract Mech 75(6):1666–1683
54. Zhou YW, Wu YF, Yun Y (2010) Analytical modeling of the bond-slip relationship at FRP-concrete interfaces for adhesively-bonded joints. Compos B Eng 41(6):423–433
55. Alhassan MA, Al-Rousan RZ, Taha HM (2020) Precise finite element modelling of the bond-slip contact behavior between CFRP composites and concrete. Constr Build Mater 240:117943
56. Yao J, Teng JG, Chen JF (2005) Experimental study on FRP-to-concrete bonded joints. Compos B Eng 36(2):99–113
57. Chen GM, Teng JG, Chen JF (2012) Process of debonding in RC beams shear-strengthened with FRP U-strips or side strips. Int J Solids Struct 49(10):1266–1282
58. Perevozchikova BV, Pisciotta A, Osovetsky BM et al (2014) Quality evaluation of the kuluevskaya basalt outcrop for the production of mineral fiber, Southern Urals, Russia. Energy Procedia 59:309–314
59. Yuan H, Teng JG, Seracino R et al (2004) Full-range behavior of FRP-to-concrete bonded joints. Eng Struct 26(5):553–565
60. Ascione L, Berardi VP, Feo L et al (2005) A numerical evaluation of the interlaminar stress state in externally FRP plated RC beams. Compos B Eng 36(1):83–90
61. Chen JF, Pan WK (2006) Three dimensional stress distribution in FRP-to-concrete bond test specimens. Constr Build Mater 20(1–2):46–58

62. Lu XZ, Jiang JJ, Teng JG et al (2006) Finite element simulation of debonding in FRP-to-concrete bonded joints. Constr Build Mater 20(6):412–424
63. Teng JG, Yuan H, Chen JF (2006) FRP-to-concrete interfaces between two adjacent cracks: theoretical model for debonding failure. Int J Solids Struct 43(18–19):5750–5778
64. Ferracuti B, Savoia M, Mazzotti C (2007) Interface law for FRP-concrete delamination. Compos Struct 80(4):523–531
65. Ascione F, Berardi VP, Feo L et al (2008) An experimental study on the long-term behavior of CFRP pultruded laminates suitable to concrete structures rehabilitation. Compos B Eng 39(7–8):1147–1150
66. Lu XZ, Ye LP, Teng JG et al (2005) Meso-scale finite element model for FRP sheets/plates bonded to concrete. Eng Struct 27(4):564–575
67. Lu XZ, Teng JG, Ye LP et al (2005) Bond-slip models for FRP sheets/plates bonded to concrete. Eng Struct 27(6):920–937
68. Ferracuti B, Savoia M, Mazzotti C (2006) A numerical model for FRP-concrete delamination. Compos B Eng 37(4–5):356–364
69. Shen DJ, Li M, Liu C et al (2021) Seismic performance of corroded reinforced concrete beam-column joints repaired with BFRP sheets. Constr Build Mater 307:124731
70. Shen DJ, Li M, Kang J et al (2021) Experimental studies on the seismic behavior of reinforced concrete beam-column joints strengthened with basalt fiber-reinforced polymer sheets. Constr Build Mater 287:122901
71. Bilotta A, di Ludovico M, Nigro E (2011) FRP-to-concrete interface debonding: experimental calibration of a capacity model. Compos B Eng 42(6):1539–1553
72. Kang THK, Howell J, Kim S et al (2012) A state-of-the-art review on debonding failures of FRP laminates externally adhered to concrete. Int J Concr Struct Mater 6:123–134
73. Zheng D, Li Q (2004) An explanation for rate effect of concrete strength based on fracture toughness including free water viscosity. Eng Fract Mech 71(16–17):2319–2327
74. Yan D, Lin G (2008) Influence of initial static stress on the dynamic properties of concrete. Cement Concr Compos 30(4):327–333
75. Buchan PA, Chen JF (2007) Blast resistance of FRP composites and polymer strengthened concrete and masonry structures—a state-of-the-art review. Compos B Eng 38(5–6):509–522
76. Hao H, Tang EKC (2010) Numerical simulation of a cable-stayed bridge response to blast loads, part II: damage prediction and FRP strengthening. Eng Struct 32(10):3193–3205
77. Saiidi MS, Johnson R, Maragakis E (2006) Strain rate effects on strength of unidirectional FRP fabrics and bond to concrete. In: Proceedings of the third international conference on FRP composites in civil engineering, Miami, FL, USA, 2006, 1315
78. Maalej M, Chhoa CY, Quek ST (2010) Effect of cracking, corrosion and repair on the frequency response of RC beams. Constr Build Mater 24(5):719–731
79. Shen DJ, Liu XZ, Li QY et al (2019) Early-age behavior and cracking resistance of high-strength concrete reinforced with dramix 3D steel fiber. Constr Build Mater 196:307–316
80. Shen DJ, Liu C, Feng ZZ et al (2019) Influence of ground granulated blast furnace slag on the early-age anti-cracking property of internally cured concrete. Constr Build Mater 223:233–243
81. Shen DJ, Liu C, Li CC et al (2019) Influence of barchip fiber length on early-age behavior and cracking resistance of concrete internally cured with super absorbent polymers. Constr Build Mater 214:219–231
82. Wu Z, Yang S, Hu X et al (2010) Analytical solution for fracture analysis of CFRP sheet-strengthened cracked concrete beams. J Eng Mech 136(10):1202–1219
83. Aydin K (2013) Free vibration of functionally graded beams with arbitrary number of surface cracks. Eur J Mech, A, Solids 42:112–124
84. Chati M, Rand R, Mukherjee S (1997) Modal analysis of a cracked beam. J Sound Vib 207(2):249–270
85. Owolabi GM, Swamidas ASJ, Seshadri R (2003) Crack detection in beams using changes in frequencies and amplitudes of frequency response functions. J Sound Vib 265(1):1–22

86. Eder R, Miller R, Baseheart T et al (2005) Testing of two 50-year-old precast post-tensioned concrete bridge girders. Precast Concr Inst J 50:90–95
87. Shifrin EI, Ruotolo R (1999) Natural frequencies of a beam with an arbitrary number of cracks. J Sound Vib 222(3):409–423
88. Bakhtiari Nejad F, Khorram A, Rezaeian M (2014) Analytical estimation of natural frequencies and mode shapes of a beam having two cracks. Int J Mech Sci 78:193–202
89. Nandwana BP, Maiti SK (1997) Detection of the location and size of a crack in stepped cantilever beams based on measurements of natural frequencies. J Sound Vib 203(3):435–446
90. Chinchalkar S (2001) Determination of crack location in beams using natural frequencies. J Sound Vib 247(3):417–429
91. Ruotolo R, Surace C (1997) DAMAGE Assessment of multiple cracked beams: numerical results and experimental validation. J Sound Vib 206(4):567–588
92. Lee YS, Chung MJ (2000) A study on crack detection using eigenfrequency test data. Comput Struct 77(3):327–342
93. Lin HP, Chang SC (2006) Forced responses of cracked cantilever beams subjected to a concentrated moving load. Int J Mech Sci 48(12):1456–1463
94. Chondros TG, Dimarogonas AD, Yao J (1998) A continuous cracked beam vibration theory. J Sound Vib 215(1):17–34
95. Patil DP, Maiti SK (2003) Detection of multiple cracks using frequency measurements. Eng Fract Mech 70(12):1553–1572
96. Yang J, Chen Y (2008) Free vibration and buckling analyses of functionally graded beams with edge cracks. Compos Struct 83(1):48–60
97. Lele SP, Maiti SK (2002) Modelling of transverse vibration of short beams for crack detection and measurement of crack extension. J Sound Vib 257(3):559–583
98. Loya J A, Rubio L, Fernández Sáez J (2006) Natural frequencies for bending vibrations of Timoshenko cracked beams. J Sound Vib 290(3):640–653
99. Szekrenyes A (2014) Coupled flexural-longitudinal vibration of delaminated composite beams with local stability analysis. J Sound Vib 333(20):5141–5164
100. Zheng DY, Kessissoglou NJ (2004) Free vibration analysis of a cracked beam by finite element method. J Sound Vib 273(3):457–475
101. Kisa M, Arif gurel M (2007) Free vibration analysis of uniform and stepped cracked beams with circular cross sections. Int J Eng Sci 45(2):364–380
102. Yang J, Chen Y, Xiang Y et al (2008) Free and forced vibration of cracked inhomogeneous beams under an axial force and a moving load. J Sound Vib 312(1):166–181
103. Miller R, Parekh K (1994) Destructive testing of deteriorated prestressed box bridge beam. Transp Res Rec 1460:37
104. Halsey JT, Miller R (1996) Destructive testing of two forty-year-old prestressed concrete bridge beams. Precast Concr Inst J 41(5):84–93
105. Ross B, Ansley M, Hamilton H (2011) Load testing of 30-year-old AASHTO Type III highway bridge girders. Precast Concr Inst J 56:152–163
106. Attanayake U, Aktan H (2011) Capacity evaluation of a severely distressed and deteriorated 50-year-old box beam with limited data. J Perform Constr Facil 25(4):299–308
107. Ismail Z, Kuan KK, Yee KS et al (2014) Examining the trend in loss of flexural stiffness of simply supported RC beams with various crack severity using model updating. Meas: J Int Meas Confederation 50:43–49
108. Kara IF, Dundar C (2009) Effect of loading types and reinforcement ratio on an effective moment of inertia and deflection of a reinforced concrete beam. Adv Eng Softw 40(9):836–846
109. Dan EB, Heinrich T, Unified procedures for predicting the deflection and centroidal axis location of partially cracked nonprestressed and prestressed concrete members. ACI J Proc 79(2):119–130
110. ACI (1995) Control of deflection in concrete structures: ACI 435R-95. American Concrete Institute, Farmington Hills
111. ACI (2014) Building code requirements for structural concrete and commentary: ACI 318-14. American Concrete Institute, Farmington Hills

112. Castel A, Gilbert RI, Ranzi G (2013) Overall stiffness reduction of cracked reinforced concrete beams due to long term effects. In:Mechanics and physics of creep, shrinkage, and durability of concrete: a tribute to Zdeňk P. Bažant, 2013, pp 443–450
113. Zanardo G, Hao H, Xia Y et al (2006) Stiffness assessment through modal analysis of an RC slab bridge before and after strengthening. J Bridg Eng 11(5):590–601
114. Capozucca R (2013) Assessment of CFRP strengthened RC beams through dynamic tests. Compos. Part B, Eng 46:69–80
115. Capozucca R, Bossoletti S (2014) Static and free vibration analysis of RC beams with NSM CFRP rectangular rods. Compos. Part B, Eng 67:95–110
116. Pascale G, Bonfiglioli B (2001) Reinforced concrete beams damaged and repaired with GFRP: dynamic testing and modelling. In: FRP composites in civil engineering, Vols. i and ii, Proceedings, 2001, pp 441–448
117. Bonfiglioli B, Pascale G, Martinez De Mingo S (2004) Dynamic testing of reinforced concrete beams damaged and repaired with fiber reinforced polymer sheets. J Mater Civ Eng 16(5):400–406
118. Camata G, Spacone E, Zarnic R (2007) Experimental and nonlinear finite element studies of RC beams strengthened with FRP plates. Compos B Eng 38(2):277–288
119. Perrella M, Berardi VP, Cricrì G (2018) A novel methodology for shear cohesive law identification of bonded reinforcements. Compos. Part B, Eng 144:126–133
120. Mostofinejad D, Mahmoudabadi E (2010) Grooving as alternative method of surface preparation to postpone debonding of FRP laminates in concrete beams. J Compos Constr 14(6):804–811
121. Skuturna T, Valivonis J (2016) Experimental study on the effect of anchorage systems on RC beams strengthened using FRP. Compos. Part B, Eng 91:283–290
122. Mostofinejad D, Khozaei K (2015) Effect of GM patterns on ductility and debonding control of FRP sheets in RC strengthened beams. Constr Build Mater 93:110–120
123. Spadea G, Bencardino F, Sorrenti F et al (2015) Structural effectiveness of FRP materials in strengthening RC beams. Eng Struct 99:631–641
124. Ascione L, Berardi VP (2011) Anchorage device for FRP laminates in the strengthening of concrete structures close to beam-column joints. Compos. Part B, Eng 42(7):1840–1850
125. Pham H, Al Mahaidi R (2004) Experimental investigation into flexural retrofitting of reinforced concrete bridge beams using FRP composites. Compos Struct 66(1):617–625
126. Skuturna T, Valivonis J (2015) The statistical evaluation of design methods of the load-carrying capacity of flexural reinforced concrete elements strengthened with FRP. Arch Civ Mech Eng 15(1):214–222
127. Ghandour AAE (2011) Experimental and analytical investigation of CFRP flexural and shear strengthening efficiencies of RC beams. Constr Build Mater 25(3):1419–1429
128. Chen GM, Zhang Z, Li YL et al (2016) T-section RC beams shear-strengthened with anchored CFRP U-strips. Compos Struct 144:57–79
129. Li A, Assih J, Delmas Y (2001) Shear strengthening of RC beams with externally bonded CFRP sheets. J Struct Eng 127(4):374–380
130. Ceroni F (2010) Experimental performances of RC beams strengthened with FRP materials. Constr Build Mater 24(9):1547–1559
131. Rizzo A, de Lorenzis L (2009) Behavior and capacity of RC beams strengthened in shear with NSM FRP reinforcement. Constr Build Mater 23(4):1555–1567
132. Al Mahmoud F, Castel A, François R et al (2009) Strengthening of RC members with near-surface mounted CFRP rods. Compos Struct 91(2):138–147
133. Teng JG, Chen JF, Smith ST et al (2003) Behaviour and strength of FRP-strengthened RC structures: a state-of-the-art review. Proc Inst Civ Eng. Struct Build 156(1):51–62
134. Bonacci JF, Maalej M (2000) Externally bonded fiber-reinforced polymer for rehabilitation of corrosion damaged concrete beams. Am Concr Inst Struct J 97(5):703–711
135. Tailjsten B (2003) Strengthening concrete beams for shear with CFRP sheets. Constr Build Mater 17(1):15–26

136. Panda KC, Bhattacharyya SK, Barai SV (2011) Shear strengthening of RC T-beams with externally side bonded GFRP sheet. J Reinf Plast Compos 30(13):1139–1154
137. China Association for Engineering Construction Standardization (2007) Technical specification for strengthening concrete structures with carbon fiber reinforced polymer laminate: CECS 146-2007. China Planning Press, Beijing (in Chinese)
138. ACI (2008) Guide for the design and construction of externally bonded FRP systems for strengthening concrete structures: 440.2R-08. American Concrete Institute, Farmington Hills
139. Leung Christopher KY, Chen Z, Lee S et al (2007) Effect of size on the failure of geometrically similar concrete beams strengthened in shear with FRP strips. J Compos Constr 11(5):487–496
140. Godat A, Qu Z, Lu XZ et al (2010) Size Effects for reinforced concrete beams strengthened in shear with CFRP strips. J Compos Constr 14(3):260–271
141. Wenwei W, Guo L (2006) Experimental study and analysis of RC beams strengthened with CFRP laminates under sustaining load. Int J Solids Struct 43(6):1372–1387
142. Pešić N, Pilakoutas K (2015) Tensile stress-strain relaxation as a failure precursor for FRP-strengthened RC beams. Compos Struct 125:530–541
143. Shen DJ, Liu XZ, Zhou BZ et al (2021) Influence of initial cracks on the frequency of a 60-year-old reinforced-concrete box beam. Mag Concr Res 73(3):121–134
144. Shen DJ, Zeng X, Zhang JY et al (2019) Behavior of RC box beam strengthened with basalt FRP using end anchorage with grooving. J Compos Mater 53(23):3307–3324
145. Shen DJ, Deng SC, Zhang JY et al (2015) Behavior of reinforced concrete box beam with initial cracks repaired with basalt fiber-reinforced polymer sheet. J Reinf Plast Compos 34(18):1540–1554
146. Zong S (2007) Experimental study on anchoring method of RC beams strengthened with BFRP. China Sci Technol Formation 20:47–48 (in Chinese)
147. Garden HN, Hollaway LC (1998) An experimental study of the influence of plate end anchorage of carbon fibre composite plates used to strengthen reinforced concrete beams. Compos Struct 42(2):175–188
148. El Maaddawy T, Soudki K (2008) Strengthening of reinforced concrete slabs with mechanically-anchored unbonded FRP system. Constr Build Mater 22(4):444–455
149. Yuan W, Guo A, Li H (2017) Seismic failure mode of coastal bridge piers considering the effects of corrosion-induced damage. Soil Dyn Earthquake Eng 93:135–146
150. Ou YC, Nguyen ND (2016) Influences of location of reinforcement corrosion on seismic performance of corroded reinforced concrete beams. Eng Struct 126:210–223
151. Choe DE, Gardoni P, Rosowsky D et al (2009) Seismic fragility estimates for reinforced concrete bridges subject to corrosion. Struct Saf 31(4):275–283
152. Zhong J, Gardoni P, Rosowsky D (2012) Seismic fragility estimates for corroding reinforced concrete bridges. Struct Infrastruct Eng 8(1):55–69
153. Zanini MA, Faleschini F, Pellegrino C (2016) Cost analysis for maintenance and seismic retrofit of existing bridges. Struct Infrastruct Eng 12(11):1411–1427
154. Gardoni P, Rosowsky D (2011) Seismic fragility increment functions for deteriorating reinforced concrete bridges. Struct Infrastruct Eng 7(11):869–879
155. Lee HS, Kage T, Noguchi T et al (2003) An experimental study on the retrofitting effects of reinforced concrete columns damaged by rebar corrosion strengthened with carbon fiber sheets. Cem Concr Res 33(4):563–570
156. Guo A, Li H, Ba X et al (2015) Experimental investigation on the cyclic performance of reinforced concrete piers with chloride-induced corrosion in marine environment. Eng Struct 105:1–11
157. Nanni A, Bradford NM (1995) FRP jacketed concrete under uniaxial compression. Constr Build Mater 9(2):115–124
158. Tastani SP, Pantazopoulou SJ (2004) Experimental evaluation of FRP jackets in upgrading RC corroded columns with substandard detailing. Eng Struct 26(6):817–829
159. Nguyen-Minh L, Vo-Le D, Tran-Thanh D et al (2018) Shear capacity of unbonded post-tensioned concrete T-beams strengthened with CFRP and GFRP U-wraps. Compos Struct 184:1011–1029

160. Nguyen-Minh L, Phan-Vu P, Tran-Thanh D et al (2018) Flexural-strengthening efficiency of CFRP sheets for unbonded post-tensioned concrete T-beams. Eng Struct 166:1–15
161. Li G (2006) Experimental study of FRP confined concrete cylinders. Eng Struct 28(7):1001–1008
162. Phan-Vu P, Tran DT, Pham TM et al (2021) Distinguished bond behaviour of CFRP sheets in unbonded post-tensioned reinforced concrete beams versus single-lap shear tests. Eng Struct 234:111794
163. Pantazopoulou SJ, Bonacci JF, Sheikh S et al (2001) Repair of corrosion-damaged columns with FRP wraps. J Compos Constr 5(1): 3–11
164. Pham TM, Hadi MNS (2013) Strain estimation of CFRP-confined concrete columns using energy approach. J Compos Constr 17(6):04013001
165. Wang LM, Wu YF (2008) Effect of corner radius on the performance of CFRP-confined square concrete columns: test. Eng Struct 30(2):493–505
166. Huynh-Xuan T, Do-Dai T, Ngo-Thanh T et al (2021) Effect of sulfate attack on reinforced concrete columns confined with CFRP sheets under axial compression. J Compos Constr 25(6):04021049
167. Eid R, Paultre P (2017) Compressive behavior of FRP-confined reinforced concrete columns. Eng Struct 132:518–530
168. Lin G, Zeng JJ, Teng JG et al (2020) Behavior of large-scale FRP-confined rectangular RC columns under eccentric compression. Eng Struct 216:110759
169. Zeng JJ, Lin G, Teng JG et al (2018) Behavior of large-scale FRP-confined rectangular RC columns under axial compression. Eng Struct 174:629–645
170. Jia J, Zhao L, Wu S et al (2020) Experimental investigation on the seismic performance of low-level corroded and retrofitted reinforced concrete bridge columns with CFRP fabric. Eng Struct 209:110225
171. Zhang XC, Li B (2020) Damage characteristics and assessment of corroded RC beam-column joint under cyclic loading based on acoustic emission monitoring. Eng Struct 205:110090
172. Yuan J, Shen D (2022) Seismic performance of corroded reinforced concrete columns strengthened with basalt fiber sheets. Appl Sci 12(11):5611
173. Zhoudao LU, Lei SU, Jiangtao YU (2011) Experimental study on the seismic behaviour of strengthened concrete column-beam joints by simulated earthquake. Procedia Eng 14:1871–1878
174. Kalogeropoulos GI, Tsonos ADG, Konstandinidis D et al (2016) Pre-earthquake and post-earthquake retrofitting of poorly detailed exterior RC beam-to-column joints. Eng Struct 109:1–15
175. Shannag MJ, Alhassan MA (2005) Seismic upgrade of interior beam-column subassemblages with high-performance fiber-reinforced concrete jackets. ACI Struct J 102(1):131
176. Ilia E, Mostofinejad D (2019) Seismic retrofit of reinforced concrete strong beam–weak column joints using EBROG method combined with CFRP anchorage system. Eng Struct 194:300–319
177. Attari N, Youcef YS, Amziane S (2019) Seismic performance of reinforced concrete beam–column joint strengthening by frp sheets. Structures. Elsevier 20:353–364
178. Bossio A, Fabbrocino F, Lignola GP et al (2015) Simplified model for strengthening design of beam–column internal joints in reinforced concrete frames. Polymers 7(9):1732–1754
179. Ghobarah A, Said A (2002) Shear strengthening of beam-column joints. Eng Struct 24(7):881–888
180. Li B, Kai Q (2011) Seismic behavior of reinforced concrete interior beam-wide column joints repaired using FRP. J Compos Constr 15(3):327–338
181. Qin LH, Li Y, Wang ZL et al (2014) Deflection calculating method of damaged concrete beams strengthened with BFRP. J Traffic Transp Eng 14(6):17–26
182. Qin LH, Li Y, Wang ZL (2015) Crack spacing and width calculation of damaged concrete beams strengthened with BFRP sheet. J Chang'an Univ (Nat Sci Ed) 35(05):89–96 (in Chinese)

183. Jiang SF, Zeng X, Shen S et al (2016) Experimental studies on the seismic behavior of earthquake-damaged circular bridge columns repaired by using combination of near-surface-mounted BFRP bars with external BFRP sheets jacketing. Eng Struct 106:317–331
184. Ma G, Chen XH, Yan L et al (2020) Effect of pre-damage on compressive properties of circular RC columns repaired with BFRP composites: testing and modeling. Compos Struct 247:112483
185. Ko H, Sato Y (2007) Bond stress–slip relationship between FRP sheet and concrete under cyclic load. J Compos Constr 11(4):419–426
186. Li B, Lim CL (2010) Tests on seismically damaged reinforced concrete structural walls repaired using fiber-reinforced polymers. J Compos Constr 14(5):597–608
187. Antoniades KK, Salonikios TN, Kappos AJ (2005) Tests on seismically damaged reinforced concrete walls repaired and strengthened using fiber-reinforced polymers. J Compos Constr 9(3):236–246
188. Antoniades K, Salonikios TN, Kappos AJ (2007) Evaluation of hysteretic response and strength of repaired R/C walls strengthened with FRPs. Eng Struct 29(9):2158–2171
189. Altin S, Anil Ö, Kopraman Y et al (2013) Hysteretic behavior of RC shear walls strengthened with CFRP strips. Compos Part B: Eng 44(1):321–329
190. Wu ZM, Hu CH, Wu YF et al (2011) Application of improved hybrid bonded FRP technique to FRP debonding prevention. Constr Build Mater 25(6):2898–2905
191. Tighiouart B, Benmokrane B, Gao D (1998) Investigation of bond in concrete member with fibre reinforced polymer (FRP) bars. Constr Build Mater 12(8):453–462
192. Okelo R, Yuan RL (2005) Bond strength of fiber reinforced polymer bars in normal strength concrete. J Compos Constr 9(3):203–213
193. Carloni C, Subramaniam KV, Savoia M et al (2012) Experimental determination of FRP–concrete cohesive interface properties under fatigue loading. Compos Struct 94(4):1288–1296
194. Davalos JF, Chen Y, Ray I (2008) Effect of FRP bar degradation on interface bond with high strength concrete. Cem Concr Compos 30(8):722–730
195. Aiello AM, Leone M, Pecce M (2007) Bond performances of FRP bars-reinforced concrete. J Mater Civ Eng 19(3):205–213
196. Cosenza E, Manfredi G, Realfonzo R (2002) Development length of FRP straight bars. Compos B Eng 33(7):493–504
197. Xiao JZ, Falkner H (2007) Bond behaviour between recycled aggregate concrete and steel bars. Constr Build Mater 21(2):395–401
198. Zhang N, Zhang Y, Gao Y et al (2019) Site amplification effects of a radially multi-layered semi-cylindrical canyon on seismic response of an earth and rockfill dam. Soil Dyn Earthq Eng 116:145–163
199. Wu YX, Zhou XH, Gao YF et al (2019) Effect of soil variability on bearing capacity accounting for non-stationary characteristics of undrained shear strength. Comput Geotech 110:199–210
200. Wu C, Oehlers DJ, Rebentrost M et al (2009) Blast testing of ultra-high performance fibre and FRP-retrofitted concrete slabs. Eng Struct 31(9):2060–2069
201. Shen DJ, Shi X, Ji Y et al (2015) Strain rate effect on bond stress–slip relationship between basalt fiber-reinforced polymer sheet and concrete. J Reinf Plast Compos 34(7):547–563
202. Shkolnik IE (2008) Influence of high strain rates on stress–strain relationship, strength and elastic modulus of concrete. Cem Concr Compos 30(10):1000–1012
203. Vos E, Reinhardt HW (1982) Influence of loading rate on bond behaviour of reinforcing steel and prestressing strands. Matériaux Constr 15:3–10
204. Huo JS, Liu JY, Lu Y et al (2016) Experimental study on dynamic behavior of GFRP-to-concrete interface. Eng Struct 20(5):371–382
205. Eligehausen R, Popov EP, Bertero VV (1982) Local bond stress-slip relationships of deformed bars under generalized excitations. In: Proceedings of the 7th European conference on earthquake engineering. Chicago, Illinois, United States, 1982, pp 69–80
206. FIB (2013) Fib model code for concrete structures 2010: fib Model Code 2010. Ernst and Sohn, Berlin

207. Ministry of Housing and Urban-Rural Development of the People's Republic of China (2010) Code for seismic design of buildings: GB 50010-2010. China Architecture and Building Press, Beijing (in Chinese)
208. Harajli MH, Hout M, Jalkh W (1995) Local bond stress-slip behavior of reinforcing bars embedded in plain and fiber concrete. Mater J 92(4):343–353
209. Harajli MH (2007) Numerical bond analysis using experimentally derived local bond laws: a powerful method for evaluating the bond strength of steel bars. J Struct Eng 133(5):695–705
210. ACI (1992) Bond under cyclic loading 408.2R-92. American Concrete Institute, Farmington Hills
211. Oh BH, Kim SH (2007) Realistic models for local bond stress-slip of reinforced concrete under repeated loading. J Struct Eng 133(2):216–224
212. Li J, Gao XL, Zhang P (2007) Experimental investigation on the bond of reinforcing bars in high performance concrete under cyclic loading. Mater Struct 40:1027–1044
213. Murcia-Delso J, Stavridis A, Shing PB (2013) Bond strength and cyclic bond deterioration of large-diameter bars. ACI Struct J 110(4):659–670
214. Lee JY, Kim KH, Kim SW et al (2014) Strength degradation of glass fiber reinforced polymer bars subjected to reversed cyclic load. Strength Mater 46:235–240
215. Wang HZ, Belarbi A (2010) Static and fatigue bond characteristics of FRP bars embedded in fiber-reinforced concrete. J Compos Mater 44(13):1605–1622
216. Akbas TT, Celik OC, Yalcin C et al (2016) Monotonic and cyclic bond behavior of deformed CFRP bars in high strength concrete. Polymers 8(6):211
217. Morita S (1973) Local bond stress-slip relationship under repeated loading. In: Proceedings, IABSE symposium on the resistance and ultimate deformability of structures (Lisbon 1973). Int Assoc Bridge Struct Eng 1973:221–227

# Chapter 2
# Bond Stress-Slip Relationship Between BFRP Sheets and Concrete Under Dynamic Loading

RC structures in civil engineering are subjected to dynamic loadings, such as blasting, impact, and earthquakes, during their service life; in such instances, the dynamic strength and fracture toughness of concrete are both higher than the corresponding static values [1, 2]. Several studies have been conducted on FRP strengthening of RC under dynamic loadings. Blast testing of FRP-retrofitted concrete slabs has been conducted [3]. The blast resistance of structures externally reinforced with FRP has also been investigated through tests and simulations [4, 5]. The bond between FRP and concrete increases with strain rate as a logarithmic function and an equation was developed through regression analysis of bond data [6]. Another study indicates that the interfacial fracture energy of FRP and concrete increases with strain rate as a logarithmic function, and strain rate is not significantly influenced by the type of bonding adhesives and FRP sheets and the effective length of FRP-concrete interface will decrease with increasing strain rate [7]. However, the model for predicting the dynamic effective bond length of the interface between FRP and concrete under dynamic loadings at different strain rates has been rarely studied to date. A reliable dynamic effective bond length model under dynamic loadings with different strain rates for the interface between FRP and concrete is fundamental for accurate modeling and understanding of debonding failures in FRP-strengthened RC structures under dynamic loading as well as for the establishment of safe and economical designs for externally bonded FRP systems.

BFRP has been used in concrete structures for external strengthening because of its superior properties, such as light weight, high temperature resistance, high tensile strength [8], freeze–thaw performance [9], and easy application. BFRP has also demonstrated good performance in acidic environments. Until recently, the dynamic performance of the interface between BFRP and concrete has not been studied in depth [8]. Hence, more work is required before general conclusions could be drawn about the interfacial properties of BFRP and concrete.

Concrete strength influences the interfacial properties of FRP and concrete [3, 10]. The dynamic interfacial properties of FRP and concrete are influenced by the dynamic



D. Shen, *Seismic Performance of Corroded Reinforced Concrete Structures Retrofitted with FRP*, https://doi.org/10.1007/978-981-99-7984-4_2

strength of concrete. A considerable amount of research on the effect of initial static loading on concrete dynamic strength has been conducted. The effect of initial static loading on dynamic compressive strength, dynamic tensile strength, and dynamic bending strength has also been investigated by experimental or numerical methods. Higher initial static loading intensity leads to lower tensile strength enhancement. The dynamic strength of concrete tends to decrease with the increase of initial static stress; however, low initial static stress slightly affects the ultimate dynamic strength. A mathematical model based on a single dimension of freedom is used to study the influence of initial static loading on the rate-dependent effect of concrete strength [1]. Initial static loading significantly affects the dynamic strength of concrete as well as the dynamic interfacial properties of FRP and concrete.

Most studies on the dynamic properties of the interface between BFRP and concrete did not consider the effect of strain rate on effective bond length [6, 7]. Virtually, several concrete structures externally bonded with BFRP would be subject to dynamic loadings, such as blasting, impact, and earthquakes. Thus, whether and how strain rate influences the effective bond length of the interface between BFRP and concrete under dynamic loading must be investigated to better understand the properties of the interface between BFRP and concrete [11–14].

## 2.1 Effective Bond Length of BFRP Sheets Bonded to Concrete Under Dynamic Loading

### *2.1.1 Experimental Program*

Commercial concrete C30 was utilized to cast the specimens, and its mix proportions are shown in Table 2.1.

The coarse aggregates consisted of gravel with a maximum size of 15 mm. River sand was utilized as fine aggregates. The concrete specimens were demolded 24 h after casting and then subjected to curing at 95% relative humidity (RH) for 28 d. Afterward, the 28 d concrete strength was determined [15]. The average concrete cubic compressive strength $f_{cu}$ was 37.5 MPa, the average axial compressive strength $f_c$ was 32.8 MPa, the concrete axial tensile strength $f_t$ was 2.62 MPa, and the compressive elastic modulus $E_c$ was $3.11 \times 10^4$ MPa. The concrete cylindrical axial compressive strength $f_c = 0.79\ f_{cu}$ was 29.6 MPa [16].

BFRP sheets with 15 mm width and 0.121 mm thickness were employed to test mechanical properties [17]. The elastic modulus of the BFRP sheet $E_{f,s}$ was 1.05 ×

**Table 2.1** Mix proportions of concrete (kg/m$^3$)

| Cement | Sand | Water | Coarse aggregates | Fly ash | Mineral powder | Admixture |
|---|---|---|---|---|---|---|
| 250 | 672 | 158 | 1010 | 37 | 37 | 5.5 |

$10^5$ MPa, the ultimate mean strain $\varepsilon_u$ was 2.18%, and the ultimate tensile strength $f_{fu,s}$ was $2.3 \times 10^3$ MPa. The tested elastic modulus, ultimate mean strain, and ultimate tensile strength mean value of the epoxy resin were $2.6 \times 10^3$ MPa, 3.47%, and 45.8 MPa, respectively.

Modified double-lap shear specimens were employed in accordance with the test method recommended by JSCE [18]. The details of the specimens are schematically shown in Fig. 2.1a. Each specimen consisted of a 510 mm long concrete block with a 100 mm × 100 mm cross-section. Similar layers of BFRP sheets with 50 mm width were bonded with epoxy resin on the two opposite sides of the concrete block along the axial direction. The BFRP bonded lengths were 50, 80, 110, and 200 mm. A reinforcing bar with a 22 mm diameter was embedded continuously through the center of both the anchorage and test blocks for tension application; two wood plates were utilized to separate the concrete blocks during casting. After curing the concrete, the reinforcing bar was cut in the middle of the two wood plates; hence, the external force was balanced by the interfacial shear force between the BFRP sheets and concrete.

Table 2.2 shows the configuration of the 21 specimens. Each specimen was assigned a label or name ("LX-DY-$n$"). The first letter in the specimen's name, L,

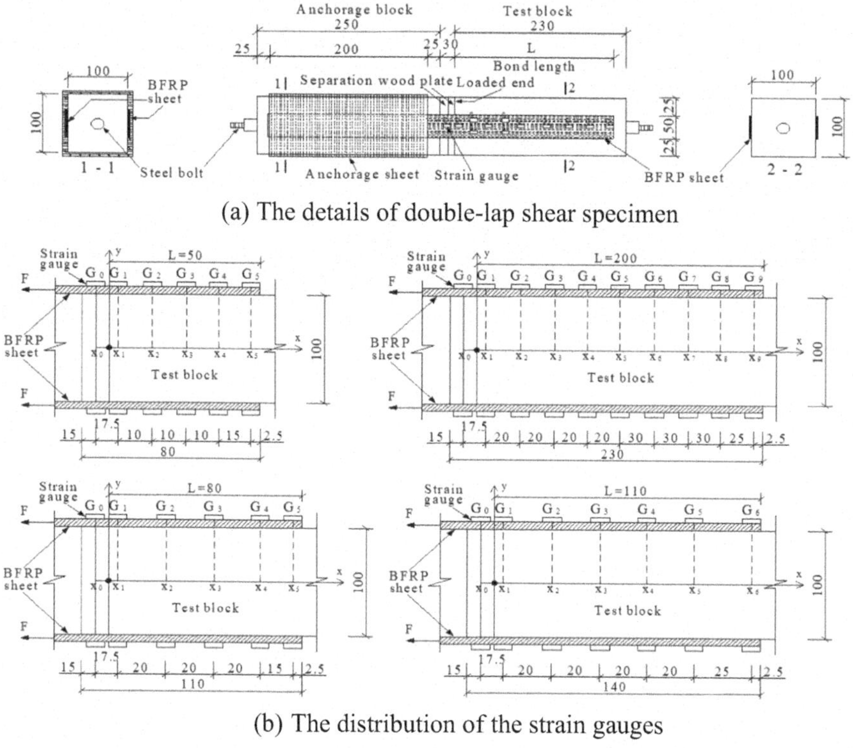


**Fig. 2.1** Schematic drawing of double-lap shear specimen (all units in mm)

refers to the bond length (i.e., L = bond length), and X indicates the bond length (i.e., 50 = 50 mm bond length). The letter D refers to dynamic loading, and Y indicates the loading rate (i.e., D0, D1, D2, and D3 indicate that the loading rates are 0.07, 0.7, 7, and 70 mm/s, respectively). The last character *n* is the progressive test number. Strain gauges with 5 mm gauge length were applied along the BFRP sheets on the side of the tested portion of the specimens to measure the axial strains during the bond test (Fig. 2.1b). The distance between the measured points of strains of the BFRP sheet is between 10 and 30 mm [19], as shown in Fig. 2.1b. Two layers of BFRP sheets were applied on the anchorage block of a specimen during the double-lap shear tests to ensure that debonding failure occurs at the test block side, as shown in Fig. 2.1a.

The application of BFRP sheets to the concrete surface involved the following steps [20]. First, the concrete surface was ground with a disk grinder to remove any stains and grit. Second, pressurized air was applied to remove dust from the surface. Third, a primer was applied followed by a resin undercoat. Lastly, a BFRP sheet with one layer was installed followed by a resin overcoat.

As illustrated in the setup scheme (Fig. 2.2), tests were performed with a servo-hydraulic testing machine (MTS 322). Two displacement transducers with a range of ± 10 mm were fixed on the concrete block on both sides of the BFRP sheets to monitor their relative displacement and reflect the global slip between BFRP sheets and concrete (Fig. 2.2a). The applied load was measured with a load cell placed between the reinforcing bars of the specimen and the test machine; uniaxial tension was applied. Load, displacement, and strain data were recorded with a DH5922 data logger at a frequency of 50 kHz. For all specimens, the 3.5 kN load was initially applied at a load rate of 3.5 kN/min. The dynamic load was then applied until failure at displacement rates of 0.07, 0.7, 7, and 70 mm/s for specimens L200-D0, L200-D1, L200-D2, and LX-D3, respectively. The loading procedure was implemented automatically under the control of a computer program. At last, three specimens were tested for each loading history case to ensure the generality of the experimental results. The strain rate of BFRP sheet was calculated from the strains of strain gauge G0, as shown in Fig. 2.1. The result of strain rates of BFRP sheet is shown in Table 2.2.

### 2.1.2 Failure Modes

Much important information on the efficiency and performance of the adhesive bonding technique can be obtained from the failure modes. Therefore, the failure modes under different strain rates must be identified. Three failure modes of FRP and concrete were observed in the shear tests: the three modes are debonding in concrete, debonding at the adhesive-concrete interface, and concrete prism failure [21]. All the specimens failed because of debonding in the concrete adjacent to the adhesive-concrete interface, in which a thin layer of concrete was attached to the BFRP strip after failure (Fig. 2.3). All of the tested double-lap specimens failed because of debonding in concrete based on the failure type classification [21]. The failure modes of the interface between BFRP and concrete were unaffected by the

**Table 2.2** Results of maximum bond stress and effective bond length of specimens L50-D3, L80-D3, L110-D3, L200-D0, L200-D1, L200-D2, and L200-D3

| Identification of specimen | Loading rate (mm/s) | Strain rate ($s^{-1}$) | $P_u$ (kN) | | Ultimate displacements (mm) | | $\tau_{max}$ (MPa) | | $L_1$ (mm) | $L_2$ (mm) | Effective bond length $Le$ (mm) | |
|---|---|---|---|---|---|---|---|---|---|---|---|---|
| | | | Individual | Ave. | Individual | Ave. | Individual | Ave. | | | Individual | Ave. |
| L50-D3-1 | | | 14.0 | | 1.639 | | 3.72 | | – | – | – | – |
| L50-D3-2 | 70 | 0.63 | 14.5 | 14.4 | 1.485 | 1.588 | 4.07 | 3.88 | – | – | – | – |
| L50-D3-3 | | | 14.7 | | 1.641 | | 3.84 | | – | – | – | – |
| L80-D3-1 | | | 14.5 | | 2.261 | | – | – | – | – | – | – |
| L80-D3-2 | 70 | 0.63 | 15.2 | 15.1 | 2.844 | 2.580 | 5.49 | 5.74 | 14.4 | 68.5 | 54.1 | 56.0 |
| L80-D3-3 | | | 15.7 | | 2.634 | | 6.00 | | 12.1 | 70 | 57.9 | |
| L110-D3-1 | | | 15.6 | | 2.607 | | 5.97 | 6.04 | 12.5 | 70.9 | 58.4 | 57.9 |
| L110-D3-2 | 70 | 0.63 | 15.7 | 15.4 | 2.869 | 3.110 | 6.11 | | 14.5 | 71.8 | 57.3 | |
| L110-D3-3 | | | 14.9 | | 3.854 | | 4.41 | | – | – | – | |
| L200-D3-1 | | | 16.2 | | 6.409 | | 5.95 | | 32.5 | 90.2 | 57.7 | |
| L200-D3-2 | 70 | 0.63 | 15.7 | 15.8 | 5.678 | 5.395 | 5.82 | 5.75 | 34.2 | 92.5 | 58.3 | 56.7 |
| L200-D3-3 | | | 15.6 | | 4.097 | | 5.49 | | 35.1 | 89.3 | 54.2 | |
| L200-D0-1 | | | 11.4 | | 2.040 | | 2.95 | 3.27 | 29.6 | 98.9 | 69.3 | 71.8 |
| L200-D0-2 | 0.07 | $0.61 \times 10^{-3}$ | 10.8 | 11.9 | 1.872 | 2.094 | 3.59 | | 30.1 | 104.3 | 74.2 | |
| L200-D0-3 | | | 13.6 | | 2.371 | | – | – | – | – | – | – |
| L200-D1-1 | | | 15.0 | | 2.412 | | 4.64 | 4.82 | 12.5 | 76 | 63.5 | 65.5 |

(continued)

**Table 2.2** (continued)

| Identification of specimen | Loading rate (mm/s) | Strain rate ($s^{-1}$) | $P_u$ (kN) | | Ultimate displacements (mm) | | $\tau_{max}$ (MPa) | | $L_1$ (mm) | $L_2$ (mm) | Effective bond length *Le* (mm) | |
|---|---|---|---|---|---|---|---|---|---|---|---|---|
| | | | Individual | Ave. | Individual | Ave. | Individual | Ave. | | | Individual | Ave. |
| L200-D1-2 | 0.7 | $0.61 \times 10^{-2}$ | 13.3 | 13.6 | 2.610 | 2.484 | 5.00 | | 13.5 | 81 | 67.5 | |
| L200-D1-3 | | | 12.5 | | 2.429 | | 3.89 | | – | – | – | |
| L200-D2-1 | | | 15.5 | | 2.904 | | 3.68 | | – | – | – | |
| L200-D2-2 | 7 | $0.47 \times 10^{-1}$ | 14.5 | 14.4 | 2.525 | 2.665 | 5.37 | 5.38 | 14.4 | 75.9 | 61.5 | 62.8 |
| L200-D2-3 | | | 13.1 | | 2.567 | | 5.39 | | 13.1 | 77.2 | 64.1 | |

*Note* $L_1$ and $L_2$ are the distance from the loaded end to the point that correspond to 10% of $\tau_{max}$, respectively

**Fig. 2.2** Test setup layout

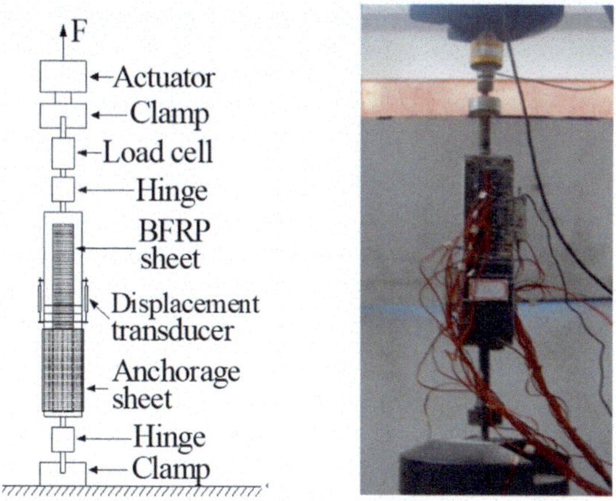


(a) loading device (b) photo of loading device

strain rate [7], which is in agreement with the test results. Bond failure typically occurred in concrete near the interface with BFRP sheets regardless of the strain rate.

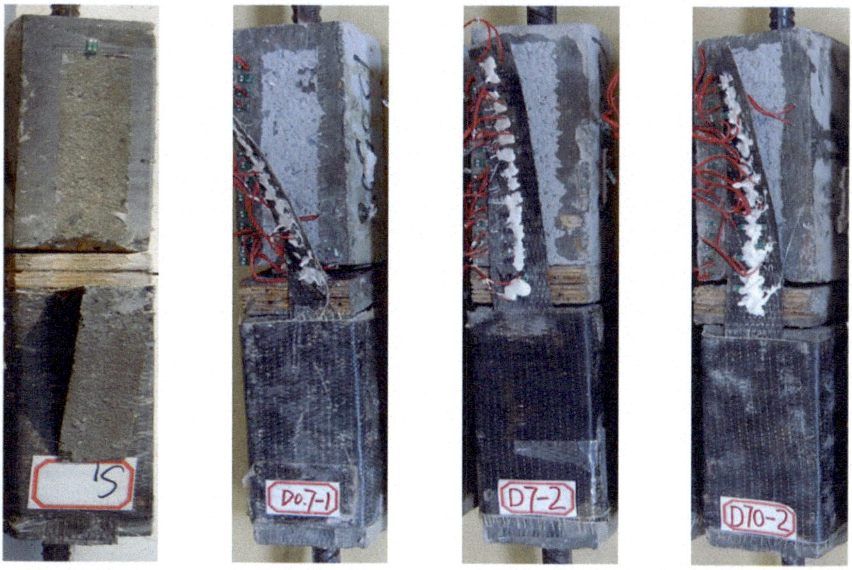


(a) L200-D0 (b) L200-D1 (c) L200-D2 (d) L200-D3

**Fig. 2.3** Failure mode of debonding in concrete of specimens L200-D0, L200-D1, L200-D2, and L200-D3

### 2.1.3 Relationship Between Load and Displacement

Figure 2.4 shows the experimental load versus global slip graphs between BFRP sheets and concrete of all specimens. The displacement in Fig. 2.4 indicates the average data of two displacement transducers bridged over the test and anchorage blocks (Fig. 2.1a). The displacement value in Fig. 2.4 contained the deformation of the BFRP sheet between the loaded end and the anchorage sheet.

Effective bond length can be experimentally determined through double-lap shear tests on BFRP with different bond lengths; by definition, the effective bond length is the bond length beyond which the ultimate load remains almost constant [22]. As shown in Fig. 2.4a–d and Table 2.2, the average ultimate loads of specimens L50-D3, L80-D3, L110-D3, and L200-D3 with bond lengths of 50, 80, 110, and 200 mm were 14.4, 15.1, 15.4, and 15.8 kN, respectively. The ultimate load and initial stiffness of double-lap shear specimens increased with the increase of bond length from 50 mm to 200 mm. The average values of the dynamic ultimate load of specimens L80-D3, L110-D3, and L200-D3 with bond lengths of 80, 110, and 200 mm were 15.43 kN, and the deviations from the corresponding individual and average values are 2.0%, 0.19%, and 2.0%, respectively. The average values of the dynamic ultimate load of specimens L80-D3, L110-D3, and L200-D3 increased at the rate of 7.1% compared with that of specimen L50-D3 with a bond length of 50 mm. A generally accepted fact in the FRP strengthening research community is that an active bond zone exists over which the majority of interfacial shear stresses are transferred from the fiber sheet to the concrete [23]. The ultimate load of specimens reaches an approximately steady case after a certain value of bond length. This length is defined as the effective bond length, beyond which no significant increase is possible; only a marginal improvement in ultimate load could occur [24]. Thus, the dynamic effective bond length of double shear specimens under 70 mm/s loading rate was between 50 and 80 mm according to the result of the ultimate load.

As shown in Fig. 2.5, the ultimate displacement of specimens L50-D3, L80-D3, L110-D3, and L200-D3 with bond lengths of 50, 80, 110, and 200 mm, respectively, increased linearly with the increase of bond length. Ultimate displacement increased at the rates of 62.5%, 95.8%, and 239.7% when the bond length was increased from 50 mm to 80, 110, and 200 mm, respectively.

To examine the sensitivity between the effective bond length and the strain loading rate, the ultimate load for the BFRP-concrete double-lap specimens at four different strain rates is shown in Table 2.2. The ultimate load and displacement were 11.9, 13.6, 14.4, and 15.8 kN and 2.094, 2.484, 2.665, and 5.395 mm, respectively, for specimens L200-D0, L200-D1, L200-D2, and L200-D3 with strain rates of $0.61 \times 10^{-3}$, $0.61 \times 10^{-2}$, $0.47 \times 10^{-1}$, and 0.63 $s^{-1}$. When the strain rate was increased from $0.61 \times 10^{-3}$ $s^{-1}$ to $0.61 \times 10^{-2}$, $0.47 \times 10^{-1}$, and 0.63 $s^{-1}$, the dynamic ultimate load increased at the rates of 14.3%, 21.0%, and 32.8%; the dynamic ultimate displacement increased at the rates of 18.6%, 27.3%, and 157.6%, respectively. The ductility of the double-lap shear specimens increased with the increase of the loading

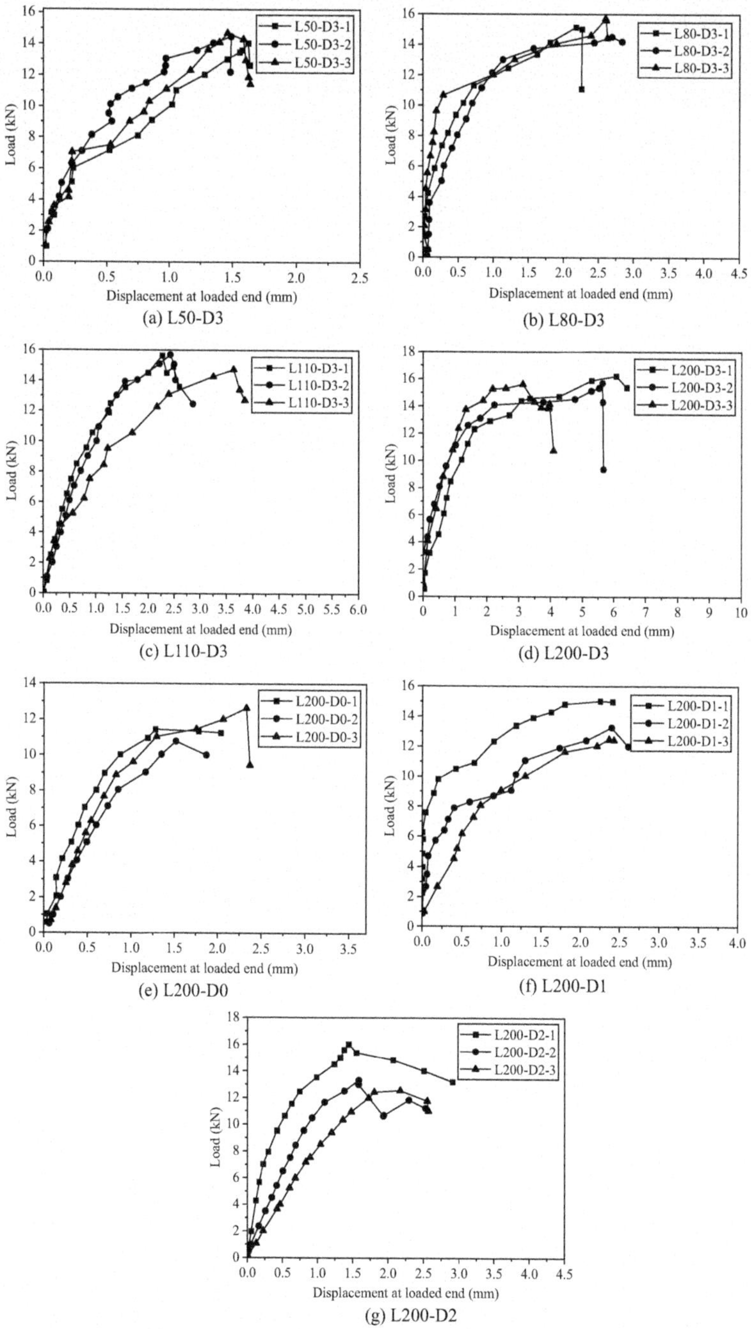


**Fig. 2.4** Load versus global slip between BFRP sheets and concrete of specimens L50-D3, L80-D3, L110-D3, L200-D0, L200-D1, L200-D2, and L200-D3

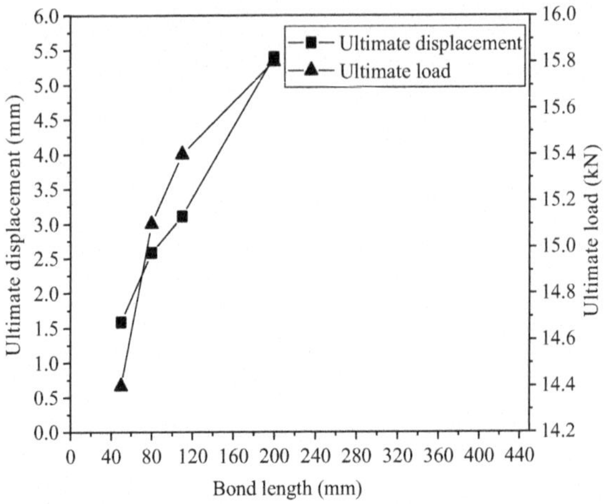


**Fig. 2.5** Relationship between ultimate displacement and load at different bond lengths

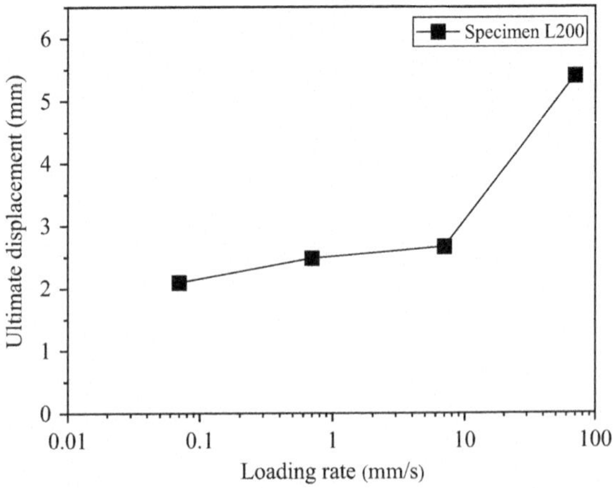


**Fig. 2.6** Relationship between ultimate displacement and loading rates

rate. As shown in Fig. 2.6, ultimate displacement increased significantly when the loading rate was higher than 7 mm/s.

### 2.1.4 Bond Strain and Bond Stress

Strain gauges were utilized to measure axial strains in the BFRP sheets during the test; the positions of the strain gauges are denoted by $x_i$. . The behavior between BFRP sheets and concrete was analyzed by plotting the experimental axial strains ($\varepsilon_i$) measured by the strain gauges during the test along the BFRP sheet (Fig. 2.1b). The strain distribution along the BFRP sheet for specimen L200-D3-1 at typical loading steps is shown in Fig. 2.7a. The strains were obtained by strain gauges mounted on the upper surface of the FRP strip. The strain distribution exhibited a descending curve from the center of specimen toward the end of sheet [25]. The strains locally deviated from the basic descending tendency and showed a zigzag curve, which is different from the theoretical relationship between the BFRP sheet strain and the distance from the loaded end [25]. These local strain deviations were due to the stress concentration caused by the aggregate embedded between the BFRP sheet and concrete. The negative strain resulted in negative bond stresses, which were not corrected. To eliminate the negative strain gradients, some of the strain values were corrected to decrease monotonically toward the sheet end, as shown in Fig. 2.7b. The BFRP sheet strains were corrected for all specimens in the same manner. The relationships between BFRP strain and distance from the loaded end of specimens under different loading rates are shown in Fig. 2.7a–i.

Figure 2.7 shows that the increase of BFRP strain was gradual until the applied load $P$ reached ultimate load $P_{\max}$. When the positions of strain gauges are larger than 50–80 mm from the loaded end, the BFRP strains are small under ultimate load $P_{\max}$. This result indicates that almost all the applied loadings were resisted in this small area. The maximum strain of BFRP for specimens L50-D3, L80-D3, L110-D3, L200-D3, L200-D0, L200-D1, and L200-D2 (Fig. 2.7a–i) was approximately 10,700, 10,900, 10,700, 10,700, 8300, 9600, and 10,500 $\mu\varepsilon$, respectively, when ultimate load $P_u$ was reached. Effective bond length is also described as the length beyond which the failure load stops increasing and FRP strain vanishes [23]. A large part of the BFRP strip near the far end still had minimal strain when the ultimate load was reached. This condition confirms the concept of effective bond length and thus indicates that increasing the bond length beyond a certain value does not further increase the bond strength [21].

Interface bond stress $\tau_{i+1/2}$, which is a local phenomenon, was determined from the following well-known relationship [26, 27], as shown in Eq. (2.1).

$$\tau_{i+1/2} = -\frac{E_f t_f (\varepsilon_{i+1} - \varepsilon_i)}{x_{i+1} - x_i} \tag{2.1}$$

where $\tau$ is bond stress; $E_f$ and $t_f$ are the elastic modulus and thickness of the BFRP sheet, respectively; and $\varepsilon_i$ is the strain measured at $x_i$ by a strain gauge. The most significant results related to the tests performed on the sheets are summarized in Fig. 2.7.

The distribution of bond stress along the length of BFRP plate is shown in Fig. 2.8. As the applied loading increased, the peak bond stress moved along the length of sheet from the loaded end, which signified debonding crack propagation.

For specimen L200-D3-1 (Fig. 2.8c), the maximum bond stress was at the distance of 12.5 mm from the loaded end when the applied load $P$ was approximately 60% of the ultimate load. Bond stress was minimal beyond the distance of approximately 30 mm from the loaded end. When applied load $P$ was 16.2 kN, the maximum bond stress was at the distance of 32.5 mm from the loaded end; bond stress was minimal beyond the distance of approximately 70 mm from the loaded end. The maximum bond stress of specimen L200-D3-1 was 5.95 MPa at a distance of 72.5 mm from

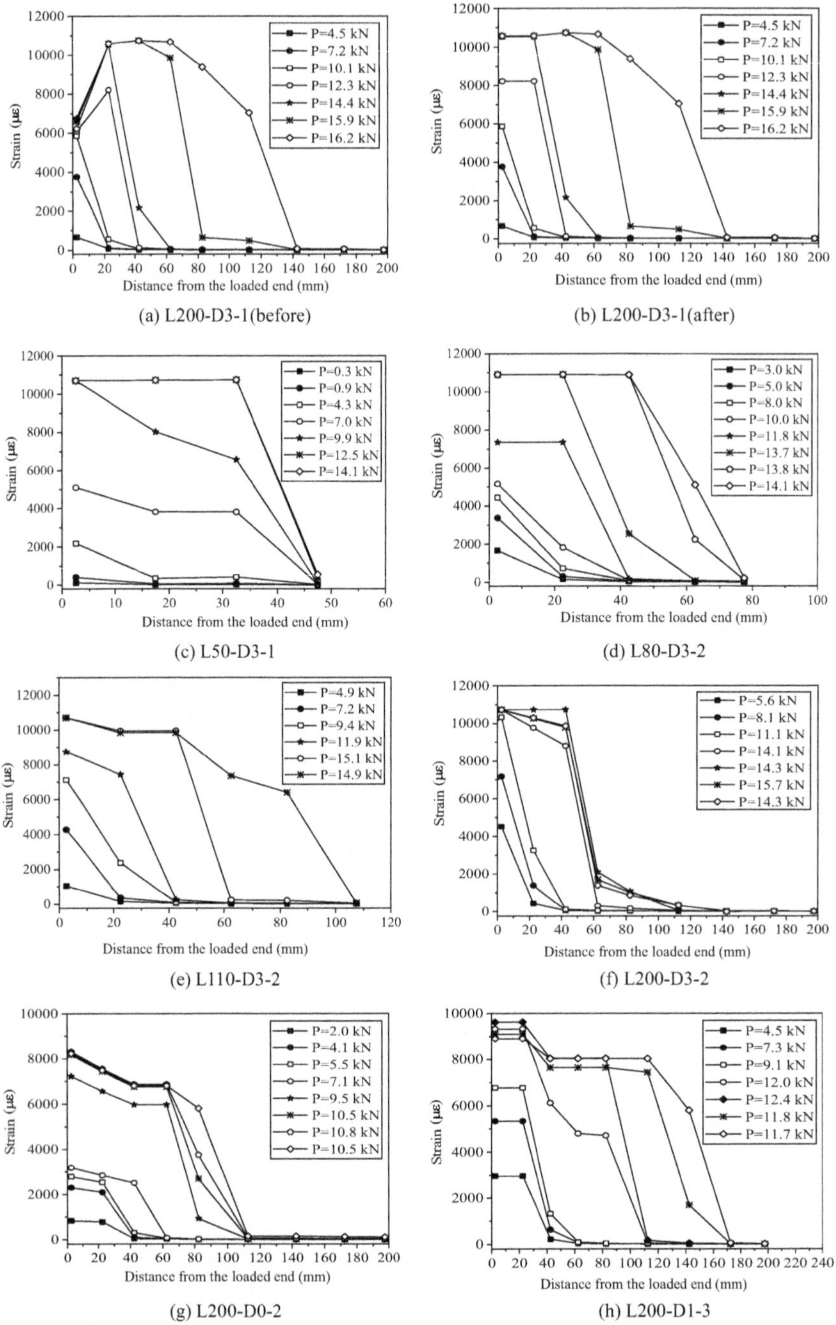


**Fig. 2.7** Relationship between distance from the loaded end and BFRP strain of specimens L50-D3-1, L80-D3-2, L110-D3-2, L200-D0-2, L200-D1-3, L200-D2-1, and L200-D3-2

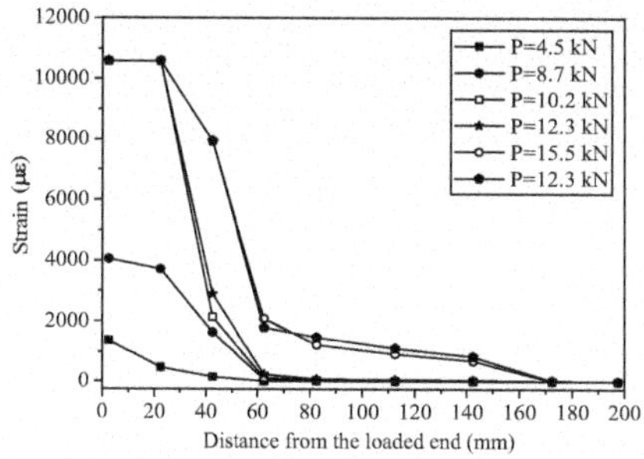


(i) L200-D2-1

**Fig. 2.7** (continued)

the loaded end. As illustrated in Fig. 2.8, the relationships between bond stress and distance from the loaded end are similar. The peak bond stress moved along the length of the sheet from the loaded end when applied load increased.

The results of the maximum bond stress $\tau_{max}$ of all specimens under different loading rates are shown in Table 2.2. The maximum bond stress of specimens L200-D0, L200-D1, L200-D2, and L200-D3 under the strain rates of $0.61 \times 10^{-3}$, $0.61 \times 10^{-2}$, $0.47 \times 10^{-1}$, and $0.63\ s^{-1}$ were 3.27, 4.82, 5.38, and 5.75 MPa, respectively. The maximum bond stress between BFRP and concrete increased with the increase of strain rate. The average measured static bond stress of the samples is 3.7 MPa for the GFRP specimens with a tensile modulus of 20.9 GPa and thickness of 1.3 mm and 4.06 MPa for the CFRP specimens with a tensile modulus of 61.5 GPa and thickness of 1.0 mm [6]. The fact that the maximum bond stresses of GFRP, CFRP, and BFRP were close confirms that the test results of BFRP were reasonable and the maximum bond stress was related to concrete strength and had no significant relation with FRP stiffness. The maximum bond stress data in Table 2.2 showed that maximum bond stress improved as the strain rate increased. This result is attributed to the sensitivity of concrete shear strength to the strain rate. The shear strength of concrete is a function of its compressive strength, and compressive strength is known to increase with the strain rate [28].

### 2.1.5 *Dynamic Effective Bond Length*

The adhesion strength of a stiffener, such as BFRP, on a substrate is generally believed to be characterized by an intrinsic length usually referred to as the effective bond length [22]. This length is defined as the length necessary to transfer the load from BFRP to the substrate. Determination of the effective bond length is of fundamental importance in the characterization of the performance of the interface between BFRP and concrete. Effective bond length has an important physical meaning, and design consequences and several proposals on how to estimate it from the characteristics

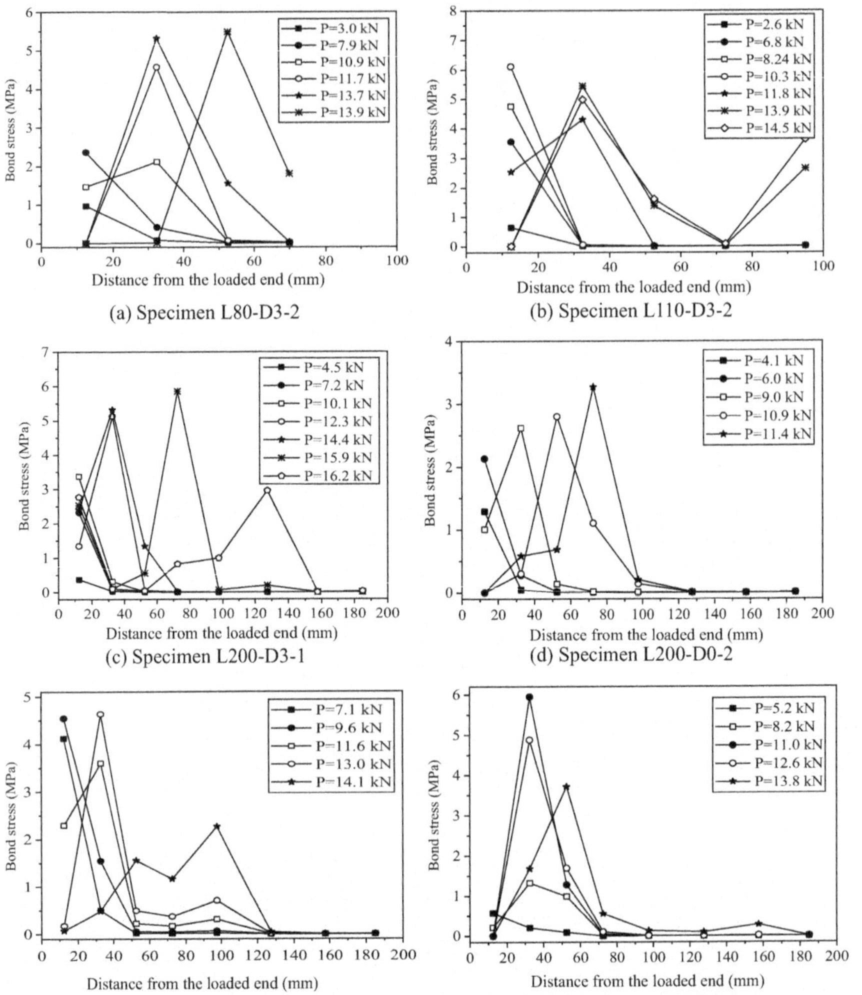

(a) Specimen L80-D3-2
(b) Specimen L110-D3-2
(c) Specimen L200-D3-1
(d) Specimen L200-D0-2
(e) Specimen L200-D1-1
(f) Specimen L200-D2-2

**Fig. 2.8** Relationship between distance from the loaded end and bond stress of specimens L80-D3-2, L110-D3-2, L200-D0-2, L200-D1-1, L200-D2-2, and L200-D3-1

of the component materials can be found in the Ref. [29]. Effective bond length is defined as the bond length over which the resultant shear contact stress is at least 97% of the ultimate strength of an infinite stiffener, such as BFRP [22, 30]. Effective bond length is the length over which strain decreases from the maximum value to a zero value, as measured from the strain profile in the stiffener by resistance strain gauges [20, 22]. Both definitions contain intrinsic ambiguities [22]. Effective bond length can be estimated from the distance between the points at which bond stresses are 10% of the maximum bond stress $\tau_{max}$ [31].

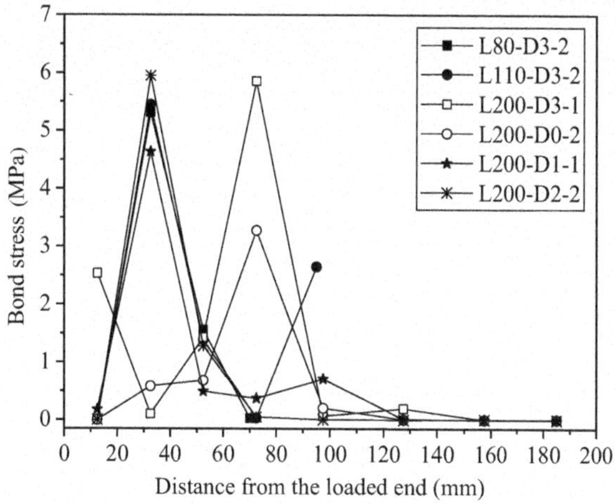


**Fig. 2.9** Relationship between bond stress and distance from the loaded end of specimens L80-D3-2, L110-D3-2, L200-D0-2, L200-D1-1, L200-D2-2, and L200-D3-1

(1) Test Result of Dynamic Effective Bond Length

The findings in Table 2.2 and Fig. 2.9 highlight the effect of strain rate on bond length. A significant decrease in bond length was observed when the strain rate increased from $0.61 \times 10^{-3}\ s^{-1}$ to $0.61 \times 10^{-2}$, $0.47 \times 10^{-1}$, and $0.63\ s^{-1}$. Given that bond length ($L_e$) is inversely proportional to concrete strength according to existing literatures [3, 16], a decreasing trend $L_e$ is expected when considering the increase of concrete strength with the increase of strain rate [2, 32].

As shown in Table 2.2, the dynamic effective bond lengths of specimens L200-D0, L200-D1, L200-D2, and L200-D3 at strain rates of $0.61 \times 10^{-3}$, $0.61 \times 10^{-2}$, $0.47 \times 10^{-1}$, and $0.63\ s^{-1}$ are 71.8, 65.5, 62.8, and 56.7 mm, respectively. When the strain rate increased from $0.61 \times 10^{-3}\ s^{-1}$ to $0.61 \times 10^{-2}$, $0.47 \times 10^{-1}$, and $0.63\ s^{-1}$, the dynamic effective bond length decreased at the rates of 8.8%, 12.5%, and 21.0%, respectively. Table 2.3 provides a summary of dynamic effective bond length and dynamic elastic modulus. The dynamic elastic modulus values were obtained from the load cell and strain gauge results.

(2) Existing Theoretical Models for Effective Bond Length

Many theoretical models have been developed based on empirical data and theories in fracture mechanics to predict the effective bond length between FRP sheets and concrete [23]. These models are commonly referred to as static effective bond length models. Ten analytical models for the evaluation of the effective bond length were analyzed by comparing the experimental value with the theoretical provisions, as shown in Table 2.4.

In ACI Code [39], CSA Code [40], and the model developed in [36], the effective bond length is inversely proportional to FRP stiffness. This condition can be attributed to the fact that the ACI Code [39] is calibrated with a few experimental observations and does not consider the cohesive nature of the concrete substrate [42]. In the model developed in [38], the effective bond length is unaffected by FRP stiffness, but the

**Table 2.3** Dynamic effective bond length and dynamic elastic modulus of specimens L200-D0, L200-D1, L200-D2 and L200-D3

| Identification of specimen | Strain rate ($s^{-1}$) | Effective bond length $L_e$ (mm) | | | | | Elastic modulus ($\times 10^5$ MPa) | | | |
|---|---|---|---|---|---|---|---|---|---|---|
| | | Model in [33] | Test results | Ratio of test results | Result of Eq. (2.11) | Ratio of Eq. (2.11) | Test results | Ratio of test results | Ratio of Eq. (2.7) | IAE (%) |
| L200-D0 | $0.61 \times 10^{-3}$ | 57.2 | 71.8 | 1 | 71.8 | 1 | 1.07 | 1 | 1 | 0 |
| L200-D1 | $0.61 \times 10^{-2}$ | 50.2 | 65.5 | 0.91 | 67.1 | 0.93 | 1.15 | 1.075 | 1.063 | 1.12 |
| L200-D2 | $0.47 \times 10^{-1}$ | 45.5 | 62.8 | 0.87 | 62.8 | 0.87 | 1.17 | 1.093 | 1.118 | 2.29 |
| L200-D3 | 0.63 | 41.0 | 56.7 | 0.79 | 57.5 | 0.80 | 1.28 | 1.196 | 1.188 | 0.67 |

**Table 2.4** Performance of effective bond length models

| References | Equation | Consideration factors | Theoretical result (mm) | IAE (%) |
|---|---|---|---|---|
| [34] | $L_e = 0.395\frac{(E_f t_f)^{0.54}}{f_{co}^{0.09}}$ | $E_f$, $t_f$, $f_c'$ | 47.9 | 33.3 |
| [35] | $L_e = \sqrt{\frac{E_f t_f}{2 f_t}}$ | $E_f$, $t_f$, $f_c'$ | 49.2 | 31.5 |
| [36] | $L_e = e^{6.13-0.580\ln(E_f t_f)}$ | $E_f$, $t_f$ | 105.2 | 46.5 |
| [16] | $L_e = \sqrt{\frac{E_f t_f}{\sqrt{f_c'}}}$ | $E_f$, $t_f$, $f_c'$ | 48.3 | 32.7 |
| [37] | $L_e = \sqrt{\frac{E_f t_f}{2 f_{ctm}}}$ | $E_f$, $t_f$, $f_{ctm}$ | 47.0 | 34.5 |
| [33] | $L_e = 1.33\frac{\sqrt{E_f t_f}}{f_t}$ | $E_f$, $t_f$, $f_t$ | 57.2 | 20.3 |
| [38] | $L_e = 100\,\text{mm}$ | – | 100.0 | 39.3 |
| [39] | $L_e = \frac{23{,}300}{(n E_f t_f)^{0.58}}$ | $E_f$, $t_f$ | 97.1 | 35.2 |
| [40] | $L_e = \frac{25{,}350}{(E_f t_f)^{0.58}}$ | $E_f$, $t_f$ | 105.6 | 47.1 |
| [41] | $L_e = \sqrt{\frac{E_f t_f}{4 f_t}}$ | $E_f$, $t_f$, $f_t$ | 34.8 | 51.5 |
| Average value | | | 69.2 | 37.2 |

where $L_e$ is the effective bond length; $t_f$ is the thickness of FRP; $E_f$ is the elastic modulus of FRP; $f_{ctm}$ is the mean tensile strength of concrete; $f_c'$ or $f_{co}$ is the cylinder axial compressive strength of concrete; $f_t$ is the axial tensile strength of concrete; and $n$ is the number of FRP layers

effective bond length of the other equations was proportional to FRP stiffness. Thus, the effective bond length equations adopted in current codes were very inconsistent. The accuracy of existing models was evaluated with integral absolute error (IAE) provided by Eq. (2.2), which is sensitive to the deviation of a model from the test results and is often utilized in model assessment [43, 44].

$$\text{IAE} = \sum \frac{\left[(\text{Expe.} - \text{Theo.})^2\right]^{1/2}}{\sum |\text{Expe.}|} \tag{2.2}$$

where Expe. and Theo. denote experimental and theoretical results, respectively. The results of the error analyses are shown in Table 2.4.

Table 2.4 shows that the average and minimum IAEs of the models are 37.2% and 20.3%, respectively. The models in the ACI Code [39], CSA Code [40], and the models developed in [36] and [38] overestimate the effective bond length up to 47.1%. The model developed in [41] underestimates the effective bond length by 51.5%. Generally, the theoretical results exhibit a relatively high scattering. This large scattering is partly attributed to the inaccuracy of the models.

Effective bond length is mainly controlled by two factors: FRP stiffness $E_f t_f$ and concrete strength $f_{co}$ [45]. Existing literature has established that effective bond length $L_e$ is proportional to FRP rigidity $E_f t_f$ [45]. The result from [16, 33, 37] shows that effective bond length $L_e$ is reversely proportional to concrete strength. Several scholars believe that effective bond length is unaffected by concrete strength [36, 38, 39]. Thus, the conclusion on the effect of concrete strength $L_e$ is somehow inconsistent [45]. The elastic modulus of FRP increases with the increase of strain rate [46, 47]. If effective bond length $L_e$ is unaffected by concrete strength and is proportional to FRP rigidity $E_f t_f$ [45], then the effective bond length $L_e$ will increase with the increase of strain rate. Table 2.2 shows that effective bond length $L_e$ decreased with the increase of strain rate. Hence, the opinion that effective bond length $L_e$ is unaffected by concrete strength is unreasonable according to the test results. Effective bond length is affected by both FRP rigidity $E_f t_f$ and concrete strength.

The model developed in [33] is the most accurate model among the existing models shown in Table 2.4. The model developed by Lu [33] underestimates effective bond length by 20.3%. An equation can be presented to predict effective bond length based on the model developed in [33], as shown in Eq. (2.3).

$$L_e = \alpha \frac{\sqrt{E_f t_f}}{f_t} \tag{2.3}$$

The constant $\alpha$ was determined from the test results and is 1.67. Thus, Eq. (2.4) can be utilized to predict the static effective bond length of the test result.

$$L_e = 1.67 \frac{\sqrt{E_f t_f}}{f_t} \tag{2.4}$$

Equation (2.4) shows that effective bond length $L_e$ is affected by the elastic modulus of FRP and the tensile strength of concrete. Equation (2.4) shows that effective bond length increases as BFRP stiffness increases. This trend was also observed in [42]. Equation (2.4) also shows that effective bond length decreases as concrete tensile strength increases.

Strain rate has a significant effect on the elastic modulus of FRP and the tensile strength of concrete [46–48]. The dynamic tensile strength of concrete increases with the increase of strain rate [48]. The reason for the increase of strength at high strain rates is believed to be caused by extremely localized yielding because of the enhanced bond among molecules at high strain rates. The relationship between concrete tensile strength and strain rate is proposed, as shown in Eq. (2.5) [48].

$$\frac{f_t^d}{f_t^s} = 1 + 0.173 \lg\left(\frac{\dot{\varepsilon}_d}{\dot{\varepsilon}_s}\right) \tag{2.5}$$

where $f_t^d$ is the dynamic tensile strength of concrete; $f_t^s$ is the static tensile strength of concrete at a strain rate of $10^{-5}$ $s^{-1}$; $\dot{\varepsilon}_d$ is the dynamic strain rate; and $\dot{\varepsilon}_s$ is the static strain rate ($0.61 \times 10^{-3}$ $s^{-1}$).

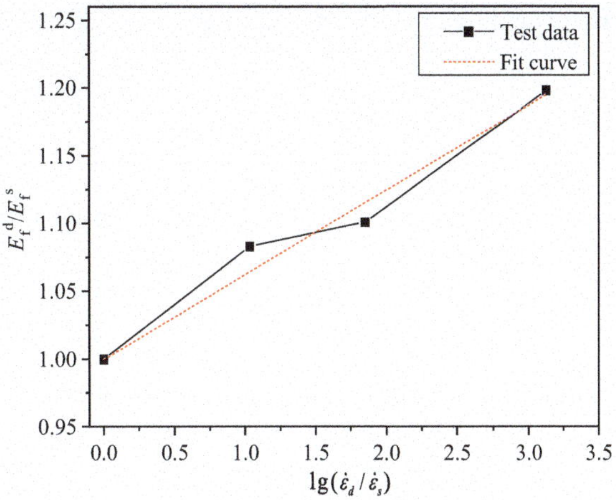


**Fig. 2.10** Fitting curve of the dynamic elastic modulus of BFRP sheet

Effective bond length is proportional to the elastic modulus of FRP [16, 33, 37]. The elastic modulus of FRP is affected by strain rate [46, 47]. shows the tested elastic modulus of BFRP under different strain rates. The elastic modulus of BFRP was 1.07 $\times$ $10^5$, 1.15 $\times$ $10^5$, 1.17 $\times$ $10^5$, and 1.28 $\times$ $10^5$ MPa when the strain rates were 0.61 $\times$ $10^{-3}$, 0.61 $\times$ $10^{-2}$, 0.47 $\times$ $10^{-1}$, and 0.63 $s^{-1}$, respectively. The dynamic elastic modulus of FRP increases with the increasing of the logarithm of the strain rate [47]. The relationship between the ratio of dynamic elastic modulus $E_f^d$ to static elastic modulus $E_f^s$ and the logarithm of the ratio of dynamic to static strain rate is shown in Fig. 2.10.

The dynamic elastic modulus of BFRP increased with the increase of the logarithm of the ratio of dynamic to static strain rate, as shown in Eq. (2.6).

$$\frac{E_f^d}{E_f^s} = 1 + \gamma \lg\left(\frac{\dot{\varepsilon}_d}{\dot{\varepsilon}_s}\right) \tag{2.6}$$

The coefficient $\gamma$ was determined through regression analysis to provide the following equation ($R^2$ value = 0.972).

$$\frac{E_f^d}{E_f^s} = 1 + 0.0625 \lg\left(\frac{\dot{\varepsilon}_d}{\dot{\varepsilon}_s}\right) \tag{2.7}$$

The values $E_f^d/E_f^s$ calculated with Eq. (2.7) were 1, 1.063, 1.118, and 1.188 when the strain rates were 0.61 $\times$ $10^{-3}$, 0.61 $\times$ $10^{-2}$, 0.47 $\times$ $10^{-1}$, and 0.63 $s^{-1}$, respectively. The results of the deviations between Eq. (2.7) and the test were 0, 1.85%, 1.54%, and 0.83%; the deviations were small. Thus, Eq. (2.7) could be utilized to predict the dynamic elastic modulus of BFRP under different strain rates.

Based on Eq. (2.4), the ratio of dynamic effective bond length to static effective bond length could be expressed as Eq. (2.8) in consideration of the strain rate on elastic modulus and concrete strength.

$$\frac{L_e^d}{L_e^s} = \frac{\sqrt{E_f^d/E_f^s}}{f_t^d/f_t^s} \tag{2.8}$$

Substitute Eqs. (2.4) and (2.7) into Eq. (2.8), as shown in Eq. (2.9).

$$\frac{L_e^d}{L_e^s} = \frac{\sqrt{1+0.0625\lg(\dot{\varepsilon}_d/\dot{\varepsilon}_s)}}{1+0.173\lg(\dot{\varepsilon}_d/\dot{\varepsilon}_s)} \tag{2.9}$$

Table 2.5 shows the relationship between the ratio of dynamic effective bond length to static effective bond length and strain rates.

The ratios of dynamic effective bond length $L_e^d$ to static effective bond length $L_e^s$ calculated with Eq. (2.9) were 1, 0.879, 0.797, and 0.716 when the strain rates were $0.61 \times 10^{-3}$, $0.61 \times 10^{-2}$, $0.47 \times 10^{-1}$, and $0.63\ s^{-1}$, respectively; the ratios of dynamic effective bond length $L_e^d$ to static effective bond length $L_e^s$ from the test results were 1, 0.91, 0.87, and 0.79 under the same strain rates. The results of the deviations between Eq. (2.9) and the test were 0, 3.4%, 8.3%, and 9.3%. The deviations were large; thus, Eq. (2.9) could not be utilized to accurately predict dynamic effective bond length. The results from Eqs. (2.5), (2.7), and (2.9) are shown in Fig. 2.11. The model for dynamic effective bond length could not be presented based on existing theoretical results. A new model must be established to predict dynamic effective bond length based on the test results.

(3) Computational Models of Dynamic Effective Bond Length

The effective bond length decreases with the increase of the logarithm of the strain rate [21]. The relationship between the ratio of dynamic effective bond length $L_e^d$ of the specimens under dynamic loading to the corresponding value $L_e^s$ of the specimens under static loading and the logarithm of the ratio of dynamic to static strain rate is shown in Fig. 2.12.

**Table 2.5** Relationship between the ratio of dynamic effective bond length to static effective bond length and strain rates

| Strain rate ($s^{-1}$) | $f_t^d/f_t^s$ | $E_f^d/E_f^s$ | Test results | Results of Eq. (2.9) | IAE of Eq. (2.9) (%) | Results of Eq. (2.11) | IAE of Eq. (2.11) (%) |
|---|---|---|---|---|---|---|---|
| $0.61 \times 10^{-3}$ | 1 | 1 | 1 | 1 | 0 | 1 | 0.0 |
| $0.61 \times 10^{-2}$ | 1.173 | 1.063 | 0.91 | 0.879 | 3.4 | 0.93 | 2.2 |
| $0.47 \times 10^{-1}$ | 1.326 | 1.118 | 0.87 | 0.797 | 8.3 | 0.87 | 0.0 |
| 0.63 | 1.521 | 1.188 | 0.79 | 0.716 | 9.3 | 0.80 | 1.3 |

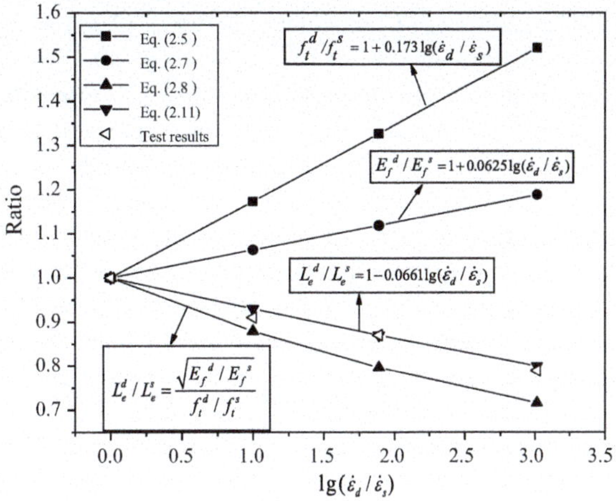


**Fig. 2.11** Effect of strain rate on effective bond length

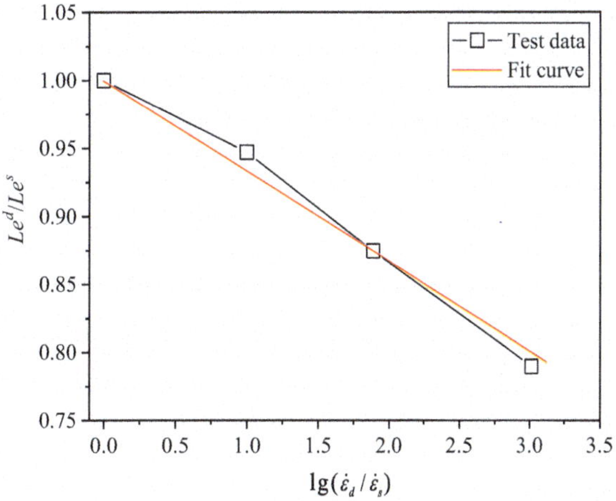


**Fig. 2.12** Relationship between effective bond length and strain rate

The relationship between the value of $L_{\mathrm{e}}^{\mathrm{d}}/L_{\mathrm{e}}^{\mathrm{s}}$ and the logarithm of the ratio of dynamic to static strain rate is linear, as shown in Eq. (2.10).

$$\frac{L_{\mathrm{e}}^{\mathrm{d}}}{L_{\mathrm{e}}^{\mathrm{s}}} = 1 + \beta \lg\left(\frac{\dot{\varepsilon}_{\mathrm{d}}}{\dot{\varepsilon}_{\mathrm{s}}}\right) \tag{2.10}$$

where $L_{\mathrm{e}}^{\mathrm{d}}$ and $L_{\mathrm{e}}^{\mathrm{s}}$ are dynamic and static bond lengths, respectively. The coefficient $\beta$ was determined through regression analysis to provide the following equation ($R^2$ = 0.990).

$$\frac{L_e^d}{L_e^s} = 1 - 0.0661 \lg\left(\frac{\dot{\varepsilon}_d}{\dot{\varepsilon}_s}\right) \tag{2.11}$$

The values $L_e^d/L_e^s$ calculated with Eq. (2.11) were 1, 0.93, 0.87, and 0.80 when the strain rates are $0.61 \times 10^{-3}$, $0.61 \times 10^{-2}$, $0.47 \times 10^{-1}$, and $0.63\ s^{-1}$, respectively. The results of the deviations between Eq. (2.11) and the test are 0, 2.2%, 0, and 1.3%, as shown in Table 2.5. The deviations were small. The results of Eq. (2.11) are also shown in Fig. 2.11. The effective bond length was proportional to BFRP rigidity $E_f t_f$ and reversely proportional to concrete tensile strength. Figure 2.11 shows that the degree of the effect of strain rate on tensile strength is larger than that of elastic modulus. Thus, dynamic effective bond length decreased with the increase of strain rate, as shown in Fig. 2.11. Equations (2.4) and (2.11) can thus be utilized to predict the dynamic effective bond length of BFRP under different strain rates. Figure 2.11 also shows that the effect of strain rate on dynamic effective bond length in Eq. (2.11) is less than that in Eq. (2.9). In Eq. (2.11), all the factors that influence dynamic effective bond length are considered; however, only the strain rate effect on both elastic modulus and concrete strength is included in Eq. (2.9). Other factors, such as the effect of strain rate on the epoxy resin, were disregarded. The ratio of effective bond length of CFRP sheet bonded to concrete under the strain rates of $1.0 \times 10^{-2}$ and $1.0 \times 10^{-1}\ s^{-1}$ to that under the strain rate of $1.0 \times 10^{-3}\ s^{-1}$ is about 0.958 and 0.923 [21], respectively. And the result from Eq. (2.11) is 0.933 and 0.866, respectively. The results of the deviations between Eq. (2.11) and the test result in [21] are 2.6% and 6.2%, respectively. The deviations were small. Thus, Eq. (2.11) can be applied to other types of FRPs, such as CFRP.

## 2.2 Bond Stress-Slip Relationship Between BFRP Sheets and Concrete Under Dynamic Loading

### *2.2.1 Experimental Program*

The material properties, test details, and the experimental program had been presented in Sect. 2.1.1. Table 2.6 shows the configuration of the 12 specimens.

### *2.2.2 Bond Stress Under Dynamic Loading*

Strain gauges were used to measure the axial strains in BFRP sheets during the test and $x_i$ indicated the positions of the strain gauges. The behavior between BFRP sheets and concrete was analyzed by plotting the experimental axial strains $\varepsilon_i$ that the strain gauges were measured during the test along the BFRP sheet (Fig. 2.1b). The strain distribution along the BFRP sheet using the typical loading steps of Specimen L200-D3-1 is shown in Fig. 2.13a. These strains were observed from the strain gauges mounted on the upper surface of the FRP strip. The strain distribution showed a descending curve from the center of the specimen toward the end of the sheet [25].

**Table 2.6** Results of maximum bond stress and ultimate load of specimens L200-D0, L200-D1, L200-D2 and L200-D3

| Identification of specimen | Loading rate (mm/s) | Strain rate ($s^{-1}$) | $P_u$ (kN) | | | Ultimate displacements (mm) | | $\tau_{max}$ (MPa) | | |
|---|---|---|---|---|---|---|---|---|---|---|
| | | | Individual | Ave | Ratio | Individual | Ave | Individual | Ave | Ratio |
| L200-D0-1 | | | 11.4 | 11.9 | 1 | 2.040 | 2.094 | 2.95 | 3.27 | 1 |
| L200-D0-2 | 0.07 | $0.61 \times 10^{-3}$ | 10.8 | | | 1.872 | | 3.59 | | |
| L200-D0-3 | | | 13.6 | | | 2.371 | | – | | |
| L200-D1-1 | | | 15.0 | 13.6 | 1.143 | 2.412 | 2.484 | 4.64 | 4.82 | 1.474 |
| L200-D1-2 | 0.7 | $0.61 \times 10^{-2}$ | 13.3 | | | 2.610 | | 5.00 | | |
| L200-D1-3 | | | 12.5 | | | 2.429 | | 3.89 | | |
| L200-D2-1 | | | 15.5 | 14.4 | 1.210 | 2.904 | 2.665 | 3.68 | | 1.645 |
| L200-D2-2 | 7 | $0.47 \times 10^{-1}$ | 14.5 | | | 2.525 | | 5.37 | 5.38 | |
| L200-D2-3 | | | 13.1 | | | 2.567 | | 5.39 | | |
| L200-D3-1 | | | 16.2 | 15.8 | 1.328 | 6.409 | 5.395 | 5.95 | 5.75 | 1.758 |
| L200-D3-2 | 70 | 0.63 | 15.7 | | | 5.678 | | 5.82 | | |
| L200-D3-3 | | | 15.6 | | | 4.097 | | 5.49 | | |

The strains deviated locally from the basic descending tendency and showed a zigzag curve, which was different from the theoretical relationship between the BFRP sheet strain and the distance from the loaded end [48]. The stress concentration from the aggregate embedding between the BFRP sheet and concrete caused these local strain deviations. The negative strain resulted in negative bond stresses that were not corrected. To eliminate the negative strain gradients, part of the strain values that compared each following value was corrected to monotonically decrease toward the sheet end, as illustrated in Fig. 2.1b. The BFRP sheet strains were corrected for all specimens in the same manner.

Figure 2.13 shows that the increase of BFRP strain was gradual until the applied load $P$ reached the ultimate load $P_{\mathrm{u}}$. The average maximum strains of BFRP for specimens L200-D0, L200-D1, L200-D2, and L200-D3 were approximately 8300, 9600, 10,500, and 10,700 $\mu\varepsilon$, respectively, when the ultimate load $P_u$ was reached. The maximum strain of BFRP increases as the strain rate increases, which agrees with the result in [21]. An inspection of the impact strain distribution graph is shown in Fig. 2.13, which reveals that the strain distributions are similar under different strain rates.

The interface bond stress $\tau_{i+1/2}$, which was a local phenomenon, was determined from the following relationship [27], as shown in Eq. (2.1). The most significant results related to the tests performed on the sheets are summarized in Fig. 2.14.

The distribution of bond stress along the length of the BFRP plate is shown in Fig. 2.14. As the applied load was increased, the peak bond stress was observed to move along the length of the sheet from the loaded end, which signified a debonding crack propagation.

The results of the maximum bond stress $\tau_{\max}$ of all specimens under different loading rates are shown in Table 2.6. The results indicate that: (1) the maximum bond stress of specimens L200-D0, L200-D1, L200-D2, and L200-D3 under the strain rates of $0.61 \times 10^{-3}$, $0.61 \times 10^{-2}$, $0.47 \times 10^{-1}$, and $0.63\ \mathrm{s}^{-1}$ were 3.27, 4.82, 5.38, and 5.75 MPa, respectively; and (2) the maximum bond stress between the BFRP and concrete increased as the strain rate increased. This result is attributed to the sensitivity of the concrete shear strength to the strain rate [6]. The shear strength of concrete is a function of its compressive strength, and the compressive strength is known to increase with the strain rate [49].

Many theoretical models have been developed based on empirical data to predict the bond stress between FRP sheets and concrete [50, 51]. These models are commonly referred to as static bond stress models. Thirteen of these models compared the experimental value with the theoretical provisions, as reported in Table 2.7.

The test result of the static maximum bond stress was 3.27 MPa. Table 2.7 showed that the static maximum bond stress obtained from these 13 models was between 0.67 MPa and 6.99 MPa. The average value of the maximum bond stress between BFRP sheets and concrete obtained from these models was 3.22 MPa. Table 2.7 showed that the average and maximum IAE of these models was 57.0% and 113.8%, respectively. The six models proposed by [31, 35, 51–53, 58] overestimated the static

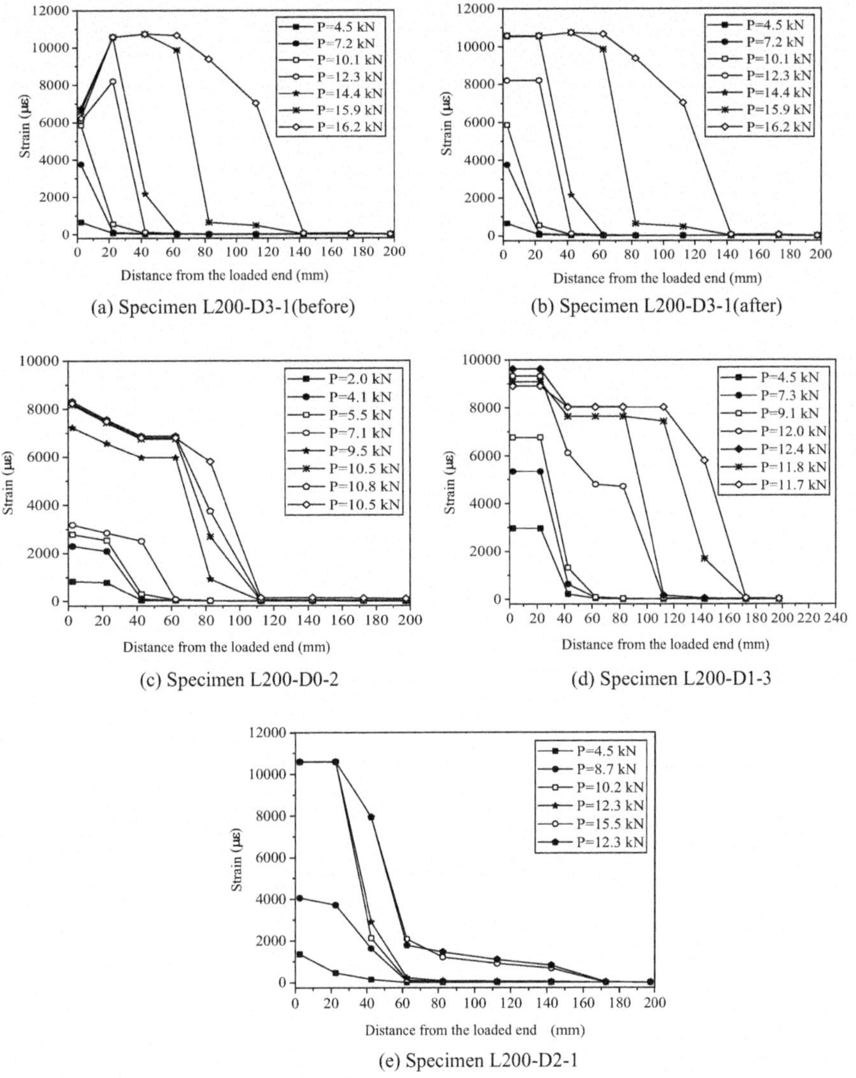


**Fig. 2.13** Relationship between distance from the loaded end and BFRP strain of specimens L200-D0-2, L200-D1-3, L200-D2-1, and L200-D3-1

maximum bond stress up to 113.8%. The seven models proposed in [36, 40, 50, 54–57] underestimated the bond stress up to 79.5%. Generally, the theoretical results had a relatively high scattering. This large scattering is partly attributed to the test method [17, 19].

Strain rate has a significant effect on the elastic modulus of FRP and the tensile strength of concrete [46–48]. The dynamic tensile strength of concrete increases as the strain rate increases [48]. Moreover, the dynamic compressive strength of

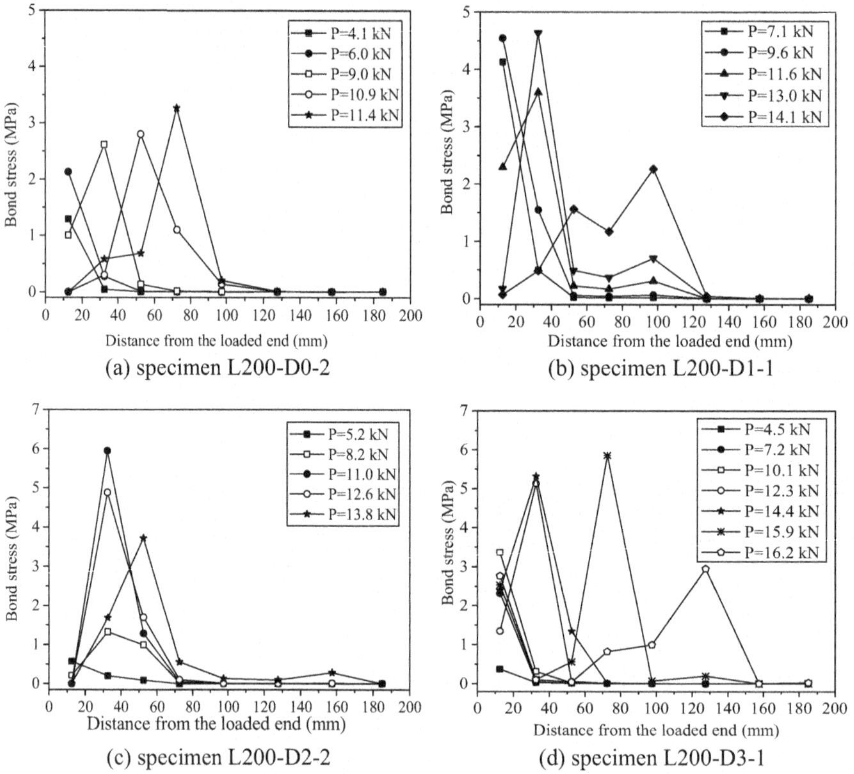


**Fig. 2.14** Relationship between distance from the loaded end and bond stress of specimens L200-D0-2, L200-D1-1, L200-D2-2, and L200-D3-1

concrete increases as the strain rate increases [60]. The increased strength at higher strain rates is attributed to the extremely localized yielding caused by the enhanced bond among the molecules at high strain rates [6]. The elastic modulus of FRP or concrete strength is included in most of the models for the evaluation of the static maximum bond stress. Thus, the strain rates influence the bond stress between BFRP sheets and concrete because the strain rates affect the elastic modulus of FRP or concrete.

The relationship between the concrete tensile strength and strain rate is shown in Eq. (2.5) [48].

The dynamic compressive strength of concrete increased as the strain rate increased [60]. The relationship between the concrete compressive strength and strain rate is proposed, as shown in Eq. (2.12).

$$\frac{f_c^d}{f_c^s} = 1 + 0.0795\,\lg\left(\frac{\dot{\varepsilon}_d}{\dot{\varepsilon}_s}\right) \tag{2.12}$$

**Table 2.7** Performance of bond stress models

| References | Equation | Consideration factors | Theoretical result (MPa) | IAE (%) |
|---|---|---|---|---|
| [35] | $\tau_{max} = 1.8 f_t$ | $f_t$ | 4.72 | 44.3 |
| [50] | $\tau_{max} = 6.13 - \ln L$ | $L$ | 3.13 | 4.3 |
| [52] | $\tau_{max} = 5.88 L^{-0.609}$ | $L$ | 0.79 | 75.8 |
| [42] | $\tau_{max} = 3.1(n_f E_f t_f)^{0.32}$ | $E_f$, $t_f$ | 6.99 | 113.8 |
| [53] | $\tau_{max} = 1.64\sqrt[4]{f_{cu}}$ | $f_{cu}$ | 4.06 | 24.2 |
| [54] | $\tau_{max} = 0.5 f_t$ | $f_t$ | 1.31 | 59.9 |
| [55] | $\tau_{max} = 110.2 \times 10^{-6} E_f t_f$ | $E_f$, $t_f$ | 1.40 | 57.2 |
| [36] | $\tau_{max} = 110.2 \times 10^{-6} E_f t_f \left(\frac{f_{co}}{42}\right)^{2/3}$ | $E_f$, $t_f$, $f_{co}$ | 1.11 | 66.1 |
| [31] | $\tau_{max} = 3.5 f_c'^{0.19}$ | $f_c'$ | 6.66 | 103.7 |
| [56] | $\tau_{max} = 2.68 \times 10^{-5} E_f t_f (f_c')^{0.2}$ | $E_f$, $t_f$, $f_c'$ | 0.67 | 79.5 |
| [57] | $\tau_{max} = 0.0182(n_f E_f t_f)^{0.5}$ | $E_f$, $t_f$ | 2.05 | 37.3 |
| [58] | $\tau_{max} = 0.165 f_c'$ | $f_c'$ | 4.88 | 49.2 |
| [51] | $\tau_{max} = 0.93 f_{co}^{0.44}$ | $f_{co}$ | 4.13 | 26.3 |
| Average value | | | 3.22 | 57.0 |

where $L$ is the effective bond length; $t_f$ is the thickness of FRP; $E_f$ is the elastic modulus of FRP; $f_{cu}$ is the cubic compressive strength of concrete; $f_c'$ or $f_{co}$ is the cylinder axial compressive strength of concrete; $f_t$ is the axial tensile strength of concrete; and $n_f$ is the number of FRP layers

where $f_c^d$ is the dynamic compressive strength of concrete and $f_c^s$ is the static compressive strength of concrete under the strain rate of $10^{-5}$ s$^{-1}$.

The dynamic elastic modulus of BFRP increases as the strain rate increases [12], as shown in Eq. 2.7.

In the models proposed in [50, 52], the bonded length $L$ is the only factor that affects the maximum bond stress; thus, these two models cannot be used to predict the strain effect on the bond stress because the bonded length is constant at different strain rates. In the models proposed in [31, 51, 53, 58], the maximum bond stress is proportional to compressive strength; the theoretical results of maximum bond stress from these models increase as the strain rates increase because the compressive strength increases with the increase of the strain rate [60]. In the models proposed in [38], the maximum bond stress is proportional to the concrete tensile strength; the theoretical results of the maximum bond stress from both models increase with the increase of the strain rate because the tensile strength increases as the strain rate increases [48]. In the models developed proposed in [55, 61], the maximum bond stress is proportional to the elastic modulus of FRP, and the bond stress of these models increases as the strain rate increases because the elastic modulus of BFRP increases with the increase of the strain rate [60]. In the models developed proposed in [36] and [56], the maximum bond stress is proportional to both the elastic modulus of FRP and concrete strength; the bond stress of both models increases as the strain rate increases because both the elastic modulus of BFRP and concrete strength increase as the strain rates increase [48]. Thus, all these models, except those proposed in [50] and [52], can lead to the conclusion that the bond stress will increase when the strain rate increases because the elastic modulus of FRP, concrete tensile, or compressive strength increases as the strain rate increases [48].

Interfacial law collects all compliance contributions at the interface level. Therefore, concrete damage at high shear stress levels until debonding must be included in the interface law [19]. The Mohr–Coulomb failure criterion for concrete maximum shear stress can be adopted and written as a function of concrete strength, as shown in Eq. (2.13).

$$\tau_{\max} \propto k_{\mathrm{b}}\sqrt{f_{\mathrm{cm}} \cdot f_{\mathrm{ctm}}} \tag{2.13}$$

where $f_{\mathrm{cm}}$ and $f_{\mathrm{ctm}}$ are the mean concrete strength in compression and tension, respectively; and $k_{\mathrm{b}}$ is a coefficient that indicates the effect of FRP reinforcement width.

The ratio of dynamic bond stress $\tau_{\max}^{\mathrm{d}}$ and static bond stress $\tau_{\max}^{\mathrm{s}}$ under different strain rates is shown in Eq. (2.14), which is based on Eq. (2.13).

$$\frac{\tau_{\max}^{\mathrm{d}}}{\tau_{\max}^{\mathrm{s}}} = \sqrt{\frac{f_{\mathrm{t}}^{\mathrm{d}} f_{\mathrm{c}}^{\mathrm{d}}}{f_{\mathrm{t}}^{\mathrm{s}} f_{\mathrm{c}}^{\mathrm{s}}}} \tag{2.14}$$

when Eqs. (2.5) and (2.12) are substituted into Eq. (2.14), the values of $\tau_{\max}^{\mathrm{d}}/\tau_{\max}^{\mathrm{s}}$, which are calculated using Eq. (2.14), are 1, 1.126, 1.235, and 1.374 when the strain rates are $0.61 \times 10^{-3}, 0.61 \times 10^{-2}, 0.47 \times 10^{-1}$, and $0.63\ \mathrm{s}^{-1}$, respectively. In addition, the results of the deviations from Eq. (2.14) and the test are 0, 23.4%, 24.9%, and 21.7%. The deviations are high; thus, Eq. (2.14) cannot be used to accurately predict the dynamic bond stress of the interface between BFRP and concrete.

The dynamic maximum bond stress increases with the increase of the strain rate [7], as shown in Eq. (2.15).

$$\frac{\tau_{\mathrm{m}}}{\tau_0} = \left(\frac{\dot{\varepsilon}}{\dot{\varepsilon}_0}\right)^{\frac{0.93}{f_{\mathrm{cu}}-17.46}} \tag{2.15}$$

where $\tau_{\mathrm{m}}$ and $\tau_0$ are the dynamic and static maximum bond stress, respectively; $\dot{\varepsilon}$ is the dynamic strain rate; $\dot{\varepsilon}_0$ is the static strain rate ($2.5 \times 10^{-5}\ \mathrm{s}^{-1}$); and $f_{\mathrm{cu}}$ is the concrete cubic compressive strength.

The values $\tau_{\max}^{\mathrm{d}}/\tau_{\max}^{\mathrm{s}}$ calculated using Eq. (2.15) are 1, 1.113, 1.223, and 1.380 when the strain rates are $0.61 \times 10^{-3}$, $0.61 \times 10^{-2}$, $0.47 \times 10^{-1}$, and $0.63\ \mathrm{s}^{-1}$, respectively. The bond stress increases by approximately 38% when the strain rate increases from $0.61 \times 10^{-3}$ to $0.63\ \mathrm{s}^{-1}$. The results of the deviations from Eq. (2.15) and the test are 0, 24.5%, 25.6%, and 21.5%. The deviations are high, so Eq. (2.15) cannot be used to accurately predict the dynamic maximum bond stress.

The dynamic bond stress increases with the increase of the strain rate [6], as shown in Eq. (2.16).

$$\frac{\tau_{\max}^{\mathrm{d}}}{\tau_{\max}^{\mathrm{s}}} = 0.052\ln(\dot{\varepsilon}) + 0.736 \tag{2.16}$$

where $\tau_{\max}^{\mathrm{d}}$ and $\tau_{\max}^{\mathrm{s}}$ are the dynamic and static maximum bond stress, respectively; and $\dot{\varepsilon}$ is the strain rate ($\mu\varepsilon$/s). The strain rates included in Eq. (2.16) range from $1.0 \times 10^{-3}\ \mathrm{s}^{-1}$ to $1.0 \times 10^{-1}\ \mathrm{s}^{-1}$.

The values $\tau_{\max}^{\mathrm{d}}/\tau_{\max}^{\mathrm{s}}$ calculated using Eq. (2.16) are 1, 1.112, 1.211, and 1.337 when the strain rates are $0.61 \times 10^{-3}$, $0.61 \times 10^{-2}$, $0.47 \times 10^{-1}$, and $0.63\ \mathrm{s}^{-1}$, respectively. Based on Eq. (2.16), the bond stress increases by approximately 34% when the strain rate increases from $0.61 \times 10^{-3}$ to $0.63\ \mathrm{s}^{-1}$. The results of the deviations from Eq. (2.16) and the test are 0, 24.6%, 26.4%, and 23.9%, respectively. The deviations are high, so Eq. (2.16) cannot be used to accurately predict the dynamic maximum bond stress of the BFRP-to-concrete interface.

Figure 2.15 shows the relationship between the ratio of the dynamic maximum bond stress $\tau_{\max}^{\mathrm{d}}$ of the specimens under dynamic loadings to the corresponding value $\tau_{\max}^{\mathrm{s}}$ of the specimen under static loading and strain rates.

The relationship between the value $\tau_{\max}^{\mathrm{d}}/\tau_{\max}^{\mathrm{s}}$ and the strain rate is shown in Eq. (2.17).

$$\frac{\tau_{\max}^{\mathrm{d}}}{\tau_{\max}^{\mathrm{s}}} = 1 + \alpha\left(\lg\left(\frac{\dot{\varepsilon}_{\mathrm{d}}}{\dot{\varepsilon}_{\mathrm{s}}}\right)\right)^{\beta} \tag{2.17}$$

where $\tau_{\max}^{\mathrm{d}}$ and $\tau_{\max}^{\mathrm{s}}$ are the dynamic and static maximum bond stress, respectively; $\dot{\varepsilon}_d$ is the dynamic strain rate; and $\dot{\varepsilon}_s$ is the static strain rate ($0.61 \times 10^{-3}\ \mathrm{s}^{-1}$). Coefficients

**Fig. 2.15** Relationship between maximum bond stress and strain rates

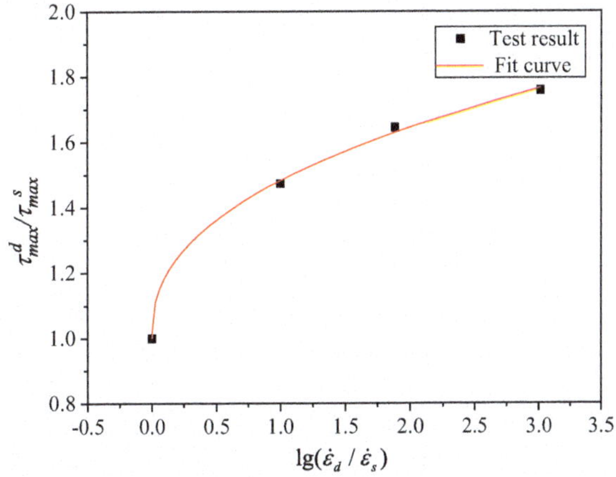

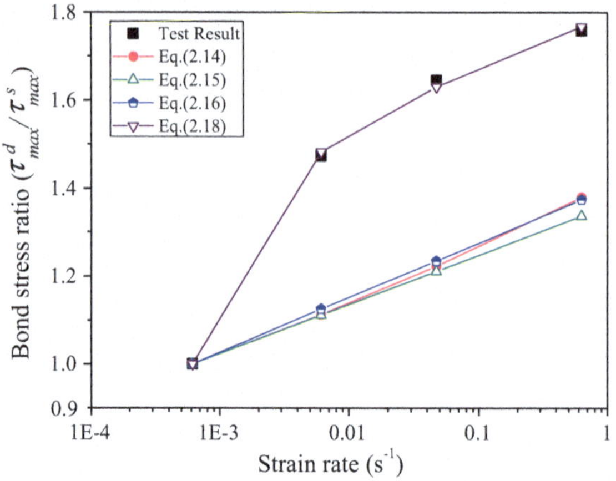


**Fig. 2.16** Strain rate effect on maximum bond stress

$\alpha$ and $\beta$ are determined from the regression analysis to provide the following equation with a $R^2$ value of 0.996.

$$\frac{\tau^d_{\max}}{\tau^s_{\max}} = 1 + 0.483\left(\lg\left(\frac{\dot{\varepsilon}_d}{\dot{\varepsilon}_s}\right)\right)^{0.418} \tag{2.18}$$

The values $\tau^d_{\max}/\tau^s_{\max}$ calculated using Eq. (2.18) are 1, 1.483, 1.631, and 1.768 when the strain rates are $0.61 \times 10^{-3}$, $0.61 \times 10^{-2}$, $0.47 \times 10^{-1}$, and 0.63 $s^{-1}$, respectively. The results of the deviations from Eq. (2.18) and the test are 0, 0.68%, 0.85%, and 0.57%. These deviations were small. Thus, Eq. (2.18) could be used to predict the dynamic bond stress of BFRP and concrete interface under different strain rates.

Figure 2.16 shows the relationship between the bond stress ratio, $\tau^d_{\max}/\tau^s_{\max}$, obtained from test results.

As shown in Fig. 2.16, the theoretical results from Eqs. (2.14), (2.15), and (2.16) underestimate the dynamic bond stress at different strain rates. Based on the test result, the relationship between the bond stress ratio, $\tau^d_{\max}/\tau^s_{\max}$, under different strain rates and the strain rates was nonlinear. Equation (2.18) was established to accurately predict the dynamic maximum bond stress under different strain rates.

### 2.2.3 *Ultimate Load Under Dynamic Loading*

Current studies suggest that the main failure mode between FRP and concrete in the shear tests is the cracking of concrete under shear, which commonly occurs a few millimeters from the adhesive-concrete interface. Therefore, the ultimate load (i.e., the maximum transferable load or bond strength) of the joint depends significantly on concrete strength.

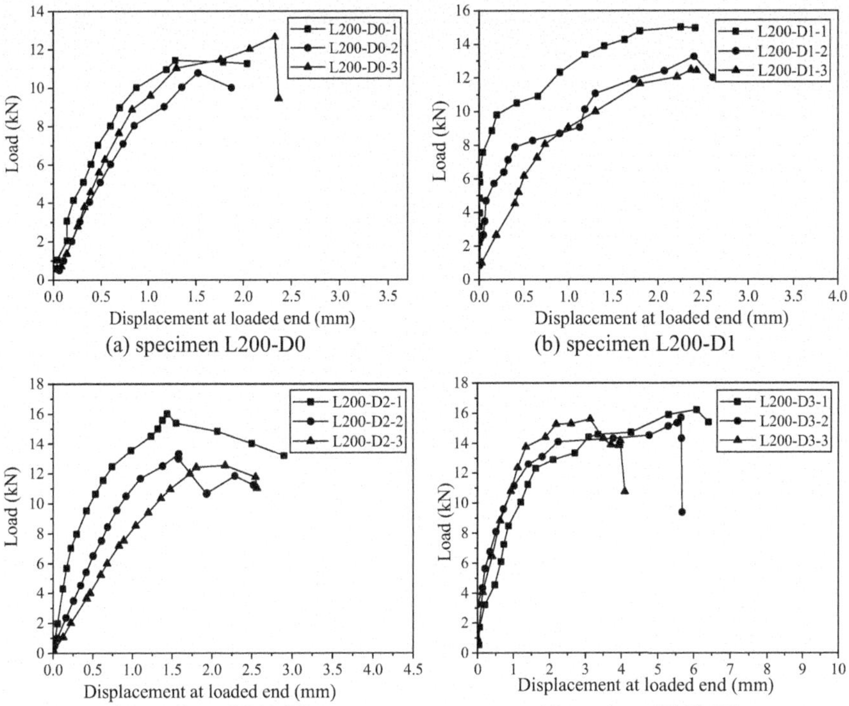


(a) specimen L200-D0

(b) specimen L200-D1

(c) specimen L200-D2

(d) specimen L200-D3

**Fig. 2.17** Load versus global slip between BFRP sheets and concrete of specimens L200-D0, L200-D1, L200-D2, and L200-D3

Figure 2.17 shows the experimental load versus global slip graphs between BFRP sheets and concrete for all specimens. The displacements in Fig. 2.17 illustrate the average data of two displacement transducers bridged over the test block and the anchorage block (Fig. 2.17a).

To examine the ultimate load sensitivity between BFRP sheets and concrete with an increasing strain rate of loading, the ultimate load for BFRP/concrete double-lap specimens at four different strain rates is shown in Table 2.6, which demonstrates that the dynamic ultimate loads are 11.9, 13.6, 14.4, and 15.8 kN for specimens L200-D0, L200-D1, L200-D2, and L200-D3 with the strain rates of $0.61 \times 10^{-3}$, $0.61 \times 10^{-2}$, $0.47 \times 10^{-1}$, and $0.63\ s^{-1}$, respectively. When the strain rates increased from $0.61 \times 10^{-3}\ s^{-1}$ to $0.61 \times 10^{-2}$, $0.47 \times 10^{-1}$, and $0.63\ s^{-1}$, the dynamic ultimate load increased at the rates of 14.3%, 21.0%, and 32.8%, respectively.

Many theoretical models had been developed to predict the ultimate load (i.e., bond strength) between FRP sheet and concrete, which were generally based on pull test results. In each of the following models, the elastic modulus, thickness, and width of the BFRP sheet are indicated as $E_f$, $t_f$, and $b_f$, respectively. Further details of the analytical formulations can be found in the original documents. The FRP-concrete joint can carry the maximum bond stress because the bonded length in the test is greater than the effective bond length, and the maximum bond stress is obtained in the following:

The models proposed in [10] are shown in Eqs. (2.19) and (2.20).

$$P_{\max} = \beta_l b_f \sqrt{2E_f t_f G_f} \tag{2.19}$$

$$G_f = 0.308\beta_w^2 \sqrt{f_t} \tag{2.20}$$

where $\beta_w$ is the width ratio factor, as shown in Eq. (2.21).

$$\beta_w = \sqrt{\frac{2.25 - b_f/b_c}{1.25 + b_f/b_c}} \tag{2.21}$$

The models proposed in [35] are shown in Eq. (2.22).

$$P_{\max} = 0.64k_w b_f \sqrt{f_t E_f t_f} \tag{2.22}$$

where $k_w$ is a geometrical factor, as shown in Eq. (2.23).

$$k_w = \sqrt{1.125\frac{2 - b_f/b_c}{1 + b_f/400}} \tag{2.23}$$

The models proposed in [50] are shown in Eq. (2.24).

$$P_{\max} = (6.13 - \ln L)b_f L \tag{2.24}$$

The models proposed in [62] are shown in Eq. (2.25).

$$P_{\max} = b_f L(5.88L^{-0.669}) \tag{2.25}$$

where $L$ is within the brackets, in cm [42].

The models proposed in [45] are shown in Eqs. (2.26) and (2.27).

$$P_{\max} = \left(0.5 + 0.08\sqrt{\frac{E_f t_f}{100 f_t}}\right) b_f L_e 0.5 f_t \tag{2.26}$$

$$L_e = 100\,\text{mm} \tag{2.27}$$

The models proposed in [63] are shown in Eqs. (2.28) and (2.29).

$$P_{\max} = 110.2 \times 10^{-6} E_f t_f b_f L_e \tag{2.28}$$

$$L_{\mathrm{e}} = e^{6.13-0.58\ln(E_{\mathrm{f}}t_{\mathrm{f}})} \tag{2.29}$$

The models proposed in [36] are shown in Eqs. (2.30) and (2.31).

$$P_{\max} = 110.2 \times 10^{-6}\left(\frac{f_{\mathrm{co}}}{42}\right)^{2/3} E_{\mathrm{f}}t_{\mathrm{f}}b_{\mathrm{f}}L_{\mathrm{e}} \tag{2.30}$$

$$L_{\mathrm{e}} = e^{6.13-0.58\ln(E_{\mathrm{f}}t_{\mathrm{f}})} \tag{2.31}$$

The models proposed in [51] are shown in Eqs. (2.32) and (2.33).

$$P_{\max} = 0.93 f_{\mathrm{co}}^{0.44} b_{\mathrm{f}} L_{\mathrm{e}} \tag{2.32}$$

$$L_{\mathrm{e}} = 0.125(E_{\mathrm{f}}t_{\mathrm{f}})^{0.57} \tag{2.33}$$

The models proposed in [64] are shown in Eqs. (2.34) and (2.35).

$$P_{\max} = (b_f + 7.4)\sqrt{2E_f t_f G_f} \tag{2.34}$$

$$G_f = 0.514 f_{co}^{0.236} \tag{2.35}$$

The models proposed in [16] are shown in Eqs. (2.36) and (2.37).

$$P_{\max} = 0.427 k_w b_f L_e \sqrt{f_{co}} \tag{2.36}$$

$$L_{\mathrm{e}} = \sqrt{\frac{E_{\mathrm{f}}t_{\mathrm{f}}}{\sqrt{f_{\mathrm{co}}}}} \tag{2.37}$$

where $k_w$ is the width ratio factor, as shown in Eq. (2.38).

$$k_{\mathrm{w}} = \sqrt{\frac{2 - b_{\mathrm{f}}/b_{\mathrm{c}}}{1 + b_{\mathrm{f}}/b_{\mathrm{c}}}} \tag{2.38}$$

Using a fracture mechanics approach, the CNR DT-200 [37] quantifies the maximum stress in the FRP sheet $\tau_{\max}$ as a function of the fracture energy $\Gamma_{FK}$ between FRP sheets and concrete. The FRP-concrete joint can carry the maximum bond stress when the bonded length is greater than the effective bond length $L_e$. The maximum bond stress is shown in Eq. (2.39).

$$\tau_{\max} = \frac{k_{\mathrm{cr}}}{\gamma_{\mathrm{f,d}}\sqrt{\gamma_{\mathrm{c}}}}\sqrt{\frac{2E_{\mathrm{f}}\Gamma_{\mathrm{FK}}}{t_{\mathrm{f}}}} \tag{2.39}$$

where $\gamma_{f,d}$ and $\gamma_c$ are the sheet and concrete safety factors, respectively. Material safety factors $\gamma_{\mathrm{f,d}} = \gamma_{\mathrm{c}} = 1$ have also been adopted in analyzing the test result.

The maximum force is finally calculated by multiplying the cross-sectional area of the sheet with the maximum bond stress obtained from Eq. (2.39). The reduction factor $k_{cr}$ is equal to 1.0 in case of end delamination.

The fracture energy between FRP sheets and concrete is calculated in Eq. (2.40).

$$\Gamma_{\mathrm{FK}} = k_{\mathrm{G}} k_{\mathrm{b}} \sqrt{f_{\mathrm{ck}} f_{\mathrm{ctm}}} \tag{2.40}$$

where $f_{ck}$ is the characteristic compressive cylinder strength of concrete; $k_G$ is the fracture energy coefficient with an average value of 0.064 [19]; and $k_b$ is a geometrical factor equal to, as shown in Eq. (2.41).

$$k_{\mathrm{b}} = \sqrt{\frac{2 - b_{\mathrm{f}}/b_{\mathrm{c}}}{1 + b_{\mathrm{f}}/b_{\mathrm{c}}}} \geq 1 \tag{2.41}$$

Thus, the ultimate load between the FRP sheet and concrete when the bonded length is greater than the effective bond length is in Eq. (2.42).

$$P_{\mathrm{max}} = b_{\mathrm{f}} \sqrt{2 E_{\mathrm{f}} t_{\mathrm{f}} k_{\mathrm{G}} k_{\mathrm{b}} \sqrt{f_{\mathrm{ck}} f_{\mathrm{ctm}}}} \tag{2.42}$$

The theoretical static (i.e., the strain rate is $0.61 \times 10^{-3}\ \mathrm{s}^{-1}$) ultimate loads between BFRP sheets and concrete obtained from these 11 models are shown in Table 2.8.

The accuracy of each analytical model was evaluated by comparing the experimental ultimate load result with the corresponding analytical predictions. The accuracy of the various models was assessed by computing the coefficient of IAE [43, 44]. IAE = 0 indicates a perfect match between the experimental and analytical ultimate loads. The test result of the static ultimate load between BFRP sheets and concrete on two opposites of the concrete was 11.9 kN, and the ultimate load on a single BFRP-concrete interface was 5.95 kN. The deviations between the experimental and theoretical results denoted as IAE0 are also shown in Table 2.8. The minimum, maximum, and average values of IAE0 for static experimental and theoretical results were 1.9%, 64.4%, and 20.9%, respectively. Generally, the theoretical results have a relatively high rate of scattering. This large scattering is partly attributed to the test method [19].

The strain rate affects the elastic modulus of FRP and concrete strength [46, 48, 60]. The ultimate load between BFRP sheets and concrete under the strain rate of $0.61 \times 10^{-3}\ \mathrm{s}^{-1}$ was considered as the static ultimate load $P^{\mathrm{s}}_{\mathrm{max}}$. When Eqs. (2.5) and (2.12) were substituted into these models, the dynamic ultimate load $P^{\mathrm{d}}_{\mathrm{max}}$ was obtained when the strain rates were $0.61 \times 10^{-2}$, $0.47 \times 10^{-1}$, and $0.63\ \mathrm{s}^{-1}$, respectively. The ratios of dynamic ultimate load $P^{\mathrm{d}}_{\mathrm{max}}$ under the strain rates of $0.61 \times 10^{-2}$, $0.47 \times 10^{-1}$, and $0.63\ \mathrm{s}^{-1}$ to the static ultimate load $P^{\mathrm{s}}_{\mathrm{max}}$ under the strain rates of $0.61 \times 10–3\ \mathrm{s}^{-1}$ are shown in Table 2.8. Table 2.6 indicated that the average test results of the value $P^{d}_{max}/P^{s}_{max}$ were 1.143, 1.210, and 1.328 when the strain rates were $0.61 \times 10^{-3}$, $0.61 \times 10^{-2}$, $0.47 \times 10^{-1}$, and 0.63 $s^{-1}$, respectively. The deviations between the experimental and theoretical results for all these 11 models when the strain rates were $0.61 \times 10^{-2}$, $0.47 \times 10^{-1}$, and 0.63 $s^{-1}$ were denoted as IAE1, IAE2, and IAE3, respectively. The deviations denoted as IAE4 were the total of IAE1, IAE2, and IAE3. Table 2.8 summarizes the results for all considered models, in which the coefficient of variation IAE is listed with the corresponding value of the ratios between the dynamic and static theoretical ultimate load value of the interface between BFRP and concrete under different strain rates.

The models proposed in [16, 35, 36, 50, 51, 61, 62] underestimated the static ultimate load at the rates of 5.7%, 1.9%, 2%, 47.3%, 5.3%, 5.4%, and 54.5%, respectively. The models proposed in [37, 45, 62, 63, 64] overestimated the static ultimate load at the rates of 3%, 16.4%, 54.5%, 23.7%, and 64.4%.

In the models proposed in [50], as well as [62], the bonded length $L$ is the only factor that can affect the ultimate load of BFRP and concrete, which the elastic modulus and concrete strength do not affect. Thus, the strain rate does not affect the ultimate load because the strain rate based on the models proposed in [50], as well as those proposed in [62], does not affect bonded length $L$. The test results show that the ultimate load increases with the increase of the strain rate, as shown in Table 2.6. Thus, the model proposed in [50] and [62] cannot be used to accurately predict the dynamic ultimate load between BFRP sheets and concrete under different strain rates.

**Table 2.8** Results of the statistical analysis in terms of coefficient of IAE and ratios

| References | Static | | Dynamic | | | | | | |
|---|---|---|---|---|---|---|---|---|---|
| | $0.61 \times 10^{-3}$ $s^{-1}$ | | $0.61 \times 10^{-2}$ $s^{-1}$ | | $0.47 \times 10^{-1}$ $s^{-1}$ | | 0.63 $s^{-1}$ | | IAE4 (%) |
| | Theoretical result (kN) | IAE0 (%) | Ratio | IAE1 (%) | Ratio | IAE2 (%) | Ratio | IAE3 (%) | |
| [61] | 5.63 | 5.4 | 1.073 | 6.2 | 1.135 | 6.2 | 1.211 | 8.8 | 21.2 |
| [35] | 5.84 | 1.9 | 1.116 | 2.3 | 1.218 | 0.6 | 1.345 | 1.3 | 4.2 |
| [50] | 3.13 | 47.3 | 1.000 | 14.3 | 1.000 | 21.0 | 1.000 | 32.8 | 68.1 |
| [62] | 9.19 | 54.5 | 1.000 | 14.3 | 1.000 | 21.0 | 1.000 | 32.8 | 68.1 |
| [45] | 6.92 | 16.4 | 1.158 | 1.3 | 1.298 | 7.3 | 1.475 | 11.1 | 19.7 |
| [63] | 7.36 | 23.7 | 1.026 | 10.3 | 1.048 | 13.4 | 1.075 | 19.0 | 42.7 |
| [36] | 5.83 | 2.0 | 1.079 | 5.6 | 1.150 | 4.9 | 1.241 | 6.6 | 17.1 |
| [51] | 5.64 | 5.3 | 1.071 | 6.3 | 1.133 | 6.4 | 1.213 | 8.7 | 21.4 |
| [64] | 9.78 | 64.4 | 1.040 | 9.0 | 1.075 | 11.2 | 1.118 | 15.8 | 36.0 |
| [16] | 5.61 | 5.7 | 1.051 | 8.1 | 1.095 | 9.5 | 1.150 | 13.4 | 31.0 |
| [37] | 6.13 | 3.0 | 1.093 | 4.3 | 1.175 | 2.9 | 1.277 | 3.8 | 11.0 |
| Average | 6.46 | 20.9 | 1.064 | 7.5 | 1.121 | 9.5 | 1.191 | 14.0 | 31.0 |

In the model proposed in [45], the effect of the strain rate on the dynamic ultimate load was overestimated at a rate of up to 11.1%. The effective bond length decreases as the strain rate increases [21], whereas the effective bond length in the model proposed in [45] is constant. Thus, the model proposed in [45] is also inadequate to accurately predict the dynamic ultimate load of BFRP and concrete under different strain rates.

The results reported in Table 2.6 show that some models are more accurate, such as those proposed in [35–37]. The IAE4 results of the models proposed in [35-37] under different strain rates are 17.1%, 4.2%, and 11.0%, respectively. The best result in terms of the IAE coefficient was from the model proposed in [35] for the static test (IAE0 = 1.9%) and for the dynamic test (IAE4 = 4.22%). All the parameters in Eq. (2.22) of the Neubauer and Rostasy [35] model could not be calibrated using the test results, whereas the test result could calibrate the parameter $k_G$ in Eq. (2.42) of the model in [37]. Based on Eq. (2.32), a $k_{\rm G}$ value can be easily obtained for each experimental test using the following expression [37], as shown in Eq. (2.43).

$$k_{\rm G} = P_{\rm max}^2 / \left( b_{\rm f}^2 2 E_{\rm f} t_{\rm f} k_{\rm b} \sqrt{f_{\rm ck} f_{\rm ctm}} \right) \tag{2.43}$$

The mean value $k_G$ is 0.060 based on the static test result. Thus, Eq. (2.42) and $k_{\rm G}$ = 0.060 can be used to accurately predict the static ultimate load of the test result.

Figure 2.18 shows the relationship between the ratio of the dynamic ultimate load $P_{\rm max}^{\rm d}$ of the specimens under dynamic loadings to the corresponding value $P_{\rm max}^{\rm s}$ of a specimen under static loading and strain rates.

The relationship between the value of $P_{\rm max}^{\rm d} / P_{\rm max}^{\rm s}$ and the strain rates is linear, as shown in Eq. (2.44).

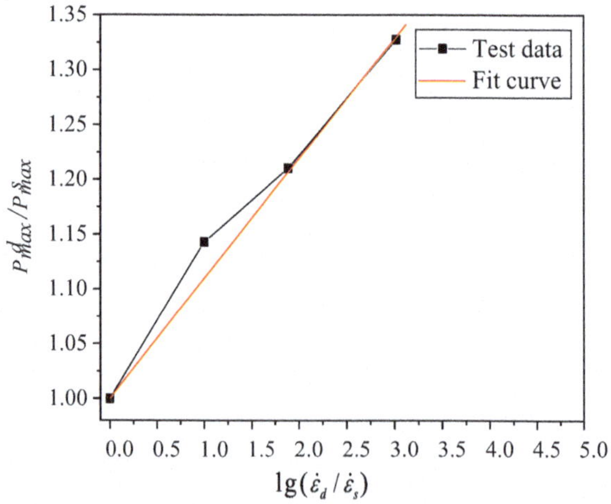


**Fig. 2.18** Relationship between ultimate load and strain rates

$$\frac{P_{\max}^{\mathrm{d}}}{P_{\max}^{\mathrm{s}}} = 1 + \gamma \lg\left(\frac{\dot{\varepsilon}_{\mathrm{d}}}{\dot{\varepsilon}_{\mathrm{s}}}\right) \tag{2.44}$$

where $P_{\max}^{\mathrm{d}}$ and $P_{\max}^{\mathrm{s}}$ are the dynamic and static ultimate loads, respectively; $\dot{\varepsilon}_{\mathrm{d}}$ is the dynamic strain rate; and $\dot{\varepsilon}_{\mathrm{s}}$ is the static strain rate ($0.61 \times 10^{-3}\ \mathrm{s}^{-1}$). The coefficient $\gamma$ was determined from the regression analysis to provide the following equation with a $R^2$ value of 0.980, as shown in Eq. (2.45).

$$\frac{P_{\max}^{\mathrm{d}}}{P_{\max}^{\mathrm{s}}} = 1 + 0.1094 \lg\left(\frac{\dot{\varepsilon}_{\mathrm{d}}}{\dot{\varepsilon}_{\mathrm{s}}}\right) \tag{2.45}$$

The values $P_{\max}^{\mathrm{d}}/P_{\max}^{\mathrm{s}}$ from Eq. (2.45) are 1, 1.109, 1.206, and 1.330 when the strain rates are $0.61 \times 10^{-3}$, $0.61 \times 10^{-2}$, $0.47 \times 10^{-1}$, and $0.63\ \mathrm{s}^{-1}$, respectively. The results of the deviations from Eq. (2.45) and the test are 0, 2.93%, 0.33%, and 0.15%, respectively. The deviations are small. Thus, Eq. (2.45) can be used to predict the dynamic ultimate load of the interface between BFRP and concrete under different strain rates. Moreover, Eqs. (2.42) and (2.45) with $k_{\mathrm{G}} = 0.060$ can be used to accurately predict the dynamic and static ultimate loads between BFRP sheets and concrete.

### 2.2.4 Bond-Slip Relationship Under Dynamic Loading

The bond-slip relationship is the constitutive model between BFRP sheets and concrete, which is essential in studying the behavior of BFRP-strengthened concrete structures. Tensile strength in the case of bond failure, strain distribution of FRP, and bond stress distributions can be obtained using the bond stress versus slip relationship model through numerical analysis [31]. The failure mode between BFRP sheets and concrete under different strain rates was debonding in the concrete layer based on test results. The strain rate did not affect the failure modes of the interface between BFRP and concrete under dynamic loading. The following assumptions were made to define the slip distribution along the BFRP sheet [26, 31]: (1) the relative displacement between the concrete and sheet at the free end of the sheet is zero; (2) the deformation of the concrete specimen that is far from the external cover is negligible concerning its BFRP counterpart; and (3) a linear variation of strains exists in the BFRP sheet between two subsequent strain gauges. The integration of the strain profile, which starts from the end strain gauge at $x = x_m$, provides the following expression for the slip at a general abscissa $x$ [26] with $x_i \le x \le x_{i+1}$ (Fig. 2.13(b)), as shown in Eq. (2.46).

$$s(x) = s(x_{i+1}) - \frac{(\varepsilon_{i+1} - \varepsilon_i)}{(x_{i+1} - x_i)} \cdot \frac{(x_{i+1} - x)^2}{2} + \varepsilon_i (x_{i+1} - x) \tag{2.46}$$

where $s(x_m) = 0$. Finally, the average slip between the two positions $x_i$ and $x_{i+1}$ was computed as Eq. (2.47).

$$s_{i+1/2} = \frac{s(x_{i+1}) + s(x_i)}{2} \tag{2.47}$$

where $x_i$ is the general strain gauge position, and $\varepsilon_i$ is the strain measured at $x_i$.

The bond length could be sufficiently long compared with the anchorage length for the validity of the assumption $s(x_m) = 0$. The post-processing procedure was performed for several BFRP strain profiles, which corresponded to different loading levels. After the calculation of all the test data, the relationships between the local average bond stress and the average slip ($\tau_{i+1/2}$ versus $s_{i+1/2}$) for all specimens are presented in Fig. 2.19.

The bond-slip curves were divided into an ascending branch and a descending branch based on the experimental data. Popovics's equation was used to calculate the bond stress [31, 64], as shown in Eq. (2.48).

$$\tau = \tau_{\max}\left[\frac{s}{s_0}\frac{n}{(n-1) + (s/s_0)^n}\right] \tag{2.48}$$

where $\tau$ is the local bond stress, in MPa; $s$ is the local slip, in mm; $\tau_{\max}$ is the maximum local bond stress, in MPa; $s_0$ is the slip corresponding to $\tau_{\max}$, in mm; and $n$ is constant.

$\tau_{\max}$ and $s_0$ were obtained directly from the experimental results. The value $n$ was calculated using the least squares method with a normalized $\tau$ and $s$ relationship. Figure 2.19 graphically shows the results of all specimens. Table 2.9 contains the fitting results of the average value of all the specimens. Maximum local bond stress ranged from 3.27 MPa to 5.75 MPa, and the slip at $\tau_{\max}$ ranged from 0.630 mm to 0.651 mm. The constant $n$ ranged from 2.98 to 3.05. The slip at $\tau_{\max}$ and constant $n$ had no clear relationship with the strain rates. The relationship between the maximum local bond stress $\tau_{\max}$ and the strain rate is shown in Eq. (2.18).

The adopted values of $s_0$ and $n$ are the average of all specimens, as shown in Table 2.9. The final results for Popovics's model are proposed, as shown in Eq. (2.49).

$$\tau = \tau_{\max}^{\mathrm{d}}\left[\frac{s}{s_0}\frac{n}{(n-1) + (s/s_0)^n}\right] \tag{2.49}$$

where $\tau_{\max}^{d}$ is the dynamic maximum local bond stress under different strain rate, in MPa; $s_0$ is the slip when the value $\tau_{\max}^{d}$ is 0.115 mm; and $n$ is 3.09.

The local bond stress versus slip of specimens under different strain rates is shown in Fig. 2.19. The relationship between the dynamic local bond stress and the slip of all specimens under different strain rates is similar. The maximum local bond stress and graph slope of the ascent stage increased as the strain rate increased. Thus, the proposed model could predict the local bond stress versus slip relationship of BFRP and concrete under different strain rates.

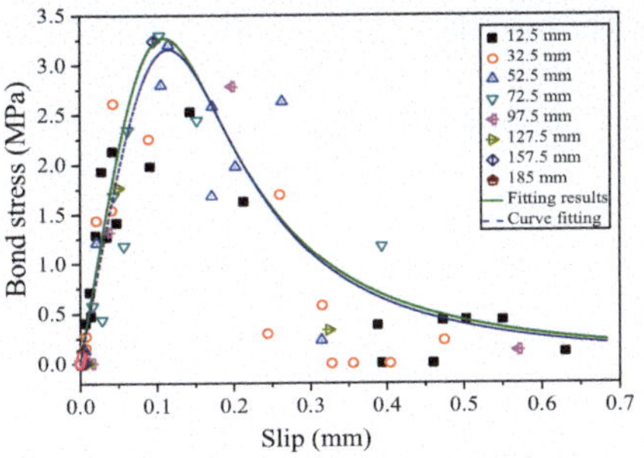


(a) specimen L200-D0

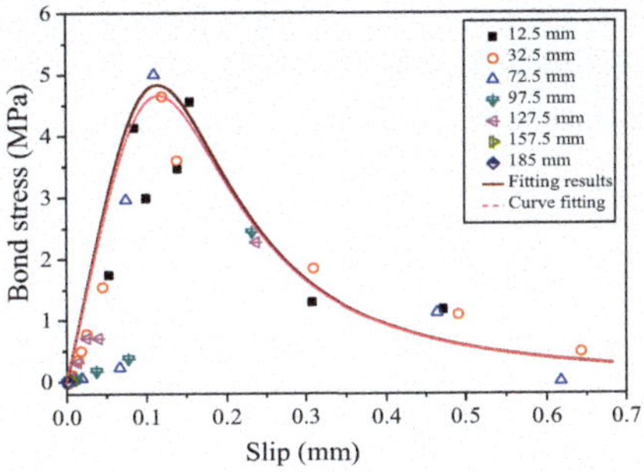


(b) specimen L200-D1

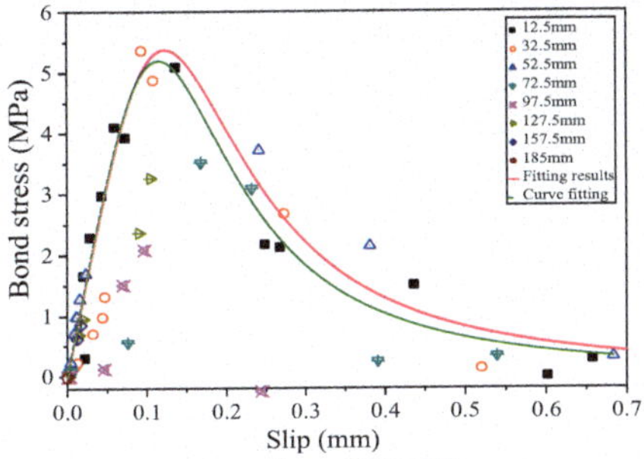


(c) specimen L200-D2

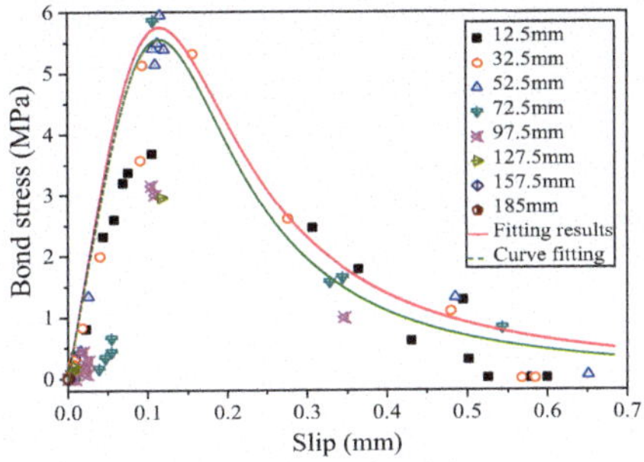


(d) specimen L200-D3

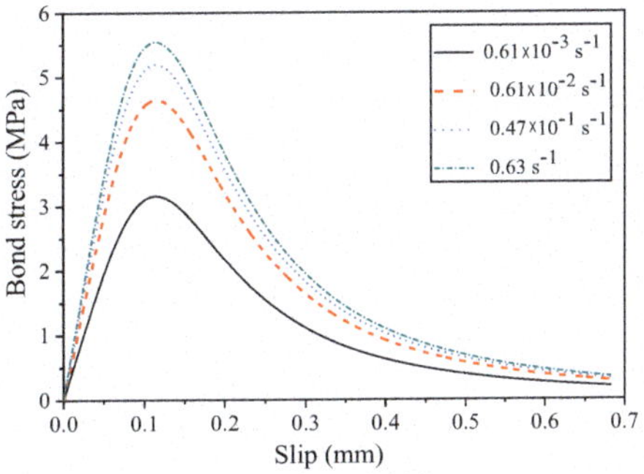


(e) specimen L200

**Fig. 2.19** Relationship between dynamic bond stress and slip under different strain rates of specimens L200-D0, L200-D1, L200-D2, and L200-D3

**Table 2.9** Test results and fitting results of the dynamic bond-slip relationship of specimens L200-D0, L200-D1, L200-D2 and L200-D3

| Identification of specimen | Strain rate ($s^{-1}$) | $\tau_{max}$ (MPa) | $s_0$ (mm) | $s_{max}$ (mm) | Regression coefficient, $n$ | Correlation coefficient, $R^2$ |
|---|---|---|---|---|---|---|
| L200-D0 | $0.61 \times 10^{-3}$ | 3.27 | 0.109 | 0.630 | 3.05 | 0.998 |
| L200-D1 | $0.61 \times 10^{-2}$ | 4.82 | 0.114 | 0.642 | 3.21 | 0.996 |
| L200-D2 | $0.47 \times 10^{-1}$ | 5.38 | 0.122 | 0.683 | 3.12 | 0.999 |
| L200-D3 | 0.63 | 5.75 | 0.115 | 0.651 | 2.98 | 0.988 |
| Average | – | – | 0.115 | 0.642 | 3.09 | – |

## 2.3 Dynamic Bond Stress-Slip Relationship Between BFRP Sheets and Concrete Under Initial Static Loading

### 2.3.1 *Experimental Program*

The material properties, test details, and the experimental program is presented in Sect. 2.1.1.

Table 2.10 shows the configuration of the 15 specimens. Each test is identified by the label "IX-n." The first alphabet of the specimen name I refers to initial static loading (i.e., I = initial static loading), X indicates the initial static loading (i.e., 0 = 0% initial static loading and 10 = 100% initial static loading), and the last number $n$ is the progressive test number. Twelve strain gauges with 5 mm gauge length were applied along the BFRP laminates on the side of the tested portion of the specimens to measure the axial strains during the bond test (Fig. 2.1b). Two layers of BFRP sheets were applied on the anchorage block of the specimen during the double-lap shear tests to ensure that debonding failure occurs at the test block side, as shown in Fig. 2.1a. Load, displacement, and strain data were recorded by a DH5922 data logger at a frequency of 50 kHz.

To determine the value of initial static loading, the mean debonding load-bearing capacity of specimen I0 under 100% initial static loading $P_{max,s}$ was determined; the average value of static debonding load-bearing capacity was 11.9 kN. The initial static loading with the proportion of 30%, 50%, and 80% for specimens I3, I5, and I8 was 3.6, 6.0, and 9.5 kN, respectively. For specimens I0, I3, I5, and I8, loads of 1.0, 3.6, 6.0, and 9.5 kN were initially applied to the specimens at a load rate of 3.5 kN/min, respectively. A dynamic load was then applied up to failure at the displacement rate of 70 mm/s, and the corresponding strain rate was 0.63 $s^{-1}$, which can represent the maximum strain rate of FRP in an earthquake [27]. A static load was applied on specimen I10 up to failure at the displacement rate of 0.07 mm/s. The loading procedure was implemented automatically under the control of a computer

**Table 2.10** Results of maximum bond stress and effective bond length of specimens I0, I3, I5, I8 and I10

| Identification of specimen | Initial static loading (%) | $\tau_{max}$(MPa) | | | $L_1$ (mm) | $L_2$ (mm) | $L_e$(mm) | | |
|---|---|---|---|---|---|---|---|---|---|
| | | Individual | Average | Ratio | | | Individual | Average | Ratio |
| I0-1 | 0 | 5.95 | 5.75 | 1 | 32.5 | 90.2 | 57.7 | 56.7 | 1 |
| I0-2 | | 5.82 | | | 34.2 | 92.5 | 58.3 | | |
| I0-3 | | 5.49 | | | 35.1 | 89.3 | 54.2 | | |
| I3-1 | 30 | 4.60 | 4.81 | 0.837 | 12.5 | 71.8 | 59.3 | 60.2 | 1.06 |
| I3-2 | | 5.22 | | | 13.1 | 74.4 | 61.3 | | |
| I3-3 | | 4.61 | | | 12.5 | 72.6 | 60.1 | | |
| I5-1 | 50 | 4.76 | 4.43 | 0.770 | 12.5 | 79.4 | 66.9 | 63.5 | 1.12 |
| I5-2 | | 4.17 | | | 12.7 | 76.2 | 63.5 | | |
| I5-3 | | 4.36 | | | 13.0 | 75.1 | 62.1 | | |
| I8-1 | 80 | 3.87 | 4.11 | 0.715 | 12.5 | 79.8 | 67.3 | 67.3 | 1.19 |
| I8-2 | | 4.21 | | | 13.3 | 80.2 | 66.9 | | |
| I8-3 | | 4.25 | | | 14.1 | 81.7 | 67.6 | | |
| I10-1 | 100 | 2.95 | 3.27 | 0.569 | 29.6 | 98.9 | 69.3 | 71.8 | 1.27 |
| I10-2 | | 3.59 | | | 30.1 | 104.3 | 74.2 | | |
| I10-3 | | – | | | – | – | – | | |

**Fig. 2.20** Failure mode of debonding in concrete

program. At least three specimens were tested for each loading history case to ensure the repeatability of the experimental results because when a test on some specimen failed, one more specimen was needed. The strain rate effect on the interface between FRP sheet and concrete was investigated [21, 27], and the dynamic data processing approach adopted in these two references was used.

### *2.3.2 Failure Modes*

All the specimens failed because of debonding in the concrete adjacent to the adhesive-concrete interface in which a thin layer of concrete was attached to the BFRP strip after failure (Fig. 2.20). Finally, all tested double-lap specimens failed because of debonding in concrete based on the failure type classification reported in [21]. Therefore, initial static loading did not affect the failure modes of the interface between BFRP and the concrete specimens under dynamic loading. The failure modes of the interface between BFRP and concrete are not affected by the strain rate or dynamic loading [21], in agreement with the test results.

### *2.3.3 Maximum Dynamic Bond Stress*

The strain distribution along the BFRP sheet at typical loading steps of specimen I0-1 is shown in Fig. 2.21a. These strains were found from the strain gauges mounted on the upper surface of the FRP strip. The strain distribution showed a descending curve from the center of the specimen toward the end of the sheet [48]. The strains locally deviated from the basic descending tendency and showed a zigzag curve, different from the theoretical relationship between the BFRP sheet strain and distance from the loaded end [48]. These local strain deviations were caused by the stress concentration because of aggregate embedding between the BFRP sheet and concrete. The negative strain resulted in negative bond stresses, which were not corrected. To eliminate negative strain gradients, part of the strain values comparing each following value was corrected to decrease monotonically toward the sheet end, as shown in Fig. 2.21b. The BFRP sheet strains were corrected for all specimens in the same manner. The relationships between the BFRP strain and distance from the loaded end of specimens under different initial static loadings are shown in Fig. 2.21b–f.

When the applied load $P$ was smaller than approximately 80% of the ultimate load $P_u$, the BFRP strain was minimal beyond the distance of approximately 50.0–70.0 mm from the loaded end, which indicates that almost all the applied load resisted with this small area. The increase of BFRP strain was gradual until the applied load $P$ reached the ultimate load $P_u$ (Fig. 2.21). The BFRP strain for specimens I0 and I10 was minimal beyond the distance of approximately 60.0 and 80.0 mm from the loaded end when the ultimate load $P_u$ was reached, which indicates that almost all the applied load resisted under this small area. The maximum strain of BFRP for specimens I0, I3, I5, I8, and I10 (Fig. 2.21b–f) was approximately $1.07 \times 10^{-4}$, $1.05 \times 10^{-4}$, $1.07 \times 10^{-4}$, $1.09 \times 10^{-4}$, and $8.30 \times 10^{-3}$ $\mu\varepsilon$, respectively, when the ultimate load $P_u$ was reached. Notably, a large part of the BFRP strip near the far end still had minimal strain when the ultimate load was reached, which confirms the concept of effective bond length, thereby indicating that increasing the bond length beyond a certain value did not further increase the bond strength [21]. Careful inspection of Fig. 2.21 revealed that local debonding near the loaded end occurred much earlier than was observed by the naked eye.

The behavior of the BFRP-to-concrete interface was analyzed by plotting the experimental axial strains $\varepsilon_i$ measured by the strain gauges during the test along the BFRP sheet (Fig. 2.1b). The interface bond stress $\tau_{i+1/2}$, which was a local phenomenon, was determined from the following well-known relationship [26, 27], as shown in Eq. (2.1). The most significant results related to the tests performed on the sheets are summarized in Fig. 2.21.

The distribution of bond stress along the length of the plate is shown in Fig. 2.22. The peak bond stress moved along the length of the sheet from the loaded end as the applied load was increased, which indicates debonding crack propagation. The maximum bond stress for specimen I0-1 (Fig. 2.22a) was at the distance of 12.5 mm from the loaded end when the applied load $P$ was approximately 60% of the ultimate load. The bond stress was minimal beyond the distance of approximately 30.0 mm from the loaded end. The maximum bond stress was at the distance of 33.0 mm from the loaded end when the applied load was $P = 14.4$ kN; the bond stress was minimal beyond the distance of approximately 70.0 mm from the loaded end. The maximum bond stress of specimen I0-1 was 5.95 MPa at a distance of 72.0 mm from the loaded end. Figure 2.22 illustrates that (1) the relationships between bond stress and distance from the loaded end are similar, and (2) the peak bond stress moved along the length of the sheet from the loaded end when the applied load was increased.

The results of the maximum bond stress of all specimens under different initial static loadings are shown in Table 2.10, which indicates that (1) the maximum bond

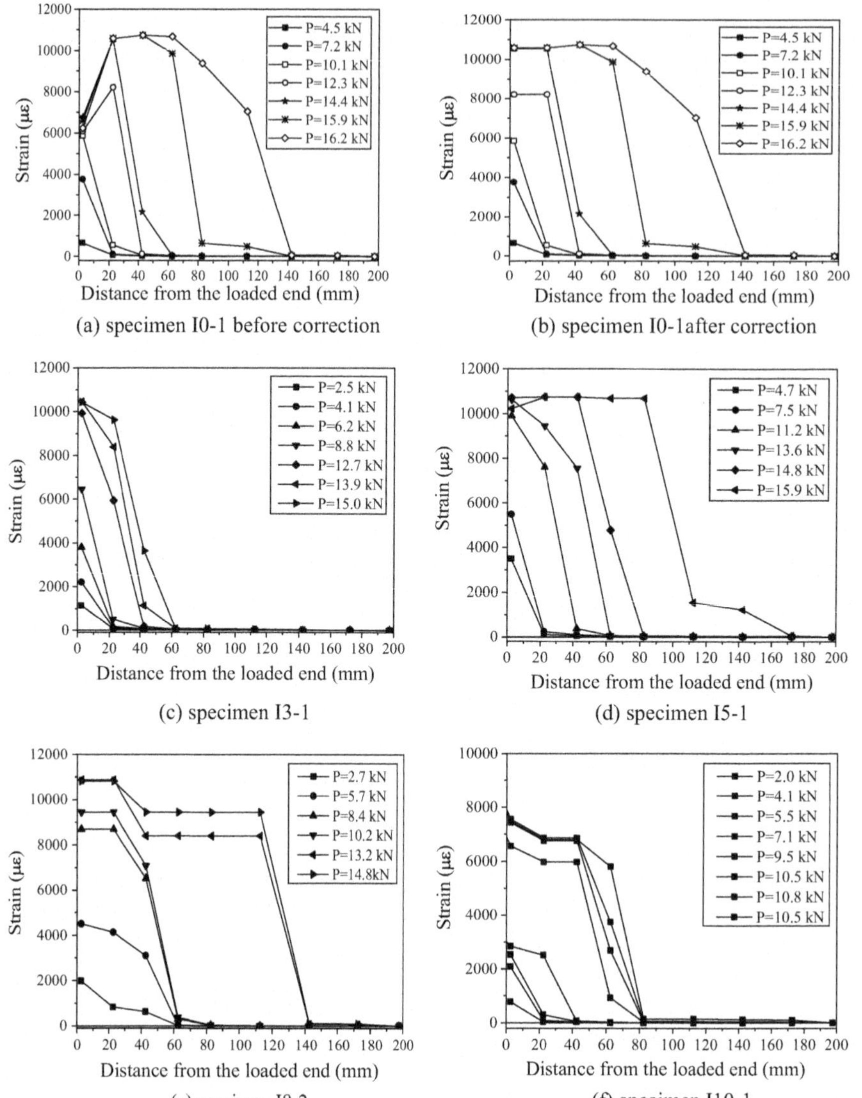


(a) specimen I0-1 before correction

(b) specimen I0-1after correction

(c) specimen I3-1

(d) specimen I5-1

(e) specimen I8-2

(f) specimen I10-1

**Fig. 2.21** Relationship between BFRP strain and distance from the loaded end of specimens I0-1, I3-1, I5-1, I8-2, and I10-1

stress between BFRP and concrete decreased with the increase of the proportion of initial static load; (2) the average value of maximum bond stress of specimen I10 under 100% static load was 3.27 MPa; (3) the average value of maximum bond stress of specimen I0 under 0% static load (i.e., dynamic loading) was 5.75 MPa; and (4) the maximum bond stress of specimens I3, I5, I8, and I10 under 30%, 50%, 80%, and

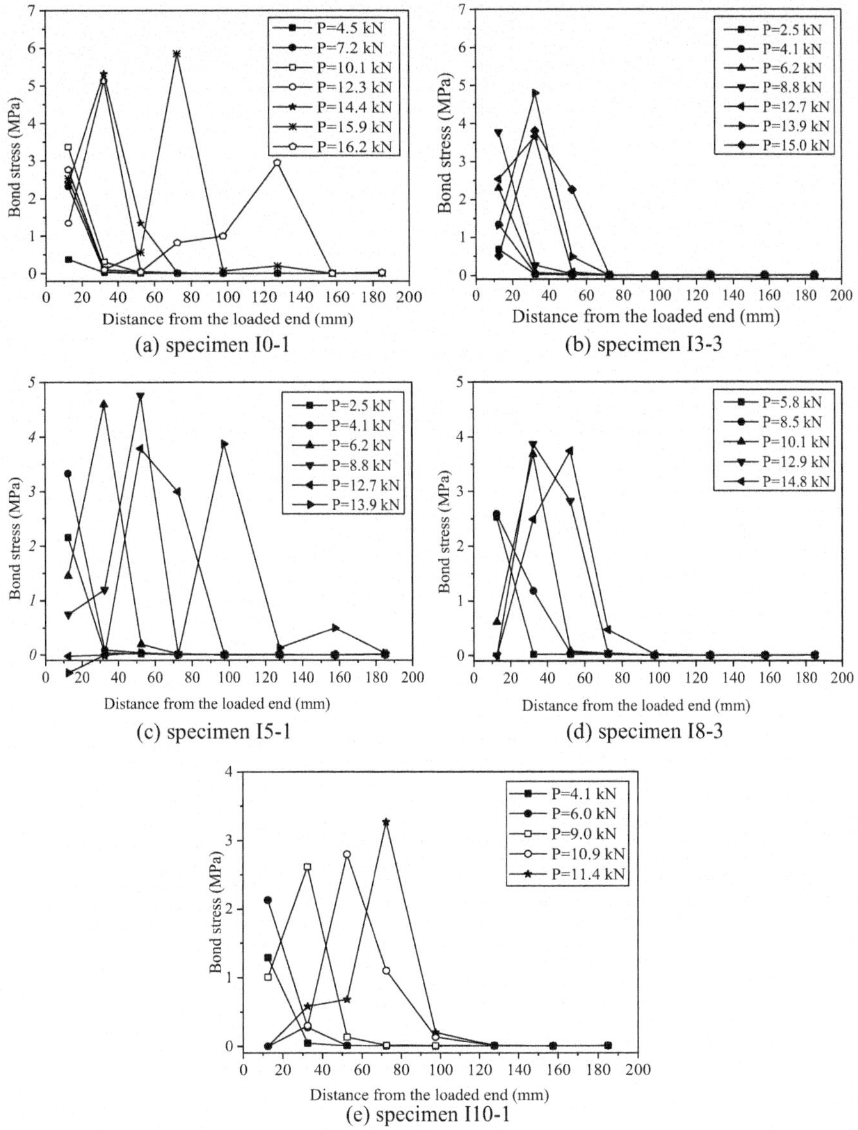


**Fig. 2.22** Relationship between distance from the loaded end and bond stress of specimens I0-1, I3-3, I5-1, I8-3 and I10-1

100% of initial static loadings was 4.81, 4.43, 4.11, and 3.27 MPa, which were lower than that of specimen I0 at the rate of 16.3%, 23.0%, 28.5%, and 43.1%, respectively.

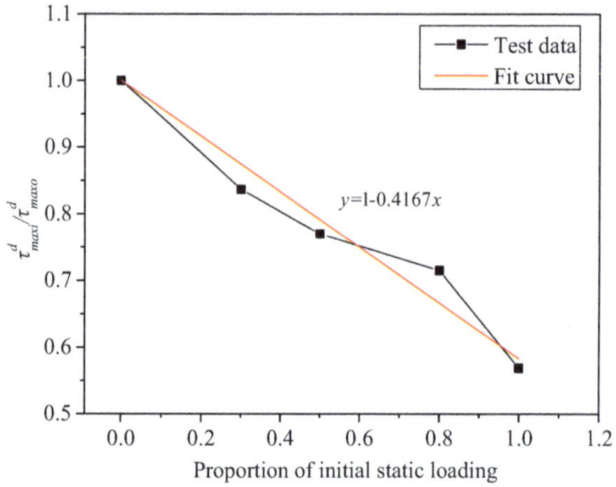


**Fig. 2.23** Relationship between maximum bond stress and initial static loading

Figure 2.23 shows the relationship between the ratio of dynamic maximum bond stress $\tau^{d}_{\max i}$ of the specimens under different initial static loadings to the corresponding value $\tau^{d}_{\max 0}$ of the specimen under 0% initial static loading and the proportion of initial static loading denoted as $P_0/P_{\max}$. The relationship between the value of maximum bond stress $\tau^{d}_{\max i}/\tau^{d}_{\max 0}$ and the proportion of the initial static loading $P_0/P_{\max}$ was linear, as shown in Eq. (2.50).

$$\frac{\tau^{d}_{\max i}}{\tau^{d}_{\max 0}} = 1 + \alpha \frac{P_0}{P_{\max}} \tag{2.50}$$

The coefficient $\alpha$ was determined from the regression analysis to provide the following equation and the $R^2$ value was 0.986, as shown in Eq. (2.51).

$$\frac{\tau^{d}_{\max i}}{\tau^{d}_{\max 0}} = 1 - 0.4167 \frac{P_0}{P_{\max}} \tag{2.51}$$

The values $\tau^{d}_{\max i}/\tau^{d}_{\max 0}$ calculated by Eq. (2.51) were 1.000, 0.875, 0.792, 0.667, and 0.583 when the proportions of the initial static loadings were 0, 30%, 50%, 80%, and 100%, respectively. The results of the deviations from Eq. (2.51) and the test were 0, 4.5%, 2.9%, 6.7%, and 2.5%; the deviations were small. Thus, Eq. (2.51) could be used to predict the dynamic bond stress of BFRP and concrete under different initial static loadings.

### 2.3.4 *Dynamic Effective Bond Length*

The effective length has an important physical meaning, and design consequences and several proposals could be found in the literature on how to estimate it from

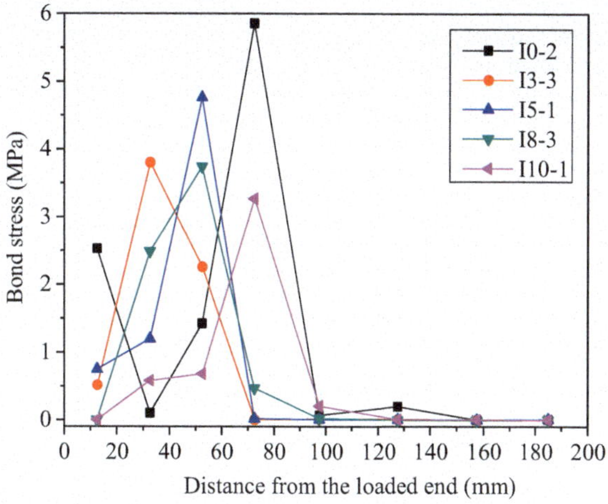


**Fig. 2.24** Relationship between bond stress and distance from the loaded end of specimens I0-2, I3-3, I5-1, I8-3, and I10-1

the characteristics of the component materials [29]. Effective bond length could be estimated by the distance between the points at which the bond stresses are 10% of the maximum bond stress $\tau_{max}$ [31]. The findings in Table 2.10 and Fig. 2.24 highlight the effect of different initial static loadings on bond length. A significant increase in the bond length was noticed when the proportion of initial static loading increased from 0 to 30%, 50%, 80%, and 100%. Given that bond length ($L_e$) is inversely proportional to concrete strength according to existing literature [16, 34], an increasing trend $L_e$ is expected when considering the decrease in concrete strength with the increase of the proportion of initial static loading [32].

As shown in Table 2.10, the dynamic effective bond length of specimen I0 under 0% initial static loading (i.e., dynamic loading) was 56.7 mm. The dynamic effective bond lengths of specimens I3, I5, I8, and I10 when applied with 30%, 50%, 80%, and 100% initial static loadings were 60.2, 63.5, 67.3, and 71.8 mm, respectively. When the initial static loading proportion increased from 0 to 30%, 50%, 80%, and 100%, the dynamic effective bond length increased at the rate of 6.2%, 12.0%, 18.7%, and 26.6%, respectively.

The relationship between $L_{ei}^{d}/L_{e0}^{d}$ and $P_0/P_{max}$ is shown in Fig. 2.25. The relationship between the value of $L_{ei}^{d}/L_{e0}^{d}$ and the proportion of initial static loading $P_0/P_{max}$ was linear, as shown in Eq. (2.52).

$$\frac{L_{ei}^{d}}{L_{e0}^{d}} = 1 + \beta \frac{P_0}{P_{max}} \tag{2.52}$$

The coefficient $\beta$ was determined from the regression analysis to provide the following equation, as shown in Eq. (2.53).

$$\frac{L_{ei}^{d}}{L_{e0}^{d}} = 1 + 0.2525 \frac{P_0}{P_{max}} \tag{2.53}$$

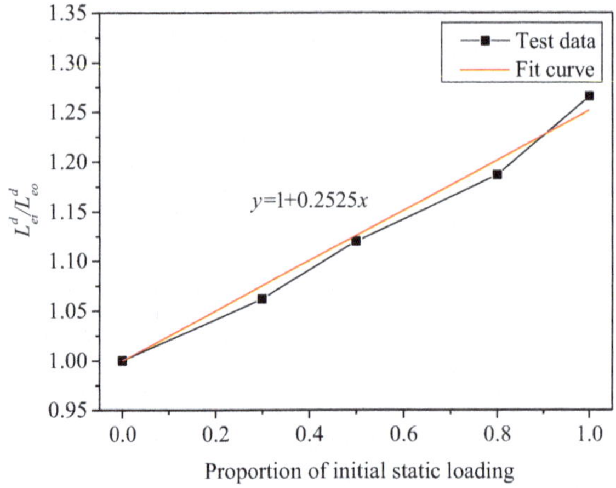


**Fig. 2.25** Relationship between dynamic effective bond length and initial static loading

The values $L^{\mathrm{d}}_{ei}/L^{\mathrm{d}}_{e0}$ calculated by Eq. (2.53) were 1.00, 1.08, 1.13, 1.20, and 1.25 when the proportions of initial static loading were 0, 30%, 50%, 80%, and 100%, respectively. The results of the deviations from Eq. (2.53) and the test were 0, 1.3%, 0.5%, 1.3%, and 1.0%, respectively; the deviations were small. Thus, Eq. (2.53) could be used to predict the dynamic effective bond length of BFRP and concrete under different initial static loadings.

The bond length ($L_{\mathrm{e}}$) was inversely proportional to concrete strength [16], as shown in Eq. (2.37).

The dynamic compressive strength of concrete decreased with the increase of initial static loading [60], as shown in Eq. (2.54).

$$\frac{f^{\mathrm{d}}_{ci}}{f^{\mathrm{d}}_{\mathrm{c}0}} = 1 - 0.181\frac{\sigma_0}{f_{\mathrm{cs}}} \tag{2.54}$$

where $\sigma_0/f_{\mathrm{cs}}$ is the proportion of initial static loading; $f^{\mathrm{d}}_{\mathrm{cd}}$ and $f^{\mathrm{d}}_{ci}$ are the dynamic compressive strength of concrete under $10^{-2}$/s strain rate and different initial static loadings, respectively.

The dynamic strength tended to decrease when the initial static stress increased. A formula that describes the relationship between initial stress and the ultimate dynamic strength was suggested, as shown in Eq. (2.55).

$$\frac{f}{f_{\mathrm{d}}} = 1.11 - 0.11e^{(f_0/f_{\mathrm{s}})^2} \tag{2.55}$$

where $f$ is the dynamic strength of concrete under different initial static stresses; $f_{\mathrm{d}}$ is the dynamic strength of concrete without initial static stresses; $f_0$ is the initial static stress; and $f_{\mathrm{s}}$ is the static strength of concrete.

Considering Eqs. (2.37) and (2.54) or Eq. (2.55), the dynamic compressive strength of concrete decreased and the dynamic effective bond length increased with the increase of the proportion of initial static loading, respectively.

### *2.3.5 Dynamic Ultimate Load*

Existing studies suggested that the main failure mode in the shear tests was cracking of concrete under shear, which occurs commonly at a few millimeters from the adhesive-concrete interface. The ultimate load (i.e., the maximum transferable load) of the joint, therefore, depended significantly on concrete strength.

The ultimate load of each double-lap shear specimen is shown in Table 2.11. Table 2.11 shows that the ultimate load decreases linearly with the increase of initial static loading proportion. The table also shows that the dynamic ultimate load of specimen I0 under 0 initial static loading (i.e., dynamic loading) was 15.8 kN. The dynamic ultimate loads of specimens I3, I5, I8, and I10 applied with 30%, 50%, 80%, and 100% initial static loading were 15.1, 14.6, 13.5, and 11.9 kN, respectively. When the proportion of initial static loading increased from 0 to 30%, 50%, 80%, and 100%, the dynamic ultimate load decreased at the rate of 4.4%, 7.6%, 14.6%, and 24.7%, respectively.

Figure 2.26 shows the relationship between the ratio of dynamic ultimate load $P^{\mathrm{d}}_{\max i}$ of the specimens under different initial static loadings to the corresponding value $P^{\mathrm{d}}_{\max 0}$ of specimen under 0 initial static loading and proportion of the initial static loading denoted as $P_0/P_{\max}$. The relationship between the value of $P^{\mathrm{d}}_{\max i}/P^{\mathrm{d}}_{\max 0}$ and the initial static loading proportion $P_0/P_{\max}$ was linear, as shown in Eq. (2.56).

$$\frac{P^{\mathrm{d}}_{\max i}}{P^{\mathrm{d}}_{\max 0}} = 1 + \gamma \frac{P_0}{P_{\max}} \tag{2.56}$$

The coefficient $\gamma$ was determined from the regression analysis to provide the following equation, as shown in Eq. (2.57).

$$\frac{P^{\mathrm{d}}_{\max i}}{P^{\mathrm{d}}_{\max 0}} = 1 - 0.2096 \frac{P_0}{P_{\max}} \tag{2.57}$$

The values $P^{\mathrm{d}}_{\max i}/P^{\mathrm{d}}_{\max 0}$ calculated by Eq. (2.57) were 1.000, 0.937, 0.895, 0.832, and 0.790 when the initial static loading proportions were 0, 30%, 50%, 80%, and 100%, respectively. The results of the deviations from Eq. (2.57) and the test were 0, 2.0%, 3.1%, 2.6%, and 4.9%; the deviations were small. Thus, Eq. (2.57) could be used to predict the dynamic ultimate load of BFRP and concrete under different initial static loadings.

**Table 2.11** Test results and fitting results of bond-slip relationship of specimens I0, I3, I5, I8 and I10

| Identification of specimen | Initial static loading (%) | Maximum ultimate load (kN) | | | $\tau_{max}$ (MPa) | $s_0$ (mm) | $s_{max}$(mm) | Regression coefficient ($n$) | Correlation coefficient ($R^2$) |
|---|---|---|---|---|---|---|---|---|---|
| | | Individual | Average | Ratio | | | | | |
| I0-1 | 0 | 16.2 | 15.8 | 1 | 5.75 | 0.115 | 0.651 | 2.98 | 0.988 |
| I0-2 | | 15.7 | | | | | | | |
| I0-3 | | 15.6 | | | | | | | |
| I3-1 | 30 | 14.3 | 15.1 | 0.956 | 4.81 | 0.111 | 0.632 | 3.40 | 0.997 |
| I3-2 | | 16.0 | | | | | | | |
| I3-3 | | 15.0 | | | | | | | |
| I5-1 | 50 | 15.9 | 14.6 | 0.924 | 4.43 | 0.122 | 0.671 | 3.35 | 0.999 |
| I5-2 | | 13.7 | | | | | | | |
| I5-3 | | 14.3 | | | | | | | |
| I8-1 | 80 | 14.3 | 13.5 | 0.854 | 4.11 | 0.117 | 0.628 | 3.23 | 0.996 |
| I8-2 | | 14.8 | | | | | | | |
| I8-3 | | 11.5 | | | | | | | |
| I10-1 | 100 | 11.4 | 11.9 | 0.753 | 3.27 | 0.109 | 0.630 | 3.05 | 0.998 |
| I10-2 | | 10.8 | | | | | | | |
| I10-3 | | 13.6 | | | | | | | |
| Average | – | – | – | – | – | 0.115 | 0.642 | 3.20 | – |

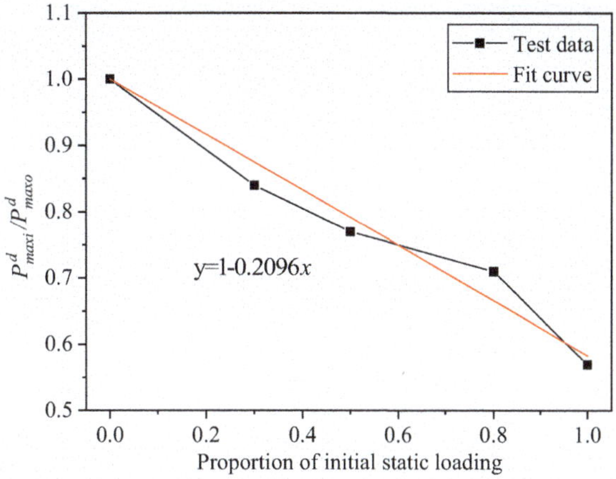


**Fig. 2.26** Relationship between dynamic ultimate load and initial static loading

### 2.3.6 *Dynamic Bond-Slip Relationship*

The bond-slip relationship is the constitutive model of the interface between BFRP sheet and concrete, which is essential to study the behavior of BFRP-strengthened concrete structures. The following assumptions were made to define the slip distribution along the BFRP sheet [31]: (1) the relative displacement between concrete and sheet at the free end of the sheet is zero; (2) the deformation of concrete specimen far from external cover is negligible concerning its BFRP counterpart; and (3) linear variation of strains exists in the BFRP sheet between two subsequent strain gauges. The integration of the strain profile, starting from the end strain gauge at $x = x_m$, provides the following expression for the slip at a general abscissa $x$, with $x_i \leq x \leq x_{i+1}$ (Fig. 2.1(b)), as shown in Eq. (2.46). Finally, the average slip between the two positions $x_i$, $x_{i+1}$ was computed as Eq. (2.47).

Bond length could be sufficiently long compared with anchorage length for the validity of assumption $s(x_m) = 0$. The post-processing procedure was performed for several BFRP strain profiles, which correspond to different loading levels. After calculating all the test data, the relationship between local average bond stress versus average slip ($\tau_{i+1/2}$ versus $s_{i+1/2}$) for all specimens is shown in Fig. 2.27. The bond-slip curves were divided into an ascending branch and a descending branch based on the experimental data. Popovics's equation was used to calculate the bond stress [31, 65], as shown in Eq. (2.48).

$\tau_{max}$ and $s_0$ are obtained directly from the experimental results. The value of $n$ is calculated by the least square method using normalized $\tau$ versus $s$ relationship. Figure 2.27 graphically shows the results of all specimens. Table 2.11 contains the fitting results of the average of all the specimens. Maximum local bond stress, $\tau_{max}$, ranged from 3.27 MPa to 5.75 MPa, and the slip at $\tau_{max}$ ranged from 0.628 mm to 0.651 mm. The constant $n$ ranged from 2.98 to 3.40. The slip at $\tau_{max}$ and constant $n$ had no clear relationship with the initial static loading proportion. The relationship

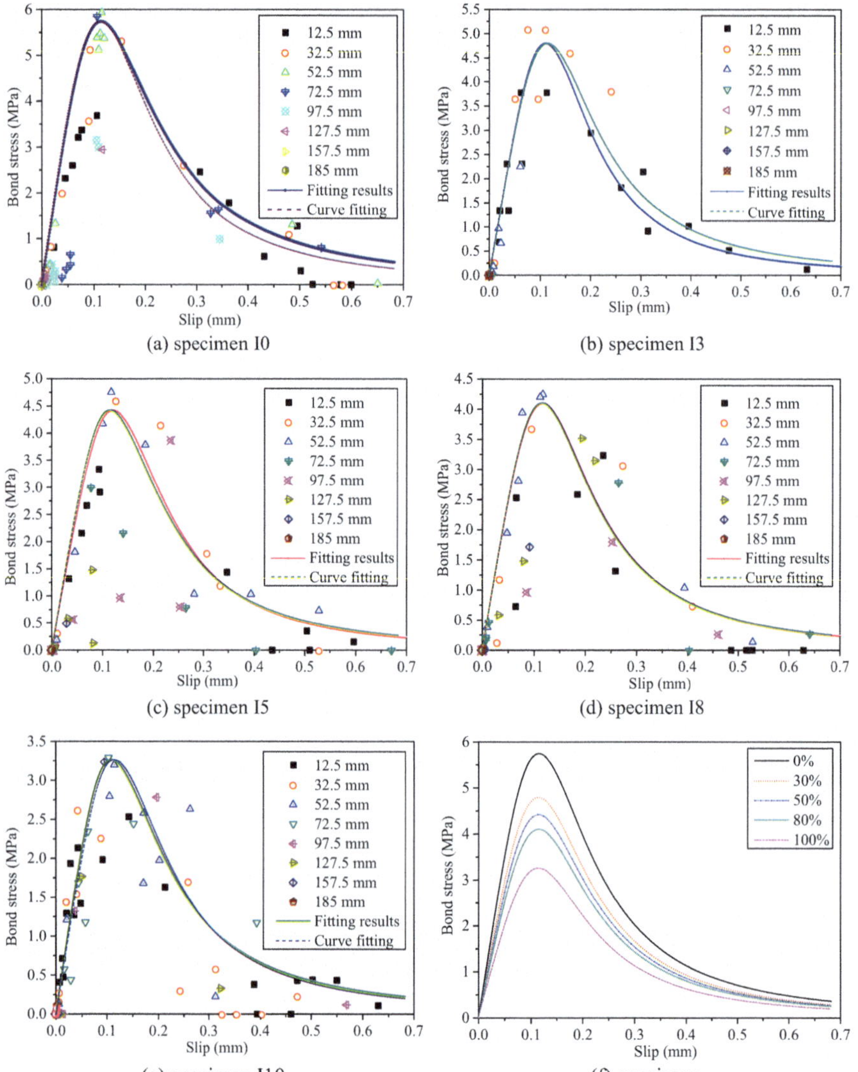


(a) specimen I0 (b) specimen I3

(c) specimen I5 (d) specimen I8

(e) specimen I10 (f) specimen

**Fig. 2.27** Relationship between dynamic bond stress and slip under different initial static loadings of specimens I0, I3, I5, I8, and I10

between maximum local bond stress $\tau_{max}$ and initial static loading proportion is shown in Eq. (2.51).

The adopted values of $s_0$ and $n$ were the average of all specimens, as shown in Table 2.11. The final results for Popovics's model are shown in Eq. (2.58).

$$\tau = \tau_{\max i}^{\mathrm{d}} \left[ \frac{s}{s_0} \frac{n}{(n-1) + (s/s_0)^n} \right] \tag{2.58}$$

where $\tau$ is local bond stress, in MPa; $s$ is local slip, in mm; $\tau_{\max i}^{\mathrm{d}}$ is dynamic maximum local bond stress under different initial static loadings, in MPa, as shown in Eq. (2.51); $s_0$ is slip at $\tau_{\max i}^{\mathrm{d}}$ with the value of 0.115, in mm; and $n$ is 3.20.

The local bond stress versus slip of specimens under different initial static loadings is shown in Fig. 2.27. Notably, the relationship between the dynamic local bond stress and the slip of all specimens under different initial static loadings was similar. The maximum local bond stress and graph slope of the ascent stage decreased with the increase of initial static loading proportion. Thus, the model presented could predict the local bond stress versus slip relationship on BFRP and concrete under different initial static loadings.

Because the dynamic bond strength between FRP and concrete is controlled by concrete properties and not the fiber type [27], the models presented can also be used to predict the performance between other FRPs and concrete, such as CFRP and GFRP when the bond length was greater than the effective bond length.

## 2.4 Summary

Investigations on the bond behavior between BFRP sheets and concrete under dynamic loading, and initial static loading were conducted. The effect of strain rate on dynamic effective bond length was also investigated. All conclusions of the results summarized were listed below:

(1) The strain rate of loading affected the dynamic effective bond length. The dynamic effective bond length decreased with the increase of strain rate. The dynamic elastic modulus of BFRP sheet increased with the increase of strain rate and the model for the dynamic elastic modulus of BFRP sheet considering strain rate was proposed. The model for the dynamic effective bond length between BFRP sheets and concrete was proposed in consideration of BFRP stiffness and concrete axial tensile strength as well as strain rate. The dynamic ultimate load did not increase when the bond length was larger than the effective bond length.

(2) The bond behavior between BFRP sheets and concrete was sensitive to strain rate. The dynamic maximum bond stress and dynamic ultimate load between BFRP sheets and concrete increased as the strain rate increased. The silp at maximum bond stress had no clear relationship with strain rate. The failure modes of the interface between BFRP sheets and concrete under dynamic loading at different strain rates were basically similar. The graphs of the dynamic local bond stress-slip relationship of all specimens under different strain rates were similar. The models for the dynamic ultimate load, bond stress, slip, and the bond stress-slip relationship between BFRP sheets and concrete were proposed in consideration of the strain rate.

(3) The initial static loading affected the bond behavior between BFRP sheets and concrete. Both the dynamic maximum bond stress and dynamic ultimate load between BFRP sheets and concrete decreased with the increase of the initial static loading proportion. The dynamic effective bond length increased with the increase of initial static loading proportion. The silp at maximum bond stress had no clear relationship with initial static loading. The models for the dynamic maximum bond stress, ultimate load, bond stress, slip, and the bond stress-slip relationship between BFRP sheets and concrete were proposed in consideration of the initial static loading proportions.

## References

1. Zheng D, Li QB (2004) An explanation for rate effect of concrete strength based on fracture toughness including free water viscosity. Eng Fract Mech 71(16–17):2319–2327
2. Yan DM, Lin G (2008) Influence of initial static stress on the dynamic properties of concrete. Cem Concr Compos 30(4):327–333
3. Wu C, Oehlers DJ, Rebentrost M et al (2009) Blast testing of ultra-high performance fibre and FRP-retrofitted concrete slabs. Eng Struct 31(9):2060–2069
4. Buchan PA, Chen JF (2007) Blast resistance of FRP composites and polymer strengthened concrete and masonry structures–a state-of-the-art review. Compos Part B: Eng 38(5–6):509–522
5. Hao H, Tang EKC (2010) Numerical simulation of a cable-stayed bridge response to blast loads, part II: damage prediction and FRP strengthening. Eng Struct 2010, 32(10):3193–3205
6. Saiidi MS, Johnson R, Maragakis E (2006) Strain rate effects on strength of unidirectional FRP fabrics and bond to concrete. In: Proceedings of the third international conference on FRP composites in civil engineering, Miami, FL, USA. 2006, 1315
7. Shi JW, Zhu H, Wu ZS et al (2012) Experimental study of the strain rate effect of FRP sheet-concrete interface. China Civ Eng J 45(12):99–107 (in Chinese)
8. Shi JW, Zhu H, Wu ZS et al (2013) Bond behavior between basalt fiber–reinforced polymer sheet and concrete substrate under the coupled effects of freeze-thaw cycling and sustained load. J Compos Constr 17(4):530–542
9. Shi JW, Zhu H, Wu ZS et al (2011) Durability of BFRP and hybrid FRP sheets under freeze-thaw cycling. Adv Mater Res 163–167:3297–3300
10. Lu XZ, Teng JG, Ye LP et al (2005) Bond-slip models for FRP sheets/plates bonded to concrete. Eng Struct 27(6):920–937
11. Shen DJ, Shi X, Ji Y, Yin FF (2015) Strain rate effect on bond stress-slip relationship between basalt fiber-reinforced polymer sheet and concrete. J Reinf Plast Compos 34:547–563
12. Yin FF (2014) Study on interfacial properties of basalt fiber reinforced polymer and concrete. Hohai University, Nanjing (in Chinese)
13. Shen DJ, Ji Y, Yin FF, Zhang JY (2015) Dynamic bond stress-slip relationship between basalt FRP sheet and concrete under initial static loading. J Compos Constr 19(6):04015012
14. Shen DJ, Shi HF, Ji Y, Yin FF (2015) Strain rate effect on effective bond length of basalt FRP sheet bonded to concrete. Constr Build Mater 82:206–218
15. Ministry of Housing and Urban Rural Development of the People's Republic of China (2002) Standard for test method of mechanical properties on ordinary concrete: GB/T 50081-2002. China Architecture and Building Press, Beijing (in Chinese)
16. Chen JF, Teng JG (2001) Anchorage strength models for FRP and steel plates bonded to concrete. J Struct Eng 127(7):784–791

17. Ministry of Housing and Urban Rural Development of the People's Republic of China (1999) Standard for test method for tensile properties of oriented fiber reinforced plastics: GB/T 3354-1999. China Architecture and Building Press, Beijing (in Chinese)
18. Japan Society of Civil Engineers (2001) Test method for bond properties of continuous fiber sheets to concrete (JSCE-E-543-2000). Research Committee on Upgrading of Concrete Structures with Use of Continuous Fiber Sheets, Tokyo, Japan
19. Bilotta A, Di Ludovico M, Nigro E (2011) FRP-to-concrete interface debonding: experimental calibration of a capacity model. Compos Part B: Eng 42(6):1539–1553
20. Ali-Ahmad M, Subramaniam K, Ghosn M (2006) Experimental investigation and fracture analysis of debonding between concrete and FRP sheets. J Eng Mech 132(9):914–923
21. Yao J, Teng JG, Chen JF (2005) Experimental study on FRP-to-concrete bonded joints. Compos Part B: Eng 36(2):99–113
22. Franco A, Royer-Carfagni G (2014) Effective bond length of FRP stiffeners. Int J Non-Linear Mech 60:46–57
23. Kang THK, Howell J, Kim S et al (2012) A state-of-the-art review on debonding failures of FRP laminates externally adhered to concrete. Int J Concr Struct Mater 6:123–134
24. Al-Zubaidy HA, Zhao XL, Al-Mahaidi R (2012) Dynamic bond strength between CFRP sheet and steel. Compos Struct 94(11):3258–3270
25. Ko H, Sato Y (2007) Bond stress-slip relationship between FRP sheet and concrete under cyclic load. J Compos Constr 11(4):419–426
26. Ferracuti B, Savoia M, Mazzotti C (2007) Interface law for FRP-concrete delamination. Steel Constr 80(4):523–531
27. Jing W, Smith ST, Qiao P (2014) Experimental investigation on FRP-to-timber bonded interfaces. J Compos Constr 18(3):A4013006
28. Kulkarni SM, Shah SP (1998) Response of reinforced concrete beams at high strain rates. Struct J 95(6):705–715
29. Silva MAG, Biscaia HC, Marreiros R (2013) Bond-slip on CFRP/GFRP-to-concrete joints subjected to moisture, salt fog and temperature cycles. Compos Part B: Eng 55:374–385
30. Chen GM, Teng JG, Chen JF (2012) Process of debonding in RC beams shear-strengthened with FRP U-strips or side strips. Int J Solids Struct 49(10):1266–1282
31. Nakaba K, Kanakubo T, Furuta T et al (2001) Bond behavior between fiber-reinforced polymer laminates and concrete. Struct J 98(3):359–367
32. Zheng D, Li Q, Wang L (2007) Rate effect of concrete strength under initial static loading. Eng Fract Mech 74(15):2311–2319
33. Lu XZ (2004) Study on FRP-concrete interface. Tsinghua University, Beijing (in Chinese)
34. Wu Z, Islam SM, Said H (2009) A three-parameter bond strength model for frp-concrete interface. J Reinf Plast Compos 28(19):2309–2323
35. Neubauer U, Rostasy FS (1997) Design aspects of concrete structures strengthened with externally bonded CFRP-plates. In: Proceedings of the seventh international conference on structural faults and repairs. ECS Publications, Edinburgh, Scotland, 1997, pp 109–118
36. Khalifa A, Gold WJ, Nanni A et al (1998) Contribution of externally bonded FRP to shear capacity of RC flexural members. J Compos Constr 2(4):195–202
37. CNR (2004) Instructions for the design, execution and control of strengthening measures through fiber-reinforced composites: CNR-DT 200/04. Italian Society Research Society
38. Yang YX, Yue QR, Hu YC (2001) Experimental study on bond performance between carbon fiber sheets and concrete. J Build Struct 22(3):36–42
39. ACI (2008) Guide for the design and construction of externally bonded FRP systems for strengthening concrete structures: ACI 440.2R-08. American Concrete Institute, Farmington Hills
40. CSA (2002) Design and construction of building components with fibre-reinforced polymers: CSA-S806. Canadian Standards Association
41. Niedermeier R (2000) Envelope line of tensile forces while using externally bonded reinforcement. Technical University of Munich, Munich

42. D'antino T, Pellegrino C (2014) Bond between FRP composites and concrete: assessment of design procedures and analytical models. Compos Part B: Eng 60:440–456
43. Girgin ZC, Anoglu N, Anoglu E (2007) Evaluation of strength criteria for very-high-strength concretes under triaxial compression. ACI Struct J 104(3):278–284
44. Wu YF, Zhou YW (2010) Unified strength model based on Hoek-Brown failure criterion for circular and square concrete columns confined by FRP. J Compos Constr 14(2):175–184
45. Wu YF, Jiang C (2013) Quantification of bond-slip relationship for externally bonded FRP-to-concrete joints. J Compos Constr 17(5):673–686
46. Huang W, Huang Z (2011) Dynamic behavior of unidirectional fiber-reinforced composites: experiments and modeling. J Reinf Plast Compos 30(1):53–59
47. Wang Z, Xia Y (1998) Experimental evaluation of the strength distribution of fibers under high strain rates by bimodal Weibull distribution. Compos Sci Technol 57(12):1599–1607
48. Huang CK, Zhao GF, Shang RJ (1997) Research on concrete strength and deformation behavior under dynamic loading and its test methods. Nanjing Hydraul Res Inst 13:17–22
49. Pandey AK (2011) An iterative approach for curvature ductility of RC beams at high strain rates. J Struct Eng: ASCE 38(4):307–317
50. Tanaka T (1996) Shear resisting mechanism of reinforced concrete beams with CFS as shear reinforcement. Hokkaido University, Sapporo
51. Japan Concrete Institute (2003) Technical report of technical committee on retrofit technology. In: Proceedings of the international symposium on the latest achievement of technology and research on retrofitting concrete structures, 2003
52. Hiroyuki Y, Wu Z (1997) Analysis of debonding fracture properties of CFS strengthened member subject to tension. The Japan Concrete Institute, pp 284–294
53. Cao SY, Chen JF, Pan JW et al (2007) ESPI measurement of bond-slip relationships of FRP-concrete interface. J Compos Constr 11(2):149–160
54. Yang YX, Yue QR, Hu YC (2001) Experimental study on bond performance between carbon fiber sheets and concrete. J Build Struct 22(3):36–42 (in Chinese)
55. Maeda T (1997) A study on bond mechanism of carbon fiber sheet. In: Proceedings of third international symposium of non-metallic (FRP) reinforcement for concrete structures, vol 1, pp 279–286
56. Sato Y, Kimura K, Kobatake Y (1997) Bond behavior between CFRP sheet and concrete (Part 1). J Struct Constr Eng 62(500):75–82
57. Lorenzis LD, Miller B, Nanni A (2001) Bond of fiber-reinforced polymer fibers to concrete. ACI Mater J 98(3):256–264
58. Ko H, Matthys S, Palmieri A et al (2014) Development of a simplified bond stress–slip model for bonded FRP-concrete interfaces. Constr Build Mater 68:142–157
59. Pellegrino C, Tinazzi D, Modena C (2008) Experimental study on bond behavior between concrete and FRP reinforcement. J Compos Constr 12(2):180–189
60. Shen DJ, Lyu XL (2010) Experimental study on the mechanical property of microconcrete in model test. China Civ Eng J 43(10):14–21
61. De Lorenzis L, Miller B, Nanni A (2001) Bond of fiber-reinforced polymer laminates to concrete. ACI Mater J 98(3):256–264
62. Yoshizawa H (1997) Analysis of debonding fracture properties of CFS strengthened RC member subject to tension. In: Non-metalic (FRP) reinforcement for concrete structures, proceedings of the third international symposium, 1997, pp 287–294
63. Maeda T (1997) A study on bond mechanism of carbon fiber sheet. In: Proceedings of third international symposium of non-metallic (FRP) reinforcement for concrete structures, 1997, vol 1, pp 279–286
64. Dai J, Ueda T, Sato Y (2007) Bonding characteristics of fiber-reinforced polymer sheet-concrete interfaces under dowel load. J Compos Constr 11(2):138–148
65. Popovics S (1973) A numerical approach to the complete stress-strain curve of concrete. Cem Concr Res 3(5):583–599

# Chapter 3
# Retrofitting of Reinforced Concrete Beam with BFRP Sheets

With the wide application of RC structures, overloading, the flaws of the construction technology, and other causes trigger local cracks were inevitably during the design life of RC structures. The load-bearing capacity of RC structures decreases after years of service because of the cracking of concrete and the steel corrosion [1, 2]. The dynamic characteristics, such as natural frequency, stiffness, damping properties, and mode shape of structure, were affected by the occurrence of cracks [3]. Thus, some measures have been utilized to prevent the concrete crack development of RC beam, among which the wrapping of fiber could reduce the average crack spacing, average crack depth, and average crack width. Consequently, the load-bearing capacity of RC beam will be increased by the application of fibers.

Many kinds of fibers such as CFRP, GFRP, AFRP, and HFRP have a positive response to increase the load-bearing capacity of RC structures. As a new type of inorganic eco-friendly high-performance fiber material, the BFRP has the advantages of high tensile stress, acid resistance, and excellent ductility [4]. In addition to the advantage of material properties, BFRP has more economic benefit than that of CFRP, which can widely be utilized in engineering. However, some novelty methods combined with BFRP should be proposed to better improve the performance of aging structures. End anchorage is an important method that can reduce the probability of FRP debonding and has been recently suggested for the prevention of the premature FRP debonding [5, 6]. End anchorage improves the performance of the strengthened beams, particularly increasing the load-bearing capacity, cracking characteristics, and also changes the failure modes [7]. There are different methods for end anchorage, the most common being the use of FRP U-strips, FRP fan anchors, steel plates with bolts [8], anchors bolts or fasteners [9], and steel clamps [10]. Grooving is widely utilized in near-surface mounted method, and the reinforcing bar is embedded in grooves cut onto the surface of the beam to be strengthened and filled with an appropriate binding agent such as epoxy paste or cement grout [11]. Results in [12] show that grooving as near-surface mounted method is able to increase the load-carrying capacities of RC beams. Steel plates have been utilized widely to



D. Shen, *Seismic Performance of Corroded Reinforced Concrete Structures Retrofitted with FRP*, https://doi.org/10.1007/978-981-99-7984-4_3

strengthen the RC beams in recent years [13]. Results reported in [14] showed that the externally bonded steel plates could greatly improve the bending stiffness and load-bearing capacity of RC beams, and to decrease the deflection at midspan and crack width of RC beams. Strengthening with FRP strips or steel plates both have certain effect on the behavior of RC beams, but both of them have the limitations. Although the FRP strips have superior mechanical properties and lightweight, the failure of flexural members strengthened with FRP strips is sudden due to its brittleness. The occurrence of cracks will weaken the stiffness of the FRP-strengthened RC beams, as reported in [15]. The steel plate has good ductility. However, the thickness of the steel plate is limited when the steel plate is used to strengthen the flexural members because of the high density of steel plate. The advantages of two materials can be comprehensively utilized when two kinds of methods are combined. The application of steel plates as well as externally bonded FRP can prevent the premature FRP debonding. The steel plate anchorage refers to the application of steel plates with anchor bolts to fix the FRP composites on the concrete structures and members [15]. The composite method of FRP strips and steel plate had better effect on the improvement of the bending resistance of beams than single method. Therefore, the application and effectiveness of the combination of BFRP strips and steel plates on the behavior of full-scale old RC beams with initial flexural cracks need to be further investigated due to its practical value and broad prospects in the actual project. The effect of the load-bearing capacity, cracking characteristics, frequency, and stiffness of the beams retrofitted by BFRP, BFRP with steel plates, and BFRP using end anchorage with grooving were analyzed.

## 3.1 Influence of Initial Cracks on the Frequency of RC Box Beam

### *3.1.1 Experimental Program*

(1) Specimens

A full-scale destructive test on a RC box beam with four initial flexural cracks was conducted, which was obtained from a bridge in Nansi Lake in Jining City, Shandong Province, China. Part of the studies has been reported in [16]. The cross-section of the test beam is shown in Fig. 3.1a. The longitudinal reinforcement of the bottom layer was consisted of four HPBs with a diameter of 12 mm and five HPBs with a diameter of 16 mm. The average depth of concrete cover was 34 mm.

Cylinder concrete samples of different sizes were obtained from the box beam and tested in compression to determine actual compressive strength and modulus of elasticity after the loading test [17]. The concrete samples in the compression area were cut into cylinders with a diameter of 100 mm and a height of 100 mm. Reinforcing bar samples were tested in tension to obtain tensile yielding strength

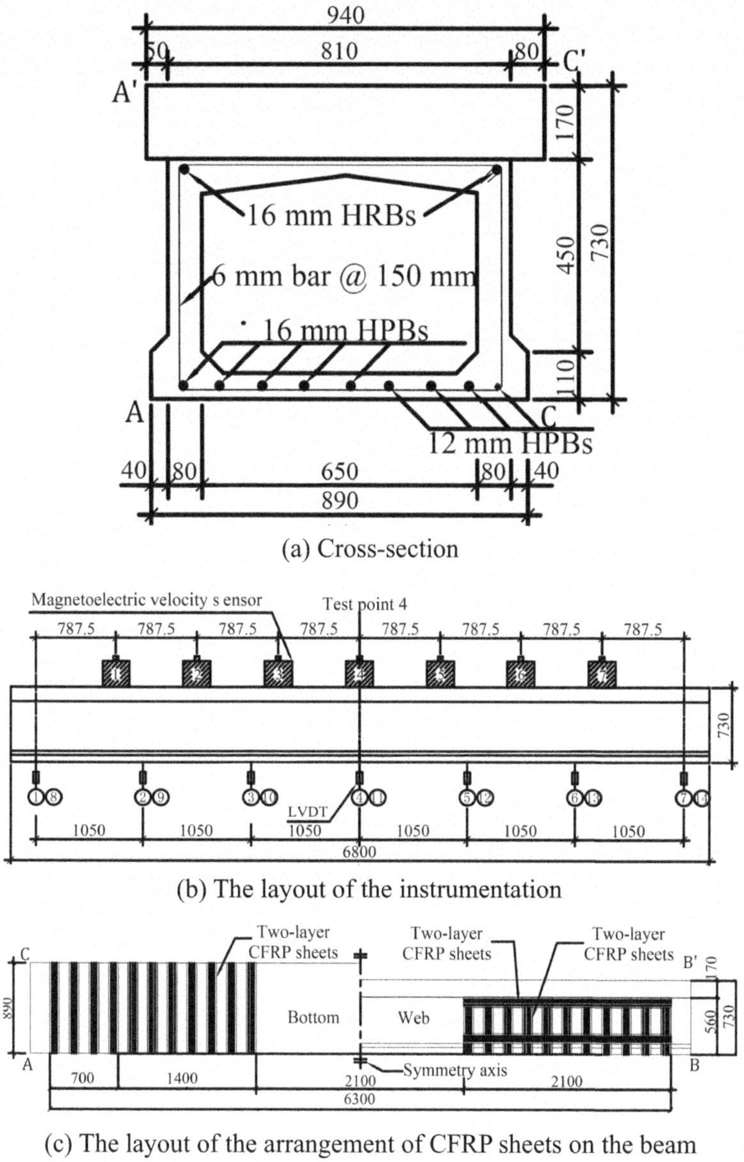


**Fig. 3.1** Specimen sizes and instrumentation layout description adapted from [16] (all units in mm)

[18]. The test results of the average yielding strength of reinforcing bar were 310 and 336 MPa for HPBs with a diameter of 12 mm and HPBs with a diameter of 16 mm, respectively. Based on the standard set out in [19], the elastic modulus of concrete was calculated by Eq. (3.1).

$$E_c = \frac{10^5}{2.2 + \frac{34.7}{f_{cu,k}}} \tag{3.1}$$

where $E_c$ is the elastic modulus of concrete; $f_{cu,k}$ is the characteristic value of the compressive strength of 150 mm × 150 mm × 150 mm concrete cube, in MPa, respectively; $f_{cu,k}$ of the specimen was tested as 52.0 MPa, and $E_c$ was calculated as 34.9 GPa. Equation (3.1) is also utilized in [16].

(2) Strengthening Scheme, Instrumentation Layout, and Load Configuration

Double layer CFRP sheets were utilized to avoid the occurrence of shear failure from the position of the concentrated load to the supports on the specimen. The FRP strengthening system has a slight effect on the stiffness of the RC beam [20]. The spacing and width of CFRP sheets were 100 mm. The tested tensile modulus, tensile strength, elongation, and thickness were $2.43 \times 10^2$ GPa, $3.5 \times 10^3$ MPa, 1.71%, and 0.167 mm, respectively. All CFRP sheets were impregnated with epoxy resin. The tested elastic modulus, ultimate mean strain, and ultimate tensile strength mean values of the epoxy resin were $2.6 \times 10^3$ MPa, 3.47%, and 45.80 MPa, respectively. CFRP strengthening was applied to increase shear resistance sufficiently and to ensure flexural failure of the retrofitted beam in accordance with ACI 440.2R-08 [21]. Figure 3.1c shows the layout of the arrangement of CFRP sheets on the beam.

The instrumentation included sensors for measuring applied load and deflection. Applied load was measured with a load cell. Deflections were measured with LVDTs. Fourteen LVDTs were placed along the length of the beams, as shown in Fig. 3.1b.

Static loading tests were conducted in four-point bending as reported in [14], as shown in Fig. 3.2. The beam was tested simply supported by pin and roller supports, with a span of 6300 mm and an overhang of 250 mm. The entire loading configuration mainly included the jack, reaction frame, steel spreader beams, and supports. The load was divided into two parts by the steel spreader beam. During the test, the deflection was applied in 0.4 mm increments from 0 to 2.0 mm, 2.0 mm increments from 2.0 to 30.0 mm, and 5.0 mm increments from 30.0 mm until failure.

(3) Modal and Vibration Test

Experimental modal analysis has been utilized successfully in the area of civil engineering to understand the response of structural systems [22]. Ambient vibration tests have received increasing attention in field and full-scale tests because of their economic efficiency and practicability. Several ambient modal identification techniques are currently available. The stochastic subspace identification method and its variants are well-known techniques that can rapidly extract modal properties [23]. As reported in [24], a Bayesian modal identification theory by utilizing the fast FFT of

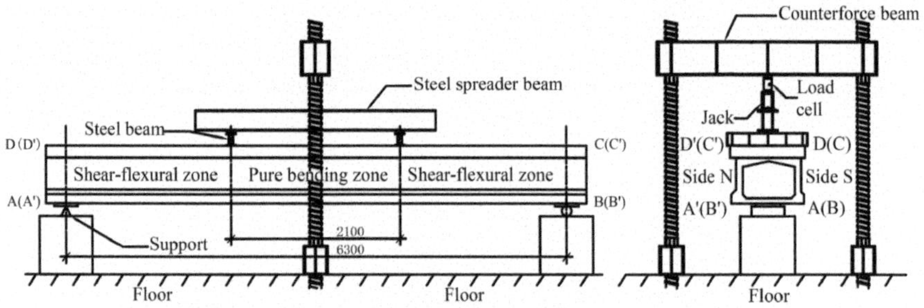


(a) Loading device

(b) Photo of loading device

**Fig. 3.2** Layout of test setup description (all units in mm)

ambient time history data was developed. Because of the advantages of modal tests and the development of modal identification techniques, experimental modal analysis is being utilized as the major technique in damage identification research [22]. As reported in [25], the crack depths of concrete structures through non-destructive test was evaluated, such as the wave transmission method. The modal test can be used to evaluate the natural frequency of the beam effectively. Therefore, this technology was utilized to evaluate the natural frequency of beams with initial flexural cracks.

The specimen was a distressed and deteriorated beam with four initial flexural cracks on the side surface after 60 years of service, as shown in Fig. 3.3. Detailed inspections were conducted before the loading test to check the flexural cracks on the side surface. The four initial flexural cracks on the specimen were located at 2151, 2845, 3449, and 4302 mm from the support, with corresponding relative depths of 243, 347, 408, and 371 mm, as shown in Fig. 3.3(a and b), respectively. These initial flexural cracks influence the vibration characteristics or responses of the RC beam, such as the decrease in natural frequency [3, 26].

A modal test was conducted on the specimen before the loading test. Seven magnetoelectric velocity sensors were arranged on top surface of the beam. Sensor data was recorded by a data acquisition system. The specific positions of sensors are shown

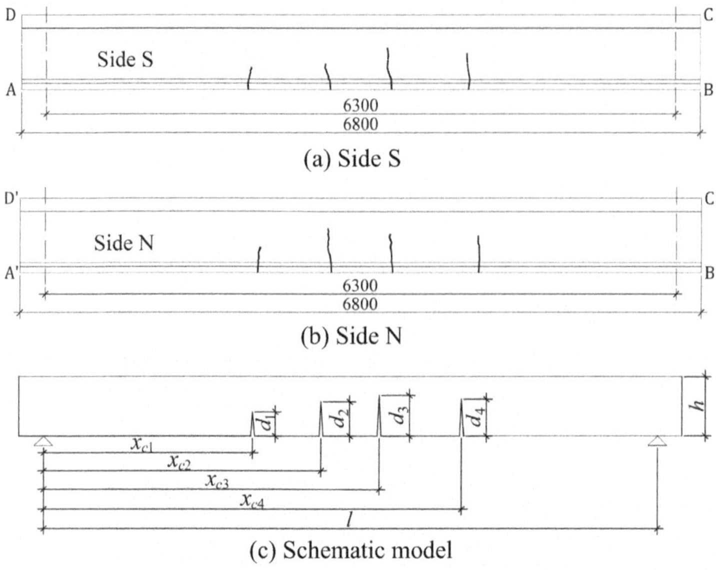


**Fig. 3.3** Initial cracks patterns (all units in mm)

in Fig. 3.1c. Appropriate hammering was utilized because natural vibration alone was insufficient to yield ideal test results. Stiffness change inferred from natural frequency data was utilized to evaluate the performance of the beam.

### *3.1.2 Test Result of Natural Frequency*

The stiffness of the cracked beam was evaluated based on the natural frequency determined by modal tests. The time-dependent velocity curve was obtained from the modal test. The amplitude–frequency curve was obtained by using the FFT method. Figure 3.4 shows the time history data and amplitude-frequency curves with test point 4 as an example. The fundamental natural frequency of the beam before CFRP strengthening with four initial flexural cracks obtained from the test was 24.43 Hz. The theoretical natural frequency of the beam without cracks was calculated as 40.08 Hz using Eqs. (3.2) and (3.3). Equation (3.2) is obtained according to the theory of vibrational science [26].

$$\omega_m = \left(\frac{n\pi}{l}\right)^2 \sqrt{\frac{E_c I_g}{\overline{m}}} \tag{3.2}$$

$$f_m = \frac{\omega_m}{2\pi} \tag{3.3}$$

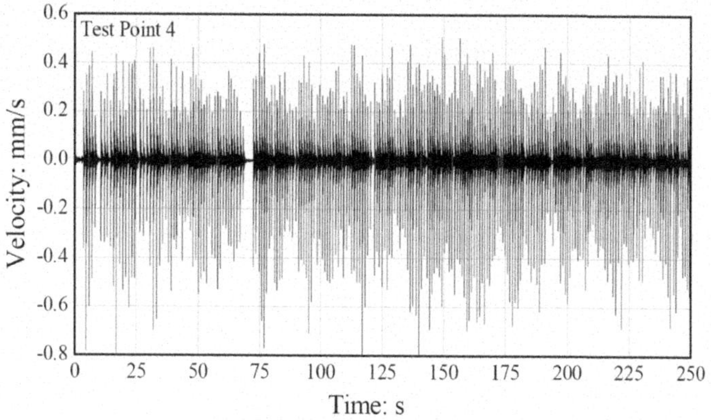

(a) Time-history curve

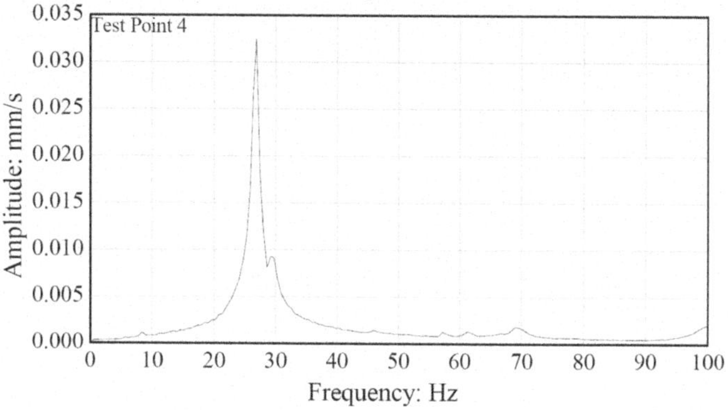

(b) Amplitude-frequency curve

**Fig. 3.4** Modal test result of the beam

where $f_m$ is the natural frequency, in Hz; $\omega_m$ is the natural angular frequency, in rad/s; $l$ is the span of the beam, in m; $E_c$ is the elastic modulus of concrete, in N/m$^2$; $I_g$ is the gross moment of inertia of a section, in m$^4$; $\overline{m}$ is the mass per unit length, in kg/m; and $n$ is the vibration mode.

The frequency results showed that the experimental natural frequency decreased by 39.0% compared with the theoretical value of the beam without cracks. Given that natural frequency $f_n$ was proportional to $\sqrt{E_c I_g}$, the decrease in stiffness $E_c I_g$ was calculated as 62.8%. The results on the decrease of natural frequency were in accordance with that in [26]. Results in [26] show that the natural frequency of RC beam at uncracked time or cracking time was 180.66 or 166.02 Hz, respectively, which decreased by 10% under monotonic loading. The decreased degree of frequency in [26] was lower than that because the number of cracks was not the same. The results obtained from modal analysis and dynamic tests show that cracking reduces frequency and stiffness [27].

The frequency results of RC beam before or after CFRP strengthening were 24.43 or 26.87 Hz, respectively, which increased by 9.1% after CFRP strengthening. The results indicated that the effect of CFRP strengthening on the frequency was slight, which was consistent with the results in [28].

### 3.1.3 Theoretical Analysis on Natural Frequency

The theoretical beam had length $l$ and a longitudinal uniform rectangular cross-section with height $h$ and width $b$; it had $N$ traversed cracks with depths of $d_1, d_2, d_3 \ldots d_n$ located at the distances $x_{c1}, x_{c2}, x_{c3} \ldots x_{cn}$ from the left support, respectively. The beam was modeled according to the Euler Bernoulli theory in $n + 1$ segments joined by $n$ rotational springs. Beam deflection was denoted as $Y_i(x, t)$ for $n + 1$ intervals of $i = 1$ for $0 < x < x_{c1}$, $i = 2$ for $x_{c1} < x < x_{c2}, \ldots$, and $i = n + 1$ for $x_{cn} < x < l$. The equation of deflective motion for each segment could be expressed as Eq. (3.4).

$$\mathrm{EI}\frac{\partial^4 Y(x,t)}{\partial x^4} + \rho A \frac{\partial^2 (x,t)}{\partial t^2} = 0 \tag{3.4}$$

Equations (3.5)–(3.6) were shown as follows:

$$\overline{x} = \frac{x}{l} \tag{3.5}$$

$$Y_{(x,t)} = y(\overline{x}, t) \tag{3.6}$$

Equation (3.4) could be changed to:

$$\frac{\mathrm{EI}}{l^4}\frac{\partial^4 y_i(\overline{x},t)}{\partial \overline{x}^4} + \rho A \frac{\partial^2 y_i(\overline{x},t)}{\partial t^2} = 0, i = 1, 2, 3, \ldots, n+1 \tag{3.7}$$

where $i = 1$ corresponds to the domain $0 < x < x_{c1}$, $i = 2$ corresponds to the domain $x_{c1} < x < x_{c2} \ldots$ and i = N + 1 corresponds to the domain $x_{cn} < x < l$; $E$ is the young's modulus, in MPa; $\rho$ is the density, in kg/mm$^3$; $A$ is the cross-sectional area, in mm$^2$; and $I$ is the moment of inertia of uniform and homogeneous beams, in mm$^4$.

The vertical deflection $y_i(\overline{x}, t)$, which is related to a section located at distance $x$ from the left end of the specimen at time $t$, could be expressed as Eq. (3.8).

$$y_i(\overline{x}, t) = X_i(\overline{x}) \cos \omega t, i = 1, 2, 3, \ldots, N+1 \tag{3.8}$$

where $X_i(\overline{x})$ is the transverse deflection of the cracked beam. The transverse deflection of the cracked beam, $X_i(\overline{x})$, is assumed to be similar to that of the uncracked beam, $u_{mc}(\overline{x})$, plus a polynomial function of $\overline{x}$.

$$X_i(\overline{x}) = u_m(\overline{x}) + B_{i,0} + B_{i,1}\overline{x} + B_{i,2}\overline{x}^2 + B_{i,3}\overline{x}^3, \overline{x}_{C,i-1} \leq \overline{x} \leq \overline{x}_{C,1} \tag{3.9}$$

where $u_m(\overline{x}) = D\sin(m\pi\overline{x})$ is the transverse deflection of the uncracked beam, $D$ is an arbitrary constant with dimensions of length and $B_{i,j}$ for $j = 0, 1, 2, 3$ are the coefficients of polynomials that have to be evaluated to satisfy the boundary and kinematic conditions.

Substituting Eq. (3.8) into Eq. (3.7) yields:

$$\frac{\mathrm{d}^4(X_i(\overline{x}))}{\mathrm{d}\overline{x}^4} - \frac{\rho A l^4}{\mathrm{EI}}\omega^2 X_i(\overline{x}) = 0, i = 1, 2, 3, \ldots, n + 1 \tag{3.10}$$

The ordinary differential equation of Eq. (3.10) had a solution by Eq. (3.11).

$$\begin{aligned} X_i(\overline{x}) = A_i \sin(m\overline{x}) + B_i \cos(m\overline{x}) + C_i h \sin(m\overline{x}) \\ + D_i h \cos(m\overline{x}), i = 1, 2, 3, \ldots, \ n + 1, \end{aligned} \tag{3.11}$$

where $A_i$, $B_i$, $C_i$, and $D_i$ for $i = 1, 2, 3, \ldots, n+1$ are the coefficients of polynomials that have to be evaluated to satisfy the boundary [Eqs. (3.13)–(3.16)] and kinematic conditions [Eqs. (3.17)–(3.20)] as follows: equality for transverse deflection, bending moment, and shear force in the cracked section. Coefficient $m$ could be expressed as Eq. (3.12).

$$\frac{\rho A l^4}{\mathrm{EI}}\omega^2 = m^4. \tag{3.12}$$

The boundary conditions of the simply supported beam were expressed as Eqs. (3.13)–(3.16).

$$X_1(0) = 0 \tag{3.13}$$

$$\ddot{X}_1(0) = 0 \tag{3.14}$$

$$X_{n+1}(l) = 0 \tag{3.15}$$

$$\ddot{X}_{n+1}(l) = 0 \tag{3.16}$$

The compatibility conditions at the point of each crack were expressed as Eqs. (3.17)–(3.20).

$$X_i(\overline{x}_{ci}) = X_{i+1}(\overline{x}_{ci}) \tag{3.17}$$

$$\ddot{X}_i(\overline{x}_{ci}) = \ddot{X}_{i+1}(\overline{x}_{ci}) \tag{3.18}$$

$$\dddot{X}_i(\overline{x}_{ci}) = \dddot{X}_{i+1}(\overline{x}_{ci}) \tag{3.19}$$

$$\dot{X}_{i+1}(\overline{x}_{ci}) - \dot{X}_i(\overline{x}_{ci}) = \frac{h}{l} f(a_i)\ddot{X}_{i+1}(\overline{x}_{ci}), i = 1, 2, 3, \ldots, n, \tag{3.20}$$

where $f(a_i)$ is the non-dimensional cracked beam sectional flexibility. It is calculated with fracture mechanics [29] and is a function of crack size. For a single-sided open crack, it was expressed as Eq. (3.21).

$$f(a_i) = 2\left(\frac{a_i}{1 - a_i}\right)^2 (5.93 - 19.69a_i + 37.14a_i^2 - 35.64a_i^3 + 13.12a_i^4),$$
$$i = 1, 2, 3, \ldots, n, \tag{3.21}$$

where $a_i$ is the non-dimensional crack-depth ratio ($a_i = d_i/h$) and $d_i$ is the depth of the crack. The initial cracking is influenced by the reinforcing bar; however, the influence degree is small. Therefore, the effect of reinforcing bar on the initial cracking was not included. The length over which the crack affects stiffness is investigated in [30]. Results show that the length over which the crack affects stiffness is 0.2 times of the length of the beam.

Rayleigh's method has been utilized by many researchers to calculate the frequency of a cracked beam. Rayleigh's method is utilized to determine natural frequencies [31]. A simplified method to evaluate the fundamental frequency for the bending vibrations of a simply supported Euler–Bernoulli beam with one crack is proposed based on Rayleigh's method [32]. A calculation method for a simply supported beam with two cracks is also proposed based on Rayleigh's method [33]. Hence, at an arbitrary mode of $m$, the maximum potential energy $U_{\max,m}$ and maximum kinetic energy $T_{\max,m}$ of the beam were set equal. Maximum kinetic energy $T_{\max,m}$ could be expressed as Eq. (3.22).

$$T_{\max,m} = \frac{1}{2}\rho Al\omega_{mc}^2\left[\int_0^{\overline{x}c1} X_1(\overline{x})^2 d\overline{x} + \int_{\overline{x}c1}^{\overline{x}c2} X_2(\overline{x})^2 d\overline{x} + \cdots + \int_{\overline{x}cn}^{1} X_{n+1}(\overline{x})^2 d\overline{x}\right] \tag{3.22}$$

where $\omega_{mc}$ is mth-natural angular frequency of the cracked simply supported beam and $\rho$ is the density of concrete, in kg/m$^3$. $U_{\max,m}$ could be expressed as Eq. (3.23).

$$U_{\max,m} = U_{1,m} + U_{2,m} \tag{3.23}$$

where $U_{1,m}$ is the potential energy caused by the bending of the beam and $U_{2,m}$ is the potential energy caused by bending moments transmitted at the crack points, respectively.

$$U_{1,m} = \frac{1}{2}\frac{\mathrm{EI}}{l^3}\left[\int_{x}^{\overline{x}c1}\left(\frac{\mathrm{d}^2X_1(\overline{x})}{\mathrm{d}\overline{x}^2}\right)\mathrm{d}\overline{x} + \int_{\overline{x}c1}^{\overline{x}c2}\left(\frac{\mathrm{d}^2X_2(\overline{x})}{\mathrm{d}\overline{x}^2}\right)^2\mathrm{d}\overline{x} + \cdots + \int_{\overline{x}cn}^{1}\left(\frac{\mathrm{d}^2X_{n+1}(\overline{x})}{\mathrm{d}\overline{x}^2}\right)^2\mathrm{d}\overline{x}\right] \tag{3.24}$$

$$U_{2,m} = \frac{1}{2}\frac{\mathrm{EI}}{l^3}\sum_{k=1}^{N}\left[\frac{h}{l}f(a_k)\left(\frac{\mathrm{d}^2X_{k+1}(\overline{x}_{ck})}{\mathrm{d}\overline{x}^2}\right)^2\right] \tag{3.25}$$

Using Eqs. (3.22)–(3.25), the *m*th-natural frequency of the simply supported beam with n cracks could be obtained. The model provided an expression for the fundamental natural frequency which was related to the depth and location of initial cracks. The model was used to investigate the change of stiffness and thus provide a simple method for detecting damage and evaluating the structure's performance. The calculated results of natural frequencies of the beam with two cracks obtained based on this method show good accuracy compared with the test results in [33].

### *3.1.4 Comparison on the Analytical Results and Test Data of Natural Frequency*

The box beam with four initial flexural cracks is shown in Fig. 3.3c. The proposed method on calculating the natural frequency of beam was extended to validate the accuracy of the test result. In this method, when the number of cracks *n* in Eq. (3.9) was four, the general expression for the mode shapes of the cracked beam was expressed as Eqs. (3.26)–(3.30).

$$X_1 = u_m(\overline{x}) + B_0 + B_1\overline{x} + B_2\overline{x}^2 + B_3\overline{x}^3, 0 \le \overline{x} \le \overline{x}_{c1} \tag{3.26}$$

$$X_2 = u_m(\overline{x}) + C_0 + C_1\overline{x} + C_2\overline{x}^2 + C_3\overline{x}^3, \overline{x}_{c1} \le \overline{x} \le \overline{x}_{c2} \tag{3.27}$$

$$X_3 = u_m(\overline{x}) + D_0 + D_1\overline{x} + D_2\overline{x}^2 + D_3\overline{x}^3, \overline{x}_{c2} \le \overline{x} \le \overline{x}_{c3} \tag{3.28}$$

$$X_4 = u_m(\overline{x}) + E_0 + E_1\overline{x} + E_2\overline{x}^2 + E_3\overline{x}^3, \overline{x}_{c3} \le \overline{x} \le \overline{x}_{c4} \tag{3.29}$$

$$X_5 = u_m(\overline{x}) + F_0 + F_1\overline{x} + F_2\overline{x}^2 + F_3\overline{x}^3, \overline{x}_{c4} \leq \overline{x} \leq 1 \tag{3.30}$$

where $B_k$, $C_k$, $D_k$, $E_k$, and $F_k$ for $k = 0, 1, 2, 3$ are the coefficients that satisfy the boundary [Eqs. (3.13)–(3.16) and Eqs. (3.31)–(3.32)] and kinematic [Eqs. (3.33)–(3.36)] conditions.

When the number of cracks $N$ in Eqs. (3.13)–(3.16) was four, the boundary conditions of the simply supported beam with four cracks could be expressed as Eqs. (3.31) and (3.32).

$$X_5(l) = 0 \tag{3.31}$$

$$\ddot{X}_5(l) = 0 \tag{3.32}$$

From Eqs. (3.17) to (3.20), the compatibility conditions at the point of each crack of the beam with four cracks were expressed as Eqs. (3.33)–(3.36)

$$X_i(\overline{x}_{ci}) = X_{i+1}(\overline{x}_{ci}) \tag{3.33}$$

$$\ddot{X}_i(\overline{x}_{ci}) = \ddot{X}_{i+1}(\overline{x}_{ci}) \tag{3.34}$$

$$\dddot{X}_i(\overline{x}_{ci}) = \dddot{X}_{i+1}(\overline{x}_{ci}) \tag{3.35}$$

$$\dot{X}_{i+1}(\overline{x}_{ci}) - \dot{X}_i(\overline{x}_{ci}) = \frac{h}{l} f(a_i)\ddot{X}_{i+1}(\overline{x}_{ci}), i = 1, 2, 3, 4 \tag{3.36}$$

When the number of cracks $N$ in Eq. (3.21) was four, $f(a_i)$ could be obtained as Eq. (3.37).

$$f(a_i) = 2\left(\frac{a_i}{1 - a_i}\right)^2 \left(5.93 - 19.69a_i + 37.14a_i^2 - 35.64a_i^3 + 13.12a_i^4\right),$$
$$i = 1, 2, 3, 4. \tag{3.37}$$

With Eq. (3.22), maximum kinetic energy $T_{max,m}$ for beams with four initial flexural cracks could be expressed as Eq. (3.38).

$$T_{\max,m} = \frac{1}{2}\rho A l \omega_{mc}^2 \left[\int_0^{\overline{x}c1} X_1(\overline{x})^2 \mathrm{d}\overline{x} + \int_{\overline{x}c1}^{\overline{x}c2} X_2(\overline{x})^2 \mathrm{d}\overline{x} + \cdots + \int_{\overline{x}cN}^{1} X_{N+1}(\overline{x})^2 \mathrm{d}\overline{x}\right] \tag{3.38}$$

With Eqs. (3.24) and (3.25), $U_{1,m}$ and $U_{2,m}$ could be expressed as Eqs. (3.39) and (3.40) for beams with four cracks, respectively.

$$U_{1,m} = \frac{1}{2}\frac{\mathrm{EI}}{l^3}\left[\int_0^{\overline{x}_{c1}}\left(\frac{\mathrm{d}^2 X_1(\overline{x})}{\mathrm{d}\overline{x}^2}\right)^2 \mathrm{d}\overline{x} + \sum_{k=1}^{3}\int_{\overline{x}_{ck}}^{\overline{x}_{c(k+1)}}\left(\frac{\mathrm{d}^2 X_{k+1}(\overline{x})}{\mathrm{d}\overline{x}^2}\right)^2 \mathrm{d}\overline{x} + \int_{\overline{x}_{c4}}^{1}\left(\frac{\mathrm{d}^2 X_5(\overline{x})}{\mathrm{d}\overline{x}^2}\right)^2 \mathrm{d}\overline{x}\right] \tag{3.39}$$

$$U_{2,m} = \frac{1}{2}\frac{EI}{l^3}\sum_{k=1}^{4}\left[\frac{h}{l}f(a_k)\left(\frac{d^2 X_{k+1}(\overline{x}_{ck})}{d\overline{x}^2}\right)^2\right] \tag{3.40}$$

By using Eqs. (3.38)–(3.40), the $m$th-natural angular frequency of the four-cracked simply supported beam could be expressed as Eq. (3.41).

$$\omega_{mc} = \sqrt{\frac{U_{\max,m}}{\frac{1}{2}\rho Al\left[\int_0^{\overline{x}_{c1}} \overline{X}_1(\overline{x})^2 d\overline{x} + \sum_{k=1}^{3}\int_{\overline{x}_{ck}}^{\overline{x}_{c(k+1)}} \overline{X}_k(\overline{x})^2 d\overline{x} + \int_{\overline{x}_{c4}}^{1} \overline{X}_5(\overline{x})^2 d\overline{x}\right]}} \tag{3.41}$$

The natural frequency of the cracked beam could be obtained with Eqs. (3.3) and (3.41). The natural frequency of the specimen with $l = 6300$ mm, $h = 730$ mm, $\rho = 2500$ kg/m$^3$, $E = 34.9$ GPa, and $b = 930$ mm was obtained through this method. The calculated natural frequency of the beam with four cracks was 25.85 Hz, which was 5.8% higher than the test result (24.43 Hz). The frequency of the beams with four cracks obtained with Eq. (3.41) showed good accuracy with the test result. The analytical method can predict natural frequency of cracked RC beam with only a small error. The results also confirmed the applicability of the calculation method for frequency of beams with $n$ cracks proposed.

### 3.1.5 Test Result of Deflection

The load-bearing capacity of RC beam is expressed as Eqs. (3.42) and (3.43) [19].

$$\alpha_1 f_c bx = f_y A_s \tag{3.42}$$

$$M_u = f_y A_s(h_0 - x/2) \tag{3.43}$$

where $\alpha_1$ is the ratio of the stress value in the rectangular stress diagram for the compression zone of concrete to the design value for the axial compressive strength of concrete; $f_c$ is the design value of the axial compressive strength of concrete, in MPa; $b$ is the width of the rectangular section or the width of a web for the inverted T-shaped section, in mm; $x$ is the depth of the concrete compression zone of the equivalent rectangular stress diagram, in mm; $f_y$ is the design value for the tensile strength of reinforcing bar, in MPa; $A_s$ is the sectional area of longitudinal reinforcement in the tension zone, in mm$^2$; $M_u$ is the ultimate flexural load-bearing capacity, in N · mm, and $h_0$ is the effective depth of a section, in mm.

The load-bearing capacity was calculated as 328.8 kN·m by using Eqs. (3.42) and (3.43). The transformed section properties of the box beam, actual concrete compressive strength, reinforcing bar tensile strength, and other parameters were also obtained from the test. The moment-deflection relationship of the beam obtained from the test is shown in Fig. 3.5. The crack patterns of the beam under ultimate load are shown in Fig. 3.6. The test result showed that the load-bearing capacity in consideration of the moment induced by self-weight (40.2 kN·m) was 342.0 kN·m. The deflections under different loads are shown in Fig. 3.7. The test results showed that before the load reached 217.3 kN·m, the deflection increased gradually. After this load, the deflection increased at a faster rate. The beam showed good ductility after the reinforcing bar began to yield.

**Fig. 3.5** Relationship between midspan moment and deflection

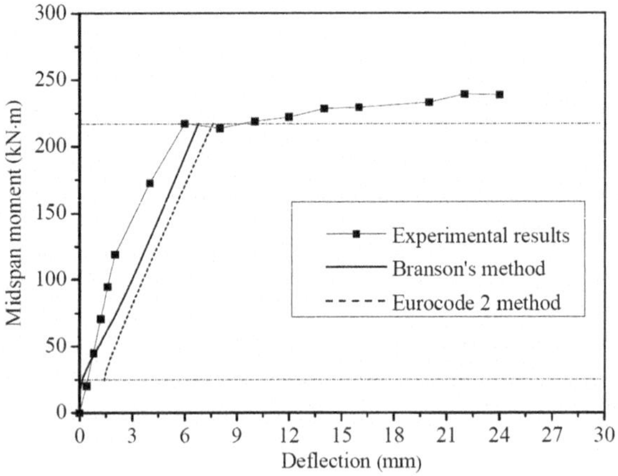


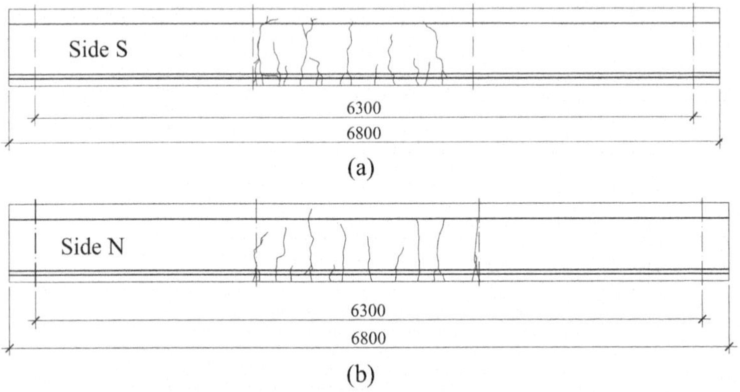


**Fig. 3.6** Crack patterns at ultimate load (all units in mm)

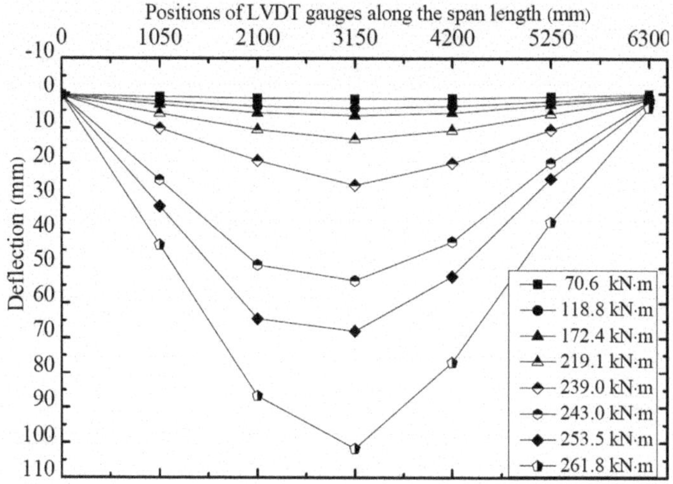


**Fig. 3.7** Relationship between deflection and applied loading

### *3.1.6 Comparison on the Analytical Result and Test Data of Deflection*

The dynamic stiffness and static stiffness are used to obtain frequency and deflection of a beam, respectively, as reported in [34]. In addition to using the frequency to calculate the dynamic stiffness to evaluate the beam performance, using the static stiffness to calculate the deflection to evaluate the beam performance is also of great value. An appropriate method for calculating the deflection after loading was necessary.

The prediction models for calculating the flexural deflection of RC beam can be classified into those derived from Branson's equation [35]. The deflection was calculated as an interpolation between a fully cracked and an uncracked state of a deformation parameter. The experimental load-midspan deflection of the tested beam was compared to Refs. [31, 36].

Elastic deflection $\delta$ for beam without initial flexural cracks can be expressed as Eq. (3.44) [37].

$$\delta = K\frac{Ml^2}{E_c I_g}, M < M_{cr} \tag{3.44}$$

where $K$ is a factor that depends on support fixity and loading conditions; $M$ is the maximum flexural moment along the span, in N·mm; $E_c$ is the elastic modulus of concrete, in MPa; and $I_g$ is the gross moment of inertia of a section, in mm$^4$.

Due to the difficulty to measure initial deflection induced by self-weight, the test result of deflection was selected to be the difference between actual deflection and

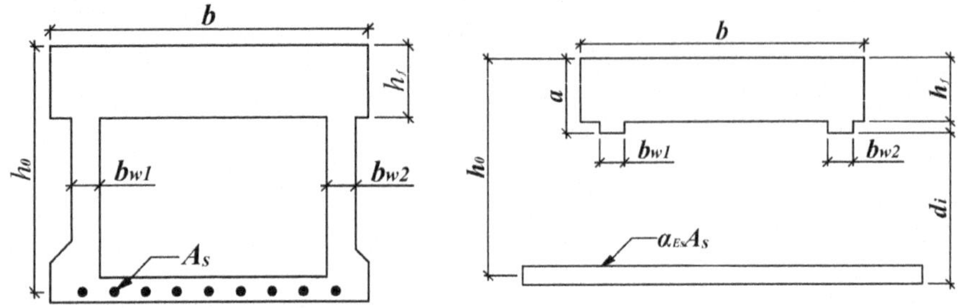


(a) Original section (b) Transformed section

**Fig. 3.8** Cracked transformed section

initial deflection. The initial deflection of intact RC beam which was induced by self-weight obtained from Eq. (3.44) was 1.3 mm.

A calculation method for the cracked moment of inertia $I_{\mathrm{cr}}$ is proposed in [37]. Regardless of the tensile zone of concrete, longitudinal reinforcements in the tension area of the cross-sectional area of the conversion can withstand the tensile stress of the concrete sectional area $\alpha_{\mathrm{Es}} A_{\mathrm{s}}$, as shown in Fig. 3.8. The cracked moment of inertia can be expressed as Eqs. (3.45)–(3.46).

$$I_{\mathrm{cr}} = \alpha_{\mathrm{Es}} A_{\mathrm{s}} (h_0 - a)^2 \frac{1}{3} b_{\mathrm{w}} (a - h_{\mathrm{f}})^3 + \frac{1}{12} b h_{\mathrm{f}}^3 + b h_{\mathrm{f}} (a - h_{\mathrm{f}}/2)^2 \tag{3.45}$$

$$\alpha_{Es} = \frac{E_s}{E_c} \tag{3.46}$$

where $\alpha_{\mathrm{Es}}$ is the modular ratio of reinforcing bar to concrete; $E_{\mathrm{s}}$ is the modulus of elasticity of reinforcing bar, in MPa; $E_{\mathrm{c}}$ is the elastic modulus of concrete, in MPa; $A_{\mathrm{s}}$ is the area of reinforcing bar, in mm$^2$; $h_0$ is the distance from the extreme compression fiber to the centroid of tension reinforcements, in mm; $a$ is the depth of the concrete compression zone, in mm; $b_{\mathrm{w}}$ is the total width of webs ($b_{\mathrm{w}} = b_{\mathrm{w1}} + b_{\mathrm{w2}}$), in mm; $h_{\mathrm{f}}$ is the flange depth in the compression zone, in mm; and $b$ is the flange width of the I-shaped section, in mm.

However, the effective moment of inertia of the cracked beam is only between $I_{\mathrm{g}}$ and $I_{\mathrm{cr}}$ [37]. $I_{\mathrm{e}}$ provides a transition between the upper and the lower bounds of $I_{\mathrm{g}}$ and $I_{\mathrm{cr}}$, respectively, as a function of the level of cracking, expressed as $M_{\mathrm{cr}}/M$. The secant stiffness during the loading procedure which is utilized to calculate the deflection is between $E_{\mathrm{c}} I_{\mathrm{g}}$ and $E_{\mathrm{c}} I_{\mathrm{cr}}$. Therefore, the effective moment of inertia $I_{\mathrm{e}}$ proposed by Branson [38] is used in [19]. Branson's method for the effective moment of inertia $I_{\mathrm{e}}$ for short-term deflections can be expressed as Eq. (3.47).

$$I_{\mathrm{e}} = \left(\frac{M_{\mathrm{cr}}}{M}\right)^3 I_{\mathrm{g}} + \left[1 - \left(\frac{M_{\mathrm{cr}}}{M}\right)^3\right] I_{\mathrm{cr}} \tag{3.47}$$

where $M_{cr}$ is the cracking moment, in N·mm, which can be expressed as Eq. (3.48).

$$M_{cr} = \frac{f_r I_g}{y_t} \tag{3.48}$$

where $f_r$ is the modulus of rupture of the concrete, in N/mm$^2$, and $y_t$ is the distance from the neutral axis to the tension face of the beam, in mm.

Deflection $\delta_c$ for cracked beams can be expressed as Eq. (3.49).

$$\delta_c = K\frac{Ml^2}{E_c I_e}, M \geq M_{cr} \tag{3.49}$$

The deviations between the deflections calculated by Branson's method and the experimental values of cracked beams were compared with those in [31]. The results showed that the Branson's method is more conservative. The deflections of unbounded partially prestressed beams before reinforcing bar obtained with Branson's method and from the test results were comparatively analyzed [39]. The deviations were in the range of − 20% to 20% in most cases.

A calculation method for the deflection of cracked beams is also proposed in Eurocode 2 [36]. Beams which are expected to crack, but may not be fully cracked, behave in a manner intermediate between the uncracked and fully cracked conditions. For beams subjected mainly to flexure, the deflection of cracked beams can be expressed as Eq. (3.50).

$$\delta_c = \zeta\delta_{II} + (1-\zeta)\delta_I \tag{3.50}$$

where $\delta_I$ and $\delta_{II}$ are the deflections calculated for the uncracked and fully cracked conditions, respectively; $\zeta$ is a distribution coefficient and can be expressed as Eq. (3.51).

$$\zeta = 1 - \beta\left(\frac{M_{cr}}{M}\right)^2 \tag{3.51}$$

where $\beta$ is a coefficient in consideration of the duration of the loading or of repeated loading on the average strain. For a single short-term loading, $\beta = 1$.

The deflections of the cracked beam obtained from Branson's method and Eurocode 2 were both close to that obtained from results. Figure 3.5 shows the relationship between the moment and deflection of the cracked beam from Branson's method, Eurocode 2, and the test results. As shown in Fig. 3.5, the slope of the experimental result changed obviously at approximately 217.3 kN·m. Hence, the reinforcing bar began to yield after reaching 217.3 kN·m. When the moment of the cracked beam induced by applied load was 217.3 kN·m, the deflection obtained from the test result was 7.3 mm. The deflections of the cracked beam based on Branson's method and Eurocode 2 were 6.8 and 7.6 mm, of which deviations were 6.8% and 4.1%, respectively, when compared with test result at the yielding point. Thus,

Branson's method is more conservative in calculating the deflection of a cracked beam; this result is consistent with the result obtained by [39] when comparing it with the results of Branson's method and Eurocode 2.

## 3.2 Behavior of RC Box Beam with Initial Cracks Retrofitted with BFRP Sheets

### 3.2.1 Experimental Program

(1) Specimens

Two full-scale test beams with initial flexural cracks identified with specimens B0 and B1. Specimen B0 represents the control test without retrofitting. Specimen B1 was strengthened with two layers of BFRP sheets on the tension face, and then subjected to the ultimate load. The cross-section of the test beams is shown in Fig. 3.1a. The average concrete cover depth of specimens B0 and B1 was 33.6 mm and 31.6 mm, respectively.

Specimen B1 was retrofitted with double-layer BFRP sheets. The BFRP sheets were applied to the tension face of the beam and had a consistent bond length of 4900 mm. The mechanical properties of FRP sheets were previously tested in the laboratory and were summarized in Table 3.1. Two layers of BFRP sheets were used as flexural reinforcement for single layer of BFRP sheet is poor for strengthening due to low modulus. Figure 3.9 shows the layout of the arrangement of BFRP sheets on the tension face of the beam.

**Table 3.1** Properties of BFRP and CFRP sheets

| | BFRP | CFRP |
|---|---|---|
| Tensile strength (MPa) | $2.3 \times 10^3$ | $3.5 \times 10^3$ |
| Tensile modulus (GPa) | $1.05 \times 10^2$ | $2.43 \times 10^2$ |
| Elongation (%) | 2.18 | 1.71 |
| Thickness (mm) | 0.121 | 0.167 |

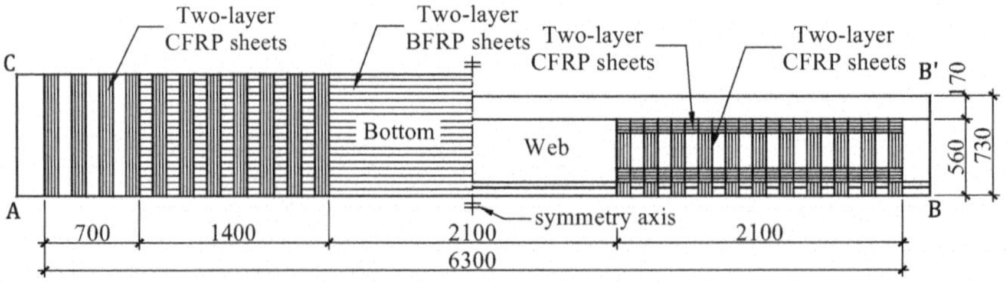


**Fig. 3.9** Details of FRP system of specimen B1 (all units in mm)

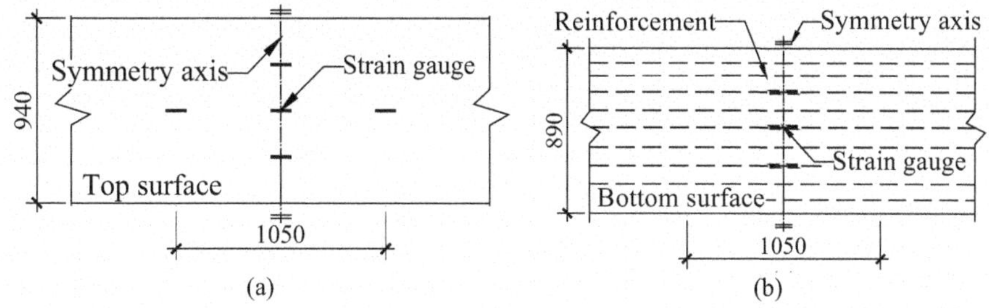


**Fig. 3.10** Instrumentation layout (all units in mm)

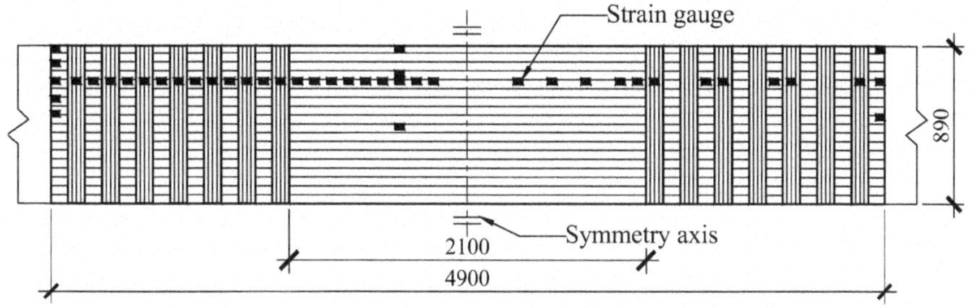


**Fig. 3.11** Strain gauge positions on BFRP sheet (all units in mm)

(2) Instrumentation Layout

Strains in the longitudinal reinforcements were measured using 3 mm uniaxial strain gauges placed on the tension reinforcements at midspan, as shown in Fig. 3.10a. Strains in the BFRP sheets were measured at various points along its length using 5 mm uniaxial strain gauges in the longitudinal direction, as shown in Fig. 3.11.

The layout of test setup included the loading system and instrumentation in this chapter was same as that shown in Fig. 3.1. The loading history and detailed arrangement and preparations on the strain gauges of tested specimens were also the same as before.

(3) Modal Tests

Specimens B0 and B1 both have four flexural cracks. The four cracks on specimen B0 were located 2150, 2850, 3450, and 4300 mm from the support, with corresponding relative depths of 245, 345, 410, and 370 mm, respectively, as shown in Fig. 3.12a. The four cracks on specimen B1 were located 2250, 2950, 3400, and 4100 mm from the support, with corresponding relative depths of 225, 370, 385, and 350 mm, respectively, as shown in Fig. 3.12b. These initial cracks have an effect on the vibration characteristics or responses of the RC beam, such as the decrease of natural frequency.

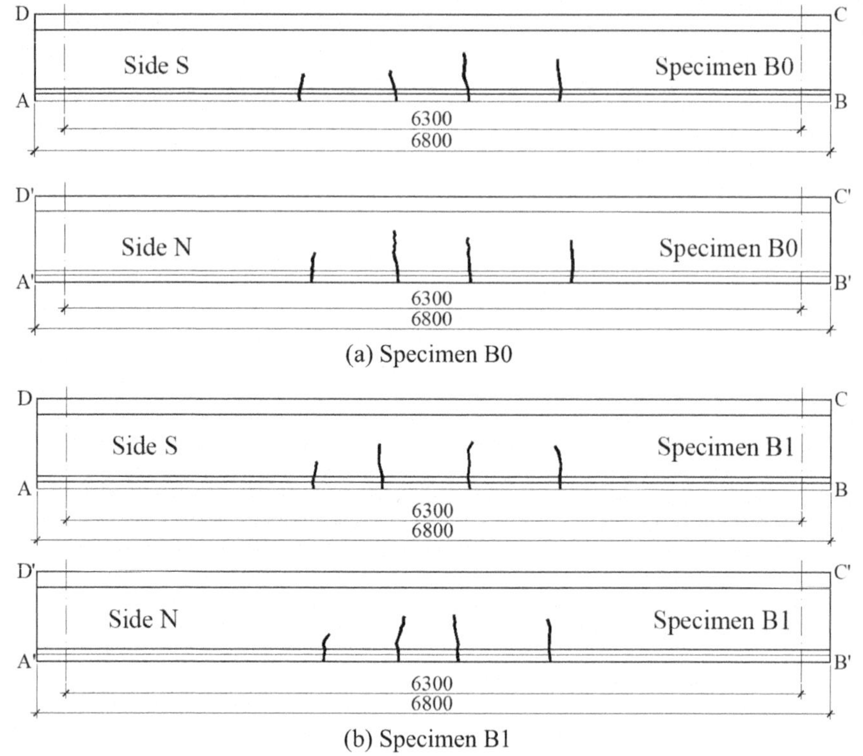


**Fig. 3.12** Initial crack patterns of specimens B0 and B1 (all units in mm)

Modal tests were conducted on specimen B0 before loading test, on specimen B1 before retrofitting, on specimen B1 after retrofitting, and on specimen B1 after loading test. Seven magnetoelectric velocity sensors were arranged along the top surface of the beam. The sensor data was received by the data acquisition system. The specific positions of the sensors are shown in Fig. 3.10a. Appropriate hammering was applied because natural vibration alone would not yield ideal test results. Stiffness change could be inferred through the natural frequency data to evaluate the health of the beam.

## 3.2.2 Load-Bearing Capacity

The moment-deflection relationship of the beams is shown in Fig. 3.13. The load-bearing capacities of specimens B0 and B1 considering the moment induced by self-weight were 342.0 and 434.9 kN·m, respectively, and a 27.2% increase associated with the retrofitting. Thus, external BFRP strengthening can also improve the load-bearing capacities of RC beams. The debonding failure mode was observed in the

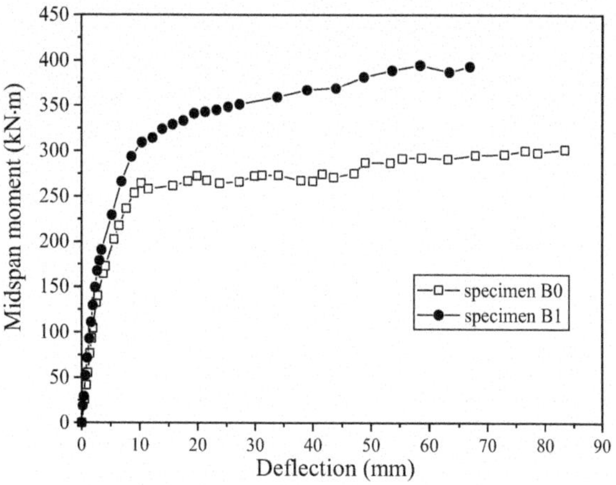


**Fig. 3.13** Relationship between midspan moment and deflection of tested specimens B0 and B1

test. As reported in [40] that the load-bearing capacities of beams strengthened with different lengths of CFRP sheets were enhanced by 33%–40% with the equivalent reinforcing ratio of 9.8%. Eight beams were tested with different damage degrees and retrofitted with different amounts of CFRP sheets, and their results showed that the load-bearing capacities were improved by 50%–87% with the equivalent reinforcing ratios of 3.4% and 5.8% [41]. CFRP sheets were also used to retrofit full-scale RC beams, and their results showed that the flexural load-bearing capacities of the beams were improved by 45% [18]. Before the sudden change in the slope of the curves, the deflections exhibited a slight difference at the same moment. However, the rate of deflection increase of specimen B0 was evidently larger than that of specimen B1 and the rate of deflection increase of specimen B0 was 30.0% larger than that of specimen B1 at the ultimate moment. The results indicated that the BFRP strengthening system has an effect on the reducing deflection of the RC beam.

The computational methods for nominal flexural strength of the section with FRP external bar were proposed in the Chinese and American standards. The calculation method of the Chinese standard was referred to [42].

If the depth of the concrete compression zone of the equivalent rectangular stress diagram x satisfies Eq. (3.52), then the flexural bearing capacity can be calculated as Eq. (3.53).

$$x \le \xi_{fb} h \tag{3.52}$$

where $\xi_{fb}$ is the relative depth of the limiting compression zone when the FRP reaches the permissible tensile strain or when concrete is crushed, and calculated as Eq. (3.54); and $h$ is the beam height, in mm.

$$M \le f_y A_s (h_0 - 0.5\xi_{fb} h) + E_f[\varepsilon_f] A_f h (1 - 0.5\xi_{fb}) \tag{3.53}$$

where $A_s$ is the sectional area of longitudinal ordinary reinforcements in the tension zone, in mm$^2$; $h_0$ is the effective depth of the section, in mm; $E_f$ is the tensile modulus of elasticity of FRP, in MPa; $[\varepsilon_f]$ is the admissible tensile strain of FRP, which can occur at $k_m\varepsilon_{cfu}$, should be smaller than two-thirds of the ultimate axial tensile strain, and can be calculated as 0.01; and $A_f$ is the area of the FRP external bar, in mm$^2$.

$$\xi_{fb} = \frac{0.8\varepsilon_{cu}}{\varepsilon_{cu} + [\varepsilon_f] + \varepsilon_i} \tag{3.54}$$

where $\varepsilon_{cu}$ is the ultimate axial strain of unconfined concrete, which can be calculated as 0.0033; and $\varepsilon_i$ is the initial strain of concrete in the tensile zone when considering the effect of secondary loading. This parameter was neglected based on the recommendation of CECS 146, and its value will be taken as zero, because the initial midspan moment was less than 20% of the load-bearing capacity of beam without retrofitting [43].

The reduction factor of the FRP bar thickness $k_m$ is considered when using multilayer FRP sheets.

$$k_m = 1 - \frac{n_f E_f t_f}{420{,}000} \le 0.9 \tag{3.55}$$

where $n_f$ is the number of plies of FRP sheets; and $t_f$ is the nominal thickness of one ply of FRP sheet, in mm.

ACI 440.2R-08 also proposes a calculation method for the capacity of beam strengthened with FRP [21]. The nominal flexural strength of the section with FRP external bar is computed as Eq. (3.56).

$$M_n = A_s f_s\left(d - \frac{\beta_1 x}{2}\right) + \psi_f A_f f_{fe}\left(d_f - \frac{\beta_1 x}{2}\right) \tag{3.56}$$

where $f_s$ is the stress in non-prestressed bar, in MPa; $d$ is the distance from the extreme compression fiber to the centroid of the tension bar, in mm; $\beta_1$ is the ratio of the depth of the equivalent rectangular stress block to the depth of the neutral axis; $x$ is the distance from the extreme compression fiber to the neutral axis, in mm; $\psi_f$ is the FRP strength reduction factor; $A_f$ is the area of FRP external bar, in mm$^2$, and $d_f$ is the effective depth of the FRP flexural bar, in mm.

The effective strain level of the FRP bar attained at failure $\varepsilon_{fe}$ can be calculated as Eq. (3.57).

$$\varepsilon_{fe} = \varepsilon_{cu}\left(\frac{d_f - x}{x}\right) \tag{3.57}$$

where $\varepsilon_{cu}$ is the ultimate axial strain of unconfined concrete.

Based on the strain level of the FRP bar, the strain level in non-prestressed bar $\varepsilon_s$ can be calculated as Eq. (3.58).

$$\varepsilon_{\rm s} = \varepsilon_{\rm fe}\left(\frac{{\rm d} - x}{{\rm d_f} - x}\right) \quad (3.58)$$

The effective stress level of the FRP bar $f_{\rm fe}$ can be derived from the strain level of the FRP by assuming a perfectly elastic behavior.

$$f_{\rm fe} = E_{\rm f}\varepsilon_{\rm fe} \quad (3.59)$$

The stress in reinforcement $f_{\rm s}$ is determined from the strain level in reinforcement $\varepsilon_{\rm s}$.

$$f_{\rm s} = E_{\rm s}\varepsilon_{\rm s} \leq f_y \quad (3.60)$$

where $E_s$ is the elastic modulus of reinforcement, in MPa.

With the strain and stress levels of the FRP and bars determined for the assumed neutral axis depth, the internal force equilibrium may be checked using Eq. (3.61).

$$x = \frac{A_{\rm s} f_{\rm s} + A_{\rm f} f_{\rm fe}}{\alpha_1 f_{\rm c}' \beta_1 b} \quad (3.61)$$

where $f_{\rm c}'$ is the specified compressive strength of concrete, in MPa; $\alpha_1$ is the multiplier of $f_{\rm c}'$ to determine the intensity of an equivalent rectangular stress distribution for concrete; and $b$ is the web width of the box beam section, in mm.

Through substitution of the parameters of FRP sheets and concrete into the calculation formulas, the ultimate flexural capacities of FRP-strengthened beam were calculated to be 450.3 and 519.4 kN·m. Compared with the experimental result of 434.9 kN·m, the calculations based on the Chinese and American standards were larger by 3.5% and 19.4%, respectively, because the factor of the initial cracks was not considered in the current standards, or the loss of capacity induced by BFRP sheets debonding. The existence of cracks has a negative effect on load-bearing capacity. Thus, the current calculation formulas, namely, Eqs. (3.53) and (3.56), were not applicable for the cracked beams retrofitted with external FRP composites.

Zhang et al. proposed a prediction method for the stiffness and deflection of retrofitted or strengthened beam using FRP composites [44]. Bending moments before and after longitudinal reinforcement yielding were expressed as functions of the tensile strain of the longitudinal reinforcement and FRP bars. The comprehensive coefficient of the variation of the average concrete strain at the upper edge of the compression zone along with the bending moment was obtained by full-range analysis. The stiffness and deflection formulas of FRP sheet-strengthened RC beams were obtained by regression analysis of the experimental data. For specimen B1, the overall moment $M$ could be calculated as Eq. (3.62).

$$M = M_{\rm s} + M_{\rm f} \quad (3.62)$$

where $M_s$ is the contribution of reinforcement to nominal flexural strength, in MPa; and $M_f$ is the contribution of the FRP sheets to nominal flexural strength, in MPa.

Through substituting the parameters into Eq. (3.62), the nominal flexural moment can be expressed as Eq. (3.63).

$$M = \sigma_s A_s \eta_s h_0 + \sigma_f A_f \eta_f h_f \tag{3.63}$$

where $\sigma_s$ is the stress of the longitudinal reinforcement, in MPa; $\sigma_f$ is the stress of the FRP bar, in MPa; $\eta_s$ and $\eta_f$ are the longitudinal and FRP bar internal force arm coefficients, respectively, with both having a value of 0.87 in the pre-yielding stage and a value of 0.95 in the post-yield stage; and $h_f$ is the distance from the extreme compression fiber to the center of FRP sheets, in mm.

The entire loading process from beginning to failure can be divided into two stages: from the beginning to reinforcing bar yielding, and from reinforcing bar yielding to the ultimate moment. In the first stage, the stress–strain relationship of reinforcing bar is linear. The nominal flexural moment can be expressed as Eq. (3.64) by introducing coefficient $a_1$, which is related to the strain of longitudinal reinforcement.

$$M = a_1 \sigma_s A_s \eta_s h_0 \tag{3.64}$$

Coefficient $a_1$ can be calculated as Eq. (3.65).

$$a_1 = 1 + \frac{\sigma_f A_f \eta_f h_f}{\sigma_s A_s \eta_s h_s} \tag{3.65}$$

The stress of the longitudinal reinforcement $\sigma_s$ in the pre-yield stage can be calculated as Eq. (3.66).

$$\sigma_s = \frac{M}{a_1 \eta_s h_0 A_s} \tag{3.66}$$

In the post-yield stage, the contribution of longitudinal reinforcement is limited. Meanwhile, the FRP bar is still linearly elastic. Thus, the nominal flexural moment can be expressed as Eq. (3.67) by introducing coefficient $a_2$, which is related to the strain of the FRP bar.

$$M = a_2 \sigma_f A_f \eta_f h_f \tag{3.67}$$

Coefficient $a_2$ can be calculated as Eq. (3.68).

$$a_2 = 1 + \frac{f_y A_s \eta_s h_0}{\sigma_f A_f \eta_f h_f} \tag{3.68}$$

The stress of the FRP bar $\sigma_f$ in the post-yield stage can be calculated as Eq. (3.69).

$$\sigma_{\mathrm{f}} = \frac{M - f_y A_{\mathrm{s}} \eta_{\mathrm{s}} h_0}{\eta_{\mathrm{f}} h_0 A_{\mathrm{f}}} \tag{3.69}$$

Based on the plane hypothesis and constant obtained by regression analysis, the stiffness formulas of FRP sheet-strengthened RC beams [44] can be expressed as Eqs. (3.70)–(3.71).

$$B_{\mathrm{s},1} = \frac{E_{\mathrm{s}} A_{\mathrm{s}} h_0^2}{\frac{\psi_{\mathrm{s}}}{a_1 \eta_{\mathrm{s}}} + \frac{8.41 \alpha_{\mathrm{E,s}} \rho_{\mathrm{s}}}{1+3.5\gamma_{\mathrm{f}}'} + 0.28} \tag{3.70}$$

$$B_{\mathrm{s},2} = \frac{E_{\mathrm{f}} A_{\mathrm{f}} h_{\mathrm{f}}^2}{\frac{\psi_{\mathrm{f}}}{a_2 \eta_{\mathrm{f}}} + 0.0081 + \frac{9.42 \alpha_{\mathrm{E,f}} \rho_{\mathrm{f}}}{1+3.5\gamma_{\mathrm{f}}'}} \tag{3.71}$$

where $\psi_{\mathrm{s}}$ and $\psi_{\mathrm{f}}$ are the uniformity coefficients of concrete and FRP strain after cracking [44], calculated using Eqs. (3.72)–(3.73), in the pre-yielding and post-yielding stages, respectively; $\alpha_{\mathrm{E,s}}$ and $\alpha_{\mathrm{E,f}}$ are the ratios of the elastic modulus of reinforcing bars and FRP bars to concrete, calculated as $\alpha_{\mathrm{E,s}} = E_{\mathrm{s}}/E_{\mathrm{c}}$ and $\alpha_{\mathrm{E,f}} = E_{\mathrm{f}}/E_{\mathrm{c}}$, respectively; and $\gamma_{\mathrm{f}}'$ is the ratio of the sectional area of the compression flange to the effective sectional area of the web, calculated using Eq. (3.76).

$$\psi_{\mathrm{s}} = \psi_{\mathrm{f}} = 1.1 - \frac{0.65 f_{\mathrm{t}}}{a_1 \rho_{\mathrm{s,te}} \sigma_{\mathrm{s}}} \tag{3.72}$$

$$\psi_s = \psi_f = 1.1 - \frac{0.65 f_t}{a_2 \rho_{f,te} \sigma_f} \tag{3.73}$$

where $f_{\mathrm{t}}$ is tensile strength of concrete, in MPa; and $\rho_{\mathrm{s,te}}$ and $\rho_{\mathrm{f,te}}$ are the ratios of the areas of longitudinal reinforcement and FRP bar to the effective cross-sectional tensile area, respectively, calculated as Eqs. (3.74)–(3.75).

$$\rho_{\mathrm{s,te}} = \frac{A_{\mathrm{s}}}{0.5bh + (b_{\mathrm{f}} - b)h_{\mathrm{f}}'} \tag{3.74}$$

$$\rho_{\mathrm{f,te}} = \frac{A_{\mathrm{f}}}{0.5bh + (b_{\mathrm{f}} - b)h_{\mathrm{f}}'} \tag{3.75}$$

where $b_{\mathrm{f}}$ is the flange width in compression zone of box beam section, in mm; and $h_{\mathrm{f}}'$ is the thickness of flange in compression zone of box beam section, in mm.

$$\gamma_{\mathrm{f}}' = \frac{(b_{\mathrm{f}} - b)h_{\mathrm{f}}'}{bh_0} \tag{3.76}$$

For beam loaded as four-point bending, the middle span deflection can be calculated as Eq. (3.77).

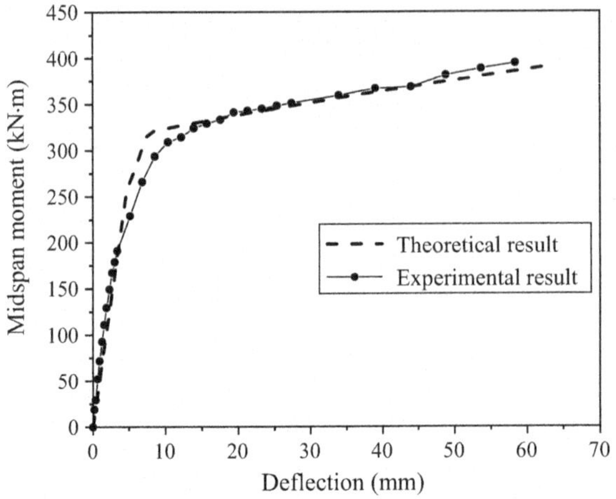


**Fig. 3.14** Experimental and theoretical deflection of specimen B1

$$\delta_{\rm i} = \frac{23}{216}\frac{Ml^2}{B_{\rm s,i}} \tag{3.77}$$

where $i = 1, 2$ in the pre-yield and post-yield stages, respectively.

$$f_{\rm n} = \frac{n^2\pi}{2l^2}\sqrt{\frac{E_{\rm c}I}{\overline{m}}} \tag{3.78}$$

where $f_{\rm n}$ is the natural frequency, in Hz; $l$ is the span of the beam, in m; $E_{\rm c}$ is the elastic modulus of concrete, in MPa; $I$ is the moment of inertia of a section, in $m^4$; $\overline{m}$ is the mass per unit length, in kg/m; and $n$ is the vibration mode.

Substituting Eqs. (3.70)–(3.71) into Eq. (3.77), the pre-yield and post-yield moment-deflection relationships were determined, as shown in Fig. 3.14. The theoretical moment-deflection relationship based on this method showed good agreement with the experimental result. Therefore, Zhang's method could be effectively used to predict the stiffness and deflection of the retrofitted beam with cracks.

### 3.2.3 *Cracking Characteristics*

The crack patterns under ultimate load of specimens B0 and B1 are shown in Fig. 3.15. Only the cracks in the constant moment region were included. The results of crack statistics and processing showed that the average crack spacing, average crack height, and average crack width of specimen B0 under ultimate load were 140, 400, and 1.12 mm, respectively. Meanwhile, the average crack spacing, average crack height, and average crack width of specimen B1 under ultimate load were 117, 319, and

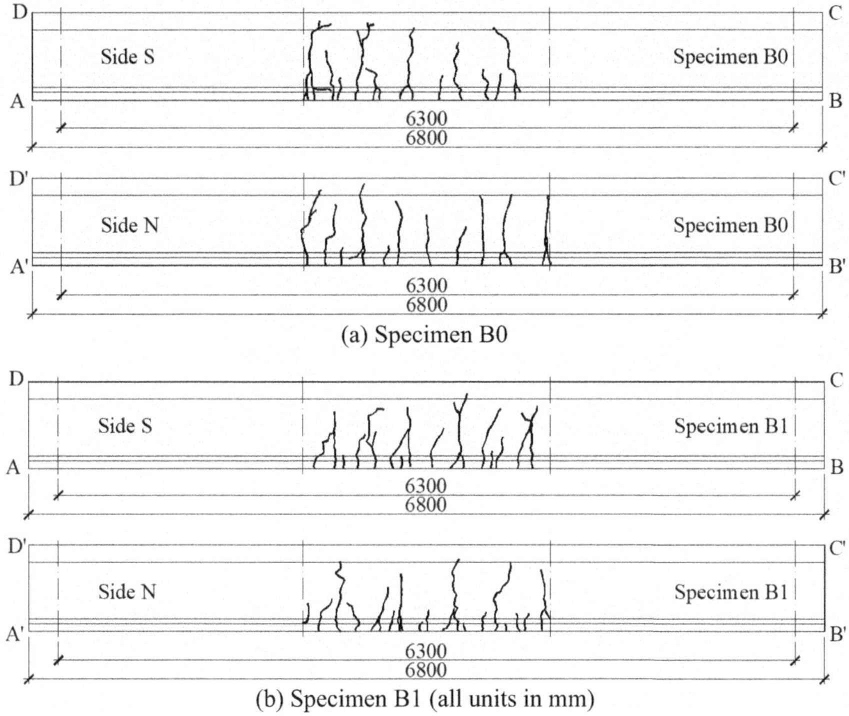


**Fig. 3.15** Crack patterns at ultimate load of specimens B0 and B1

0.97 mm, 16.4%, 20.3%, and 13.4% lower than the beam without retrofitting, respectively. Figure 3.16 shows that the maximum crack widths of specimen B1 were smaller than those of specimen B0 at different loading levels. In particular, after a load of 150 kN·m, the growth rate of the crack width of specimen B0 was evidently higher than that of specimen B1.

Thus, the effectiveness of the BFRP retrofitting system on crack control was evident. The fracture distribution of the FRP sheet-retrofitted beam is dispersive, which is different from the concentrated cracks on the beam without retrofitting. Beams strengthened with GFRP sheets show improved cracking characteristics with smaller crack widths compared with conventional RC beams [45].

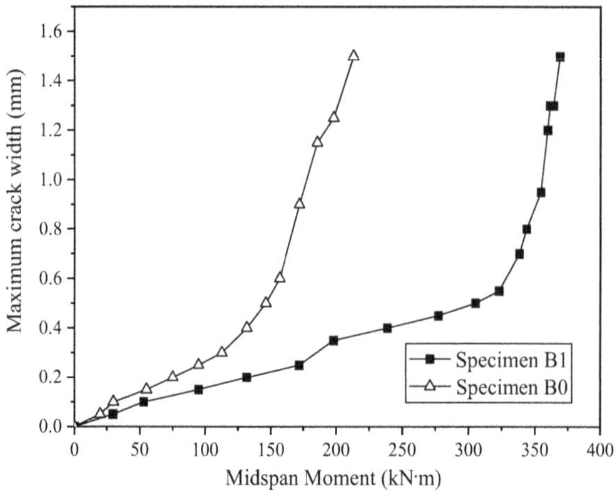


**Fig. 3.16** Comparison of maximum crack widths between specimens B0 and B1

### *3.2.4 Stiffness of Specimens*

A procedure for assessing stiffness variations caused by damage to retrofit of RC beams was presented. This procedure was based on modal analysis and testing. Modal tests were conducted before retrofitting, after retrofitting, and after loading. The stiffness of cracked beams with and without external strengthening was evaluated based on the natural frequency determined by modal tests. The time-dependent velocity curve was obtained using the modal test data. With the FFT method, the amplitude–frequency curve was obtained. Figures 3.17 and 3.18 show the time history data and amplitude–frequency curves, respectively, considering test point 4 as an example. The fundamental natural frequency and damping ratio could be obtained from the curve. The experimental natural frequencies of B0 before loading, B1 before retrofitting, B1 after retrofitting, and B1 after loading were 23.44, 24.43, 26.38, and 17.59 Hz, respectively. The theoretical natural frequency of specimens B0 and B1 without cracks was calculated as 40.0 and 38.9 Hz, respectively, using Eq. (3.79) [26].

$$f_{\mathrm{n}} = \frac{n^2\pi}{2l^2}\sqrt{\frac{E_{\mathrm{c}}I}{\overline{m}}} \tag{3.79}$$

where $f_{\mathrm{n}}$ is the natural frequency, in Hz; $l$ is the span of the beam, in m; $E_{\mathrm{c}}$ is the elastic modulus of concrete, in MPa; $I$ is the moment of inertia of a section, in m$^4$; $\overline{m}$ is the mass per unit length, in kg/m; and $n$ is the vibration mode.

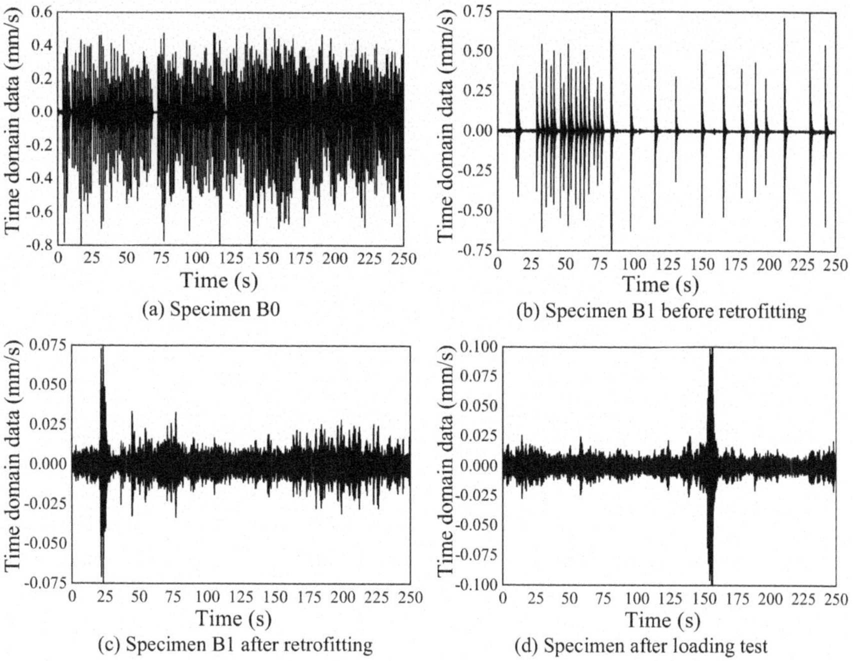


**Fig. 3.17** Time history curves of specimens B0 and B1

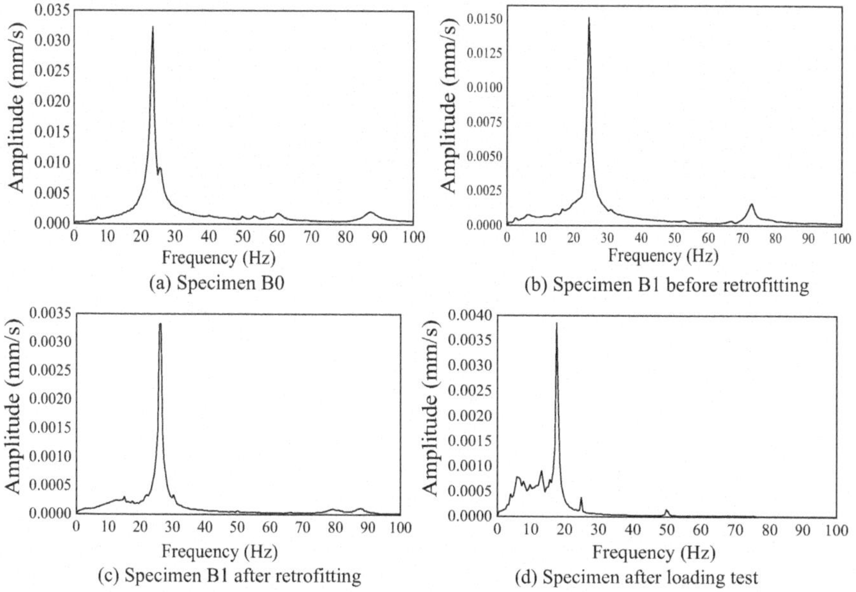


**Fig. 3.18** Amplitude-frequency curves of specimens B0 and B1

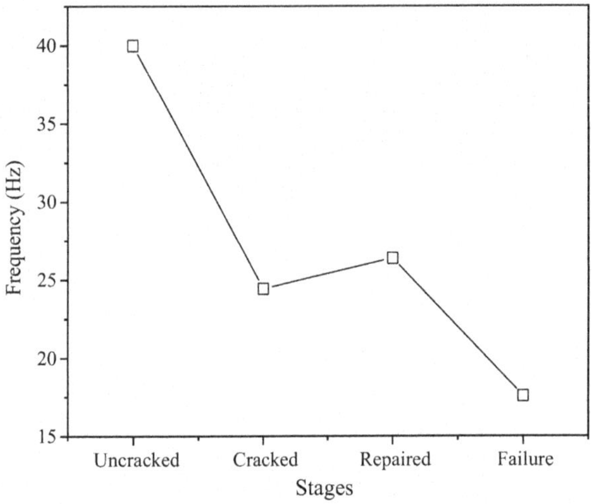


**Fig. 3.19** Frequency of specimen B1 at different stages

Given that the natural frequency $f_n$ is proportional to $\sqrt{E_c I}$, the decrease of stiffness $E_c I$ could be calculated as 65.7% and 60.6% for specimens B0 and B1, respectively. The stiffness increased at the rate of 16.6% after retrofitting and decreased at the rate of 55.5% after loading because of severe damage to the section. Figure 3.19 shows the natural frequency of specimen at different stages. The FRP strengthening system caused the natural frequency and overall stiffness to increase slightly. The experimental results reported in [46] show that the load-induced cracking caused a decrease of natural frequency by approximately 10% because of a reduction in the $E_c I$ of the beams [26]. Results reported in [46] obtain by modal analysis and dynamic tests show that cracking results in a reduction of stiffness, whereas the application of the FRP composite results in an increase of stiffness. Results reported in [47] that the stiffness of CFRP sheet-strengthened RC beam is increased by approximately 6.0% compared with that of the beam without strengthening. Results reported in [20] indicated that the FRP strengthening system only has a slight effect on the flexural stiffness of the RC beam.

### *3.2.5 Strain Distribution in BFRP Sheets*

The composite behavior of the BFRP-strengthened beam was investigated by examining the strain distribution in the sheet. Figure 3.20 shows the strain distribution from midspan to the end of the sheet at various load levels. Figure 3.20 shows that the strain distribution along the sheet could be divided into three distinct zones: (1) de-stressed zone, (2) bond development zone, and (3) full composite zone. The strains in the de-stressed zone were observed to be approximately equal to zero. The strains in the sheet increased linearly from zero to a strain value of 4320 $\mu\varepsilon$ in the bond development zone. The observations indicated that strain at the transition point

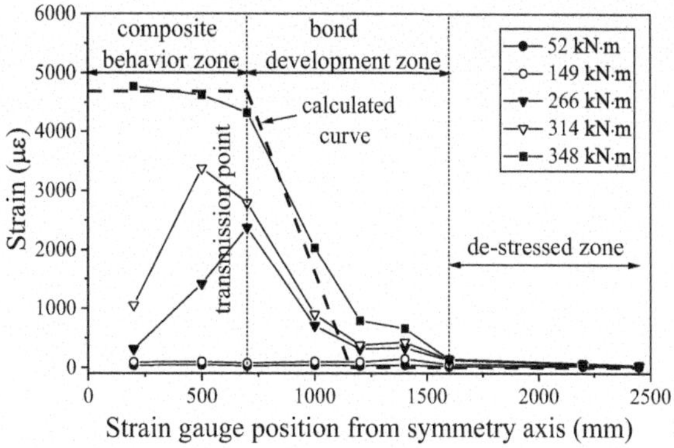


**Fig. 3.20** Strain distribution of BFRP sheet

between the bond development zone and the full composite zone, if limited to a certain critical value, may prevent ripping failures. The strain at the transition point $\varepsilon_{\mathrm{ft}}$ is expressed as Eq. (3.80).

$$\varepsilon_{\mathrm{ft}} = \frac{1}{E_{\mathrm{c}}} \frac{M_{\mathrm{t}}}{I_{\mathrm{c}}} (d_{\mathrm{f}} - x) \tag{3.80}$$

where $M_t$ is the applied bending moment at the transition point, in kN·m; and $I_{\mathrm{c}}$ is the equivalent moment of inertia of cracked section, in m$^4$.

Figure 3.20 shows that the strain distribution along the sheet can be calculated when the strain at the transition point $\varepsilon_{\mathrm{ft}}$ and the total bond development length $l_{\mathrm{dev}}$ are obtained. The total bond development length $l_{\mathrm{dev}}$ to be used in practice is calculated as follows Eq. (3.81).

$$l_{\mathrm{dev}} = c + \frac{d_{\mathrm{f}}}{2} + \frac{4.61}{\lambda} \tag{3.81}$$

where $c$ is the thickness of concrete cover, in mm; and $\lambda$ is a constant that can be calculated using Eq. (3.82).

$$\lambda = \sqrt{\frac{1}{E_{\mathrm{f}} t_{\mathrm{f}}} \frac{G_{\mathrm{a}} G_{\mathrm{c}}}{G_{\mathrm{c}} t_{\mathrm{a}} + G_{\mathrm{a}} c}} \tag{3.82}$$

where $G_{\mathrm{a}}$ and $G_{\mathrm{c}}$ are the shear modulus of the adhesive and concrete, respectively, in MPa; and $t_{\mathrm{a}}$ is the thickness of the adhesive layer, in mm.

As the specimen B1 is cracked, the shear modulus $G_{\mathrm{c}}$ of concrete can be calculated as Eq. (3.83).

$$G_{\mathrm{c}} = \frac{0.4 E_{\mathrm{c}}}{2(1 + \nu_{\mathrm{c}})} \tag{3.83}$$

where $\nu_c$ is the Poisson's ratio of concrete.

The shear modulus $G_a$ of adhesive can be calculated as Eq. (3.84).

$$G_{\mathrm{a}} = \frac{E_{\mathrm{a}}}{2(1 + \nu_{\mathrm{a}})} \tag{3.84}$$

where $E_{\mathrm{a}}$ is the elastic modulus of adhesive, in MPa; and $\nu_{\mathrm{a}}$ is the Poisson's ratio of adhesive.

In Fig. 3.20, the calculated strain distribution, assuming that the concrete section was cracked, is shown as a dashed line. The figure shows that the result of the theoretical calculation is similar to the experimental result. Therefore, the ripping failure can be prevented by calculating the transition strain. The strain distribution along the BFRP sheet can also be predicted by calculating the total bond development length and the transition strain.

## 3.3 Behavior of Reinforced Concrete Box Beam Retrofitted with BFRP Using End Anchorage in Grooving

### *3.3.1 Experimental Program*

(1) Specimens

Three full-scale test RC beams with initial flexural cracks were removed. The average concrete cover depths of specimen B2 was 30.0 mm, respectively.

The specimens B0 were same as that shown in Sect. 3.2.1. Specimen B2 was anchored with grooving. In practical engineering, end anchorage was sometimes not allowed to settle in side surface of the beam, a new end anchorage with grooving could be used. End anchorage with grooving required chiseling grooves at the bottom of the beam at a distance of 940 mm from the supports. The sectional dimension of the grooves was 30 mm × 30 mm [48]. The details of the grooves are shown in Fig. 3.21.

Two-layer BFRP strips were used for flexural reinforcements for single layer of BFRP strips was poor for strengthening according to calculation.

Figure 3.22 shows the layout of the arrangement of FRP strips on the bottom surface and side surface of RC beams.

(2) Instrumentation Layout and Test Setup

In order to stick strain gauges to reinforcing bar, part of concrete cover was scrapped. Strains in the BFRP strips were measured at various points along its length using 5 mm uniaxial strain gauges in the longitudinal direction, as shown in Fig. 3.23.

The layout of test setup included the loading system and instrumentation was same as that shown in Fig. 3.1. The loading history and detailed arrangement and preparations on the strain gauges of tested specimens were also same as before.

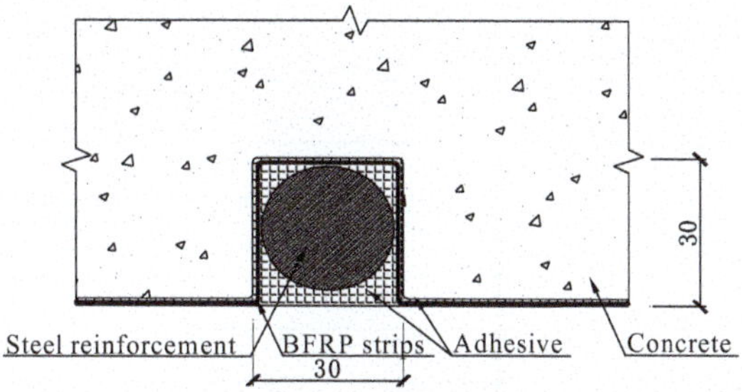


(a) Schematic diagram of groove

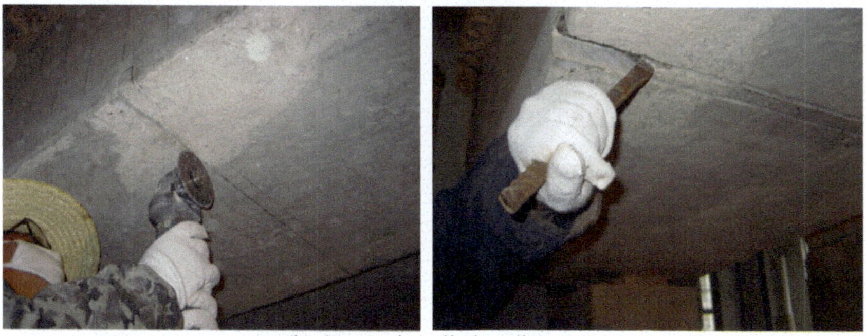

(b) Groove installation diagram

**Fig. 3.21** Details of grooves (all units in mm)

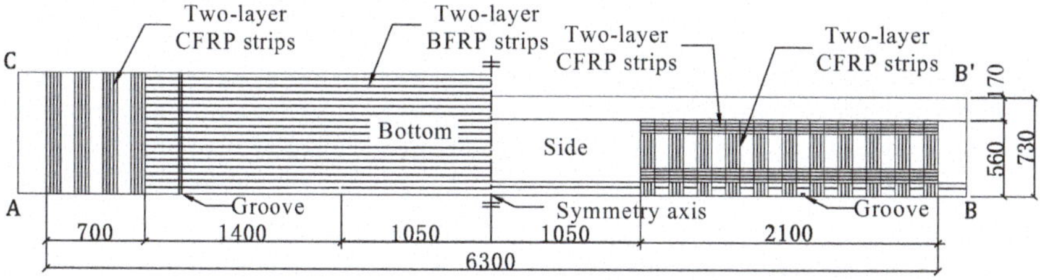


**Fig. 3.22** Details of FRP system of specimen B2 (all units in mm)

(3) Modal Tests

Experimental modal analysis has been recently utilized successfully in the area of civil engineering to understand the response of structural systems [10]. Ambient vibration tests are the focus of increasing attention in field tests and full-scale tests because of economics and practicability. Several ambient modal identification techniques are currently available. Stochastic subspace identification method and its variants are well-known techniques capable of rapidly extracting modal properties [22]. A Bayesian modal identification theory using fast Fourier transform of ambient time history data was developed [23]. Owing to the advantages of modal tests and the development of modal identification techniques, experimental modal analysis

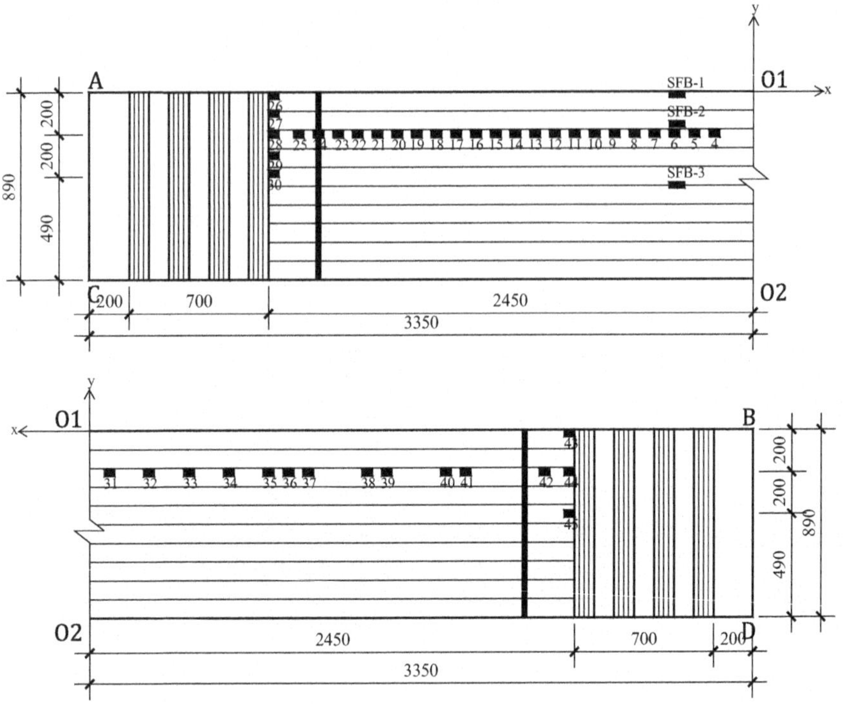


**Fig. 3.23** Strain gauge positions on BFRP strips in specimen B2 (all units in mm)

is being used as the major technique for damage identification research. The crack depths of concrete structures with non-destructive testing, such as the wave transmission method, were evaluated [25]. Modal tests using ambient vibration response are conducted to determine the change in natural frequency of the bridge [49]. Therefore, this technology was used to evaluate the natural frequency of beams with initial flexural cracks strengthened with BFRP strips. The specimens were distressed and deteriorated beams with initial flexural cracks on the side surface after 60 years of service. Detailed inspections were conducted before the loading test to check the flexural cracks on the side surface. Specimens B0 and B1 both had four initial flexural cracks and specimen B2 had three initial flexural cracks. These initial flexural cracks had an effect on the vibration characteristics or responses of the RC beam, such as the decrease of natural frequency.

For specimen B0, modal tests were conducted before loading test and after loading test. For specimens B1 and B2, modal tests were conducted before retrofitting, after retrofitting, and after loading test. Seven magnetoelectric velocity sensors were arranged along the top surface of the beam. The sensor data was received by the data acquisition system. Appropriate hammering was applied because natural vibration alone would not yield ideal test results. Stiffness change could be inferred through the natural frequency data to evaluate the health of the beam.

### 3.3.2 Failure Mode and Cracking Characteristics

Reference specimen B0 failed in a flexural manner. As the sustaining load, the initial flexural cracks were initiated from the bottom of the beam in pure bending zone. At 90.0 kN load, these initial cracks were widened and extended upward and new flexural cracks formed. As the applied load was further increased to 208.6 kN, the maximum crack width was increased to 1.5 mm. The ultimate load of specimen B0 was measured as 287.4 kN. The photograph of specimen B1 after failure is shown in Fig. 3.24a.

Specimen B1 was strengthened with double-layer BFRP strips anchored with U-strips and failed by debonding of CFRP and BFRP strips in the end zone of the beam. The width of initial flexural cracks was constant and the maximum crack width was 0.1 mm before the load was increased to 49.7 kN. New flexural cracks had formed as the applied load was increased to 68.3 kN. As the load further increased, the old flexural cracks propagated and more flexural cracks formed in pure bending zone of the beam. These cracks widened and the maximum crack width was increased to 1.5 mm at a load of 331.7 kN. After the load of 342.1 kN, the CFRP U-strips began to debond from the side surface of the beam and then the BFRP strips soon debonded from the concrete cover in the end zone of the beam. The ultimate load of specimen B1 was measured as 375.9 kN. The photograph of specimen B1 after failure is shown in Fig. 3.24b.

Specimen B2 was strengthened with double-layer BFRP strips anchored with grooving and failed by debonding of BFRP strips in pure bending zone of the beam. The first new flexural crack was observed at a load of 135.2 kN. The width of the new crack was 0.05 mm and the maximum crack width of initial cracks was 0.2 mm. As the load was increased to 367.9 kN, the maximum crack width was increased to 1.5 mm. Meanwhile, the specimen made a noise just like that the BFRP strips were straightened out. However, the debonding was not observed and the possible reason was that the FRP debonding had already taken place, but the end anchorage with grooving made the BFRP strips close to the concrete cover in the end zone of the beam. As the load was further increased to 383.5 kN, CFRP U-strips also began to debond from the side of the beam. At the same time, BFRP strips also began to the concrete cover in pure bending zone of the beam. The debonding occurred near the cracks. This failure mechanism is identified as a kind of "intermediate crack-induced debonding." [50] When a new flexural crack was formed in the concrete, the tensile stresses released by the cracked concrete were transferred to BFRP strips. As a result, high local interfacial stresses between BFRP strips and concrete were induced near the crack. As the applied loading increased further, the tensile stresses developed in the strips and hence the interfacial stresses between BFRP strips and concrete near the crack also increased. When these stresses reached critical values, debonding initiated at the crack and then propagated toward to whole BFRP strips. The debonding failure resulted from the flexural cracks. The ultimate load of specimen B2 was measured as 415.5 kN. The photograph of specimen B2 after failure is shown in Fig. 3.24c.

(a) B0

(b) B1

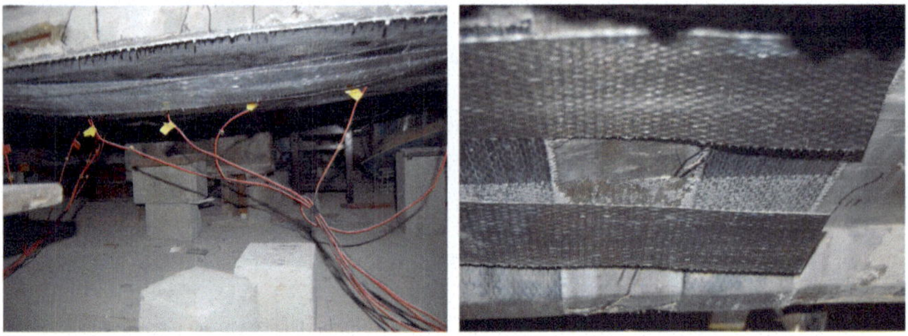

(c) B2

**Fig. 3.24** Details of specimens of B0, B1, and B2 in failure

The crack patterns under ultimate load of specimens B0, B1, and B2 are shown in Figs. 3.15 and 3.25. Only the cracks in the pure bending zone were included. The average crack spacing, average crack depth, and average crack width of specimen B0, B1, and B2 are shown in Table 3.2. BFRP-strengthened beams can effectively postpone the development of cracks. Meanwhile, the average crack spacing, average crack depth, and average crack width of specimen B2 under ultimate load were 105, 310, and 0.91 mm, which were 25.0%, 22.5%, and 18.8% or 10.3%, 2.8%, and 6.2% lower than those of specimen B0 or B1, respectively. The specimen B2 had a better performance in the confinement of cracks compared with that of specimen B1. The maximum crack width of specimen B2 was lower than that of specimens B0, and B1 at different loading levels, as shown in Fig. 3.26.

After the load of 150 kN·m, the growth rate of the crack width of specimen B1 was evidently higher than that of specimen B2. Therefore, the control of end anchorage with grooving on crack amount, crack spacing, average crack depth, and average crack width was more effective than that of end anchorage with U-strips.

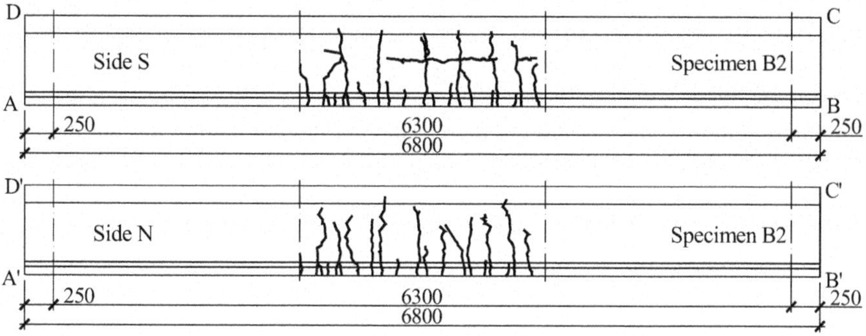


**Fig. 3.25** Crack patterns of specimen B2 at ultimate load (all units in mm)

**Table 3.2** Properties of cracks under ultimate load of specimens of B0, B1, and B2

| Identification of specimen | Average crack spacing (mm) | Average crack depth (mm) | Average crack width (mm) |
|---|---|---|---|
| B0 | 140 | 400 | 1.12 |
| B1 | 117 | 319 | 0.97 |
| B2 | 105 | 310 | 0.91 |

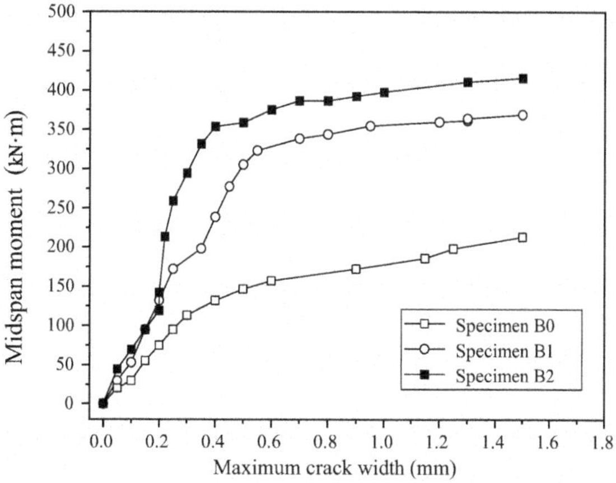


**Fig. 3.26** Relationship between midspan moment and maximum crack width of specimens of B0, B1, and B2

The capacity of restraining development of cracks for specimen B2 was also slightly better than that for specimen B1.

### 3.3.3 Ductility of Specimens

The ductility includes deflection ductility and curvature ductility [5]. To relate the definitions of structural ductility to the physical behavior of the beams, the load–deflection and moment–curvature curves at the midspan section of beams are shown in Figs. 3.27 and 3.28. The curvature at midspan was evaluated from the measured strains over the depth of the beam. The data in Figs. 3.27 and 3.28 was utilized to quantify ductility. The following definitions of structural ductility were adopted:

Deflection ductility was calculated by Eq. (3.85).

$$\mu_{\Delta} = \frac{\Delta_{\mu}}{\Delta_{y}} \tag{3.85}$$

Curvature ductility was calculated by Eq. (3.86).

$$\mu_{\phi} = \frac{\phi_{\mu}}{\phi_{y}} \tag{3.86}$$

where $\Delta_{\mu}$ is the midspan deflection at ultimate failure; $\Delta_{y}$ is the midspan deflection at yielding of reinforcing bars; $\phi_{\mu}$ is the curvature at midspan section at ultimate failure; and $\phi_{y}$ is the curvature at midspan section at yielding of reinforcing bars, respectively.

The ductility indices were calculated for all beams tested, and the results were shown in Table 3.3. The data was expressed in terms of ratios of the ductility of the

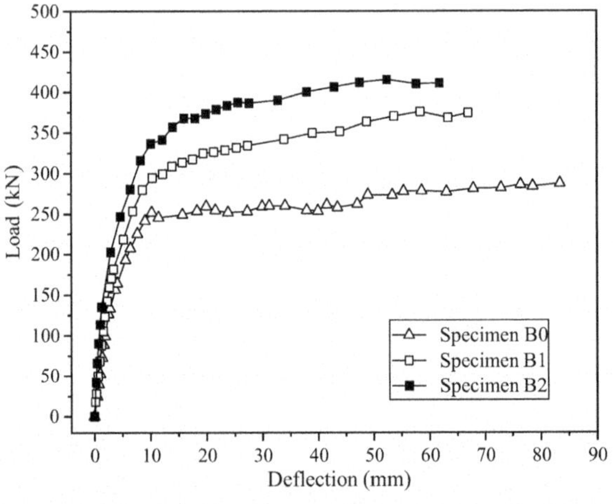


**Fig. 3.27** Relationship between load and deflection of specimens B0, B1, and B2

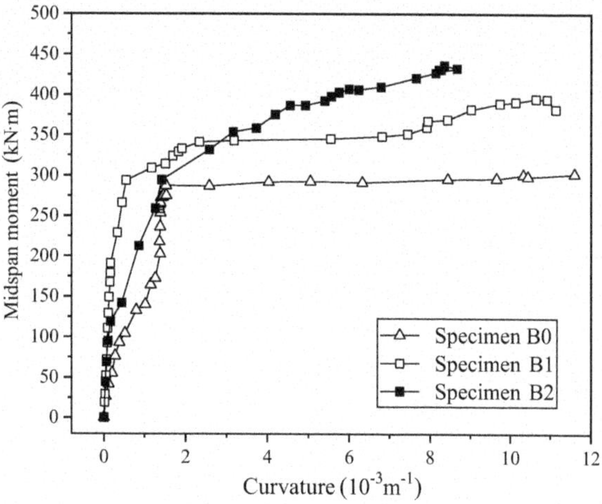


**Fig. 3.28** Relationship between midspan moment and curvature of specimens B0, B1, and B2

**Table 3.3** Ductility indices of specimens B0, B1, and B2

| Identification of specimen | Deflection ductility (mm) | Deflection ductility ratio | Curvature ductility (mm) | Curvature ductility ratio |
|---|---|---|---|---|
| B0 | 9.2 | 1.00 | 8.4 | 1.00 |
| B1 | 5.6 | 0.61 | 4.9 | 0.59 |
| B2 | 2.7 | 0.29 | 3.3 | 0.40 |

strengthened beams with end anchorage to those of the reference beam. These results had two significant implications. First, the bond slip between the BFRP strips and concrete substrate could lead to substantial losses in ductility, as shown by specimens B1, and B2, as reported in [51]. The deflection ductilities of specimens B1 and B2 were only 29%–61% of that of specimen B0. A distinct loss in ductility was observed in the strengthened beams. Secondly, the deflection ductility of specimen B2 was 52% lower than that of specimen B1. For specimen B2, more ductility was sacrificed in order to increase the load-bearing capacity.

### 3.3.4 Load-Bearing Capacity of Specimens

According to ACI 440.2R-08 [21], the debonding strain of externally bonded FRP strips $\varepsilon_{fd}$ can be calculated as Eq. (3.87).

$$\varepsilon_{fd} = 0.41\sqrt{\frac{f'_c}{nE_f t_f}} \le 0.9\varepsilon_{fu} \tag{3.87}$$

where $f_c'$ is the specified compressive strength of concrete, in MPa; $n$ is the number of plies of FRP strips, in MPa; $E_f$ is the tensile modulus of elasticity of FRP strips, in MPa; $t_f$ is the nominal thickness of one ply of FRP strips, in mm; and $\varepsilon_{fu}$ is the design rupture strain of FRP strips.

Specimens B0, B1, and B2 served for more than 60 years in the same working conditions. The load-bearing capacity of the RC beam strengthened with BFRP strips could be calculated from Eq. (3.56). The experimental load-bearing capacities considering the moment induced by self-weight were 342.0, 434.9, and 476.5 kN·m, which increased by 27.2%, and 39.3%, for specimens B0, B1, and B2, respectively. Test results showed that the load-bearing capacity of specimen B2 was higher than that of specimen B1. The theoretical load-bearing capacities of specimens B1 and B2 were both 519.4 kN·m. The experimental and calculated load-bearing capacities are shown in Table 3.4. For specimens B1, and B2, the theoretical load-bearing capacities were 19.4%, and 9.0% higher than the experimental values. The possible reason is that the loss of capacity induced by BFRP strips debonding was not considered in standard [21]. The present calculation method of RC beams strengthened with CFRP strips obtained from standard [21] considered the failure mode of rupture of FRP strips and plane cross-section assumption. The calculation method was not applicable for the BFRP-strengthened beams with failure of BFRP debonding.

Tomas proposed a calculation method for the load-bearing capacities of RC beams strengthened with different anchorages [52]. The calculation method takes into account the effect of the bond slip on the load-bearing capacities of FRP-strengthened beams. The suggested method for calculating RC flexural structures strengthened with external FRP bars is based on the theory of built-up bars. The following assumptions are made: the beam is consisted of two bars: concrete and FRP; bars are with partial shear connection.

The load-bearing capacities, $M_R$, of the BFRP-strengthened beams were calculated by Eq. (3.88).

$$M_R = k \cdot M_n \tag{3.88}$$

The coefficient $k$ that evaluates the rigidity of the bond between BFRP strips and concrete was calculated by applying Eq. (3.89).

**Table 3.4** The load-bearing capacities of specimens B0, B1, and B2

| Identification of specimen | Experimental $M_{exp}$ | Calculated (ACI) $M_{ACI}$ | Calculated (Toma) $M_{Toma}$ | $M_{exp}/M_{ACI}$ | $M_{exp}/M_{Toma}$ |
|---|---|---|---|---|---|
| B0 | 342.0 | 330.3 | – | 1.04 | – |
| B1 | 434.9 | 519.4 | 444.3 | 0.84 | 0.98 |
| B2 | 476.5 | 519.4 | 491.3 | 0.92 | 0.97 |

*Note* $M_{exp}$ means the load-bearing capacities obtained from experiments; $M_{ACI}$ means the load-bearing capacities calculated by ACI 440.2R-08; and $M_{Toma}$ means the load-bearing capacities calculated by Tomas' method

$$k = 1 - \frac{\mathrm{ch}(\lambda(0.5l - l_a))}{\lambda l_a \mathrm{ch}(0.5\lambda l)} \mathrm{sh}(\lambda l_a) \tag{3.89}$$

where $l$ is the length of the beam span, in mm; $l_a$ is the distance from support to the concentrated force, in mm; and the value $\lambda$ assessing the stiffness of the bond was calculated by Eqs. (3.90)–(3.92).

$$\lambda^2 = \mu\gamma \tag{3.90}$$

$$\mu = \frac{b_c G_{w,eff}}{a} \tag{3.91}$$

$$\gamma = \frac{1}{E_c A_{c,eff}} + \frac{1}{E_f A_f} + \frac{a^2}{E_c I_{c,eff}} \tag{3.92}$$

where $A_{c,eff}$ is the area of effective RC cross-section, in mm$^2$; $I_{c,eff}$ is the moment of inertia of effective RC cross-section, in mm$^4$; $a$ is the distance from the weight center of the concrete and the BFRP strips, in mm; $b_c$ is the width of concrete beam, in mm; and $G_{w,eff}$ is the characteristic of the bond shear stiffness, in N/mm.

The moment of inertia of a concrete cross-section with cracks was calculated by Eq. (3.93).

$$I_{c,eff} = \frac{b \cdot x^2}{3} \tag{3.93}$$

The characteristic of the bond shear stiffness $G_{w,eff}$ is received from experimental research results and can be calculated using Eq. (3.94).

$$G_{w,eff} = \frac{3k_1\beta}{E_f A_f} E_c \tag{3.94}$$

$$\beta = \sqrt{\frac{2.5 - b_f/b_c}{1.5 + b_f/b_c}} \tag{3.95}$$

where $\beta$ is coefficient which evaluates the influence of the width of external bars on the operation of strengthened RC members; $b_f$ is the width of the external BFRP strips, in mm; $k_1$ is coefficient which evaluates the type of anchorage of external bars; and $k_1$ is also received from experimental research results.

The calculated load-bearing capacities of specimens B1 and B2 in consideration of the stiffness obtained by the result of bond performance were 444.3, and 491.3 kN·m, of which deviations were 2.16%, and 3.11%, when compared with experimental results, respectively, as shown in Table 3.4. The theoretical load-bearing capacities obtained from Tomas' method showed good accuracy with the experimental results. This method was semi-theoretical and semi-experimental, and it could be effectively utilized to calculate the load-bearing capacities of BFRP-strengthened beams with different anchorages.

The moment-deflection relationship of the beams is shown in Fig. 3.27. Before the sudden change in the slope of the curves, the deflections exhibited a slight difference at the same moment. The rate of deflection increase of specimen B2 was evidently lower than that of specimen B1 and the rate of deflection increase of specimen B2 was 30.0% lower than that of specimen B0 at the ultimate load. Therefore, end anchorage with grooving had a better performance in deflection control than that with U-strips.

### 3.3.5 *Stiffness and Natural Frequency*

The time history curves and amplitude-frequency curves for four-point bending test of specimens B0 and B2 are shown in Figs. 3.29 and 3.30, respectively. The theoretical natural frequency of specimens B0, B1, and B2 without cracks was calculated as 40.0, 38.9, and 38.9 Hz, respectively.

The experimental natural frequency before retrofitting, after retrofitting, after loading, and calculated natural frequency before retrofitting are shown in Table 3.5. Given that the natural frequency $f_n^2$ is proportional to $E_cI$, the decrease of stiffness $E_cI$ after 60 years of service could be calculated as 65.7%, 60.6%, and 61.0% for

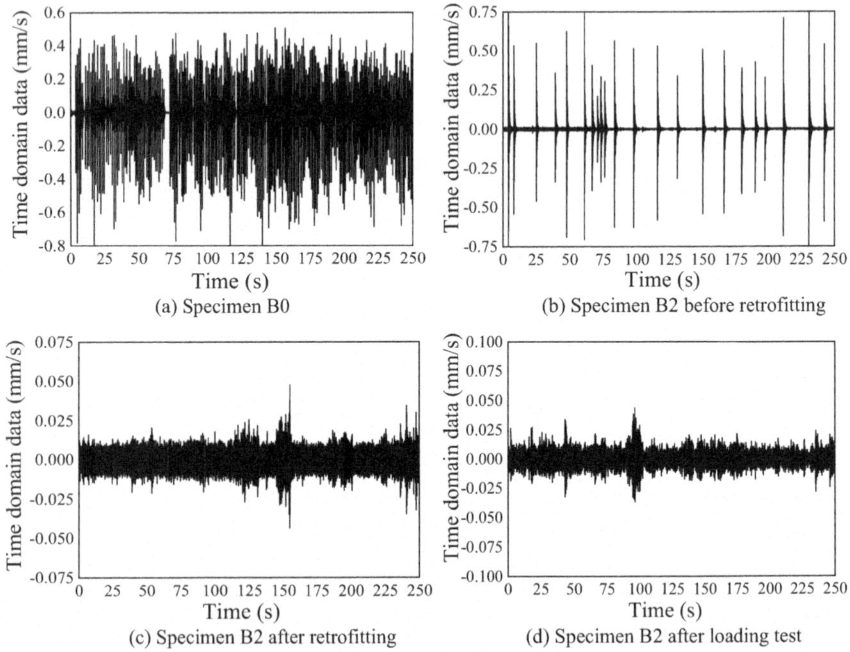


**Fig. 3.29** Time history curves of specimens B0 and B2

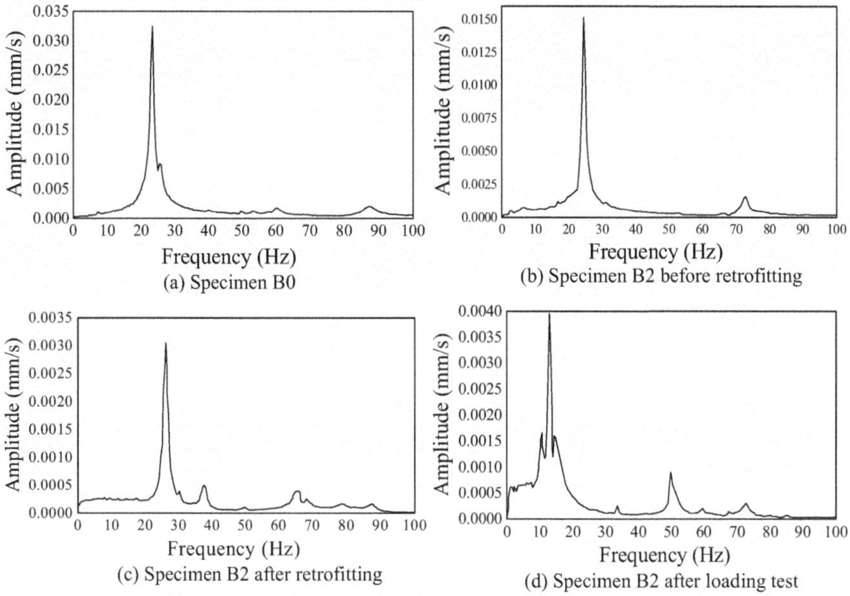


**Fig. 3.30** Amplitude-frequency curves of specimens B0 and B2

specimens B0, B1, and B2, respectively. The stiffness of specimen B1 increased by 16.6% after retrofitting, and decreased by 55.5% after loading because of severe damage to the section. Meanwhile, the stiffness of specimen B2 increased by 22.2% after retrofitting and decreased by 54.7% after loading. Results in [47] show that the stiffness of CFRP-strengthened RC beam is increased by approximately 6.0% compared with that of beam without strengthening. Results in [26] show that the load-induced cracking causes a decrease of natural frequency by approximately 10% because of a reduction in the $E_c I$ of the beams. The natural frequency of specimen B2 was almost the same as that of specimen B2 at different stages, as shown in Fig. 3.31. Therefore, grooving in the section of the RC box beam didn't decrease the stiffness of specimen B2.

**Table 3.5** Natural frequency of specimens B0, B1, and B2

| Identification of specimen | Experimental natural frequency (Hz) | | | Calculated natural frequency (Hz) |
|---|---|---|---|---|
| | Before retrofitting $f_{exp}$ | After retrofitting $f_{exp}$ | After loading $f_{exp}$ | Before retrofitting $f_{calc}$ |
| B0 | 23.44 | – | 21.10 | 40.0 |
| B1 | 24.43 | 26.38 | 17.59 | 38.9 |
| B2 | 24.30 | 26.87 | 18.08 | 38.9 |

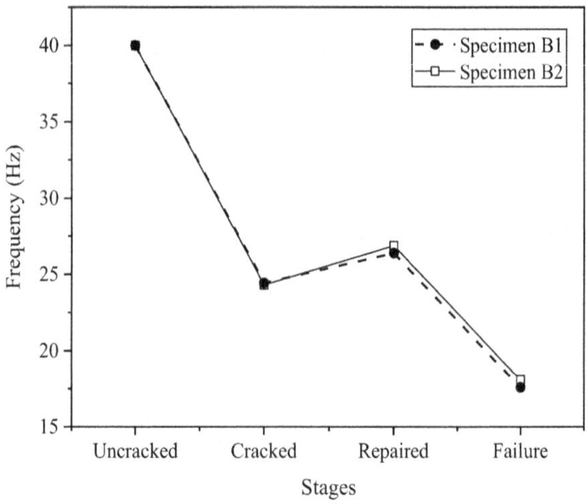


**Fig. 3.31** Frequency of specimens B1 and B2 at different stages

### *3.3.6 Relationship Between Load and Strain*

The relationships between load and longitudinal strains in the BFRP strips for specimens B1 and B2 obtained from 8 strain gauges showed a slight tensile relaxation in the end anchorage zone during the final loading stages (see Fig. 3.32). A closer look at the load-strain histories recorded by five gauges at the zone of end anchorage for specimens B1, and B2 (see Fig. 3.33) confirmed that the partial relaxation began under the load of 275.5, and 350.4 kN. The anchorage strain relaxation in the BFRP strips for specimen B2 (see Fig. 3.33b) occurred at about 27.3% higher load level than that of specimen B1 (see Fig. 3.33a), because end anchorage with grooving delayed and prevented further propagation of the horizontal strips-end cracks through concrete. Therefore, the end anchorage with grooving effectively delayed the tensile relaxation and the debonding of BFRP strips. It could be assumed that rapid strain relaxation in end anchorage zone of the bonded FRP strips was an omen of debonding failure. Results in [53] show that the anchorage tensile relaxation in the FRP strips occurs at about 10% higher load level in comparison with the beam strengthened without using U-strips.

The FRP-strengthened beams are likely to suffer from FRP debonding. ACI 440-2R.08 [21] prescribes the relationship for evaluation of debonding strain in the FRP strips was calculated by Eq. (3.87).

For specimens B1 and B2, the effective strain level of the BFRP bars attained at failure, $\varepsilon_{fe}$, was calculated as 2079 $\mu\varepsilon$ using Eq. (3.87) and the design value of $\varepsilon_{fd}$ was calculated as 1585 $\mu\varepsilon$ using Eq. (3.87). The theoretical failure modes of specimen B1 and B2 were both BFRP debonding in the pure bending zone because $\varepsilon_{fe} > \varepsilon_{fd}$. However, the specimen B1 was premature damage with FRP debonding in the end region of beam. When the experimental strain in BFRP strips was increased to the debonding strain, the experimental load-bearing capacity of specimen B2 was 367.9

**Fig. 3.32** Tensile strains measured along the BFRP strips of specimens B1 and B2

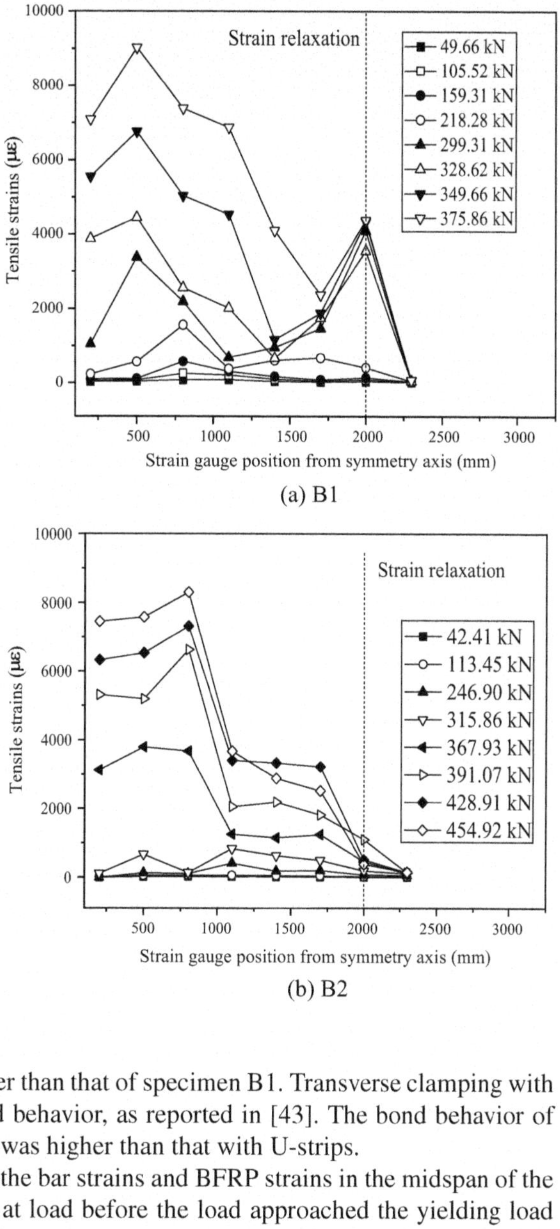


kN·m, which was 19.2% higher than that of specimen B1. Transverse clamping with FRP layers can improve bond behavior, as reported in [43]. The bond behavior of end anchorage with grooving was higher than that with U-strips.

For specimens B1 and B2, the bar strains and BFRP strains in the midspan of the beams were almost the same at load before the load approached the yielding load (see Fig. 3.34). The yielding load of specimens B1, and B2 were 294.5, and 357.2 kN, respectively. After yielding, the strains in BFRP strips increased more rapidly than strains in bar. The possible reason was that the bar had begun to debond from the concrete cover. Therefore, for specimen B2, the BFRP strips remained coordinated with bar at a higher load level than specimen B1.

**Fig. 3.33** Strains in the BFRP strips anchorage zone of specimens B1 and B2

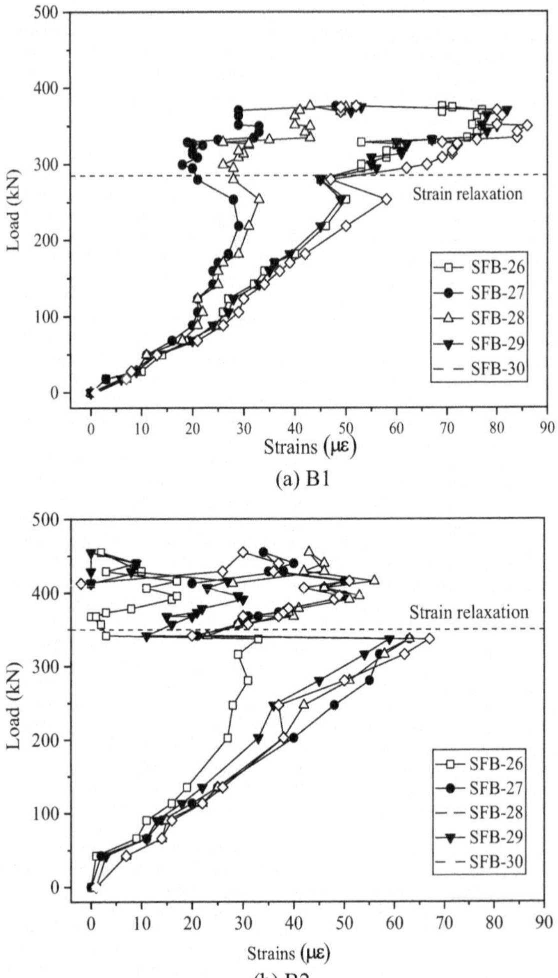

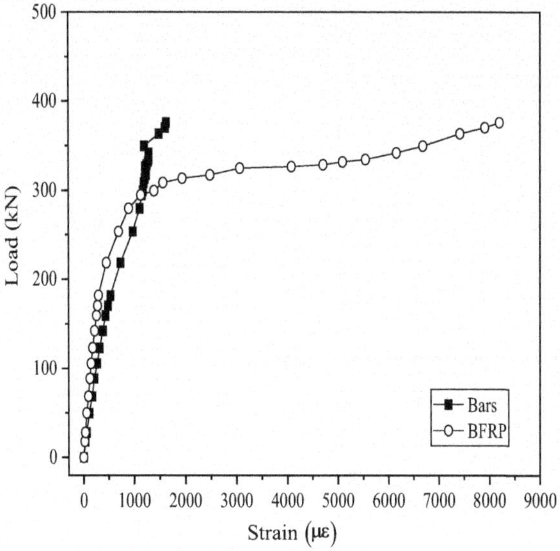


(a) B1

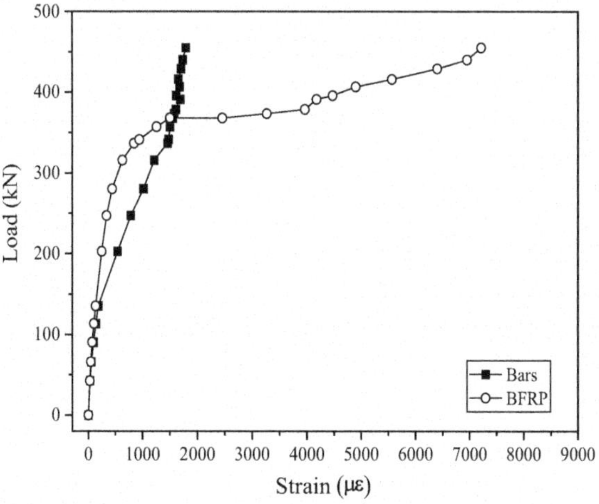


(b) B2

**Fig. 3.34** Relationship between load and strain of BFRP and bar of specimens B1 and B2

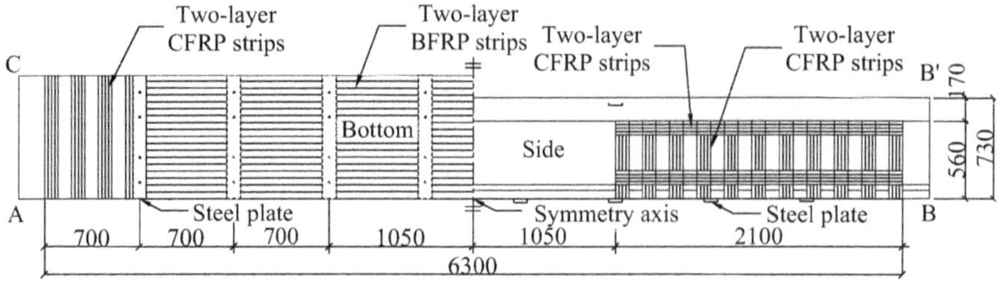


Fig. 3.35 Details of FRP system of specimens B3 (all units in mm)

## 3.4 Behavior of Reinforced Concrete Box Beam Retrofitted with BFRPs Using Steel Plate Anchorage

### *3.4.1 Experimental Program*

(1) Description of Specimens

Three full-scale RC box beams with initial flexural cracks were utilized as test specimens. The test specimens were designated as specimens B0, B1, and B3 respectively. Figure 3.1a demonstrates the cross-section of these full-scale test specimens.

The specimens B0 and B1 were same as that shown in B0 and B1 of Sect. 3.2.1. Figure 3.35 demonstrates the layout of the arrangement of CFRP and BFRP strips of Specimen B3.

At the end of the FRP U-strips, the CFRP strips were set horizontally as the trim strips. The steel plates were placed above the BFRP strips of specimen B3 as anchorages, as demonstrated in Fig. 3.35.

This anchorage was started from the end of BFRP strips. Eight steel plates were utilized in specimen B3, and the spacing between two steel plates was 700 mm. The anchorage device included anchor bolts and steel plates, and the size of steel plate was 900 mm × 100 mm × 10 mm. Each steel plate had four reserved holes, the diameter of which was 12 mm, as demonstrated in Fig. 3.36.

The process of anchoring steel plate was mainly divided into four steps: drilling holes at the bottom, cleaning holes, fixing anchor bolts, and fixing steel plates, as demonstrated in Fig. 3.37. The FRP U-strips and steel plates were both symmetrically along the center line of test specimen [54].

The surface slurry and oil stain on the bonding surfaces of CFRP and BFRP strips were removed in advance through polishing. The alcohol was utilized to clean up the surfaces of CFRP and BFRP strips after polishing to ensure that bonding between concrete and FRP strips was effective.

(2) Instrumentation Layout and Test Setup

Figure 3.38a demonstrates that the 5 mm uniaxial strain gauges were utilized to obtain the strains at various points along the length of BFRP strips in the longitudinal

**Fig. 3.36** Details of steel plate anchorage (all units in mm)

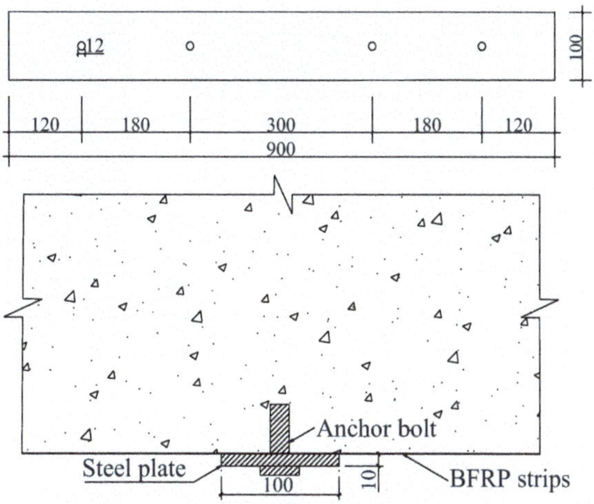

(a) Diagram

(b) Actual photo

direction. The strain in the steel plates was measured by the same kind of 5 mm gauges, as demonstrated in Fig. 3.38b.

The layout of test setup included the loading system and instrumentation was same as that shown in Fig. 3.1. The loading history and detailed arrangement and preparations on the strain gauges of tested specimens were also same as before.

(3) Modal Tests

Three flexural cracks were found on specimen B3, the positions and forms of which were demonstrated in Fig. 3.39. The vibration characteristics and responses of RC

(a) Drill holes at the bottom (b) Clean hole

(c) Fix anchor bolt (d) Fix steel plate

**Fig. 3.37** Process of steel plate anchorage

beams were affected by these initial flexural cracks, i.e., the decrease of natural frequency.

The modal test for specimen B0 was divided into two stages, which was conducted before and after loading test, respectively. The modal test for specimens B1 and B3 was divided into three stages, which was conducted before retrofitting, after retrofitting, and after loading, respectively. Seven magnetoelectric velocity sensors were set along the top surface of test specimens, the specific positions of which were demonstrated in Fig. 3.34a. The data acquisition system was utilized to obtain the sensor data. The ideal test results were obtained based on the combination of the natural vibration and appropriate hammering. The natural frequency data could be utilized to refer the change of stiffness, which played an important role in the evaluation of the health of the beam.

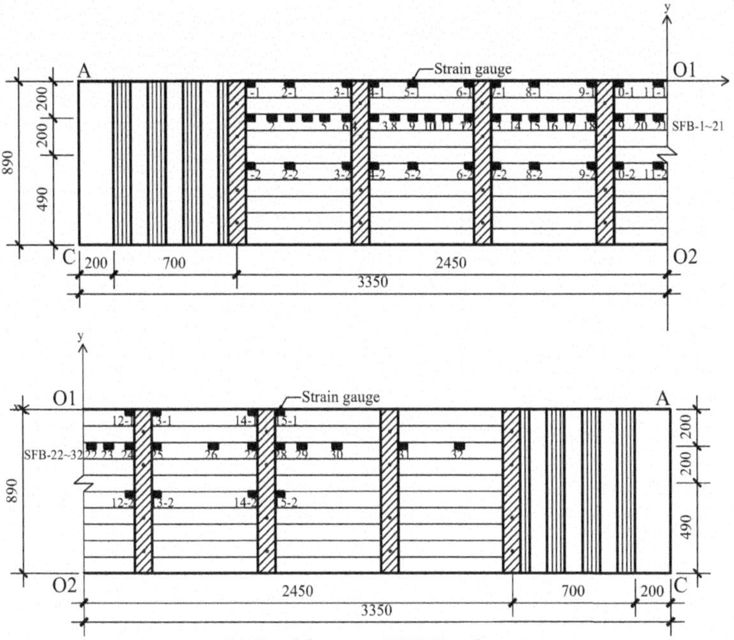


(a) Positions on BFRP strips

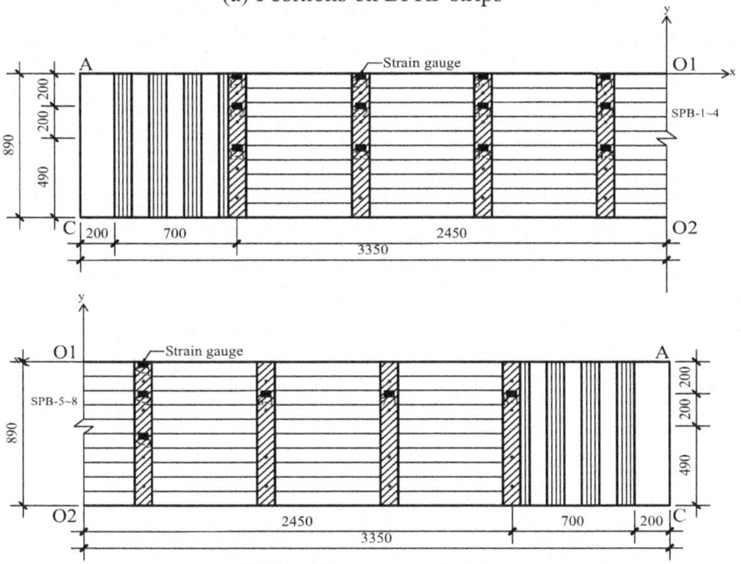


(b) Positions on steel plates

**Fig. 3.38** Strain gauge positions in specimen B3 (all units in mm)

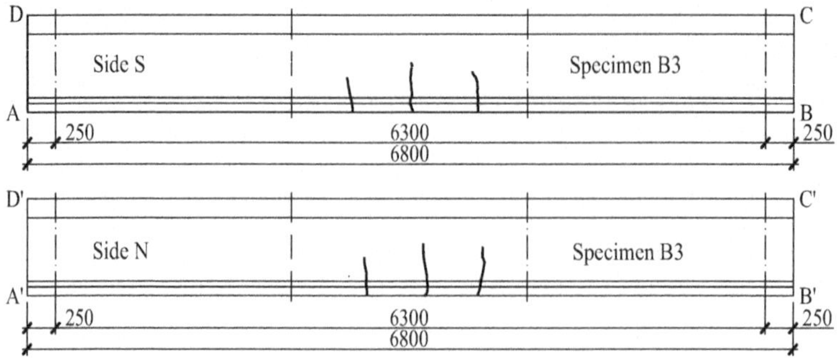


**Fig. 3.39** Initial Crack patterns of specimen B3 (all units in mm)

### *3.4.2 Analysis on Cracking Characteristics*

The cracking characteristics of specimens B0 and B1 were same as that shown in B0 and B1 of Sect. 3.3.2. For specimen B3, steel plates were utilized to anchor BFRP strips. Specimen B3 failed when the BFRP strips in pure bending zone of test specimen debonded. Under the load of 49.3 kN, the change of initial crack width was negligible, and no new cracks occurred at that moment. When the load further increased to 127.3 kN, new flexural cracks occurred, the width maximum of which was 0.05 mm. The maximum crack width was 1.5 mm when the load further increased to 376.9 kN. The BFRP strips were straightened out at this time, which made a great sound as a result of the slip failure. However, FRP debonding was not observed at this moment. This phenomenon may be due to that the FRP debonding had already occurred in fact, but the steel plates anchored the BFRP strips on the adhesive interface in pure bending zone of test specimen. Under the load of 385.1 kN, debonding at the BFRP strips and adhesive interface was occurred in pure bending zone of the test RC box beam, the position of which was near the cracks. This failure mechanism was considered as a kind of intermediate crack-induced debonding as demonstrated in [49]. The tensile stress released by cracked concrete was transferred to the BFRP strips because of the occurrence of new flexural cracks in concrete. This phenomenon induced the high level of local interfacial stresses at the interface of BFRP strips and concrete surface.

The tensile stress developed in the strips as the load increased, and the interfacial stress increased simultaneously. When the values of these stresses reached critical values, debonding occurred near the cracks, and then spread to other parts of BFRP strips. The ultimate load of specimen B3 was 421.5 kN. Figure 3.40 demonstrates the actual photos of specimen B3 after failure.

The average crack width, average crack depth, and average crack spacing of specimen B3 under ultimate load were 21.4%, 25.8%, and 30.0% or 9.3%, 6.9%, and 16.2% lower than those of specimens B0 or B1, respectively. Compared with specimen B0, the development of cracks was postponed effectively by the application of BFRP for specimens B1 and B3. This phenomenon indicated that the BFRP strips

(a) Left side (b) Right side

**Fig. 3.40** Details of specimen B3 in failure

had an evident influence on the control of cracks. The results demonstrated in [45] show that the FRP-strengthened RC beams had better cracking characteristics and smaller crack widths than unretrofitted RC beams. Thus, the results obtained based on specimen B1 and B3 could be considered as reasonable. The reason may be that FRP composites could carry the load throughout the loading stage, which could significantly reduce the development of cracks, as demonstrated in [55].

Figure 3.41 demonstrates the relationship between maximum crack width and midspan moment of three test specimens. Compared with specimens B0 and B1, the maximum crack width of specimen B3 was the lowest at different levels of load, as demonstrated in Fig. 3.41. The crack widths of specimen B1 increased more quickly than that of specimen B3 when the moment exceeded 150 kN·m. The performance of specimen B3 in the confinement of cracks was better than that of specimen B1, and the steel plates utilized in specimen B3 could restrain the development of cracks more effectively than FRP U-strips.

**Fig. 3.41** Relationship between midspan moment and maximum crack width of specimens B0, B1, and B3

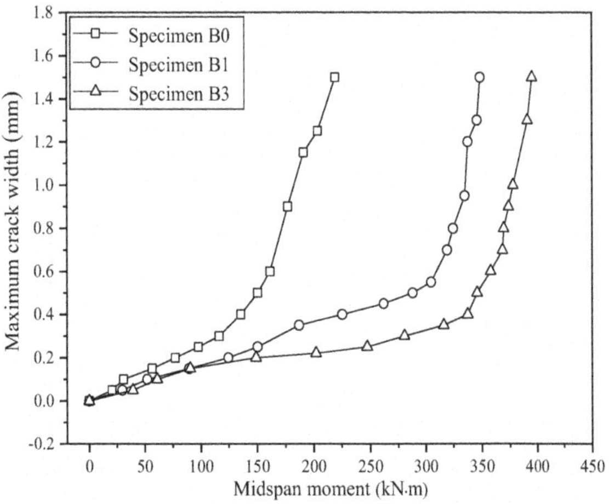

### 3.4.3 Analysis on Ductility

Figure 3.42 demonstrates that the load–deflection curves of test specimens were divided into two stages. The deflections of specimens B1 and B3 were small at the elastic stage (first stage) because the effect of BFRP strips was not fully exerted. As the load continued to increase, the cracks occurred in the tensile zone of test specimens. The beams were at the elastoplastic stage (second stage) at this moment, the deflections of specimens B1 and B3 changed greatly. The BFRP strips fully exerted its effect, which effectively inhibited the cracking of concrete and delayed the development of the cracks. The load–deflection curves of specimens B1 and B2 showed a good linear relationship at the first stage, which indicated that the BFRP strips and steel plates were uniformly loaded at this moment. The midspan deflection of specimen B1 was smaller than that of specimen B3 at the same load, as demonstrated in Fig. 3.42.

The slopes of load–deflection curves changed abruptly at the moment when the bar yielded. Therefore, the load at the yielding of bar could be determined by the sudden change of slopes of load–deflection curves. The midspan deflection at ultimate failure or yielding of bar of specimens B0, B1, and B3 were 83.4, 58.1, and 49.1 mm or 9.1, 10.4, and 11.9 mm, and the corresponding load were 287.4, 375.9, and 421.5 kN or 241.6, 301.7, and 334.6 kN, respectively. The curvature at ultimate failure or yielding of bar for specimens B0, B1, and B3 were $11.7 \times 10^{-3}$, $9.3 \times 10^{-3}$, and $8.5 \times 10^{-3}$ $m^{-1}$ or $1.4 \times 10^{-3}$, $1.9 \times 10^{-3}$, and $2.4 \times 10^{-3}$ $m^{-1}$, respectively.

The ductility indices of specimens B0, B1, and B3 were calculated, and the corresponding results are demonstrated in Table 3.6.

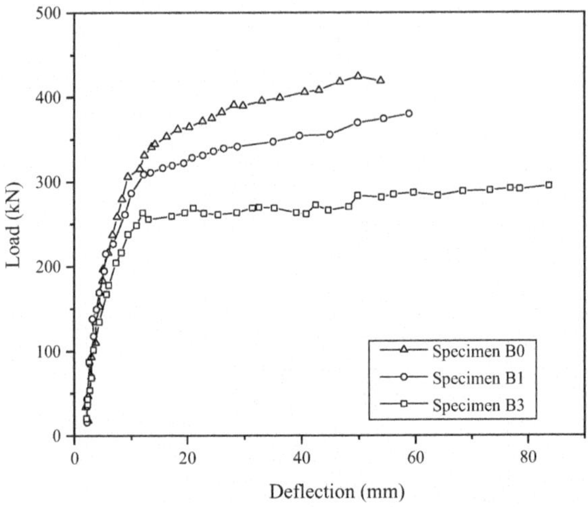


**Fig. 3.42** Relationship between load and deflection of specimens B0, B1, and B3

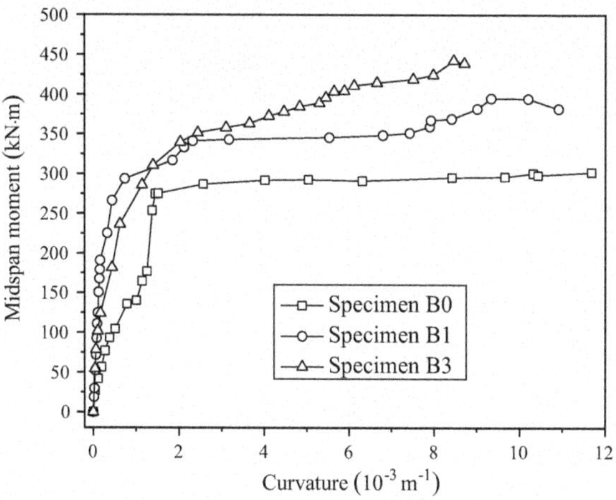


**Fig. 3.43** Relationship between midspan moment and curvature of specimens B0, B1, and B3

**Table 3.6** Ductility indices of specimens B0, B1, and B3

| Specimen | Deflection ductility (mm) | Deflection ductility ratio | Curvature ductility (mm) | Curvature ductility ratio |
|---|---|---|---|---|
| B0 | 9.2 | 1.000 | 8.4 | 1.000 |
| B1 | 5.6 | 0.609 | 4.9 | 0.583 |
| B3 | 4.1 | 0.446 | 3.5 | 0.417 |

Compared with specimen B0, the deflection ductility or curvature ductility of specimens B1 and B3 decreased by 39.1% and 55.4% or 41.7% and 58.3%, respectively. A distinct loss in ductility was occurred in both specimens B1 and B3. This phenomenon may be due to the existence of bond slip between the BFRP strips and concrete substrate, as demonstrated in [51]. The deflection ductility and curvature ductility of specimen B3 were 26.8% and 28.6% lower than the ductility of specimen B1, respectively. This meant that specimen B3 sacrificed more ductility to improve its load-bearing capacity.

### *3.4.4 Analysis on Load-Bearing Capacity*

All test specimens were in service for 60 years under normal service condition. The theoretical load-bearing capacity of three test specimens was obtained through Eq. (3.56), while the experimental value was obtained through experimental tests. Table 3.7 demonstrates the results of experimental and theoretical load-bearing capacities for specimens B0, B1, and B3, respectively.

**Table 3.7** The load-bearing capacities of specimens B0, B1, and B3

| Specimen | Experimental $M_{\mathrm{exp}}$ | Calculated (ACI) $M_{\mathrm{ACI}}$ | Calculated (Tomas) $M_{\mathrm{Toma}}$ | $M_{\mathrm{exp}}/M_{\mathrm{ACI}}$ | $M_{\mathrm{exp}}/M_{\mathrm{Toma}}$ |
|---|---|---|---|---|---|
| B0 | 342.0 | 330.3 | – | 1.035 | – |
| B1 | 434.9 | 519.4 | 444.3 | 0.837 | 0.979 |
| B3 | 486.4 | 519.4 | 507.8 | 0.936 | 0.958 |

*Note* $M_{\mathrm{exp}}$ means the load-bearing capacities obtained from experimental tests; $M_{\mathrm{ACI}}$ means the load-bearing capacities calculated by ACI 440.2R-08 [21]; $M_{\mathrm{Toma}}$ means the load-bearing capacities calculated by Tomas' method

The moment caused by self-weight for specimens B0, B1, and B3 was 40.2, 45.2, and 43.8 kN·m, respectively. The application of BFRP increased the experimental load-bearing capacity of RC box beams effectively. The results demonstrated in [56] show that the experimental load-bearing capacity of BFRP-strengthened beams is 15.4% higher than that of unretrofitted beam. Meanwhile, the load-bearing capacity of beam increased more by the application of the steel plates than FRP U-strips. This phenomenon can be explained that the steel plates could effectively prevent the stress concentration at the edge of the anchorage when compared with the traditional FRP U-strips. The theoretical load-bearing capacity of specimens B0, B1, and B3 calculated by Eq. (3.56) was 330.3, 519.4, and 519.4 kN·m, respectively. The experimental values of specimens B1 and B3 were 19.4%, and 6.8% lower than the corresponding theoretical values, respectively. This phenomenon may be due to the following reasons: the existence of initial cracks would decrease the load-bearing capacity of RC box beam, and the effect of cracks was not considered in the current standard ACI 440.2R-08 [21]. Meanwhile, the loss of capacity resulted from the debonding of BFRP strips and the effect of the anchorage on the increase of capacity were not considered. Therefore, the current calculation method raised in ACI 440.2R-08 [21], namely, Eq. (3.56), was not suitable for the cracked beams using anchorages.

The load-bearing capacity of FRP-strengthened RC beams using different kinds of anchorages can be obtained through the calculation method proposed by Tomas [52]. The effect of the bond slip is considered in this method. The theory of built-up bars is the theoretical basis of the suggested method which is utilized to calculate the FRP-strengthened RC flexural structures. The following condition is assumed in the calculation: the composition of the beam includes two bars with partial shear connection, FRP strips, and concrete.

$$M_{\mathrm{n}} = A_{\mathrm{s}} f_{\mathrm{s}}\left(d - \frac{x}{2}\right) + A_{\mathrm{f}} f_{\mathrm{f}}\left(d_{\mathrm{fe}} - \frac{x}{2}\right) \tag{3.96}$$

where $M_{\mathrm{n}}$ is the load-bearing capacity of BFRP-strengthened beams when the bond between BFRP strips and concrete is stiff, in kN·m.

Considering the bond performance, the load-bearing capacities of specimens B1 and B3 calculated by Eqs. (3.88) and (3.96) were 444.3 and 507.8 kN·m, respectively. The calculated values showed that steel plates had better performance on the increase of the load-bearing capacity than FRP U-strips. Compared with the experimental values, the deviation between experimental values and calculated value based on Eq. (3.88) for specimens B1 and B3 was 2.1% and 4.2%, respectively. The theoretical load-bearing capacity calculated by the calculation method proposed by Tomas demonstrated good accuracy with the experimental values. The nature of semi-theoretical and semi-experimental determined that the method proposed by Tomas was effective and accurate.

### *3.4.5 Stiffness and Natural Frequency*

The natural frequency plays an important role in examining stiffness change and damage of RC beams in the dynamic tests [57]. The application of FRP composites on the strengthened RC beams affects the dynamic behaviors of the beams [58]. A procedure on the basis of the modal analysis and tests was proposed to evaluate the stiffness variation induced by the cracks and strengthening of RC beams. The modal tests were divided into three stages for specimens B1 and B3, which included before retrofitting, after retrofitting, and after loading, respectively. The natural frequency was obtained through the modal tests, which could be utilized to evaluate the stiffness of the unstrengthened or strengthened beams with cracks. The curve of time-dependent velocity was determined based on the data of modal tests. The curve of amplitude-frequency was determined through the fast Fourier transform (FFT) method, as demonstrated in [16]. Figures 3.44 and 3.45 demonstrate the curves of time history and amplitude-frequency for specimens B0 and B3, respectively.

Assuming that all test beams were uncracked, the theoretical natural frequency of three specimens calculated by Eq. (3.78) was 40.0, 38.9, and 38.9 Hz, respectively. Table 3.8 demonstrates the test results of experimental natural frequency at three different stages and the calculated values for specimens B0, B1, and B3, respectively.

Considering that the natural frequency was proportional to the stiffness, the stiffness could be determined from the natural frequency, as reported in [59]. After 60 years of service, the stiffness of specimens B0, B1, and B3 decreased by 41.5%, 37.3%, and 36.8%, respectively. The stiffness of specimens B1 and B3 increased by 8.2% and 8.6% after retrofitting, respectively. This phenomenon indicated that the application of BFRP increased the stiffness of the cracked RC beams. Compared with the stiffness after retrofitting, the stiffness after loading test for specimens B1 and B3 decreased by 33.3% and 30.8% due to the severe damage to the section of the beams. The results demonstrated in [26] show that the cracking induced by loading decreases the natural frequency because the value of $E_c I$ decreases. The values of natural frequency of specimens B1 and B3 were almost the same after retrofitting. The results demonstrated in [59] show that different kinds of anchorages have little influence on the initial stiffness of strengthened beams. However, the stiffness after loading test of specimen B3 was 0.8 Hz larger than that of specimen B1. This phenomenon could be explained that the steel plates started to function earlier than FRP U-strips, and the stiffness of steel plates was larger than that of FRP U-strips.

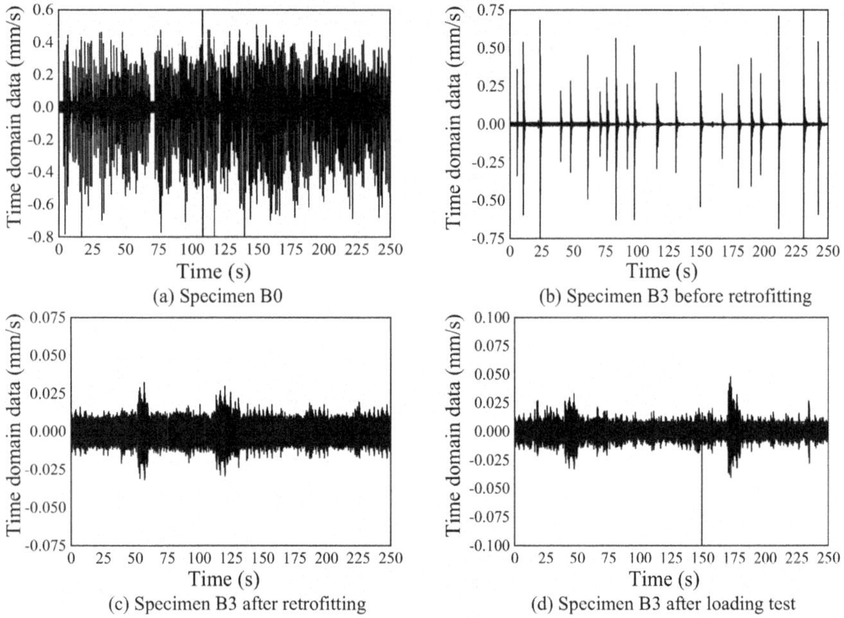


**Fig. 3.44** Time history curves of specimens B0 and B3

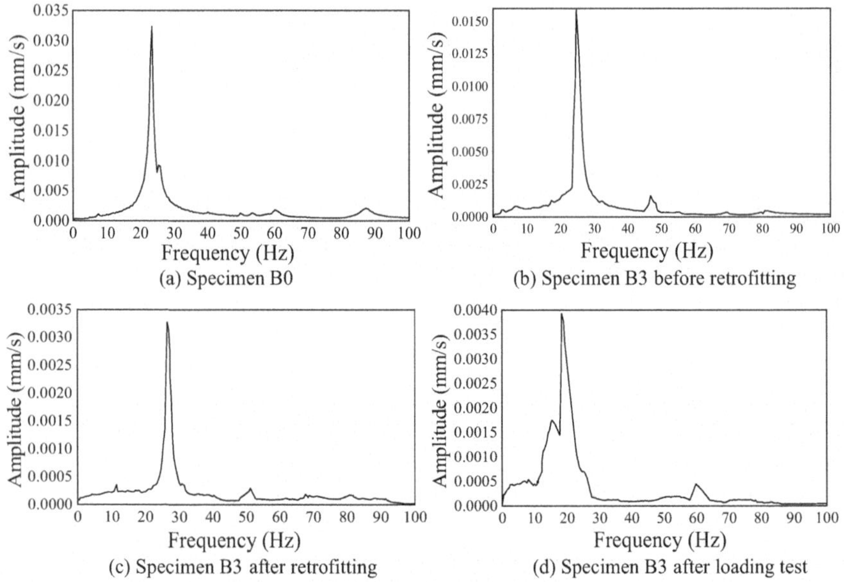


**Fig. 3.45** Amplitude-frequency curves of specimens B0 and B3

**Table 3.8** Natural frequency of specimens B0, B1, and B3

| Specimen | Experimental natural Frequency (Hz) | | | Calculated natural frequency (Hz) |
|---|---|---|---|---|
| | Before retrofitting $f_{exp}$ | After retrofitting $f_{exp}$ | After loading test $f_{exp}$ | Before strengthening $f_{calc}$ |
| B0 | 23.4 | – | 21.1 | 40.0 |
| B1 | 24.4 | 26.4 | 17.6 | 38.9 |
| B3 | 24.5 | 26.6 | 18.4 | 38.9 |

### *3.4.6 Relationship Between Load and Strain*

The stress–strain relationship is significant parameter of the structural characteristics, which should be studied and quantified in the investigations [60]. The failure of FRP-strengthened beams will be caused by the FRP bending [59]. For specimens B1 and B3, the value of $\varepsilon_{fe}$ was 2079 με, and the design value of $\varepsilon_{fd}$ was 1585 με. The theoretical failure mode of the beam is determined from the relationship between $\varepsilon_{fe}$ and $\varepsilon_{fd}$. The calculated value of $\varepsilon_{fe}$ was larger than design value of $\varepsilon_{fd}$, which determined that the theoretical failure mode of specimens B1 and B3 was FRP bonding. The difference was that specimen B1 was premature damage because of the occurrence of FRP debonding in the end zone of beam. When the strains in BFRP strips reached the debonding value, the corresponding load of specimen B3 was 13.6% higher than that of specimen B1. Therefore, BFRP-strengthened beam using steel plate anchorage had better performance on the bond behavior than that using FRP U-strips anchorage.

Figure 3.46 demonstrates the relationship between the load and the strains in BFRP strips and bars. The strain in steel plates was less than that in FRP U-strips, which indicated that the steel plate anchorage could effectively prevent the BFRP cloth from slipping. The anchor bolts would not be pulled out when the test beam was damaged, so that the BFRP strips can fully exert its tensile strength. At the initial stage, the strains in BFRP strips and bars were in linear changes, which indicated that BFRP strips and bars were both in the elastic stage at this time, as demonstrated in Fig. 3.46.

The strains in BFRP strips and bars of two test specimens were almost the same before the load reached the yielding value. The yielding load of specimen B1 was 294.5 kN, and the strain in BFRP strips and bars was 1152 and 1130 με, respectively. The yielding load of specimen B3 was 366.1 kN, and the strain in BFRP strips and bars was 1477 and 1446 με, respectively. This phenomenon indicated that the BFRP strips and bars bonded well and worked together to carry the load-bearing capability before yielding load, and no debonding phenomenon occurred at this moment.

The strain in BFRP strips was slightly larger than that in bars, which indicated that there is no slippage between BFRP strips and concrete cover at this stage. When the load exceeded the yielding value, the bars yielded, and the strains in BFRP developed more quickly. The phenomenon could be explained that the bar started to debond

**Fig. 3.46** Relationship between load and strain of BFRP and bar of specimens B1 and B3

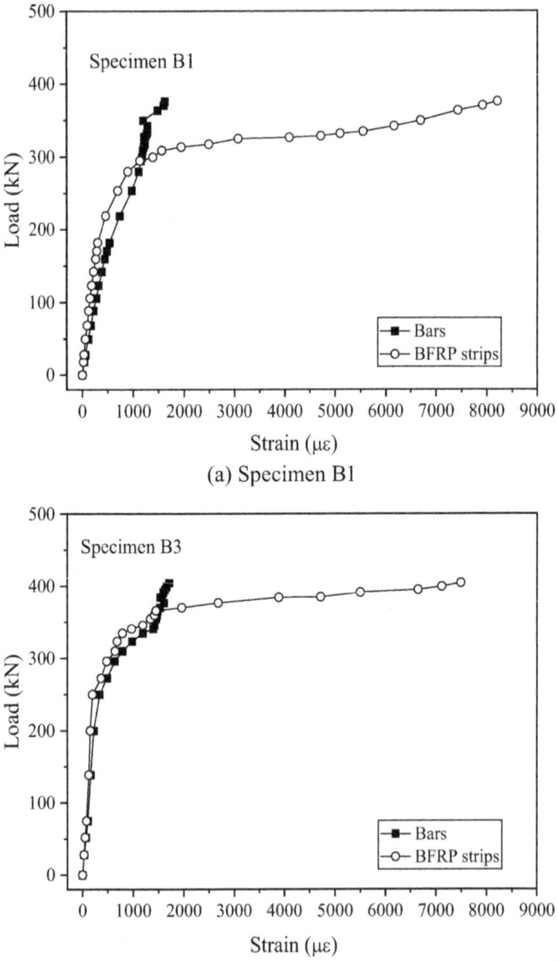


(a) Specimen B1

(b) Specimen B3

from concrete, and BFRP strips carried more load. The BFRP strips of specimen B3 remained coordinated with bars at a higher load level than specimen B1. This phenomenon indicated that the steel plate anchorage was more effective to postpone BFRP bonding of BFRP-strengthened beams than FRP U-strips.

## 3.5 Summary

Investigations on the fundamental natural frequency of deteriorated RC box beam with initial flexural cracks and performance of deteriorated RC box beams retrofitted with BFRP sheets by three types of anchorage, such as U-strips anchorage, end anchorage with grooving, and steel plate anchorage, were conducted. The major results obtained from this chapter were summarized as follows.

(1) Initial flexural cracks significantly influenced the fundamental natural frequency and stiffness of deteriorated RC box beam due to section weakening. The load-bearing capacity, stiffness, and natural frequency of cracked RC box beam decreased compared with those of theoretical value of uncracked beam. The model for the natural frequency of the deteriorated RC box beam was proposed in consideration of depths and locations of cracks.
(2) The behavior of deteriorated RC box beam with initial flexural cracks could be improved by retrofitting with BFRP sheets by end anchorage of U-strips. The load-bearing capacity and stiffness of the retrofitted RC box beam increased compared with those of the RC box beam without retrofitting. BFRP retrofitting had an obvious confinement effect on crack development of concrete. The average crack spacing, width, and depth of RC box beam decreased after retrofitting with BFRP.
(3) End anchorage with grooving had significant effect on the load-bearing capacity of the deteriorated RC box beam with initial flexural cracks. The end anchorage with grooving of deteriorated RC box beam could postpone the occurrence of BFRP debonding, confine crack development of concrete, and decrease the crack spacing. The load-bearing capacity of deteriorated RC box beam retrofitted with BFRP sheets by end anchorage with grooving was larger than that of the beam by U-strips. A model based on built-up bars was used to calculate the load-bearing capacity of BFRP-retrofitted RC box beam.
(4) Retrofitting with BFRP by the steel plate anchorage could improve the behavior of deteriorated RC box beams. The steel plates anchorage had a better effect on improving the load-bearing capacity of beams than FRP U-strips. The steel plate anchorage could postpone the occurrence of initial debonding and delay or prevent the further propagation of the debonding of BFRP sheets.

## References

1. Wu Z, Yang S, Hu X et al (2010) Analytical solution for fracture analysis of CFRP sheet-strengthened cracked concrete beams. J Eng Mech 136(10):1202–1219
2. Shen DJ, Yang Q, Jiao Y et al (2017) Experimental investigations on reinforced concrete shear walls strengthened with basalt fiber-reinforced polymers under cyclic load. Constr Build Mater 136:217–229
3. Aydin K (2013) Free vibration of functionally graded beams with arbitrary number of surface cracks. Eur J Mech-A/Solids 42:112–124

4. Shen DJ, Li M, Kang JC et al (2021) Experimental studies on the seismic behavior of reinforced concrete beam-column joints strengthened with basalt fiber-reinforced polymer sheets. Constr Build Mater 287:122901
5. Mostofinejad D, Mahmoudabadi E (2010) Grooving as alternative method of surface preparation to postpone debonding of FRP laminates in concrete beams. J Compos Constr 14(6):804–811
6. Skuturna T, Valivonis J (2016) Experimental study on the effect of anchorage systems on RC beams strengthened using FRP. Compos. Part B, Eng 91:283–290
7. Mostofinejad D, Khozaei K (2015) Effect of GM patterns on ductility and debonding control of FRP sheets in RC strengthened beams. Constr Build Mater 93:110–120
8. Spadea G, Bencardino F, Sorrenti F et al (2015) Structural effectiveness of FRP materials in strengthening RC beams. Eng Struct 99:631–641
9. Ascione L, Berardi VP (2011) Anchorage device for FRP laminates in the strengthening of concrete structures close to beam-column joints. Compos. Part B, Eng 42(7):1840–1850
10. Pham H, Al-Mahaidi R (2004) Experimental investigation into flexural retrofitting of reinforced concrete bridge beams using FRP composites. Compos Struct 66(1):617–625
11. Rizzo A, de Lorenzis L (2009) Behavior and capacity of RC beams strengthened in shear with NSM FRP reinforcement. Constr Build Mater 23(4):1555–1567
12. Al-Mahmoud F, Castel A, François R et al (2009) Strengthening of RC members with near-surface mounted CFRP rods. Compos Struct 91(2):138–147
13. Garden HN, Hollaway LC (1998) An experimental study of the influence of plate end anchorage of carbon fibre composite plates used to strengthen reinforced concrete beams. Compos Struct 42(2):175–188
14. Rogers RA, Wotherspoon L, Scott A et al (2012) Residual strength assessment and destructive testing of decommissioned concrete bridge beams with corroded pretensioned reinforcement. Precast/Prestressed Concr Inst J 2012:100–118
15. Qin Z, Tian Y, Li G et al (2019) Study on bending behaviors of severely pre-cracked RC beams strengthened by BFRP sheets and steel plates. Constr Build Mater 219:131–143
16. Shen DJ, Deng SC, Zhang JY et al (2015) Behavior of reinforced concrete box beam with initial cracks repaired with basalt fiber-reinforced polymer sheet. J Reinf Plast Compos 34(18):1540–1554
17. China Association for Engineering Construction Standardization (2007) Technical specification for testing concrete strength with drilled core: CECS 03-2007. China Building Industry Press, Beijing, (in Chinese)
18. China Iron and Steel Association (2010) Metallic materials-tensile testing-part 1: method of test at room temperature: GB/T 228.12010. Standards Press of China, Beijing (in Chinese)
19. Ministry of Housing and Urban Rural Development of the People's Republic of China (2010) Code for design of concrete structures: GB 50010-2010. China Architecture and Building Press, Beijing (in Chinese)
20. Oehlers DJ, Liu IST, Ju G et al (2004) Moment redistribution in continuous plated RC flexural members. Part 2: flexural rigidity approach. Eng Struct 26(14): 2209–2218
21. ACI (2008) Guide for the design and construction of externally bonded FRP systems for strengthening concrete structures: 440.2R-08. American Concrete Institute, Farmington Hills
22. Catbas FN, Aktan AE (2000) Modal analysis as a bridge health monitoring tool. Adv Technol Struct Eng 1–10
23. Brincker R, Zhang LM, Andersen P (2001) Modal identification of output-only systems using frequency domain decomposition. Smart Mater Struct 10(3):441–445
24. Yuen KV, Katafygiotis LS (2003) Bayesian fast Fourier transform approach for modal updating using ambient data. Adv Struct Eng 6(2):81–95
25. Kim JH, Kwak HG, Min J (2010) Characterization of the crack depth in concrete using self-compensating frequency response function. NDT E Int 43(5):375–384
26. Maalej M, Chhoa CY, Quek ST (2010) Effect of cracking, corrosion and repair on the frequency response of RC beams. Constr Build Mater 24(5):719–731

27. Bonfiglioli B, Pascale G, Martinez De Mingo S (2004) Dynamic testing of reinforced concrete beams damaged and repaired with fiber reinforced polymer sheets. J Mater Civ Eng 16(5):400–406
28. Capozucca R, Nilde Cerri M (2002) Static and dynamic behaviour of RC beam model strengthened by CFRP-sheets. Constr Build Mater 16(2):91–99
29. Tada H, Paris PC, Irwin GR (1973) The stress analysis of cracks handbook. Del Research Corporation
30. Fu C, Wang Y, Tong D (2018) Stiffness estimation of cracked beams based on nonlinear stress distributions near the crack. Math Prob Eng 2018:1–12
31. Rao SVKM, Dilger WH (1993) Evaluation of short-term deflections of partially prestressed concrete members. Am Concr Inst Struct J 89(1):71–78
32. Fernández-Sáez J, Rubio L, Navarro C (1999) Approximate calculation of the fundamental frequency for bending vibrations of cracked beams. J Sound Vib 225(2):345–352
33. Bakhtiari-Nejad F, Khorram A, Rezaeian M (2014) Analytical estimation of natural frequencies and mode shapes of a beam having two cracks. Int J Mech Sci 78:193–202
34. Zheng T, Ji T (2012) An approximate method for determining the static deflection and natural frequency of a cracked beam. J Sound Vib 331(11):2654–2670
35. Barris C, Torres L, Comas J et al (2013) Cracking and deflections in GFRP RC beams: an experimental study. Compos. Part B, Eng 55:580–590
36. BSI (2001) Design of concrete structures-part 1-1: general rules and rules for buildings: EN 992-2:2001. British Standards Institution, London
37. ACI (1995) Control of deflection in concrete structures: ACI 435R-95. American Concrete Institute, Farmington Hills
38. Branson DE, Trost H (1982) Unified procedures for predicting the deflection and centroidal axis location of partially cracked nonprestressed and prestressed concrete members. J Proc 79(2):119–130
39. Du J, Au Francis TK (2014) Moment of inertia of cracked sections and deflections for UPPC beams. Eng Mech 31(2):170–176
40. Nguyen DM, Chan TK, Cheong HK (2001) Brittle failure and bond development length of CFRP-concrete beams. J Compos Constr 5(1):12–17
41. Benjeddou O, Ouezdou MB, Bedday A (2007) Damaged RC beams repaired by bonding of CFRP laminates. Constr Build Mater 21(6):1301–1310
42. Panda KC, Bhattacharyya SK, Barai SV (2011) Shear strengthening of RC T-beams with externally side bonded GFRP sheet. J Reinf Plast Compos 30(13):1139–1154
43. China Association for Engineering Construction Standardization (2007) Technical specification for strengthening concrete structures with carbon fiber reinforced polymer laminate: CECS 146-2007. China Planning Press, Beijing (in Chinese)
44. Pu Z, Hong ZHU, Shaoping M et al (2011) Calculation of sectional stiffness and deflection of FRP sheets strengthened reinforced concrete beams. J Build Struct 32(4):87–94 (in Chinese)
45. Tan KH, Saha MK (2008) Cracking characteristics of RC beams strengthened with FRP system. J Compos Constr 12(5):513–521
46. Pascale G, Bonfiglioli B (2001) Reinforced concrete beams damaged and repaired with GFRP: Dynamic testing and modelling. FRP Compos Civ Eng i and ii, Proceedings 441–448
47. Chen CF, Huang PY (2003) An analysis on bending stiffness of the strengthened reinforced concrete beam with carbon fiber-sheet. South China Univ Technol: Nat Sci Ed 31(1):10–13 (in Chinese)
48. Wang W (2014) Study on performance of 60-year old reinforced concrete beams retrofitted With FRP sheet. Hohai University, Nanjing (in Chinese)
49. Siringoringo DM, Fujino Y (2008) System identification of suspension bridge from ambient vibration response. Eng Struct 30(2):462–477
50. Teng JG, Smith ST, Yao J et al (2003) Intermediate crack-induced debonding in RC beams and slabs. Constr Build Mater 17(6):447–462
51. Spadea G, Swamy RN, Bencardino F (2001) Strength and ductility of RC beams repaired with bonded CFRP laminates. J Bridg Eng 6(5):349–355

52. Skuturna T, Valivonis J (2014) Design method for calculating load-carrying capacity of reinforced concrete beams strengthened with external FRP. Constr Build Mater 50:577–583
53. National Bureau of Standards of China (1999) Test method for tensile properties of orientation fiber reinforced polymer matrix composite materials: GB/T 3354-1999. Standards Press of China, Beijing (in Chinese)
54. Du J (2014) Research on performance of 60 ages beams rehabilitated with fiber reinforced polymer and steel plate. Hohai University, Nanjing (in Chinese)
55. Wu G, Wang HT, Wu ZS et al (2012) Experimental study on the fatigue behavior of steel beams strengthened with different fiber-reinforced composite plates. J Compos Constr 16(2):127–137
56. Zong S (2007) Experimental study on anchoring method of RC beams strengthened with BFRP. China Sci Technol Formation 20:47–48 (in Chinese)
57. Capozucca R (2009) Static and dynamic response of damaged RC beams strengthened with NSM CFRP rods. Compos Struct 91(3):237–248
58. Shen DJ, Yang Q, Huang CB et al (2019) Tests on seismic performance of corroded reinforced concrete shear walls repaired with basalt fiber-reinforced polymers. Constr Build Mater 209:508–521
59. Shen DJ, Zeng X, Zhang JY et al (2019) Behavior of RC box beam strengthened with basalt FRP using end anchorage with grooving. J Compos Mater 53(23):3307–3324
60. Zanardo G, Hao H, Xia Y et al (2006) Stiffness assessment through modal analysis of an RC slab bridge before and after strengthening. J Bridg Eng 11(5):590–601

# Chapter 4
# Retrofitting of Corrosion-Damaged Reinforced Concrete Columns with BFRPs

The increase in corrosive substances in the atmospheric environment causes the corrosion of RC structures. The research on the seismic performance of corroded structures is important for the evaluation and retrofitting of corroded RC structures [1–3]. Relevant experimental research mainly focused on the quantitative influence of reinforcing bar corrosion on the seismic performance of RC columns. The effect of corrosion rates and axial compression ratios on the seismic performance of corroded RC columns has been studied in [4]. Research on the reinforced concrete column with an axial compression ratio of 0.23 had been studied, to analyze the effect of corrosion rate on the seismic performance of the corroded RC columns, as reported in [5]. The effect of corrosion rate on the seismic performance of RC columns under artificial climate corrosion also has been studied in [6]. The above studies have not achieved relatively consistent results. Existing RC structures in densely populated areas are particularly vulnerable to earthquakes. The hostile environment may cause severe reinforcing bar corrosion and durability issues due to material deterioration [7–11]. The seismic performance of RC structures under earthquake excitations is negatively impacted [12, 13]. The primary lateral load-bearing components of RC buildings are columns, which are crucial to the overall seismic performance of structures. According to earlier research, corroded RC columns perform worse than uncorroded ones in terms of load-bearing capacity, ductility, and energy dissipation [14–16].

FRP is frequently used in the seismic rehabilitation of RC structures [17]. Numerous experimental investigations on the seismic behaviors of FRP retrofitted RC columns have been conducted in recent years [18–20]. The tests demonstrate that the external bonded FRP could greatly enhance the seismic behaviors of the retrofitted columns [21–23]. Researches have been done in [24–30] to examine the impact of FRP retrofitting on the seismic behavior of corroded RC columns. The FRP retrofitting method can effectively stop the spread of shear cracks [24]. Additionally, external bonded FRP can increase the load-bearing, ductility, and energy dissipation capability of columns [25, 28]. With the addition of the layer of FRP, flexibility was



D. Shen, *Seismic Performance of Corroded Reinforced Concrete Structures Retrofitted with FRP*, https://doi.org/10.1007/978-981-99-7984-4_4

improved to a greater extent [26, 29]. The effect of FRP retrofitting on the seismic behavior of corroded RC columns has been studied recently in [27–30]. Several codes and requirements, including ASCE [31], ACI 440 [32], JSCE [33], FIB [34], TR55 of the UK Concrete Society [35], and GB 50608-2010 [36], have introduced the FRP retrofitting design. However, there is no explicit determination of the deformation capacity for FRP retrofitted corroded specimens in these standards and rules. Engineers are unable to assess the performance of corroded RC column after retrofitting with FRP. Although the seismic performance of corroded RC columns repaired with CFRP has been studied, that of repaired with BFRP is lacking. To study the seismic performance of corroded RC columns repaired with BFRP, the cyclic loading test of RC columns was conducted. The effect of the layers of BFRP sheets on the seismic performance indexes, such as hysteretic performance, load-bearing capacity, stiffness, and ductility of the RC columns repaired with BFRP sheets, was compared and analyzed [37].

## 4.1 Seismic Performance of Corrosion-Damaged Columns

### *4.1.1 Experimental Program*

(1) Specimen Design

The specimens were constructed with the 1/2-scale and tested under cyclic loading referenced by the Chinese Standard JGJ 50010-2015 [38]. Three reinforced concrete columns were designed and fabricated with corrosion rate of 0, 8%, and 15%, respectively. The axial compression ratio of the specimens was 0.2. The corrosion process was applied to the reinforcing bars. The corrosion rate was determined by the mass loss. The concrete strength grade was C30 with a slump of $16 \pm 2$ cm. The thickness of the protective layer of the concrete was set as 30 mm. Moreover, commercial concrete was applied. The water-to-binder ratio of concrete in all tested specimens was 0.48, and the mixed proportions by weight of cement, sand, coarse aggregates, fly ash, ground granulated basalt furnace slag, and superplasticizer were 210, 719, 1112, 70, 70, and 4.18 kg/m$^3$, respectively. Three longitudinal reinforcements with a 16 mm diameter of grade III steel were arranged on each side with symmetrical reinforcement. The stirrups were made of grade I steel with a diameter and spacing of 8 mm and 80 mm, respectively. The stirrups at both ends of the specimen were properly densified with a spacing of 50 mm. The geometric size of the specimens and the repairing form are shown in Fig. 4.1.

(2) Accelerated Corrosion Test

The wet electrification method was adopted to accelerate corrosion. By putting each specimen into the corresponding corrosion test tank, the corrosion was carried out with a constant voltage and current source. The specific steps were as follows.

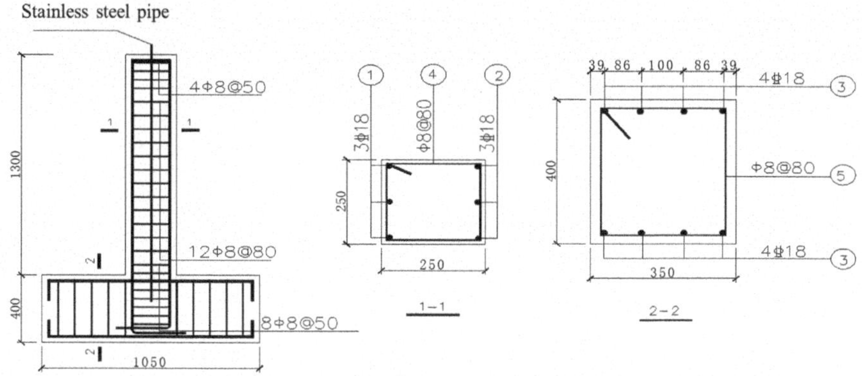


**Fig. 4.1** Geometric size of column and arrangement of reinforcements

Firstly, the corrosion test tank was built. After that, specimens were separately placed into the tank and the water was injected. Then the salt was added to make a 5% NaCl solution. The corrosion rate refers to the mass loss rate. To improve the speed of reinforcement corrosion, high concentrations of NaCl solution were adopted. If water leakage was found, it could be remedied in time.

The specimens were immersed in the tank for seven days to let the chloride ion enter the concrete and then started energizing. The positive pole of the constant voltage and current source was connected with the hook welded on the specimen. The negative electrode was connected with the stainless-steel tube through a wire, while the stainless-steel tube was put into NaCl solution to act as the cathode. The accelerated corrosion test of the specimen is shown in Fig. 4.2.

**Fig. 4.2** Accelerated corrosion of specimens

**Table 4.1** Material properties of main reinforcements of specimens Z1, Z2, and Z3

| Specimens | Actual corrosion rate | Yielding strength (MPa) | Ultimate strength (MPa) |
|---|---|---|---|
| Z1 | 0 | 440 | 640 |
| Z2 | 8.4% | 399 | 587 |
| Z3 | 15.1% | 376 | 547 |

(3) Material Properties

The measured cubic compressive strength of concrete curing at 95% relative humidity (RH) for 28 d was 40.34 MPa. The yielding strength and ultimate strength of the uncorroded longitudinal reinforcement are 440 and 640 MPa, respectively. A balance with an accuracy of 0.1 g was used to weigh the longitudinal reinforcements to obtain the weight per unit length of 1.58 g/mm. After the test of the corroded specimen was completed, the main reinforcements were removed away from the failure section of the damaged specimen. The main reinforcements were de-rusted with steel brushes to investigate the mechanical properties to obtain the yielding and ultimate strength, as shown in Table 4.1.

(4) Test Setup and Loading Method

The load–displacement mixed control loading is adopted in the test, and the loading device is shown in Fig. 4.3. Dimensions and reinforcements of all tested specimens were identical, as depicted in Fig. 4.1. The vertical load of the specimen was applied through the reaction frame on the test bench. Two 25-ton hydraulic jacks were installed on the reaction frame and placed upside down. Two steel plates with a thickness of 35 mm that clamp the rollers were placed below the jack. The horizontal cyclic loading was applied by a 500 kN hydraulic actuator (stroke of ± 150 mm), as shown in Fig. 4.4.

The axial compression ratio was set as 0.2. The vertical load was applied first and then the horizontal cyclic loading was applied after the vertical load was stable. Before the test, it was determined that the preload cyclic loading test should be carried out twice and the preload value should not exceed 30% of the calculated value of the cracking load. Before reaching the yielding load, the horizontal load control was used first and the cycle was staged with a load of 10 kN in each stage. After reaching the yielding load, the cyclic loading was controlled according to the top displacement of the column in multiple increments of 4 mm and each stage was cycled three times. The loading procedure is shown in Fig. 4.5.

### *4.1.2 Failure Modes*

When the load of uncorroded specimen Z1 reached 40, 50, and 60 kN, the first three bending transverse cracks would appear at 10, 25, and 50 cm from the bottom of the

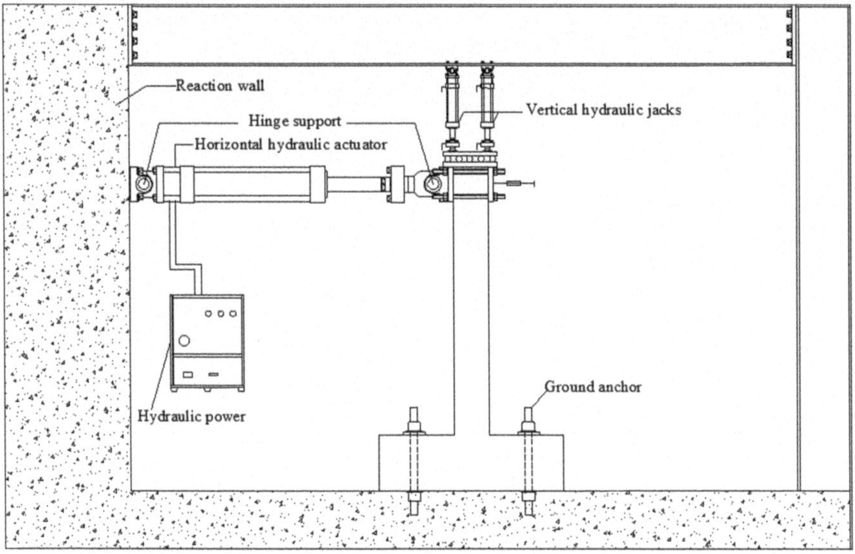


**Fig. 4.3** Scheme of the test set-up

**Fig. 4.4** Photo of the test set-up

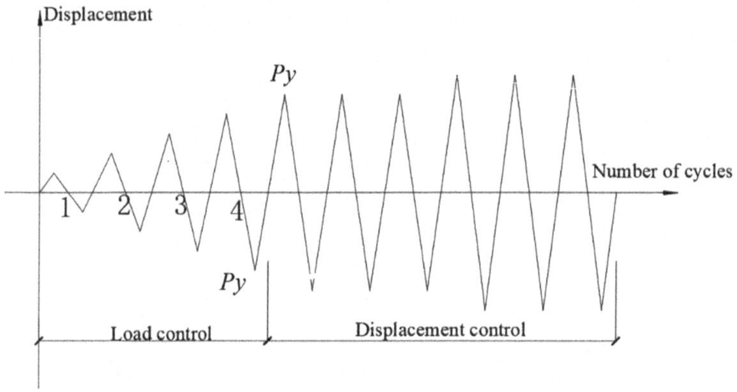


**Fig. 4.5** Loading procedure

specimen, respectively. When the load was 80 kN, the bending shear crack appeared 35 cm away from the bottom of the specimen and developed along the bending crack. With the increase of the shear force, the bending shear cracks in the parallel plane of the bottom of the specimen continued to develop, penetrate, and finally staggered into a network. After the specimen reached the ultimate load, the concrete near the bottom would be crushed and peeled off in the vertical shear plane, resulting in a decrease in load-bearing capacity. The plastic hinge was formed at the bottom of the column and the specimen was damaged. With the increase in loading cycles, the horizontal cracks gradually appeared and continued to widen. The cracks were mainly distributed in the height range of about 350–500 mm from the bottom of the specimen. The horizontal cracks gradually developed obliquely until the forward and reverse cracks were connected. Two or three cracks at the bottom of the specimen developed into main cracks. The core concrete was divided into several pieces, which were crushed and destroyed, as shown in Fig. 4.6.

For specimen Z2, the original longitudinal cracks continued to develop under horizontal cyclic loading. The failure process of corroded specimen was similar to that of the uncorroded specimen, while the bending transverse crack and bending shear crack appeared when the load reached 30 and 60 kN, respectively. Due to the corrosion of reinforcing bars, the bonding performance between the reinforcing bar and the concrete degraded. When the damage occurred, a large area of protective layer fell off. The longitudinal reinforcement exposed with the yielding of compressed reinforcements, as shown in Fig. 4.7. Short vertical cracks appeared only when the specimen was close to the failure load.

For specimen Z3, the original longitudinal cracks continued to develop under horizontal cyclic loading. The failure process of corroded specimen was similar to that of the uncorroded specimen, while the bending transverse crack and bending shear crack appeared when the load reached 25 and 50 kN, respectively. Due to the corrosion of reinforcing bars, the bonding performance between the reinforcing bar and the concrete degraded. When the damage occurred, a large area of protective layer fell off. The longitudinal reinforcement was exposed with the yielding of compressed

**Fig. 4.6** Final failure mode of specimen Z1 ($\rho_w = 0$)

reinforcement bars, as shown in Fig. 4.8. Short vertical cracks appeared only when the specimen was close to the failure load.

### *4.1.3 Hysteretic Capacity*

Figures 4.9, 4.10 and 4.11 are the load–displacement hysteresis curves of specimens. The hysteresis curve was a straight line without a hysteresis loop under small loads. The displacement at the top of the column was small and the specimen was in the elastic stage. Before the specimen yielded, the concrete underwent plastic deformation, but the overall deformation was small. The slope of the loading curve changed slightly and there was a small amount of residual deformation after the component was unloaded.

After the specimen yielded, the displacement-controlled loading began. The load-bearing capacity would continue to increase with a slow growth rate. In addition, it was found that before the load reached the limit; the latter-stage load curve passed through the maximum point of the load–displacement in the former-stage load curve. Compared the loading curves of all stages, the slope of the latter curve was smaller than that of the former curve, indicating that the stiffness of the specimen was degraded under cyclic loading.

**Fig. 4.7** Final failure mode of specimen Z2 ($\rho_w = 8\%$)

When the specimen was unloaded, the initial unloading curve was relatively steep and the recovered deformation was very small. The slope of the unloading curve decreased with the increase in unloading times, indicating that the unloading stiffness was degraded.

Figures 4.9, 4.10 and 4.11 are the hysteresis curves of the specimens with corrosion rate of 0, 8%, and 15% under the same axial compression ratio, respectively. The hysteresis curve of the uncorroded specimen was much fuller than that of the corroded specimen, and the hysteresis loop area was larger, indicating that the energy-dissipation capacity and ductility of specimens decreased after being corroded. The hysteresis curve of the uncorroded specimen Z1 was symmetrical, while the curve of the corroded specimens Z2 and Z3 were asymmetric due to the serious and uneven degree of corrosion.

### *4.1.4 Skeleton Curves*

The envelope curve formed by the peak point in the first cycle at each loading stage of the load–displacement, and curve shall be taken as the skeleton curve of specimens.

**Fig. 4.8** Final failure mode of specimen Z3 ($\rho_w = 15\%$)

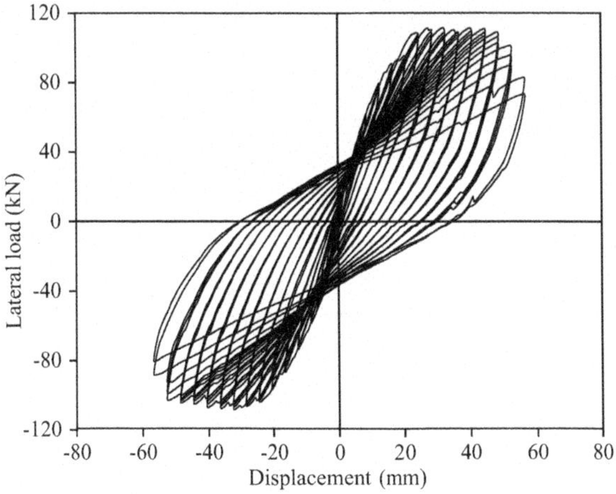


**Fig. 4.9** Hysteretic curve Z1 ($\rho_w = 0$)

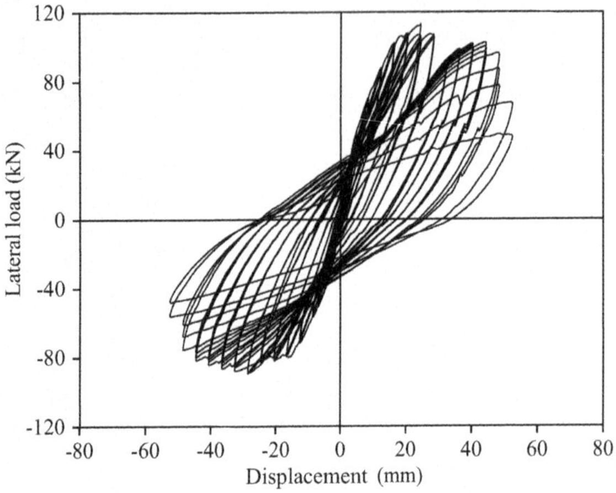


**Fig. 4.10** Hysteretic curve Z2 ($\rho_w = 8\%$)

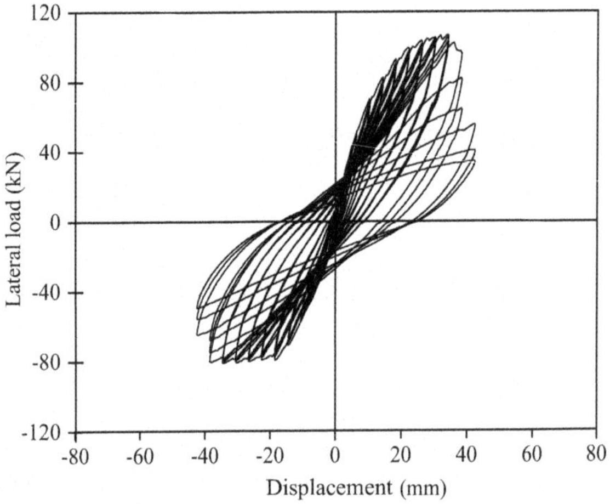


**Fig. 4.11** Hysteretic curve Z3 ($\rho_w = 15\%$)

The strength, stiffness, and ductility of the specimen can be more clearly reflected in the skeleton curve. The skeleton curves of specimens are shown in Fig. 4.12.

As shown in Fig. 4.12, the load-bearing capacity and ductility of corroded specimen Z2 were significantly lower than those of uncorroded specimen Z1. The load-bearing capacity and ductility of specimen Z2 with corrosion rate of 8% were significantly lower than that of specimen Z3 with corrosion rate of 15%.

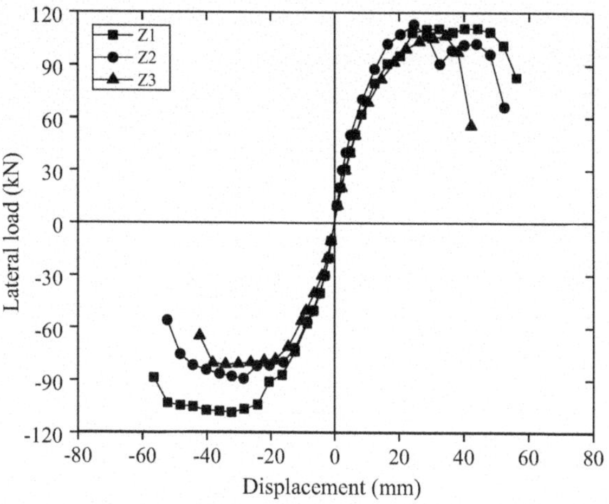


**Fig. 4.12** Skeleton curve of corrosion-damaged specimens Z1, Z2, and Z3

## *4.1.5 Ductility and Load-Bearing Capacity*

$P_u$ is the ultimate load of the specimen, and the failure load was taken as 0.85 $P_u$. $\Delta_u$ is the ultimate displacement of the specimen; $\Delta_y$ is the yielding displacement; $u = \Delta_u/\Delta_y$ is the displacement ductility ratio. The values of load and displacement at each stage are listed in Table 4.2.

The results of specimens Z1, Z2, and Z3 are compared, as shown in Table 4.2. Compared with the uncorroded specimen Z1, the load-bearing capacity and ductility of Z2 with a corrosion rate of 8% decreased by 10.1% and 14.8%, respectively. Compared with the uncorroded specimen Z1, the load-bearing capacity and ductility of Z3 with a corrosion rate of 15% decreased by 14.9% and 22.2%, respectively. Therefore, the reinforcing bars with severe corrosion (> 8%) would lead to a significant decrease in the load-bearing capacity and ductility of the specimens. The ductility of the corroded specimen decreased more obviously than the load-bearing capacity.

**Table 4.2** Characteristic loads and ductility of specimens Z1, Z2, and Z3

| Specimens | $P_u$ (kN) | $\Delta_y$ (mm) | $\Delta_u$ (mm) | $U$ | 0.85 $P_u$ (kN) |
|---|---|---|---|---|---|
| Z1 | 109.58 | 14.43 | 54.27 | 3.76 | 93.14 |
| Z2 | 98.50 | 12.51 | 46.22 | 3.69 | 83.73 |
| Z3 | 93.20 | 12.86 | 42.21 | 3.28 | 79.22 |

## 4.2 Seismic Performance of Corrosion-Damaged Columns Retrofitted with BFRP Sheets

### *4.2.1 Specimen Design and Fabrication*

(1) Specimen Design

The reinforced concrete columns were strengthened with BFRP sheets. The specimens were numbered Z1, Z2, Z4, and Z5, respectively. The specific structural parameters are shown in Table 4.3.

(2) Retrofitting Schemes of RC Column

Before retrofitting, the rust on the surface of the corroded concrete column was polished clean. The construction site is shown in Fig. 4.13. Before sticking the FRP sheets, several preparations needed to be finished. Firstly, the crushed concrete and the dust on the surface of the specimens were removed and cleaned up, respectively. Secondly, the cracks of the concrete were rehabilitated by the epoxy cement mortar and the rehabilitated specimens needed to be cured for at least one day. When the epoxy cement mortar was hardened, the surfaces of the specimens were polished smooth. Subsequently, utilizing the mixed epoxy resins to bond the BFRP sheets and the coatings brushed on the BFRP surface were periodically inspected. The FPR sheets were wrapped around the reinforced concrete columns along the longitudinal FRP direction. At last, additional concentration was taken to guarantee the effectiveness of the coatings, while the epoxy resins needed to be carefully cared for over several days, as shown in Fig. 4.13.

(3) Material Properties

After the test of the corroded specimen was completed, the main reinforcements were removed away from the failure section of the damaged specimen. The main reinforcements were de-rusted with steel brushes to investigate the mechanical properties to obtain the yielding and ultimate strength, as shown in Table 4.4. The high-performance unidirectional BFRP sheet was used and the fiber impregnation adhesive

**Table 4.3** Main parameters of hysteretic specimens Z1, Z2, Z3, Z4, and Z5

| Specimens | Designed corrosion rate | Axial compression ratio | Repairing form |
|---|---|---|---|
| Z1 | 0 | 0.2 | – |
| Z2 | 8% | 0.2 | – |
| Z3 | 15% | 0.2 | – |
| Z4 | 8% | 0.2 | One layer of BFRP sheets |
| Z5 | 8% | 0.2 | Two layers of BFRP sheets |

(a) A roll of fiber sheets

(b) Retrofitting columns with BFRP sheets

**Fig. 4.13** Sticking the BFRP sheets

**Table 4.4** Material properties of main reinforcements of specimens Z1, Z2, Z3, Z4, and Z5

| Specimens | Actual corrosion rate | Yielding strength (MPa) | Ultimate strength (MPa) |
|---|---|---|---|
| Z1 | 0 | 440 | 640 |
| Z2 | 8.4% | 399 | 587 |
| Z3 | 15.1% | 376 | 547 |
| Z4 | 8.3% | 402 | 547 |
| Z5 | 7.6% | 415 | 562 |

**Table 4.5** Technical indexes of BFRP sheet

| Number | Test items | Test results |
|---|---|---|
| 1 | Tensile strength (MPa) | 2303 |
| 2 | Elastic modulus (MPa) | $1.05 \times 10^5$ |
| 3 | Elongation (%) | 2.18 |

TLS-503 was selected as the binder. The mechanical properties of the BFRP sheet are shown in Table 4.5. The measured results of colloidal properties and bonding properties are shown in Table 4.6.

### 4.2.2 *Failure Modes*

For the specimens repaired with BFRP sheets, since the specimens were wrapped, the development of cracks could not be observed during the loading process. However, with the continuous application of horizontal cyclic loading, the sound of crackling was constantly heard in the BFRP sheets. It can be seen from the last few cycles before

**Table 4.6** Test results of colloid and bonding properties

| Number | Test items | | Test results | | Test results |
|---|---|---|---|---|---|
| | | | Class A | Class B | |
| 1 | Colloidal properties | Tensile strength (MPa) | $\geq 40$ | $\geq 30$ | 45.79 |
| | | Elastic modulus (MPa) | $\geq 2.5 \times 10^3$ | $\geq 1.5 \times 10^3$ | $2.6 \times 10^3$ |
| | | Elongation (%) | $\geq 1.5$ | | 3.47 |
| 2 | Bonding properties | Normal bonding strength with concrete (MPa) | $\geq 2.5$ (C30 concrete failure) | | 5.87 |

the end of the test that the BFRP sheet near the root of the column bulged, indicating that the concrete was crushed. However, the concrete was still well wrapped in the BFRP sheet, as shown in Figs. 4.14 and 4.15, which means that the BFRP can play an important role in restraining concrete. The corrosion rate of specimen Z4 was 8% and specimen Z4 was repaired with one layer of FRP sheet. A slight sound of cracking could be heard when the lateral load was 50 kN, which could be attributed to the breaking of hardened resins. As the test proceeded, the sounds were loud and continuous. However, no evident cracks of the concrete at the surface were observed in the first stage. In the displacement-control stage, BFRP sheets fractured at the corners and bulged at the core area when the displacement was 32 and 36 mm, respectively. Due to the crushing of concrete at the core area, the BFRP sheets were entirely bulged and delaminated from the surface when the displacement was 44 mm, as well as the load-bearing capacity decreased. The test was terminated when the displacement was 66 mm, and the shearing failure mode could be found in Fig. 4.14.

The corrosion rate of specimen Z5 repaired with two layers of sheets was 8%. A slight sound of cracking could be heard when the lateral load was 60 kN, which could be attributed to the breaking of hardened resins. As the test proceeded, the sounds were loud and continuous. However, no evident cracks of the concrete at the surface were observed in the first stage. In the displacement-control stage, BFRP sheets fractured at the corners and bulged at the core area when the displacement was 36 and 40 mm, respectively. Due to the crushing of concrete at the core area, the BFRP sheets were entirely bulged and delaminated from the surface when the displacement was 48 mm, as well as the load-bearing capacity decreased. The test was terminated when the displacement was 72 mm, and the shearing failure mode could be found in Fig. 4.15.

**Fig. 4.14** Z4 ($\rho_w = 8\%$, one layer)

**Fig. 4.15** Z5 ($\rho_w = 8\%$, two layers)

### 4.2.3 Hysteretic Response

Figure 4.10 shows hysteretic curves of the corroded but unrepaired specimen Z2. Corroded specimen Z4 with corrosion rate of 8% was repaired with one layer of BFRP sheet, as shown in Fig. 4.16. Corroded specimen Z5 was repaired with two layers of BFRP sheets, as shown in Fig. 4.17. The hysteretic curves of Z4 and Z5 are much fuller than that of Z2, and the area of the hysteretic loop is larger. The results showed that the load-bearing capacity, energy-dissipation capacity, and ductility of specimens were greatly improved after repairing with BFRP sheets.

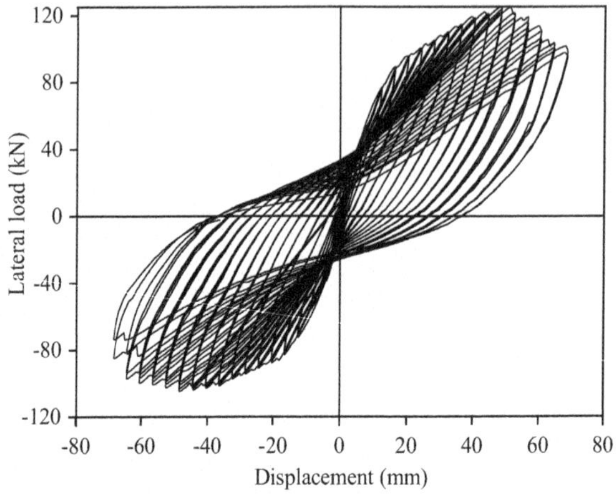


**Fig. 4.16** Hysteretic curve Z4 ($\rho_w = 8\%$, one layer)

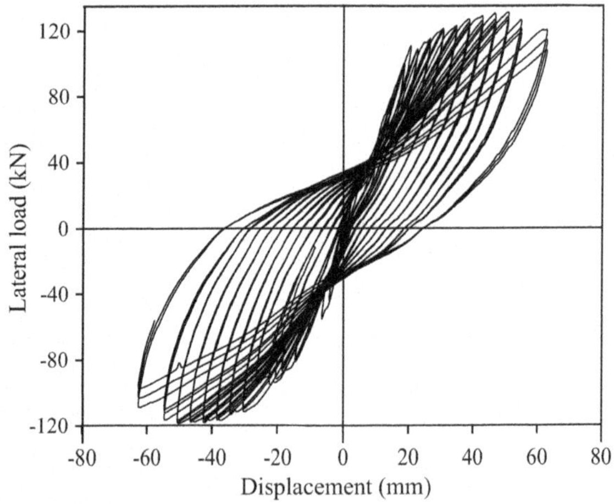


**Fig. 4.17** Hysteretic curve Z5 ($\rho_w = 8\%$, two layers)

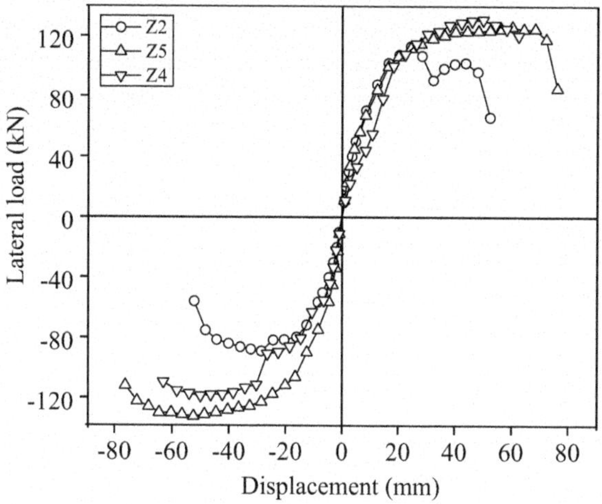


**Fig. 4.18** Skeleton curve of corrosion-damaged specimens retrofitted

### 4.2.4 *Skeleton Curves*

The skeleton curves of specimens are shown in Fig. 4.18. The corroded and repaired specimens Z4 and Z5 have a larger load-bearing capacity and ductility than the unrepaired specimen after retrofitting. Therefore, the seismic behavior of columns is significantly improved by using BFRP for retrofitting.

### 4.2.5 *Ductility and Load-Bearing Capacity*

The values of load and displacement at each stage are listed in Table 4.7.

The load-bearing capacity of the retrofitted specimens Z4 and Z5 with a corrosion rate of 8% increased by 4.1% and 13.99%, respectively, and the ductility increased by 6.58% and 10.95%, respectively, compared with those of the uncorroded specimen Z1. The load-bearing capacity of the repaired specimens Z4 and Z5 with corrosion rate of 8% increased by 16.1% and 20.6%, respectively, and the ductility increased by 24.14% and 30.27%, respectively, compared with those of specimen Z2 with

**Table 4.7** Characteristic loads and ductility of specimens Z1, Z2, Z4, and Z5

| Specimens | $P_u$ (kN) | $\Delta_y$ (mm) | $\Delta_u$ (mm) | $u$ | 0.85 $P_u$ (kN) |
|---|---|---|---|---|---|
| Z1 | 109.58 | 14.43 | 54.27 | 3.76 | 93.14 |
| Z2 | 98.50 | 12.51 | 46.22 | 3.69 | 83.73 |
| Z4 | 114.04 | 14.52 | 57.84 | 3.98 | 96.93 |
| Z5 | 124.90 | 15.09 | 60.21 | 3.99 | 106.17 |

a corrosion rate of 8%. The load-bearing capacity and ductility of the specimen were greatly improved due to the wrapping of the BFRP sheet. With the increase of repairing layers, the restraint effect of the BFRP sheet on the specimen also increased, and the improving effect on the load-bearing capacity and ductility of reinforced specimens was also more obvious. This indicated that utilizing BFRP sheets could retrofit the ductility and load-bearing capacity of corroded to the level of the uncorroded specimen.

## 4.3 Summary

Experimental investigations on the seismic performance of corroded RC columns without retrofitting or retrofitted with BFRP sheets were conducted. The major results obtained from this chapter were summarized as follows.

(1) The reinforcement corrosion affected the seismic performance of RC columns. The brittle failure mode of the corroded RC columns was observed. The severe spalling of concrete were observed in test indicating the bond behavior between reinforcing bars and concrete deteriorated. The load-bearing capacity, cumulative energy dissipation, and ductility of columns decreased with the increase of corrosion rate.

(2) The seismic performance of corroded RC columns improved by retrofitting with BFRP sheets. The use of BFRP sheets to confine the concrete in the corroded RC columns improved the bond behavior between reinforcing bars and concrete. The ductile failure mode of the corroded RC columns retrofitted with BFRP sheets was also observed. The stiffness, displacement ductility ratios, and cumulative energy dissipation of corroded RC columns retrofitted with BFRP sheets increased compared with those of corroded RC columns. The techniques of retrofitting with BFRP sheets for improving the seismic performance of corroded RC columns were developed.

## References

1. Choe G, Shinohara Y, Kim G et al (2020) Concrete corrosion cracking and transverse bar strain behavior in a reinforced concrete column under simulated marine conditions. Appl Sci 10(5):1794
2. Ou YC, Nguyen ND (2016) Influences of location of reinforcement corrosion on seismic performance of corroded reinforced concrete beams. Eng Struct 126:210–223
3. Zheng A, Li S, Zhang D et al (2021) Shear strengthening of RC beams with corrosion-damaged stirrups using FRP grid-reinforced ECC matrix composites. Compos Struct 272(1):114229
4. Chen XX, Niu DT, Wang XM (2005) The force-restoring model of corroded reinforced concrete members with flexure and compression. J Xi'an Univ Archit Technol (Nat Sci Ed) 37:155–159 (in Chinese)

5. Gong JX, Zhong WQ, Zhao GF (2004) Experimental study on low-cycle behavior of corroded reinforced concrete member under eccentric compression. Build Struct 25:92–104
6. Enright MP, Dan MF (1998) Probabilistic analysis of resistance degradation of reinforced concrete bridge beams under corrosion. Eng Struct 20(11):960–971
7. Elghazy M, Refai AE, Ebead U et al (2017) Effect of corrosion damage on the flexural performance of RC beams strengthened with FRCM composites. Compos Struct, S0263822317317695
8. Su MN, Zeng C, Li WQ et al (2019) Flexural performance of corroded continuous RC beams rehabilitated by ICCP-SS. Compos Struct 232:111556
9. Wang ZH, Li L, Zhang YX et al (2019) Bond-slip model considering freeze-thaw damage effect of concrete and its application. Eng Struct 201:109831
10. Dong W, Li W, Vessalas K et al (2020) Piezoresistivity deterioration of smart graphene nanoplate/cement-based sensors subjected to sulphuric acid attack. Compos Commun 23
11. Yu XH, Dai KY, Li YS (2021) Variability in corrosion damage models and its effect on seismic collapse fragility of aging reinforced concrete frames. Constr Build Mater 295:123654
12. Yalciner H, Sensoy S, Eren O (2015) Seismic performance assessment of a corroded 50-year-old reinforced concrete building. J Struct Eng 141(12):05015001
13. Liu X, Jiang H, He L (2017) Experimental investigation on seismic performance of corroded reinforced concrete moment-resisting frames. Eng Struct 153:639–652
14. Yang SY, Song XB, Jia HX et al (2016) Experimental research on hysteretic behaviors of corroded reinforced concrete columns with different maximum amounts of corrosion of rebar. Constr Build Mater 121:319–327
15. Ying M, Yi C, Gong J (2012) Behavior of corrosion damaged circular reinforced concrete columns under cyclic loading. Constr Build Mater 29:548–556
16. Kyda B, Dgl A, Xhy A (2021) Experimental investigation on the seismic performance of corroded reinforced concrete columns designed with low and high axial load ratios. J Build Eng 44
17. Sun FJ, Pang SH, Zhang ZW et al (2020) Retrofitting seismically damaged steel sections encased concrete composite walls using externally bonded CFRP strips. Compos Struct 236:111927
18. Raza S, Khan M, Menegon SJ et al (2019) Strengthening and repair of reinforced concrete columns by jacketing: state-of-the-art review. Sustainability 11(11):3208
19. del Vecchio C, di Ludovico M, Prota A et al (2016) Modelling beam-column joints and FRP strengthening in the seismic performance assessment of RC existing frames. Compos Struct 142:107–116
20. Ma R, Xiao Y, Lib KN (2000) Full-scale testing of a parking structure column retrofitted with carbon fiber reinforced composites. Constr Build Mater 14(2):63–71
21. Galal K, Arafa A, Ghobarah A (2005) Retrofit of RC square short columns. Eng Struct 27(5):801–813
22. Wang D, Huang L, Yu T et al (2017) Seismic performance of CFRP-retrofitted large-scale square RC columns with high axial compression ratios. J Compos Constr 21(5).
23. Wang D, Wang Z, Yu T et al (2018) Seismic performance of CFRP-retrofitted large-scale rectangular RC columns under lateral loading in different directions. Compos Struct 192:475–488
24. Lee HS, Kage T, Noguchi T et al (2003) An experimental study on the retrofitting effects of reinforced concrete columns damaged by rebar corrosion strengthened with carbon fiber sheets. Cem Concr Res 33(4):563–570
25. Li X, Zhao ZH, Liang YS et al (2016) Cyclic behaviour of earthquake- and corrosion-damaged RC columns. Mag Concr Res, pp 1–17
26. Leng LY, Zhang PF (2014) Research on CFRP strengthening corroded reinforced concrete columns in seismic performance. Appl Mech Mater 477–478:681–685
27. Bousias SN, Triantafillou TC, Fardis MN et al (2004) Fiber-reinforced polymer retrofitting of rectangular reinforced concrete columns with or without corrosion. ACI Struct J 101(4)

28. Rajput AS, Sharma UK (2019) Seismic upgrade of corroded confined reinforced concrete columns using composite materials. ACI Mater J 116(5):37–48
29. Rajput AS, Sharma UK, Engineer K (2018) Seismic retrofitting of corroded RC columns using advanced composite materials. Eng Struct 181:35–46
30. A X L, B Y L (2018) Experimental study of seismic behavior of partially corrosion-damaged reinforced concrete columns strengthened with FRP composites with large deformability. Constr Build Mater 191:1071–1081
31. ASCE (2017) Seismic evaluation and retrofit of existing buildings: ASCE 41-17: American Society of Civil Engineers. Reston
32. ACI (2004) Guide for the design and construction of externally bonded FRP systems for strengthening concrete structures: ACI 440-3R-04. American Concrete Institute, Farmington Hills
33. JSCE (2001) Test method for bond properties of continuous fiber sheets to concrete: JSCE-E-543-2000). Japan Society of Civil Engineers, Tokyo, Japan
34. Fib (2006) fib Model code for retrofitting of concrete structures by externally bonded FRPs with emphasis on seismic applications 2006: fib Model Code 2006[S]. Lausanne, Switzerland, Berlin
35. Arya C, Clarke JL, Kay EA et al (2002) TR 55: Design guidance for stengthening concrete structures using fibre composite materials: a review. Eng Struct 24(7):889–900
36. Ministry of Housing and Urban Rural Development of the People's Republic of China (2010) Standard for technical code for infrastructure application of FRP composites: GB 50608-2010. China Architecture and Building Press, Beijing (in Chinese)
37. Yuan JJ (2013) Experimental study on retrofitting measures of multi-aged reinforcement concrete columns. Hohai University, Nanjing (in Chinese)
38. Ministry of Housing and Urban Rural Development of the People's Republic of China (2010) Standard for code for design of concrete structures: GB 50010-2010. China Architecture and Building Press, Beijing (in Chinese)

# Chapter 5
# Retrofitting of Corrosion-Damaged Reinforced Concrete Beam-Column Joints with BFRP Sheets

The beam-column joint designed as the critical region of RC structures suffers high shear stress and large shear deformations during earthquake, which may result in the brittle failure after the reinforcing bars corroded [1]. Results reported in [1] indicate that the load-bearing capacity, stiffness, and energy dissipation of corroded joints significantly decrease attributed to the reinforcement corrosion. Results reported in [2] indicate that the damaged characteristics of corroded beam-column joints can be evaluated by the acoustic emission technique, and the initial diagonal cracks of corroded specimens occur earlier than those of reference specimen. Tendency of b-values obtained from the acoustic emission of corroded specimens is more stable than that of reference specimen, which indicates that more microcracks and the reappearance of old cracks occur with the decrease of the load-bearing capacity. Indeed, the load-bearing capacity of corroded RC beams [3, 4], columns, and shear walls [5] can be increased by repairing with BFRP, together with the seismic behavior of pre-damaged beam-column joints [6] can also be improved. Therefore, investigations on the corroded beam-column joints retrofitted with BFRP sheets for improving the seismic performance and durability of corroded beam-column joints are concerned and significant.

Recently, frequent earthquakes have caused great damages to RC structures. It could be due to the deficiency of seismic design concepts, inadequate designs, or increasing seismic intensity that resulted in the deterioration of seismic performance of aged RC building or structures, as reported in [7]. That might imperil people's life and property. Especially the aged RC structures are seriously injured under the moderate, severe, or destructive earthquake damages, bringing a great challenge to the seismic design and post-earthquake retrofitting [8]. The beam-column joints are considered to be the weaker regions in RC structures when suffered high shearing stress and serious inelastic deformations under the earthquake [9]. Phenomena from the 1999 Kocaeli (Turkey) and 1999 Chi-Chi (Taiwan, China) earthquakes exhibit brittle failures in the joint, which possibly results in the collapse of whole buildings



D. Shen, *Seismic Performance of Corroded Reinforced Concrete Structures Retrofitted with FRP*, https://doi.org/10.1007/978-981-99-7984-4_5

[10, 11]. Therefore, researches on the seismic performances of RC beam-column joints are needed.

Firstly, the corrosion-damaged beam-column joints were accomplished by the accelerating corrosion test and tested under the cyclic loadings. The seismic performances of corrosion-damaged beam-column joints were investigated [12, 13]. Secondly, the corrosion-damaged beam-column joints were retrofitted with BFRP sheets in certain methods. The seismic performances of corrosion-damaged beam-column joints retrofitted with BFRP sheets were discussed and compared with those of uncorroded joints [12]. Thirdly, the beam-column joints were pre-damaged in interlayer displacement angles after the accelerating corrosion test to simulate the moderately, severely, and destructively damaged levels of earthquake. Then, the corrosion-damaged and pre-damaged joints were retrofitted with BFRP sheets and the models for predicting the shear strength of the joint core areas considering the influences of earthquake damage, reinforcement corrosion, and BFRP retrofitting were proposed [14, 15]. Lastly, the seismic performance of uncorroded beam-column joints retrofitted with BFRP sheets in different methods considering the influence of the layers and patterns of BFRP sheets, as well as RC slab was investigated [16].

## 5.1 Seismic Performance of Corrosion-Damaged Beam-Column Joints

### *5.1.1 Experimental Program*

(1) Specimens and Materials

Four RC interior beam-column joints designed as specimens BC-0-BC-3 were fabricated and tested. Specimens BC-0-BC-3 were the corroded joints with the corrosion rates of 0, 3%, 9%, and 15%, respectively. Ordinary concrete utilized in the experimental test was offered by the commercial concrete, the components proportions of which are shown in Table 5.1. According to Chinese Standard GB/T 50081-2019 [17], the measured cubic compressive strength $f_{cu}$, and elastic modulus $E_c$ of concrete were 39.3, and $3.24 \times 10^4$ MPa, respectively. The dimension and reinforcement form of specimens are shown in Fig. 5.1. Every specimen was divided into five parts as follows: (1) left beam, (2) right beam, (3) upper column, (4) lower column, and (5) joint core area. The length of these beams and columns was 1225, 1225, 800, and 525 mm, respectively. Each beam of tested specimens, with the cross-section dimension of 200 mm × 300 mm, was composed of 6 ribbed bars of 14 mm diameter applied for longitudinal reinforcements and closed plain bars of 6 mm diameter applied for stirrups with a space of 100 mm (50 mm in encryption area). Each column of tested specimens, with the cross-section dimension of 250 mm × 250 mm, was composed of 8 ribbed bars of 16 mm diameter applied for longitudinal reinforcements and closed plain bars of 6 mm diameter applied for stirrups with a space of

75 mm (50 mm in encryption area). The average values of yielding strength of ribbed and plain bars were 335, and 235 MPa, respectively.

(2) Accelerating Corrosion Test

Reinforcement corrosions are commonly obtained by two methods, i.e., natural corrosion and electrochemical corrosion [18, 19]. The natural corrosion of reinforcing bars in the laboratory will last a too long time (continuously about several years or longer), which is impractical for a laboratory with limited space [19]. Therefore, the electrochemical corrosion with an appropriately and constantly direct current is applied for accelerating laboratory testing, and has been proved to have a similar result as natural corrosion [19–21].

**Table 5.1** Mix proportions of concrete (kg/m$^3$)

| Cement | Sand | Water | Coarse aggregates | Fly ash | Ground granulated basalt furnace slag | Superplasticizer |
|---|---|---|---|---|---|---|
| 212 | 721 | 168 | 1123 | 70 | 70 | 4.22 |

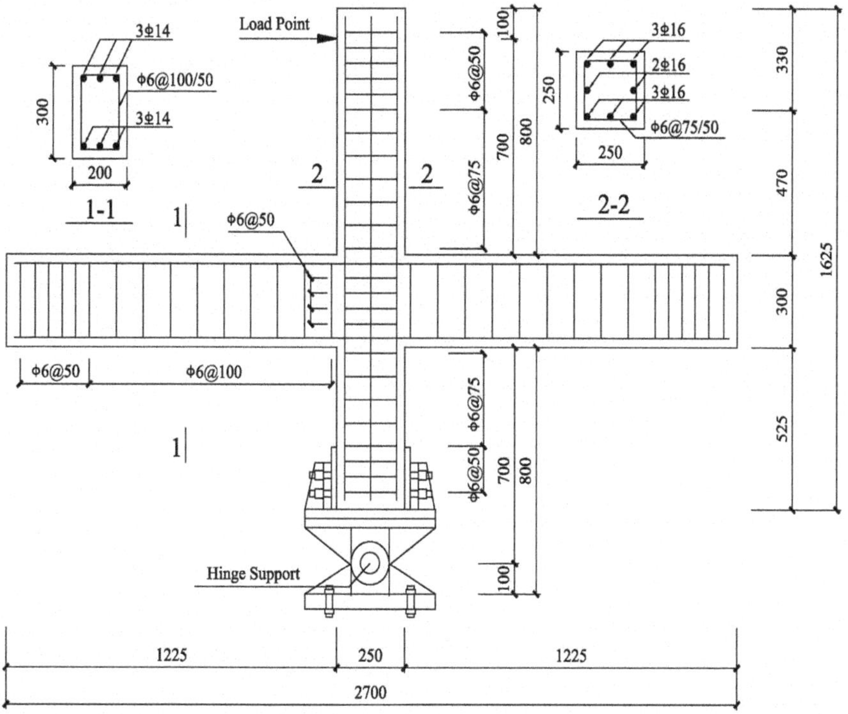


**Fig. 5.1** Geometry and reinforcement details of all specimens (all units in mm)

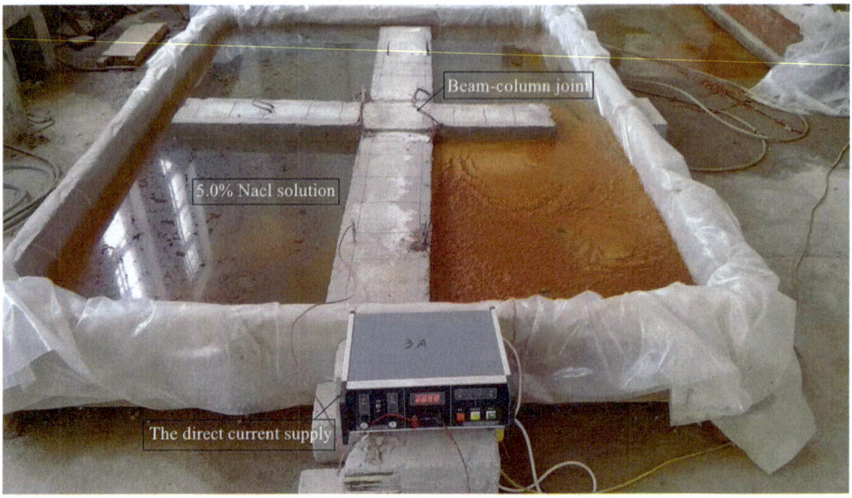


**Fig. 5.2** Setup of accelerated corrosion process

The corroded beam-column joints were conducted by the electrochemical corrosion to accelerate the corrosion test. Specimens BC-1, BC-2, and BC-3 were separately arranged in the NaCl solution with a concentration of 5.0% in a sink for 7 days, as shown in Fig. 5.2. Then, stainless-steel pipes set as cathode were arranged in the NaCl solution and "U" type embedded anchors connecting to reinforcing bars were set as anode in the specimen. Conducting wires were applied for connecting the anode, cathode, and the direct current supply.

Specimens with various corrosion rates were conducted in the same method. An adjustable constantly direct current ranging from 1 A to 3 A was applied for accelerating corrosion test. In the process of testing, electric current flowed out from the positive pole of the direct current supply over the concrete permeated by NaCl solution, the stainless-steel pipe, and finally returned to the negative pole of the direct current supply. Faraday's law is utilized to calculate the detailed time for the accelerated test as follows, which is proven to be feasible and reasonable in a sense [1, 22].

$$\Delta_m = \frac{MIt}{ZF} \tag{5.1}$$

where $\Delta_m$ is the logical mass loss of reinforcing bars, in g; $M$ is the atomic weight of the iron, which is taken as 56 g/mol; $I$ is the current intensity, in A; $t$ is the duration of corrosion, in s; $Z$ is the ionic charge, which is taken as 2; and $F$ is the Faraday's constant, which is taken as 96,500 A s.

Besides, the mass loss of corroded reinforcing bar can also be calculated by the corrosion rate $\rho_{\text{w}}$ as Eq. (5.2).

$$\Delta_m = \frac{\pi}{4} d^2 L \rho \rho_w \tag{5.2}$$

where $L$ is the cumulative connecting length of reinforcing bars in concrete, in mm; $\rho$ is the density of steel, in g/mm$^3$; and $d$ is the diameter of reinforcing bars, in mm.

The duration of corrosion $t$ is derived by Eqs. (5.1) and (5.2) as Eq. (5.3).

$$t = \frac{\pi d^2 L \rho \rho_w Z F}{4 M I} \tag{5.3}$$

Consequently, the parameters of $t$ and $I$ could be applied to evaluate the corrosion rate of reinforcing bars. The duration of specimens with the target corrosion rates of 3%, 9%, and 15% was 37, 55, and 61.8 days when the reinforcing bars were impressed with a current intensity of 1, 2, and 3 A, respectively.

(3) Test Setup and Procedure

The layout of test setup included the loading system and instrumentation is shown in Fig. 5.3. Column and beams were in vertical and horizontal positions, respectively. The lower column was connected to a hinge support, which was fixed on the steel girder by high strength steel bolts on the ground. Top of the beam was also connected to the steel girder by a hinge, which was connected with the load cell and links. The links were applied for allowing freely horizontal displacement, but restraining the vertical displacement.

Tested specimens were subject to a constant axial load offered by two vertical jacks and a quasi-static load offered by a horizontal hydraulic actuator on the top of columns. The actuator, which was arranged on the reaction walls, had a 500 kN-capacity with a stroke of 500 mm when fully extended. Researches on beam-column joint tests [23–26] indicate that most tests suffer the axial load ratios less than 0.5. For contrasting the test results with previous results, the actual axial load ratio of 0.2 was applied for all specimens.

Figure 5.4 shows the horizontal loading history at both push and pull cycles. According to Chinese Standard JGJ/T 101-2015 [27], the loading history of tested specimens consisted of two parts.

The first stage was controlled by load, which was taken $\pm 0.167\ P_y$ as per step until reaching to $\pm\ P_y$. Then the second stage was controlled by displacement, which took $\pm\ 0.5\ \Delta_y$ as per step and started from $\Delta_y$. Test was terminated at $\pm\ 0.85\ P_{\max}$ ($P_{\max}$ was the peak load) according to the Chinese Standard JGJ/T 101-2015 [27]. During the test, test data was collected by the collection system where the load–displacement curves were gathered and the yielding load was obtained. Displacements of beams and column measured by the LVDTs were applied for calculating the deformations, as shown in Fig. 5.3a. Reinforcement strains were obtained by using the strain gauges, and several preparations were needed for protecting the strain gauges and guaranteeing the results. Firstly, the resistance value of strain gauge was measured by a multimeter, which was usually 120 Ω. The strain gauges with resistance variations within ± 0.5 Ω were chosen for use. Secondly, the surface of reinforcing bars was

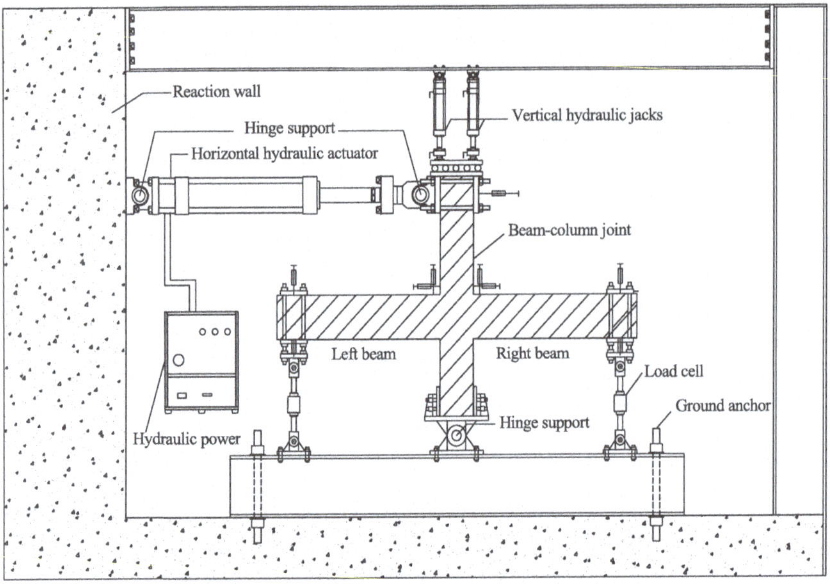


(a) Schematic drawing

(b) Photo

**Fig. 5.3** Test setup of the loading system

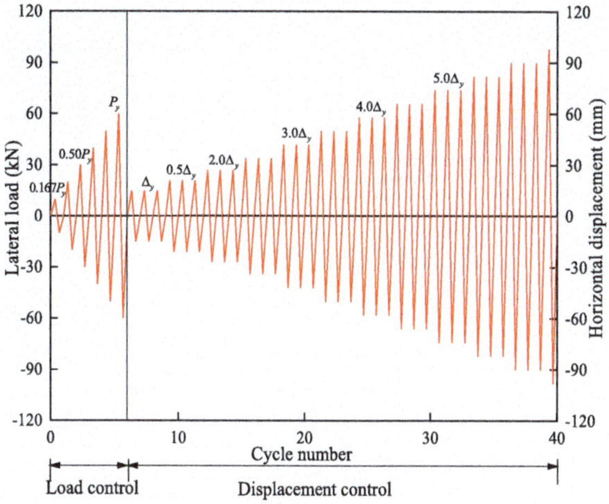


**Fig. 5.4** Horizontal loading history

polished smoothly by the trowel and sandpaper, and the rust was cleaned out by the cotton balls with alcohol before sticking the strain gauges. At this time, the resistance value of strain gauge was examined again. Thirdly, moisture barriers made of silicone rubber and gauze were applied for sealing up the strain gauges stuck on the reinforcing bars from the water and air. For checking out the effectiveness, the insulation and resistances of strain gauges were examined by the megameter and multimeter after the reinforcing bars were immersed into the water for at least 24 h, respectively. The workability of strain gauges was proven to be effective only when the insulated resistance value between the strain gauge and the reinforcing bars is more than $2 \times 10^8$ Ω [28] and the resistance value of the strain gauge is 120 Ω within a certain resistance variation, respectively. Detailed arrangement and preparations on the strain gauges are shown in Fig. 5.5.

### 5.1.2 Crack Pattern and Failure Modes

(1) Reference Specimen BC-0

The final failure of reference specimen BC-0, which was directly tested to failure without any retrofitting, is described in Fig. 5.6a. When the specimen was controlled by the force, micro-flexural cracks were noticed at the lateral load of 40 kN. These cracks began to widen with the test proceeded, and cross-inclined cracks could be noticed on the core area at the lateral load of 60 kN. When the specimen was controlled by the displacement, the previous cracks progressed and widened with a maximum width of 0.75 mm at the displacement of 20 mm (1.18% drift). Simultaneously, penetrated cracks occurred on the beams. The number of cracks on the surface of tested beam-column joints increased with the increasing displacement. The crushing and spalling of the concrete in the core area within a specific region were observed, while the maximum width of cracks was 2.0 mm at the displacement of 52 mm (3.06% drift). A considerable spalling of concrete in the core area and exposing of

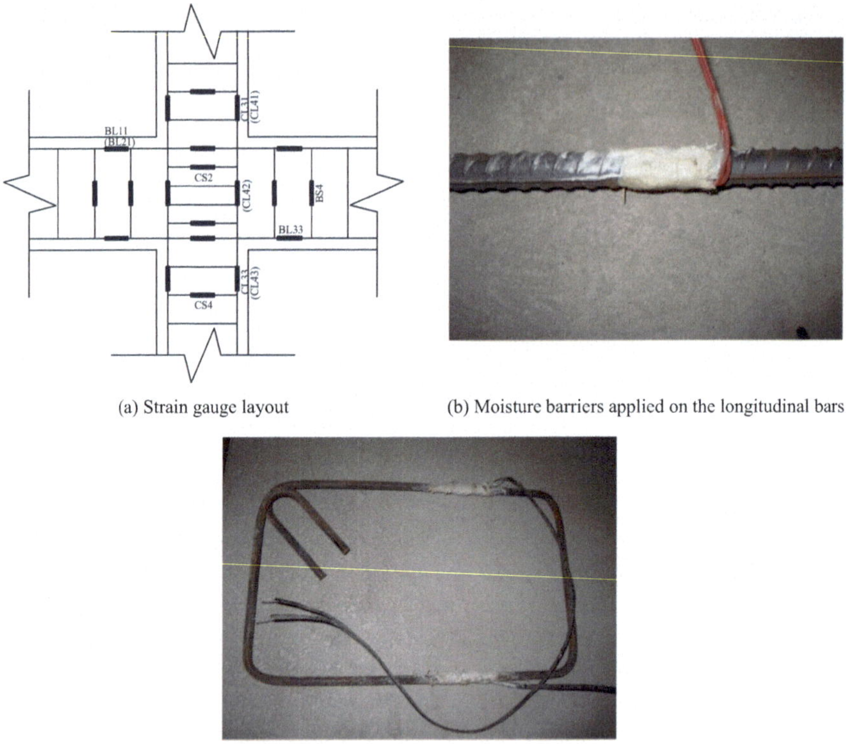


(a) Strain gauge layout

(b) Moisture barriers applied on the longitudinal bars

(c) Moisture barriers applied on the stirrup

**Fig. 5.5** Strain gauge layout and preparations on strain gauges

reinforcing bars were observed, whereas the load-bearing capacity of the specimen BC-0 started to decline at the displacement of 68 mm (4.00% drift). The test was ceased when the load-bearing capacity of the specimen BC-0 decreased by 15% of the peak load in Fig. 5.6.

(2) Specimen BC-1

When the specimen BC-1 was controlled by the force, microcracks could be found nearby the column of the right beam at the load of 30 kN. The inclined cracks in the core area could be found when the load was 55 kN. The widths of these cracks increased to 0.30 mm when the load was 65 kN. In the displacement-control stage, the width of cracks at the corner of column increased to 0.50 mm with penetrating to the beam when the displacement was 24 mm. A considerable spalling of concrete in the core area and exposing of reinforcing bars could be observed when the displacement was 58 mm. When the displacement was 68 mm, the concrete in the core area is completely peeled off with the exposing of the aggregates. The left beam appeared to detach from the column when the displacement was 76 mm, while the load-bearing capacity of the specimen decreased by 15% of the peak load in Fig. 5.7.

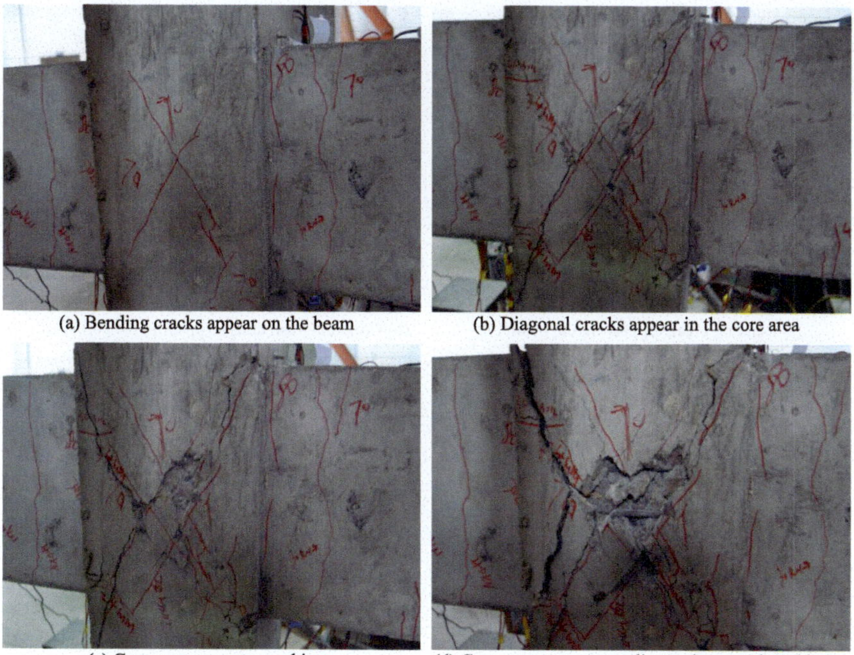

(a) Bending cracks appear on the beam (b) Diagonal cracks appear in the core area

(c) Core area concrete crushing (d) Core area concrete peeling and exposed steel bars

**Fig. 5.6** Failure process diagram of specimen BC-0

(3) Specimen BC-2

The microcracks on the right beam at the distance from the column could be observed and a small crack appeared along the longitudinal reinforcements direction at the severely corroded part of the left beam when the load was 30 kN. The inclined cracks in the core area could be found when the load was 65 kN and the displacement was 26 mm. In the displacement-control stage, the widths of the cracks in the core area spaned to 1.25 mm with the limitedly spalling area of concrete when the displacement was 36 mm. As the displacement increased to 52 mm, a considerably spalling area of concrete in the core area was observed with the exposing of longitudinal bars in the column. When the displacement was 64 mm, the load-bearing capacity of the specimen began to decrease with the limitedly spalling area of concrete in the core area. It could be seen that the test was ceased when the serious spallings of concrete in the core area and the exposings of corroded stirrups and longitudinal reinforcements of columns were observed at the displacement was 80 mm. At this moment, the load-bearing capacity dropped to 85% of the peak load with one stirrup ruptured in Fig. 5.8.

(4) Specimen BC-3

In the load-control stage, microcracks were observed on the beams with the distance of 100 mm from the column when the load was 5 kN. The widths of these cracks

(a) Bending cracks appear on the beam (b) Cross diagonal cracks appear in the core area

(c) Diagonal Cracks spread to the end of the beam (d) Crushed concrete, exposed corroded steel bars

**Fig. 5.7** Failure process diagram of specimen BC-1

were 0.05 mm. As the load at the end of the column increased, cracks on the beams develop rapidly. The cracks could be observed at the corners of right beam and the width of cracks at the corners of left beam was 0.2 mm when the loads were 19.3 and – 45.0 kN, respectively. In the displacement-control stage, the width of cracks at the core area was 0.15 mm when the displacement was 23 mm. The spalling of concrete and exposing of reinforcing bars were observed when the displacement was 50 mm, and corroded stirrup fractured at the core area when the displacement was 52 mm. The test was ceased when the load-bearing capacity of specimen dropped by 15% of the peak load at the displacement was 78 mm in Fig. 5.9.

## *5.1.3 Hysteretic Response and Load-Bearing Capacity*

Figure 5.10 shows the load–displacement hysteretic loops of all specimens. The encircled area of the hysteretic loop can be utilized for estimating the energy dissipation of tested specimen during the loading process [29].

To study the degradation of ductility and load-bearing capacity, the envelope curves of all tested specimens are shown in Fig. 5.11, where the envelopes were taken by averaging the maximum absolute values in the process of push and pull

(a) Diagonal cracks appear in the core area of the node

(b) Concrete at the inclined crack is crushed

(c) The concrete is further crushed and peeled off

(d) Exposed corroded steel bars

**Fig. 5.8** Crack propagation of specimen BC-2

for every cycle. The displacement and lateral load exhibited a linear relationship, which indicated that specimens were in elastic stage at the early stage. An obviously long yielding platform was found after a steady hardening up to the peak load, following by a slowly softening after the peak load. The skeleton curves of corroded specimens BC-1-BC-3 are not symmetrical, and one important reason is that the corrosion rate of the beams on both sides of the corroded specimens is not the same. When the load at the end of the column increased to a large extent, previous loading had already caused certain damage to the specimen. The damage to the specimen is more severe under pull loading than under push stage. Besides, the effect of reinforcement corrosion on the skeleton curve of the specimens is consistent. The skeleton curves of corroded specimens BC-1-BC-3 with the corrosion rate of 3%, 9%, and 15% were enclosed by that of uncorroded specimen BC-1. As the corrosion rate of reinforcing bars increased, the yielding platform of the skeleton curve of corroded specimens shortened with the considerably decreased, exhibiting brittle failure modes.

Ductility is also utilized to evaluate the seismic behaviors of RC structures under the earthquake [29, 30]. The yielding displacement $\Delta_y$ corresponding to the yielding load was calculated by the method reported in [31]. As shown in Fig. 5.12, $P_y$ is the yielding load and $\Delta_y$ is the yielding displacement, which are calculated depended

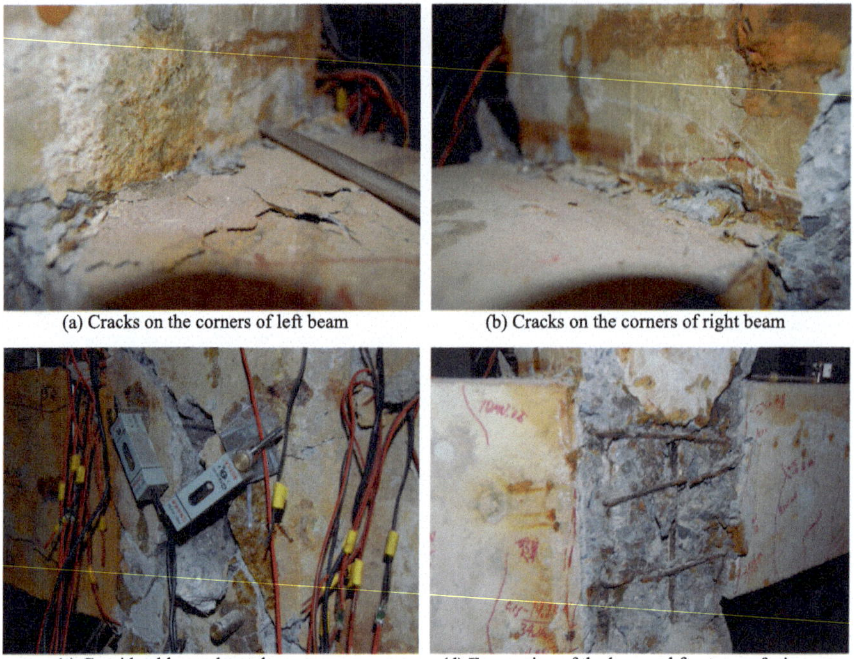

(a) Cracks on the corners of left beam (b) Cracks on the corners of right beam

(c) Considerable cracks at the core area (d) Exposuring of the bars and fractures of stirrups

**Fig. 5.9** Failure mode of specimen BC-3

on the assumption that the area of shadow zone $S_{OBD}$ equals to the area of shadow zone $S_{ABC}$; $P_u$ is the ultimate load, which is characterized as 85% of the peak lateral load $P_{\max}$ and $\Delta_u$ is the ultimate displacement.

Table 5.2 shows the test results of all specimens. The load-bearing capacity and displacement of slightly corroded specimen BC-1 held over 90% of those of reference specimen BC-0, which illustrated that the hysteretic loop was limitedly affected by the corrosion of reinforcing bars. Similar tendency in the variation of hysteretic curves of corroded specimens is also reported in [5]. However, corroded specimens BC-2 and BC-3 showed a considerable decrease in the load-bearing capacity as well as displacement, holding around 85% of those of reference specimen BC-0. Displacements of corroded specimens were possibly recovered to that of reference specimen after repairing with BFRP sheets, while only both the load-bearing capacity and displacement of specimen with the corrosion rate of 3% could be recovered to those of reference specimen.

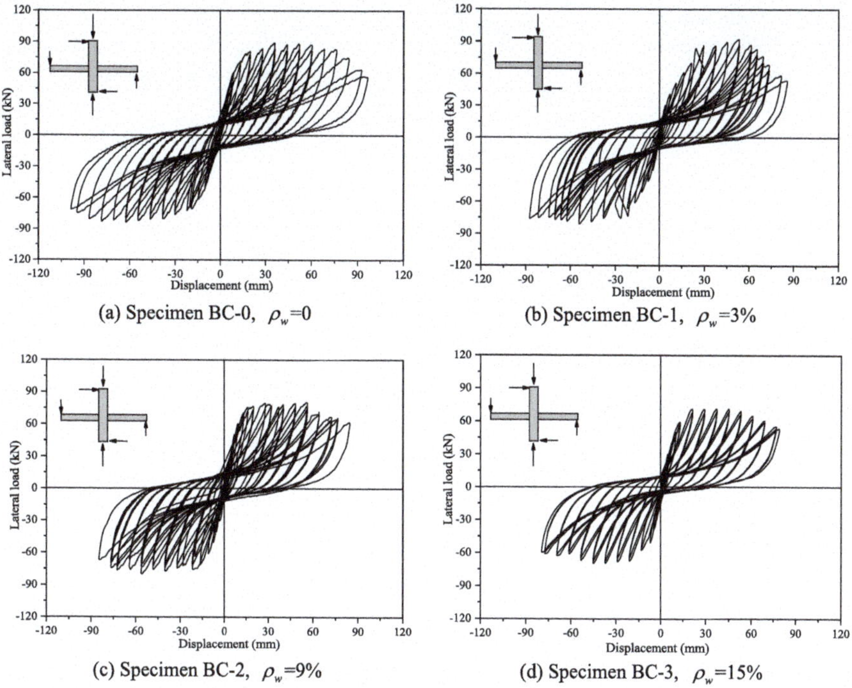


(a) Specimen BC-0, $\rho_w$=0

(b) Specimen BC-1, $\rho_w$=3%

(c) Specimen BC-2, $\rho_w$=9%

(d) Specimen BC-3, $\rho_w$=15%

**Fig. 5.10** Load–displacement hysteretic curves of all specimens

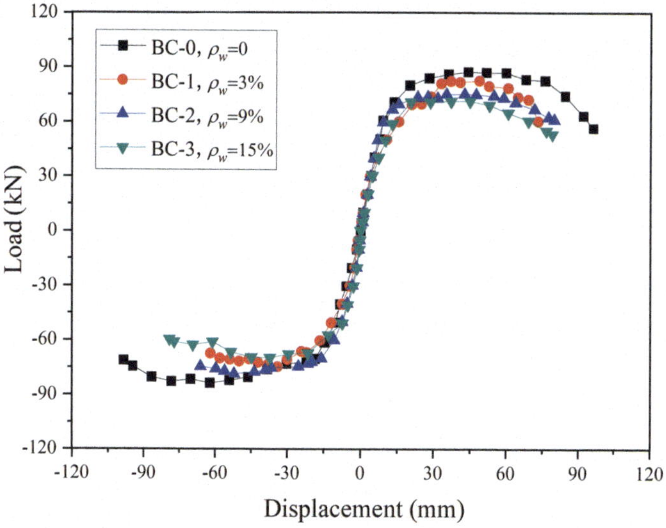


**Fig. 5.11** Envelope curves of hysteretic loops for reference and retrofitted specimens

Fig. 5.12 Definition of yielding displacement and lateral load

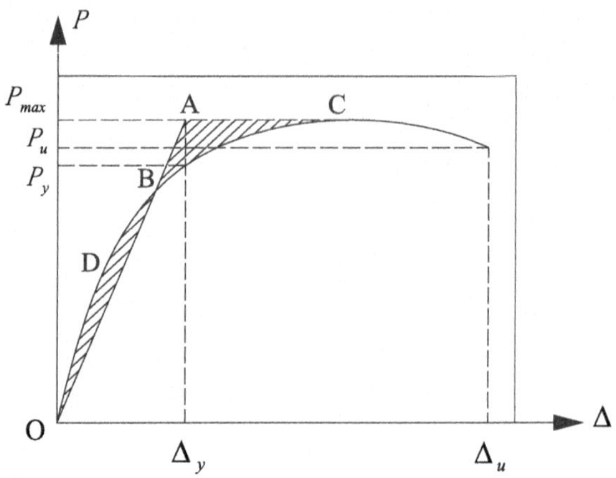


**Table 5.2** Test results of all specimens

| Specimen | Corrosion rate (%) | $P_y$ (kN) | | $P_u$ (kN) | | $\Delta_y$ (mm) | | $\Delta_u$ (mm) | |
|---|---|---|---|---|---|---|---|---|---|
| | | Value | Ratio of corroded or retrofitted specimens to specimen BC-0 (%) | Value | Ratio of corroded or retrofitted specimens to specimen BC-0 (%) | Value | Ratio of corroded or retrofitted specimens to specimen BC-0 (%) | Value | Ratio of corroded or retrofitted specimens to specimen BC-0 (%) |
| BC-0 | 0 | 70.8 | 100 | 85.6 | 100 | 14.8 | 100 | 82.3 | 100 |
| BC-1 | 3 | 67.6 | 84.5 | 79.4 | 92.8 | 13.5 | 91.2 | 70.5 | 85.7 |
| BC-2 | 9 | 62.7 | 88.6 | 76.9 | 89.8 | 12.5 | 84.5 | 63.1 | 76.7 |
| BC-3 | 15 | 51.7 | 73.0 | 70.6 | 82.5 | 12.3 | 83.1 | 69.3 | 84.2 |

*Note* $P_y$ is yielding strength; $P_u$ is ultimate strength; $\Delta_y$ is yielding displacement; and $\Delta_u$ is ultimate displacement

### *5.1.4 Relationship Between Half-Cell Potentials, Corrosion Rate, and Load-Bearing Capacity*

(1) Experimental Program

The experimental program involved four specimens (denoted as BC-1, BC-2, BC-3, and BC-4). Designed concrete strength is 30 MPa. Considering the reinforcement corrosion, corrosion rates of specimens BC-0, BC-1, BC-2, and BC-3 are 0, 3%, 9% and 15%, respectively. Detail of specimen is shown in Fig. 5.13.

An adjustable direct current supply providing 24 V constant potential at 0 to 3 A was used for the accelerated corrosion process. The direct current was applied to the main reinforcement embedded in the concrete, using the reinforcing bar as the

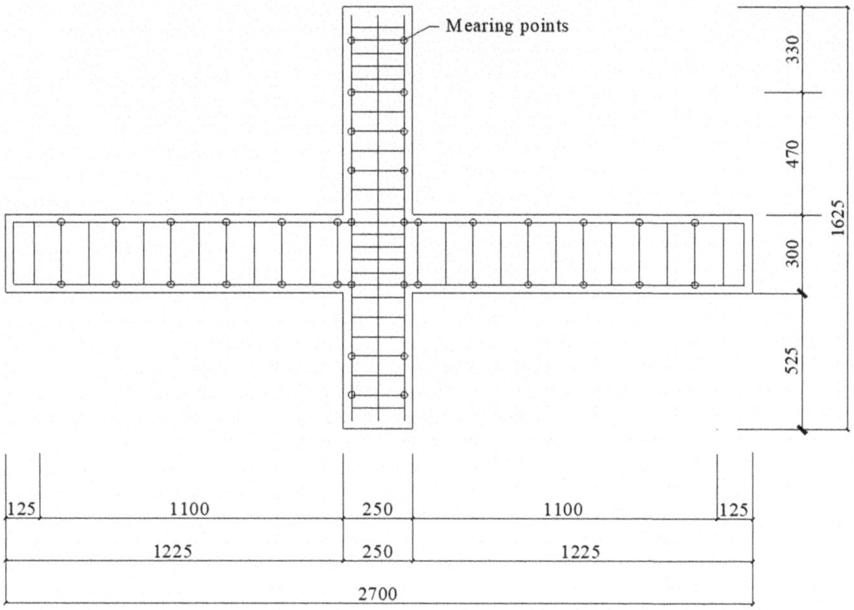


**Fig. 5.13** Geometry and reinforcement details of specimen (all units in mm)

anode and a copper plate as the cathode. The current was passed from the reinforcing bars to the copper plate placed inside of 5.0% salt solution (NaCl). The setup for the accelerated corrosion process is shown in Fig. 5.2.

According to Faraday's law [32], the duration time t of exposure can be evaluated by as Eq. (5.4).

$$t = \frac{\pi D^2 L \gamma \rho_w F}{2MI} = 0.05515 \frac{L\rho_w}{I} \quad (5.4)$$

where $t$ is the duration time of exposure and $I$ is the average magnitude of electrical current and $M$ is the molar mass for iron (56 g/mol for iron) and $F$ is the Faraday's constant and taken equal to 96,500 A·s and $D$ is the diameter of the reinforcing bars in mm; $L$ is the total length of reinforcing bars (22,400 mm), $\gamma$ is the density of the iron and equal to $7.86 \times 10^{-3}$ g/mm$^3$, and $\rho_w$ is corrosion rate, in %.

When corrosion rate of specimen BC-1 was 3%, current was 1 A, the duration time of exposure was 37 days; when corrosion rate of specimen BC-2 was 9%, current was 3 A, the duration time of exposure was 37 days; when corrosion rate of specimen BC-3 was 15%, and current was 3 A, the duration time of exposure was 61 days.

The half-cell potential technique can be used to locate corroding reinforcing bars without disrupting the concrete cover to the reinforcing bars. A copper–copper sulfate ($Cu$-$CuSO_4$) reference electrode is selected for half-cell potential measurements. The methodology consists of comparing the half-cell potential of the steel with a reference potential, and depending on the potential reading to estimate the probability

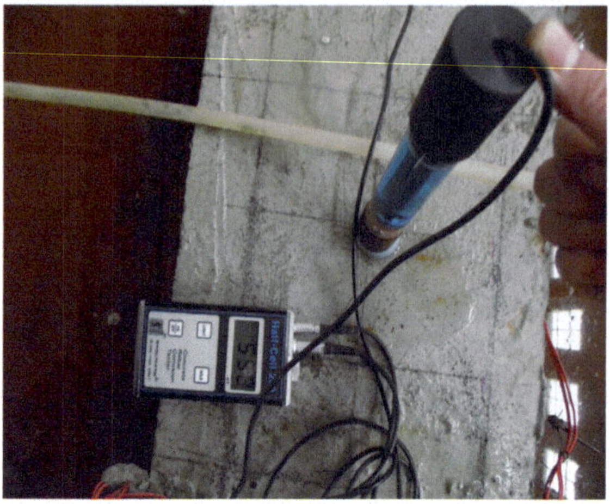

**Fig. 5.14** Test on the measuring points of beam-column joint

of corrosion. The measurements are taken by connecting the reference electrode to the positive terminal and the prestressing strand to the negative terminal of a voltmeter [33].

Measuring points on the lateral surface of the specimens are shown in Fig. 5.13, and half-cell potential at the surface of concrete joint is measured by a portable corrosion meter (HC200), which is illustrated in Fig. 5.14.

The loading shown in Fig. 5.4 included one loading cycle at each load control and two cycles at each displacement control. Figure 5.3 shows the experimental setup. The bottom of column was connected with a hinge support, and one double-acting hydraulic actuator was attached to the other side. The actuator, which was fixed on the reacting walls, has a capacity of 500 kN with a stroke of 500 mm when fully extended. The top of column was bolted to two vertical hydraulic jacks.

(2) Test Result and Analysis

Half-cell potentials of BC-0 are shown in Table 5.3. The maximum value is – 31 mV, minimum value is – 159 mV, and average value is – 132 mV. It shows that potentials are uniform except one point.

Half-cell potentials of BC-1 are shown in Table 5.4. The maximum value was – 406 mV, minimum value was – 774 mV, and average value was – 510 mV. Values

**Table 5.3** Half-cell potentials of specimen BC-0 (mV)

| Test results obtained from the points at the surface of specimen in Fig. 5.13 | | | | | | | | | | |
|---|---|---|---|---|---|---|---|---|---|---|
| – 134 | – 125 | – 138 | – 145 | – 129 | – 134 | – 119 | – 150 | – 108 | – 139 | – 136 |
| – 140 | – 136 | – 134 | – 140 | – 135 | – 130 | – 130 | – 143 | – 135 | – 133 | – 141 |
| – 129 | – 134 | – 130 | – 159 | – 31 | – 121 | – 128 | – 140 | – 131 | – 128 | – 127 |
| – 131 | – 139 | – 145 | – 128 | – 132 | – 136 | – 137 | – 125 | – 146 | – 132 | – 128 |

were more uniform than that of BC-1. The half-cell potential decreased at 378 mV and at the degree of 286.4%.

Half-cell potentials of BC-2 are shown in Table 5.5. The maximum value was – 534 mV, minimum value was – 760 mV, and average value was – 614 mV. Values were more uniform than that of BC-0. The half-cell potential decreased at 482 mV and at the decrease degree of 365.2%.

Half-cell potentials of BC-3 are shown in Table 5.6. The maximum value was – 594 mV, minimum value was – 683 mV, and average value was – 630 mV. Values were more uniform than that of BC-1. The half-cell potential decreased at 482 mV and at decrease the degree of 377.3%.

Half-cell potentials were used to evaluate specimens' corrosion rate, which were based on American standard ASTM C876-91 [33]. As shown in Table 5.4, the potentials of BC-0 are more positive than – 0.20 VCSE. It illustrates that BC-0 is over 90% probabilities that no reinforcement corrosion is occurring in that area. As shown in Tables 5.4–5.6, the potentials of BC-1, BC-2, and BC-3 much less than – 0.35 VCSE. It illustrates that all of them are over 90% probabilities that reinforcement corrosion is occurring in that area. Figure 5.15 shows that half-cell potential decreases with the increase of corrosion rate.

**Table 5.4** Half-cell potentials of specimen BC-1 (mV)

| Test results obtained from the points at the surface of specimen in Fig. 5.13 | | | | | | | | | | |
|---|---|---|---|---|---|---|---|---|---|---|
| – 439 | – 440 | – 454 | – 463 | – 431 | – 528 | – 561 | – 570 | – 609 | – 640 | – 632 |
| – 406 | – 419 | – 441 | – 444 | – 432 | – 622 | – 509 | – 530 | – 540 | – 542 | – 568 |
| – 437 | – 774 | – 467 | – 471 | – 443 | – 562 | – 473 | – 500 | – 536 | – 527 | – 450 |
| – 452 | – 462 | – 454 | – 540 | – 543 | – 504 | – 600 | – 602 | – 462 | – 458 | – 490 |

**Table 5.5** Half-cell potentials of specimen BC-2 (mV)

| Test results obtained from the points at the surface of specimen in Fig. 5.13 | | | | | | | | | | |
|---|---|---|---|---|---|---|---|---|---|---|
| – 760 | – 730 | – 712 | – 651 | – 573 | – 595 | – 600 | – 619 | – 607 | – 622 | – 612 |
| – 706 | – 715 | – 677 | – 600 | – 534 | – 549 | – 573 | – 577 | – 575 | – 569 | – 580 |
| – 714 | – 699 | – 675 | – 603 | – 537 | – 538 | – 549 | – 565 | – 570 | – 569 | – 566 |
| – 689 | – 692 | – 685 | – 665 | – 556 | – 575 | – 550 | – 543 | – 573 | – 580 | – 581 |

**Table 5.6** Half-cell potentials of specimen BC-3 (mV)

| Test results obtained from the points at the surface of specimen in Fig. 5.13 | | | | | | | | | | |
|---|---|---|---|---|---|---|---|---|---|---|
| – 632 | – 640 | – 615 | – 627 | – 604 | – 636 | – 632 | – 650 | – 677 | – 664 | – 667 |
| – 640 | – 641 | – 683 | – 610 | – 612 | – 617 | – 600 | – 642 | – 660 | – 616 | – 610 |
| – 594 | – 596 | – 652 | – 654 | – 652 | – 657 | – 640 | – 655 | – 654 | – 660 | – 605 |
| – 603 | – 603 | – 604 | – 615 | – 615 | – 606 | – 604 | – 614 | – 618 | – 619 | – 620 |

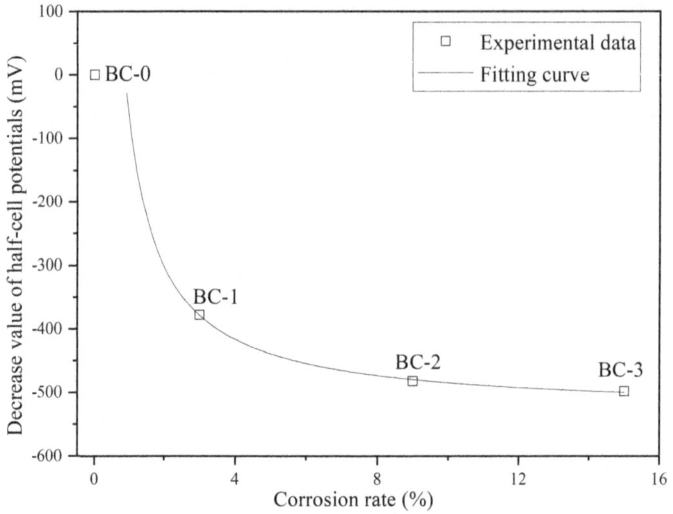


**Fig. 5.15** Relationship between half-cell potentials and corrosion rate

Quantitative nonlinear relationship is between the half-cell potential and corrosion rate. Half-cell potential is obtained from Eq. (5.5):

$$\Delta E_{\text{coor}} = \frac{455.8}{\rho_w} - 530.3 \tag{5.5}$$

where $\rho_w$ is the corrosion rate, in %; $\Delta E_{\text{corr}}$ is the decrease value of half-cell potentials, in mV. According to Eq. (5.5), when corrosion rate of beam-column joint is 3%, 9%, and 15%, decrease of half-cell potential are − 378.4, − 479.7, and − 499.9 mV, respectively. Compared with experimental data, decrease of half-cell potential changes at − 0.4, 2.3, and − 1.9 mV. So within the corrosion rate of 0 and 15%, we can use Eq. (5.5) to analyze the relationship between half-cell potentials and corrosion rate.

Load-bearing capacity of specimens is attained by cyclic loading. As shown in Fig. 5.16, load-bearing capacity decreases with the increase of corrosion rate. Generally, there is obvious linear relationship between them. Herein, Eq. (5.6) presents their relationships:

$$R(\rho) = 1 - 0.0119\rho_w \tag{5.6}$$

where $\rho_w$ is the corrosion rate, in %; $R(\rho)$ is ratio of load-bearing capacity of corroded specimen to normal specimen. When corrosion rate of specimens is 3%, 9%, and 15%, the load-bearing capacity rates are 96.4%, 89.3%, and 82.2%, respectively. Compared with experimental data, load-bearing capacity rate changes at 3.7%, −

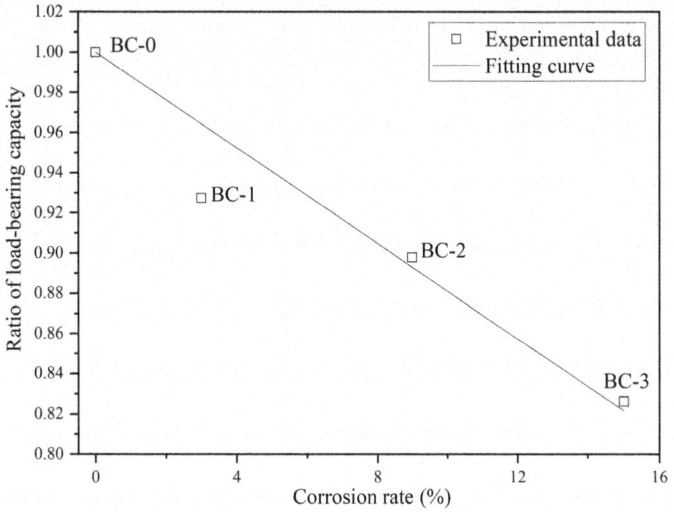


**Fig. 5.16** Relationship between corrosion rate and load-bearing capacity

0.5%, and – 0.4%, respectively. So within the corrosion rate of 0 and 15%, Eq. (5.6) is applied to analyze the relationship between corrosion rates and load-bearing capacity.

Figure 5.17 shows relationship between half-cell potentials and load-bearing capacity. Load-bearing capacity decreases with the decrease of half-cell potentials.

Compared with BC-0, the ratio of load-bearing capacity of corroded specimen to normal specimen is at the rate of 92.7%, 89.8%, and 82.6%, respectively. It is

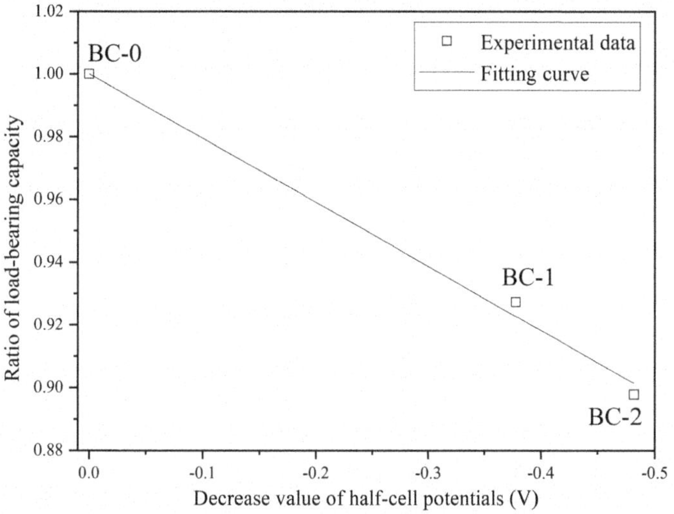


**Fig. 5.17** Relationship between half-cell potentials and load-bearing capacity

concluded that within the corrosion rates of 0 and 9%, half-cell potentials and load-bearing capacity have obvious linear relationship. Load-bearing capacity is given as Eq. (5.7)

$$R(\rho) = 1 + 0.2\Delta E_{\text{corr}} \tag{5.7}$$

where $\Delta E_{\text{corr}}$ is the decrease value of half-cell potentials, in V; $R(\rho)$ is ratio of load-bearing capacity of corroded specimen and normal specimen, in %. When $\Delta E_{\text{corr}}$ are − 0.378 V and − 0.482 V, $R(\rho)$ are 92.4% and 90.4%, respectively. Compared with experimental data, load-bearing capacity rate changes at − 0.3% and 0.6%, respectively. Within the corrosion rate of 0 and 9%, Eq. (5.7) could be used to analyze the relationship between half-cell potentials and load-bearing capacity.

Compared with BC-0, decrease rate of the half-cell potentials and load-bearing capacity is at the rate of 7.3%, 10.2%, and 17.4%, respectively. We could find that within the corrosion rate of 0 and 9%, decrease rate of half-cell potentials and load-bearing capacity have obvious linear relationship plotted in Fig. 5.18. Decrease rate of load-bearing capacity could be obtained as Eq. (5.8):

$$\alpha = 0.027\beta \tag{5.8}$$

where $\alpha$ is decrease rate of load-bearing capacity, in %; $\beta$ is decrease rate of the half-cell potentials, in %. When $\beta$ are 286.4% and 365.2%, $\alpha$ are 7.73% and 9.86%. Compared with experimental data, load-bearing capacity rate changes at 0.47% and − 0.36%. Within the corrosion rate of 0 and 9%, Eq. 5.5 was used to analyze the relationship between decrease rates of half-cell potentials and that of load-bearing capacity.

### 5.1.5 *Stiffness Degradation*

Depending on the theory of peak-to-peak stiffness reported in [34], stiffness of tested specimens was calculated by the slope of maximum load point of the first cycle in the loading process. The peak-to-peak secant stiffness is calculated as Eq. (5.9) [35]:

$$K_i = \frac{|+F_i| + |-F_i|}{|+X_i| + |-X_i|} \tag{5.9}$$

where $K_i$ is the secant stiffness at the $i$ cycle, in kN/mm; $+F_i$ and $-F_i$ are the peak loads at the $i$ cycle of the push and pull process, respectively, in kN; $+X_i$ and $-X_i$ are the correlative displacements of the peak loads at the $i$ cycle, respectively, in mm.

The comparisons of stiffness degradation of reference and retrofitted specimens are shown in Fig. 5.19 and Table 5.7. The initial stiffness of specimens BC-1 to BC-3 with the corrosion rates of 3%, 9%, and 15% of 8.0, 7.9, and 7.4 decreased by 2.4%, 3.7%, and 9.8% compared to that of uncorroded specimen BC-0 of 8.2 kN/mm,

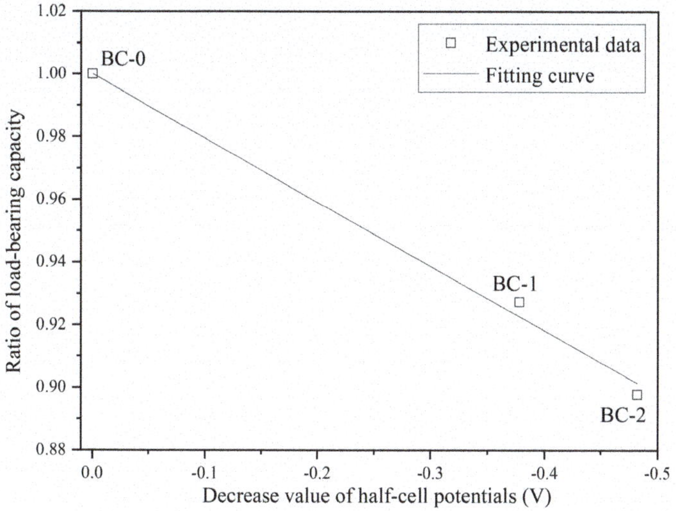


**Fig. 5.18** Relationship between decrease rate of the half-cell potentials and load-bearing capacity

respectively. The initial stiffness of the specimen decreased as the corrosion rate of reinforcing bars increased. Table 5.7 shows that the stiffness of corroded specimens BC-1 and BC-2 held over 95% of that of reference specimen, which illustrates that the secant stiffness degradation was limitedly influenced by the reinforcement corrosion with a low rate. However, severely corroded specimen BC-3 had a considerable decrease in stiffness, holding about 85% of that of reference specimen.

**Fig. 5.19** Secant stiffness degradation of all tested specimens

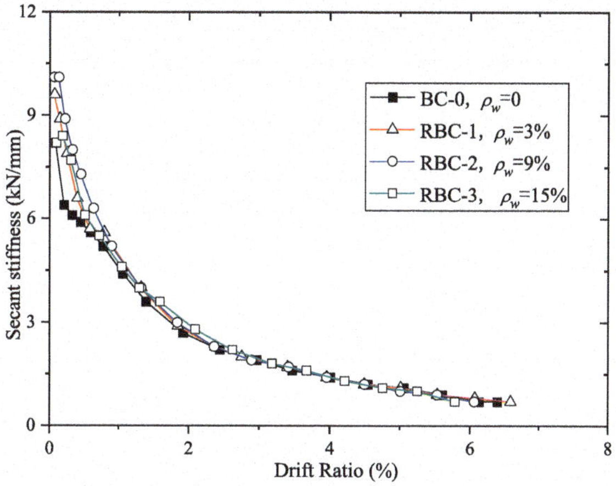

**Table 5.7** Test results of secant stiffness

| Specimen | Corrosion rate | $K_0$ (kN/mm) | | $K$ (kN/mm) | |
|---|---|---|---|---|---|
| | | Value | Ratio of corroded or retrofitted specimens to specimen BC-0 (%) | Value | Ratio of corroded or retrofitted specimens to specimen BC-0 (%) |
| BC-0 | 0 | 8.2 | 100 | 0.7 | 100 |
| BC-1 | 3% | 8.0 | 97.6 | 0.7 | 100 |
| BC-2 | 9% | 7.9 | 96.3 | 0.7 | 100 |
| BC-3 | 15% | 7.4 | 90 | 0.6 | 85.7 |

*Note* $K_0$ is the initial secant stiffness; and $K$ is the final secant stiffness

### *5.1.6 Ductility*

The displacement ductility and energy dissipation were always adopted to quantify the elastic–plastic deformation of RC members, as reported in [36]. The displacement ductility ratio is characterized as $\mu = \Delta_u/\Delta_y$, and the drift can be characterized as $\theta_u = \Delta_u/h_w$.

Table 5.8 shows that the displacement ductility ratios of corroded specimens BC-1, BC-2, and BC-3 only decrease by less than 10% of that of reference specimen, which illustrates that the displacement ductility was limitedly affected by the reinforcement corrosion. However, the drifts of corroded specimens BC-1, BC-2, and BC-3 decreased considerably, holding about 80% of that of reference specimen.

**Table 5.8** Test results of ductility, drift, and energy dissipation

| Specimen | Corrosion rate (%) | $\mu$ | | $\theta_u$ | | $Q$(kN·m) | |
|---|---|---|---|---|---|---|---|
| | | Value | Ratio of corroded or retrofitted specimens to specimen BC-0 (%) | Value | Ratio of corroded or retrofitted specimens to specimen BC-0 (%) | Value | Ratio of corroded or retrofitted specimens to specimen BC-0 (%) |
| BC-0 | 0 | 5.54 | 100 | 1/19 | 100 | 30.1 | 100 |
| BC-1 | 3 | 5.19 | 93.7 | 1/22 | 85.7 | 29.1 | 96.7 |
| BC-2 | 9 | 5.04 | 91.0 | 1/24 | 76.7 | 24.7 | 82.1 |
| BC-3 | 15 | 5.01 | 90.4 | 1/22 | 84.2 | 19.6 | 65.1 |

*Note* $\mu$ is displacement ductility ratio; $\theta_u$ is drift; and $Q$ is cumulative dissipated energy

### 5.1.7 Energy Dissipation Capacity

In the process of cyclic loading, energy dissipation characterized as the region enclosed inside the hysteretic curve in each cycle could be utilized to identify the ability of RC structures to dissipate the input energy from the earthquake. The energy dissipation capacity of specimens can be evaluated by two parameters, namely equivalent viscous damping ratio $\xi_{eq}$ [37] and cumulative energy dissipation $Q$ [38] calculated by summing up the energy dissipation obtained from the whole curves during the test. The equivalent viscous damping ratio $\xi_{eq}$ can be calculated as Eq. (5.10) [37, 39].

$$\xi_{eq} = \frac{1}{2\pi}\frac{S_{ABCD}}{S_{OAE} + S_{OCF}} \tag{5.10}$$

where $S_{ABCD}$ is the area of energy dissipated at the $i$ cycle; $S_{OAE}$ and $S_{OCF}$ are the energy consumed by a liner elastic body equivalent to the specimen, characterized as the area of triangle zone $OAE$ and $OCF$, respectively, as shown in Fig. 5.20. Results reported in [40] illustrate that the RC members with the high equivalent viscous damping ratio have a better energy dissipation capacity.

Figure 5.21 shows a comparison of the equivalent viscous damping ratios of reference and retrofitted specimens. The equivalent damping coefficient of the non-corroded specimen BC-0 is relatively large, and its energy dissipation capacity is strong. As the degree of component corrosion increases, the equivalent viscous damping coefficient of the specimen gradually decreases, indicating that the hysteresis curve is "pinching," and the energy dissipation capacity of the specimen decreases with the increase of steel corrosion rate.

When the displacement was 10 mm, the equivalent viscous damping coefficients of specimens BC-1 to BC-3 of 0.079, 0.072, and 0.112 are 26.9%, 33.3%, and 3.7% lower than that of specimen BC-0 of 0.108, respectively. When the displacement

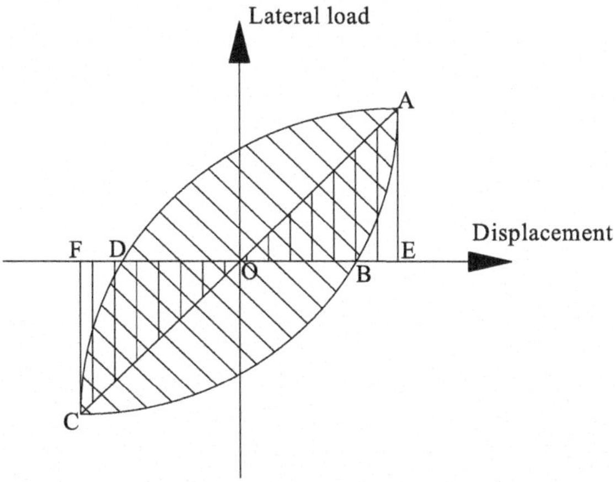


**Fig. 5.20** Evaluation of the energy dissipation for each cycle

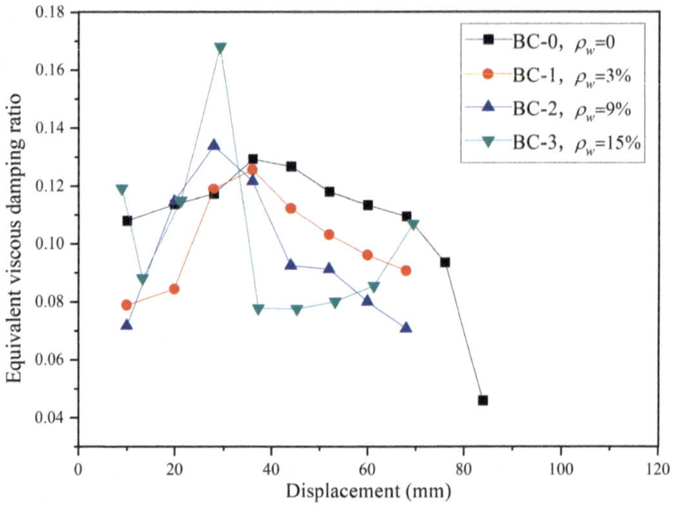


**Fig. 5.21** Equivalent viscous damping ratio $\xi_{eq}$ of all specimens

was 60 mm, the equivalent viscous damping coefficients of specimens BC-1 to BC-3 of 0.096, 0.080, and 0.085 are 15.0%, 29.2%, and 24.7% lower than that of specimen BC-0 of 0.113, respectively. In the later stage, the equivalent viscous damping coefficient sharply decreased as the displacement increased.

The cumulative energy dissipation of reference and retrofitted specimens is shown in Fig. 5.22. The cumulative energy dissipation of specimens with the corrosion rates of 3% and 9% held over 80% of that of reference specimen, which illustrated that the reinforcement corrosion exhibited a limited influence on the energy dissipation capacity, as shown in Table 5.8. But the energy dissipation capacity of severely corroded specimen BC-3 decreased considerably, holding around 65% of that of reference specimen.

### *5.1.8 Strains Profiles of Reinforcing Bars*

Methods of utilizing the strains of reinforcing bars to analyze the bond behavior of corroded RC joints or seismic-damaged RC beam-column joints retrofitted with BFRP sheets have been proposed in [41–43]. Figure 5.5 shows the strain gauges layout of tested specimens. Forty strain gauges were arranged on the longitudinal reinforcements in beam, longitudinal bars in column, beam stirrups, and column stirrups, which were designed with a prefix of "BL," "CL," "BS," and "CS," respectively. The suffix of "11" was referred to the serial number of reinforcing bars in concrete. For example, "BL11" and "BL21" were designed as the first strain gauge of first, and second longitudinal reinforcements in beam, respectively. Several strain gauges were

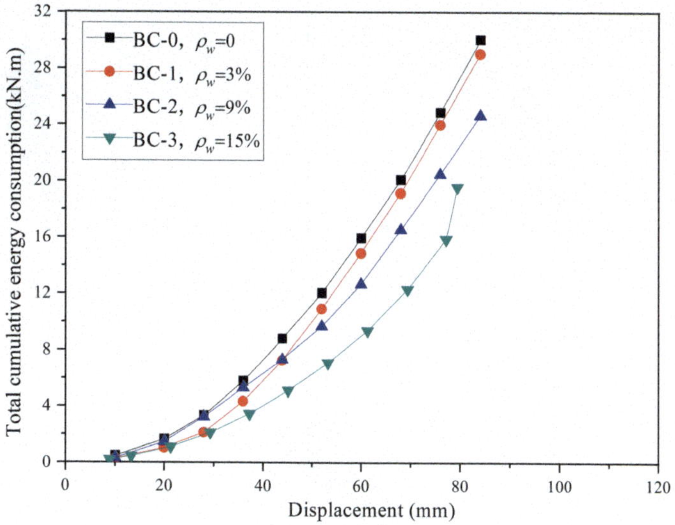


**Fig. 5.22** Cumulative energy dissipation of all test specimens at each loading cycle

invalid during the test, which were possibly associated with the following reasons: (a) strain gauges were damaged during the pouring process; (b) several strain gauges were destroyed by the electric current under the NaCl solution.

Load-strain loops of longitudinal reinforcements in beam of specimens BC-0, BC-1, and BC-3 are shown in Fig. 5.23a–c. The curve of reference specimen BC-0 showed an obvious "*K*" shape, while specimens with the corrosion rates of 3% and 15% exhibited vague and uneven curves. The bond behavior of corroded longitudinal reinforcements in beam of specimens BC-1 and BC-3 was poorer than that of reference specimen BC-0, which made the longitudinal reinforcements slip at the medium loading stage with larger strains. Strains of corroded reinforcing bars of specimens BC-1 and BC-3 changed from compressive strains to tensile strains earlier than those of BC-0 in the pull stage, as Fig. 5.23a–c. That might attribute to the fact that the bending moment on both sides of the joint reversed, resulting in the reinforcing bars pulling on one side and compressing on the other side.

When one side of the reinforcing bars yielded, the resultant load at both ends of the reinforcing bars passing through the core area was greater than the yielding load of the reinforcing bars. This extra force could only be transferred by the bond, producing high bonding stress and resulting in the reinforcing bars slipping under the cyclic loading. Revised deformations increased with the deteriorations of the bond behavior. Therefore, the tensile stress on the one side stayed steady while the compressive strains changed to tensile strains on the other side of the reinforcing bar.

Strains of longitudinal reinforcements in uncorroded column were lower than those in the beam under the cyclic loading, as shown in Figs. 5.23a and 5.24a, which coincided with the failure mode of "strong column-weak beam." Strains of corroded longitudinal reinforcements in column of specimens BC-1 and BC-3 with corrosion

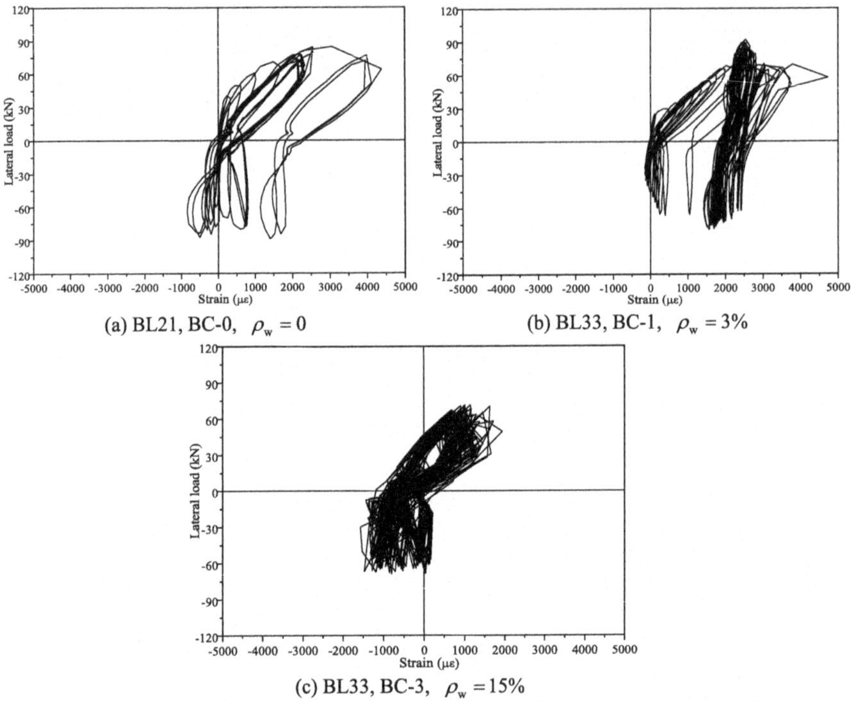


**Fig. 5.23** Load-strain loops of longitudinal reinforcements in beam

rates of 3% and 15% were larger than those of specimen BC-0 at the early loading stage, and these strains increased when the lateral load limitedly increased, as shown in Fig. 5.24a–c. That might attribute to the cross-section area of reinforcing bar decreased and the corrosion products increased after the reinforcing bars corroded, resulting in the microcracks generating in the concrete and the bond behavior deteriorating, as reported in [44]. The deterioration of bond behavior of corroded reinforcing bars results in the decrease of the region area and the instability of the load-strain loop at each cycle [41].

As shown in Fig. 5.25a–c, the strains of corroded stirrups in beam of specimens BC-2 and BC-3 were larger than those of the reference specimen BC-0 in the joint region possibly owing to the reinforcement corrosion, which contributed to larger stresses and strains. The bond behavior between corroded reinforcing bars and concrete decreased at early stage as the corrosion rate increased from 9% to 15%. Comparing with unretrofitted specimens, the bond behavior or strains of retrofitted corroded stirrups were improved or decreased, as shown in Fig. 5.25b–d. The BFRP sheets exhibited a good performance in increasing the bond behavior and decreasing the strains of corroded stirrups in beam.

The load-strain loops of stirrups in column are shown in Fig. 5.26. The strains in the stirrup adjacent to the joint core area from the strain gauge CS4 were significantly

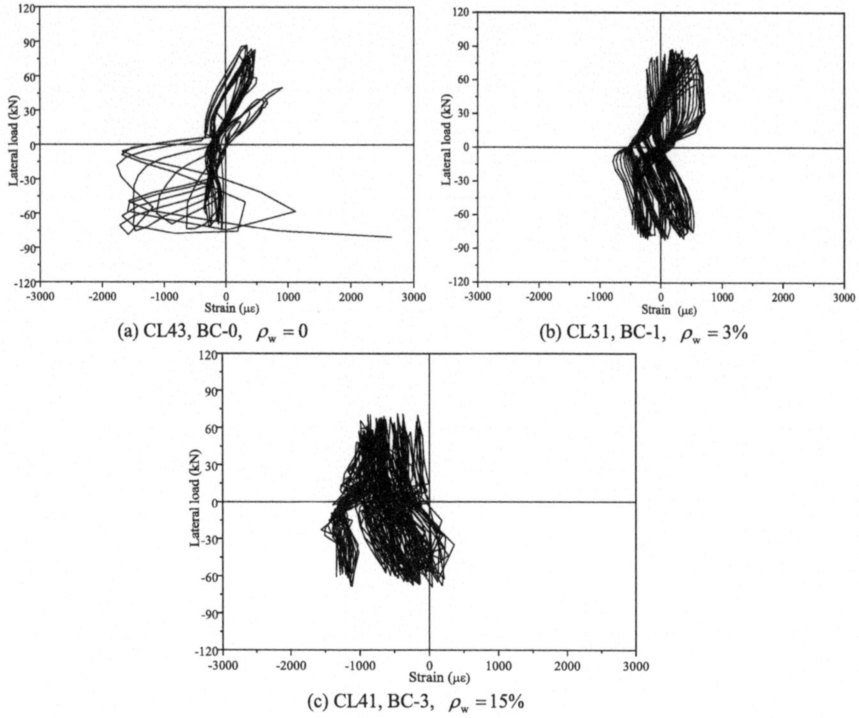


**Fig. 5.24** Load-strain loops of longitudinal reinforcements in column

lower than those in the joint core area from the strain gauge CS2. In the push and pull stage, the concrete at the core area of the joint suffered more alternate stress contributed to cracking and spalling, and the prematurely slipping of corroded stirrups in column at the early stage.

## 5.2 Seismic Performance of Corrosion-Damaged Beam-Column Joints Retrofitted with BFRP Sheets

### *5.2.1 Experimental Program*

(1) Specimens and materials

Four RC interior beam-column joints designed as specimens BC-0, RBC-1, RBC-2, and RBC-3 were fabricated and tested. Specimens BC-0 was the reference joint and RBC-1, RBC-2, and RBC-3 were the corroded joints with the corrosion rates of 3%, 9%, and 15%, respectively, retrofitted with BFRP sheets. Ordinary concrete utilized in the experimental test was offered by the commercial concrete, the components

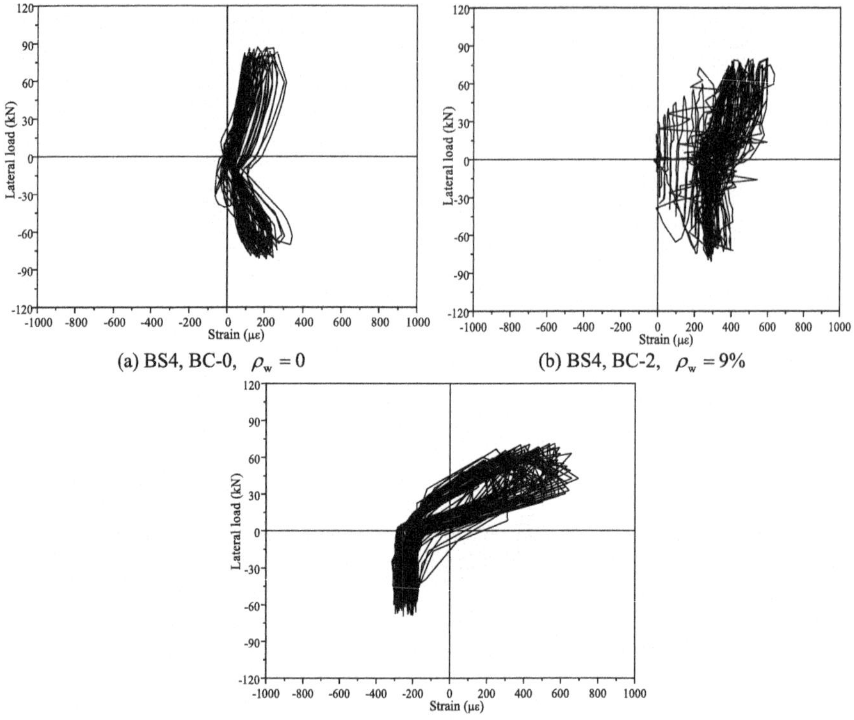


(a) BS4, BC-0, $\rho_w = 0$

(b) BS4, BC-2, $\rho_w = 9\%$

(c) BS4, BC-3, $\rho_w = 15\%$

**Fig. 5.25** Load-strain loops of stirrups in beam

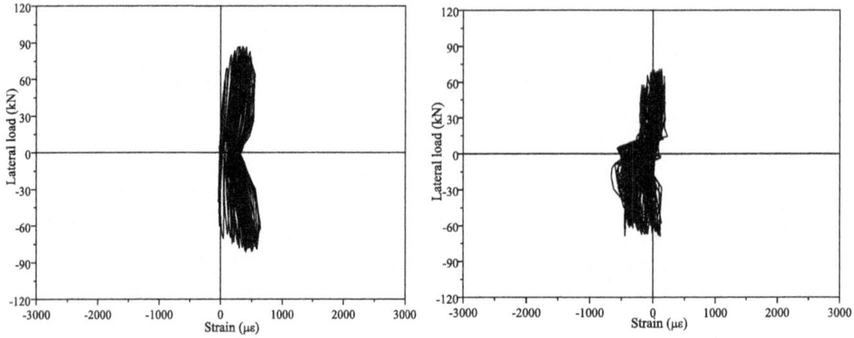


(a) CS4, BC-1, $\rho_w = 0$

(b) CS2, BC-3, $\rho_w = 15\%$

**Fig. 5.26** Load-strain loops of stirrups in column

proportions of which are shown in Table 5.1. Test results of cubic compressive strength $f_{cu}$ and elastic modulus $E_c$ of concrete were 39.3, and 3.24 × $10^4$ MPa, respectively. The dimension and reinforcement form of specimens BC-0, RBC-1, RBC-2, and RBC-3 are same as the descriptions shown in Fig. 5.1. The average values of yielding strength of ribbed and plain bars were 335, and 235 MPa, respectively.

(2) Accelerating Corrosion Test

The setup of accelerated corrosion process of specimens with the corrosion rates of 3%, 9%, and 15% was as same as that shown in Fig. 5.2. And detailed duration of corrosion $t$ could also be evaluated by Eq. (5.3), and the duration of specimens with the target corrosion rates of 3%, 9%, and 15% was 37, 55, and 61.8 days when the reinforcing bars were impressed with a current intensity of 1, 2, and 3 A, respectively.

(3) Method of Retrofitting

Considering several methods reported in [11, 23, 24, 40], corroded beam-column joints were retrofitted by the detailed arrangements, as shown in Fig. 5.27. Firstly, for restraining the cracks observed on the beams and columns and improving the flexural capacities of the column and beam, one floor of L-type BFRP sheet with 200 mm width was wrapped at the corners. Secondly, for confining the concrete of the core area, one layer of consecutive BFRP sheet with 250, and 300 mm width was applied on the surfaces of column, and beam, respectively. Thirdly, one layer of BFRP sheet with 300 mm width was wrapped around the column to anchor those sheets. Fourthly, one layer of BFRP sheet with 100 and 50 mm width was applied on the beams to increase the shear strength and restrain the crack development. Lastly, an additional layer of X-type BFRP strip with 50 mm width was applied to increase the shear strength of the core area. Before applying the BFRP sheets, the surface of joints should be polished smoothly, and corners around the beams as well as the column needed to be grounded with a 20 mm radius, as reported in [5]. Mixed epoxy resins were applied for sticking the BFRP sheets to the joint surface, where the coatings brushed on the BFRP sheet surface were periodically inspected, and then attention was paid to ensure the effectiveness of coatings. Several days were taken for epoxy resins to reach full care.

A thickness of 0.121 mm BFRP sheets was applied for the tested specimens, and the average value of tensile elastic modulus, ultimate average strain, and ultimate tensile strength were 1.05 × $10^5$ MPa, 2.2%, and 2.30 × $10^3$ MPa, respectively. The repairing arrangements tried to offer an effective restraint by BFRP sheets, and improve the deteriorated seismic behaviors of beam-column joints, which were caused by reinforcement corrosion, to prevent the RC buildings from non-ductile failure under earthquake.

(4) Test Setup and Procedure

The layout of test setup included the loading system and instrumentation was the same as that shown in Fig. 5.3. And the loading history and detailed arrangement and preparations on the strain gauges of tested specimens were also same as that shown in Figs. 5.4 and 5.5.

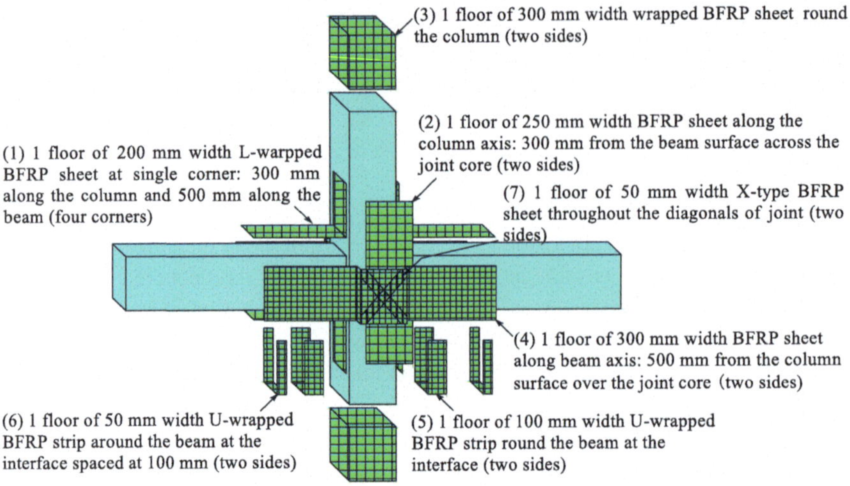


**Fig. 5.27** Repairing arrangement for corroded specimens with different corrosion rates (dimensions in mm)

## 5.2.2 *Crack Pattern and Failure Modes*

All retrofitted specimens presented a commonly similar failure mode and exhibited a development of failure from the delaminating of BFRP sheets to local bucking of stirrups and crushing of the core concrete.

In the load-control stage, cracking sounds were firstly heard at the lateral load of 40 kN momently, longer, and continuously for specimens RBC-1, RBC-2, and RBC-3, respectively. The lateral load when cracks were first visible on the surfaces for specimens RBC-2 and RBC-3 was 60 and 40 kN, respectively, while no cracks were observed on the specimen RBC-1 in this stage. These initial cracks were epoxy cracks at the interface of joint, as show in Fig. 5.28a. As the test proceeding, these cracks continued to propagate rapidly with the sounds becoming continuously and loudly. In the end, flexural cracks and local bulging of BFRP sheets occurred at the interface, as shown in Fig. 5.28b.

In the displacement-control stage, these cracks expanded and continuous sounds from the breaking of the epoxy resins were heard. BFRP sheets on the interface of beam suddenly began to rupture with the concrete spalling at the lateral displacement of 36.0 (2.36% drift), 36.0 (2.36% drift), and 32.0 mm (2.10% drift) for specimens RBC-1, RBC-2, and RBC-3, respectively, as shown in Fig. 5.28c. Due to the deficiency of the BFRP confinement and deterioration of bond behavior between concrete and the corroded reinforcing bars, the concrete cover at the edge of beams spalled and then the BFRP sheets in the joint region bulged at the lateral displacement of 72.0 (4.72% drift), 68.0 (4.46% drift), and 56.0 mm (3.67% drift) for specimens RBC-1-RBC-3, respectively. After the concrete cover spalled, specimens deteriorated directly with local buckling of corroded stirrups and wrapping delaminating

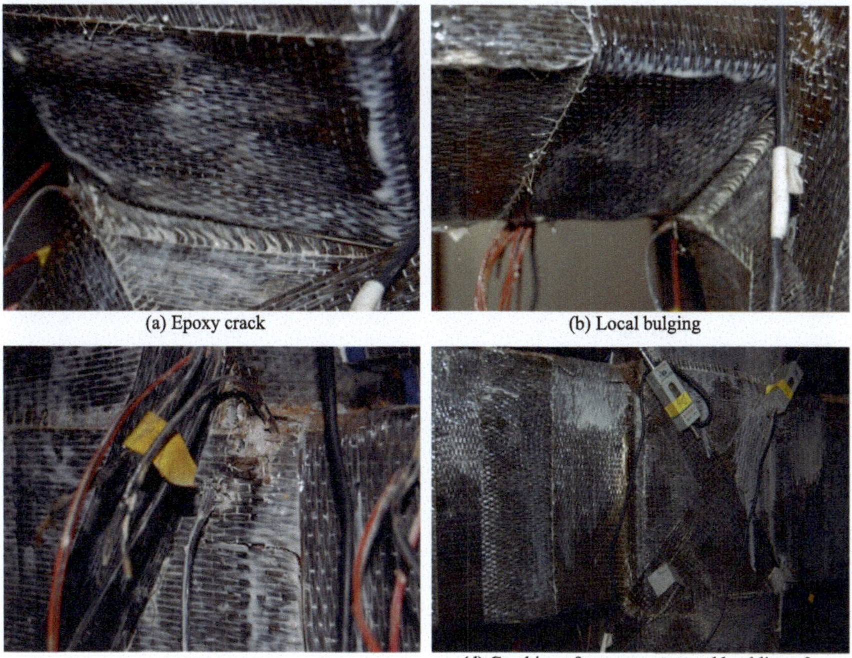

(a) Epoxy crack (b) Local bulging (c) Rupture of BFRP sheets (d) Crushing of core concrete and buckling of corroded stirrup

**Fig. 5.28** Observed failure of retrofitted specimens

of BFRP sheets, as shown in Fig. 5.28d. BFRP sheets provided an effective confinement to the compressed concrete in the joint core areas of the retrofitted specimens, contributing to the delaying of concrete spalling and crushing as well as the stirrups buckling. However, large cracks and buckling occurred near the joint core areas in contrast with reference specimen, which was primarily related to the continuous deficiency of BFRP confinement during the test.

### *5.2.3 Hysteretic Response and Load-Bearing Capacity*

Figure 5.29 shows the load–displacement hysteretic loops of all specimens. The encircled area of the hysteretic loop can be utilized for estimating the energy dissipation of tested specimen during the loading process [29]. The hysteretic curve of reference specimen was roughly full and inverted-S typed without obvious pinching phenomenon, whereas the hysteretic responses of the retrofitted specimens were commonly improved contrasting with those of the corresponding unretrofitted specimens. FRP sheets should availably promote the hysteretic response of the retrofitted

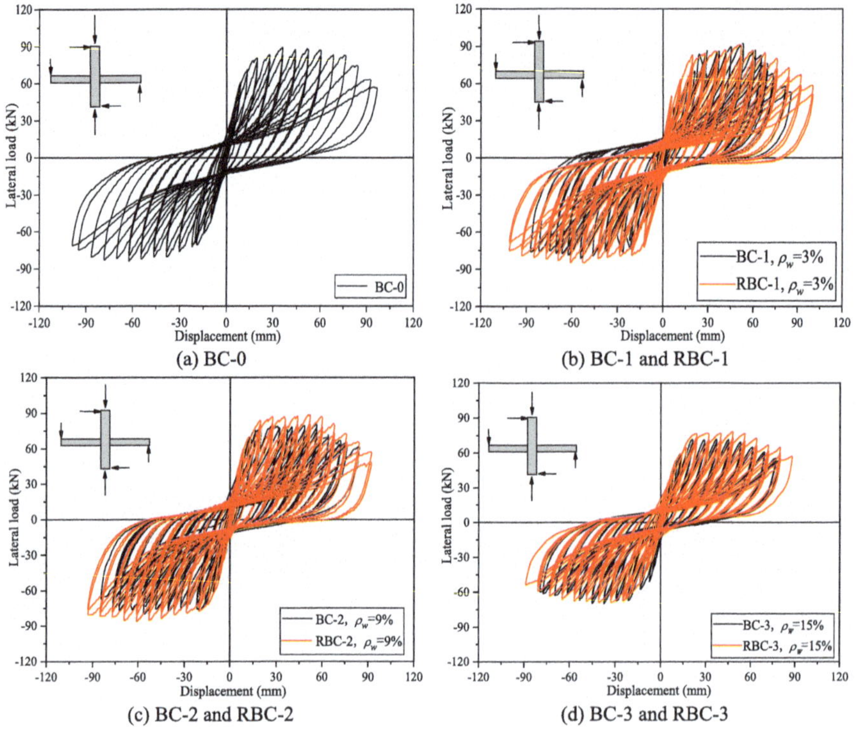


**Fig. 5.29** Load–displacement hysteretic curves of all specimens

specimen, though actual restraint provided by FRP sheets was, in a sense, inefficient in enlarging the internal area of hysteretic curves with the corrosion rate of 15%.

To study the degradation of ductility and load-bearing capacity, the envelope curves of all tested specimens are shown in Fig. 5.30, where the envelopes were taken by averaging the maximum absolute values in the process of push and pull for every cycle. The displacement and lateral load exhibited a linear relationship, which indicated that specimens were in elastic stage at the early stage.

An obviously long yielding platform was found after a steady hardening up to the peak load, following by a slowly softening after the peak load. The envelope curve of reference specimen was enclosed by those of retrofitted specimens, indicating that the effect of BFRP sheets on increasing the load-bearing capacities of retrofitted specimens was effectively achieved or slightly insufficient except for the specimen RBC-3. Due to the corrosion inhomogeneity, the load-bearing capacity of specimen RBC-3 with the corrosion rate of 15% was less than that of reference specimen BC-0 in the negative direction cycles.

Ductility is also utilized to evaluate the seismic behaviors of RC structures under the earthquake [29, 30]. The yielding displacement $\Delta_y$ corresponding to the yielding load was calculated by the method reported in [31]. As shown in Fig. 5.12, $P_y$ is the yielding load and $\Delta_y$ is the yielding displacement, which are calculated depended

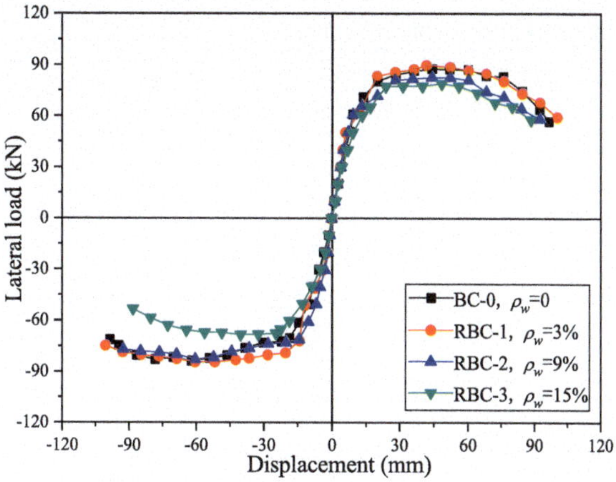


**Fig. 5.30** Envelope curves of hysteretic loops for reference and retrofitted specimens

on the assumption that the area of shadow zone $S_{OBD}$ equals to the area of shadow zone $S_{ABC}$; $P_u$ is the ultimate load, which is characterized as 85% of the peak lateral load $P_{\max}$ and $\Delta_u$ is the ultimate displacement. Table 5.9 shows the test results of all specimens. The load-bearing capacity and displacement of slightly corroded specimen BC-1 held over 90% of those of reference specimen BC-0, which illustrated that the hysteretic loop was limitedly affected by the corrosion of reinforcing bars. However, corroded specimens BC-2 and BC-3 showed a considerable decrease in the load-bearing capacity as well as displacement, holding around 85% of those of reference specimen BC-0. Displacements of corroded specimens were possibly recovered to that of reference specimen after repairing with BFRP sheets, while only both the load-bearing capacity and displacement of specimen with the corrosion rate of 3% could be recovered to those of reference specimen.

## 5.2.4 *Stiffness Degradation*

Depending on the theory of peak-to-peak stiffness reported in [34], stiffness of tested specimens was calculated by the slope of maximum load point of the first cycle in the loading process. The peak-to-peak secant stiffness is calculated as Eq. (5.9).

A comparison of stiffness degradation of reference and retrofitted specimens is shown in Fig. 5.31. The initial stiffness of specimens BC-0, RBC-1, RBC-2, and RBC-3 was 8.2, 9.6, 10.1, and 8.2 kN/mm, which was 17.1%, 23.2%, and 0 higher than that of specimen BC-0, respectively. The initial stiffness of corroded specimens retrofitted with BFRP sheets was higher than that of reference specimen, which were possibly attributed to the cured FRP sheets and the increased initial stiffness of corroded specimens. Similar conclusion of the initial stiffness of corroded members is also reported in [45]. Then the degradation rates of the stiffness of retrofitted specimens were lower than that of reference specimen at the early stage, and gradually

**Table 5.9** Test results of all specimens

| Specimen | Corrosion rate (%) | $P_y$(kN) | | $P_u$(kN) | | $\Delta_y$(mm) | | $\Delta_u$(mm) | |
|---|---|---|---|---|---|---|---|---|---|
| | | Value | Ratio of corroded or retrofitted specimens to specimen BC-0 (%) | Value | Ratio of corroded or retrofitted specimens to specimen BC-0 (%) | Value | Ratio of corroded or retrofitted specimens to specimen BC-0 (%) | Value | Ratio of corroded or retrofitted specimens to specimen BC-0 (%) |
| BC-0 | 0 | 70.8 | 100 | 85.6 | 100 | 14.8 | 100 | 82.3 | 100 |
| BC-1 | 3 | 67.6 | 84.5 | 79.4 | 92.8 | 13.5 | 91.2 | 70.5 | 85.7 |
| BC-2 | 9 | 62.7 | 88.6 | 76.9 | 89.8 | 12.5 | 84.5 | 63.1 | 76.7 |
| BC-3 | 15 | 51.7 | 73.0 | 70.6 | 82.5 | 12.3 | 83.1 | 69.3 | 84.2 |
| RBC-1 | 3 | 64.1 | 90.5 | 88.1 | 102.9 | 13.2 | 89.2 | 84.2 | 102.3 |
| RBC-2 | 9 | 62.0 | 87.6 | 83.1 | 97.1 | 11.7 | 79.1 | 80.3 | 97.6 |
| RBC-3 | 15 | 60.2 | 85.0 | 73.6 | 86.0 | 13.4 | 90.5 | 84.1 | 102.2 |

had a similar tendency. Table 5.10 shows that the stiffness of corroded specimens BC-1 and BC-2 held over 95% of that of reference specimen, which illustrates that the secant stiffness degradation was limitedly influenced by the reinforcement corrosion with a low rate. However, severely corroded specimen BC-3 had a considerable decrease in stiffness, holding about 85% of that of reference specimen. Stiffness of specimens with the corrosion rates of 3%, 9%, and 15% could be recovered to that of reference specimen.

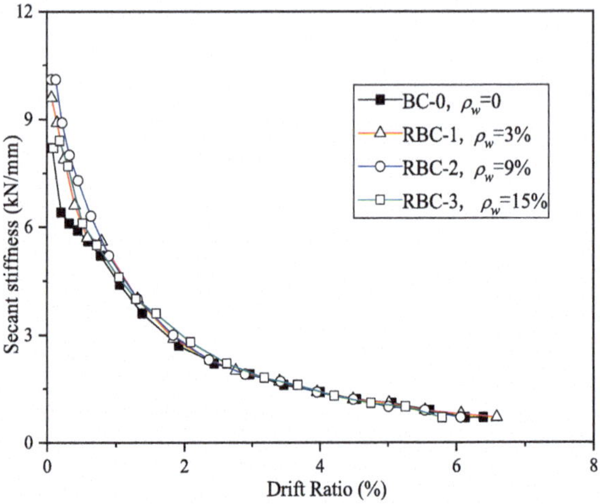


**Fig. 5.31** Secant stiffness degradation of reference and retrofitted specimens

**Table 5.10** Test results of secant stiffness

| Specimen | Corrosion rate (%) | $K_0$(kN/mm) | | $K$(kN/mm) | |
|---|---|---|---|---|---|
| | | Value | Ratio of corroded or retrofitted specimens to specimen BC-0 (%) | Value | Ratio of corroded or retrofitted specimens to specimen BC-0 (%) |
| BC-0 | 0 | 8.2 | 100 | 0.7 | 100 |
| BC-1 | 3 | 8.0 | 97.6 | 0.7 | 100 |
| BC-2 | 9 | 7.9 | 96.3 | 0.7 | 100 |
| BC-3 | 15 | 7.4 | 90 | 0.6 | 85.7 |
| RBC-1 | 3 | 9.6 | 117.1 | 0.7 | 100 |
| RBC-2 | 9 | 10.1 | 123.2 | 0.7 | 100 |
| RBC-3 | 15 | 8.2 | 100 | 0.7 | 100 |

*Note* $K_0$ is the initial secant stiffness, $K$ is the final secant stiffness

### 5.2.5 Ductility

The displacement ductility and energy dissipation were always adopted to quantify the elastic–plastic deformation of RC members, as reported in [36]. The displacement ductility ratio is characterized as $\mu = \Delta_u/\Delta_y$, and the drift can be characterized as $\theta_u = \Delta_u/h_w$.

Table 5.11 shows that the displacement ductility ratios of corroded specimens BC-1, BC-2, and BC-3 only decrease by less than 10% of that of reference specimen, which illustrates that the displacement ductility was limitedly affected by the reinforcement corrosion. However, the drifts of corroded specimens BC-1, BC-2, and BC-3 decreased considerably, holding about 80% of that of reference specimen. The displacement ductility ratio and drift of corroded specimens could be recovered to those of reference specimen, while the drift of specimen with the corrosion rate of 9% could not be recovered to that of reference specimen as expected, which were possibly associated with the following reasons: (a) the corrosion heterogeneity of reinforcing bars; (b) asymmetry of the loading device; and (c) the inefficiency of BFRP sheets offered to confine the joint area.

### 5.2.6 Energy Dissipation Capacity

In the process of cyclic loading, energy dissipation characterized as the region enclosed inside the hysteretic curve in each cycle could be utilized to identify the ability of RC structures to dissipate the input energy from the earthquake. The energy dissipation capacity of specimens can be evaluated by two parameters, namely equivalent viscous damping ratio $\xi_{eq}$ [37] and cumulative energy dissipation $Q$ [38] calculated by summing up the energy dissipation obtained from the whole curves during

**Table 5.11** Test results of ductility, drift, and energy dissipation

| Specimen | Corrosion rate (%) | $\mu$ | | $\theta_u$ | | $Q$(kN·m) | |
|---|---|---|---|---|---|---|---|
| | | Value | Ratio of corroded or retrofitted specimens to specimen BC-0 (%) | Value | Ratio of corroded or retrofitted specimens to specimen BC-0 (%) | Value | Ratio of corroded or retrofitted specimens to specimen BC-0 (%) |
| BC-0 | 0 | 5.54 | 100 | 1/19 | 100 | 30.1 | 100 |
| BC-1 | 3 | 5.19 | 93.7 | 1/22 | 85.7 | 29.1 | 96.7 |
| BC-2 | 9 | 5.04 | 91.0 | 1/24 | 76.7 | 24.7 | 82.1 |
| BC-3 | 15 | 5.01 | 90.4 | 1/22 | 84.2 | 19.6 | 65.1 |
| RBC-1 | 3 | 6.61 | 119.3 | 1/18 | 102.3 | 33.3 | 110.6 |
| RBC-2 | 9 | 6.84 | 123.5 | 1/19 | 97.6 | 31.5 | 104.7 |
| RBC-3 | 15 | 6.30 | 113.7 | 1/18 | 102.2 | 29.0 | 96.3 |

*Note* $\mu$ is displacement ductility ratio; $\theta_u$ is drift; and $Q$ is cumulative dissipated energy

the test. The equivalent viscous damping ratio $\xi_{eq}$ can be calculated as Eq. (5.10) [37, 39].

Figure 5.32 shows a comparison of the equivalent viscous damping ratios of reference and retrofitted specimens. The equivalent viscous damping ratios of tested specimens increased at the early stage, and then decreased at the last few cycles as the displacement increased. However, when the drift ratio was 1.8%, the equivalent viscous damping ratios of specimens BC-0, RBC-1, RBC-2, and RBC-3 were 11.7%, 14.6%, 14.6%, and 13.4%, which were 24.8%, 24.8%, and 14.5% higher than that of reference specimen BC-0, respectively. Equivalent viscous damping ratios of retrofitted specimens were larger than that of the reference specimen before the specimen reached to the peak load, which was generally associated with the following reasons: (a) hardened epoxy resin with BFRP sheets leading to increase the brittleness of the retrofitted specimens; (b) corroded products filling with the space between the reinforcing bars and concrete. Similar tendency in the variation of equivalent viscous damping ratios of tested specimens is also reported in [46, 47]. At the middle stage, equivalent viscous damping ratios of retrofitted specimens were lower than that of reference specimen. However, the viscous damping ratios of specimens BC-0, RBC-1, RBC-2, and RBC-3 were 4.6%, 10.1%, 10.5%, and 10.5%, which were 119.6%, 128.3%, and 128.3% higher than that of reference specimen BC-0 at the drift ratio of 5.5%, respectively.

The significant decrease of the viscous damping ratio of BC-0 might attribute to the significant spalling of the concrete in core area and the yielding of reinforcing bars occurred when the drift ratio was 5.5%, while the corroded retrofitted specimens were confined by the BFRP sheets. With the application of BFRP sheets, the spalling of concrete and buckling of the reinforcing bars were delayed, resulting in the higher viscous damping ratios of retrofitted specimens at the later loading stage.

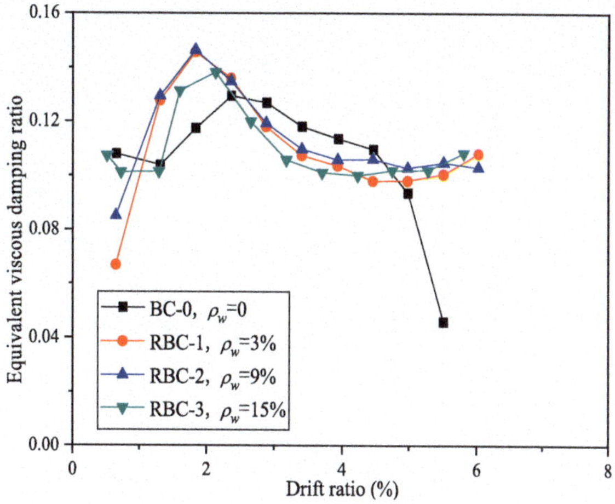


**Fig. 5.32** Equivalent viscous damping ratio $\xi_{eq}$ of all specimens

The cumulative energy dissipation of reference and retrofitted specimens is shown in Fig. 5.33. The cumulative energy dissipation of specimens with the corrosion rates of 3% and 9% held over 80% of that of reference specimen, which illustrated that the reinforcement corrosion exhibited a limited influence on the energy dissipation capacity, as shown in Table 5.8. But the energy dissipation capacity of severely corroded specimen BC-3 decreased considerably, holding around 65% of that of reference specimen. The total cumulative energy dissipation of retrofitted specimens BC-0, RBC-1, RBC-2, and RBC-3 was 30.1, 33.3, 31.5, and 29.0 kN·m, which was 10.6%, 4.7% higher, or 3.7% lower than that of reference specimen BC-0, respectively. The energy dissipation capacities of corroded specimens with slight corrosion rates of 3% and 9% were fully recovered to that of reference specimen after repairing with BFRP sheets, except the severely corroded specimen with the corrosion rate of 15%.

### 5.2.7 *Strains Profiles of Reinforcing Bars*

The bond behavior between corroded reinforcing bars and concrete or strains of longitudinal reinforcements in the beam of specimen RBC-1 retrofitted with BFRP sheets was better or lower than those of specimen BC-1 with the corrosion rate of 3%, as shown in Fig. 5.34b, d. The effect of BFRP sheets on confining the strains of corroded reinforcing bars and the instability of load-strain loops decreased as the corrosion rate increased, as shown in Fig. 5.34d–f.

After the corroded specimens were retrofitted with BFRP sheets, the strains of longitudinal reinforcements in column of specimens RBC-1 and RBC-3 were lower than those of specimens BC-1 and BC-3 in the push stage, indicating that the corroded reinforcing bars were confined by the BFRP sheets and bond behavior of corroded

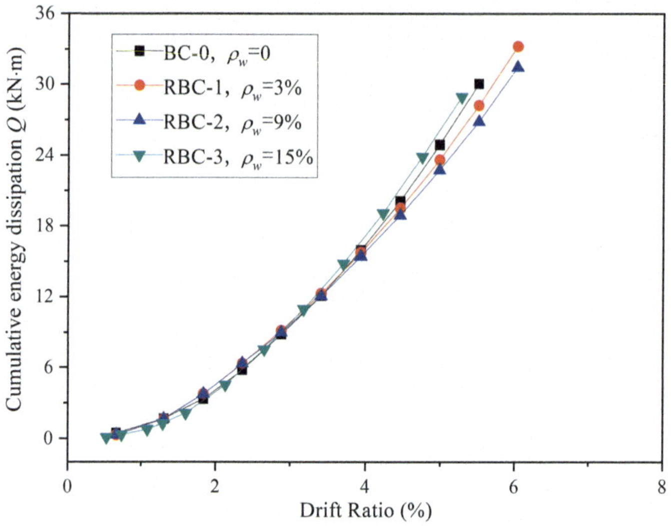


**Fig. 5.33** Cumulative energy dissipation of all tested specimens at each loading cycle

specimens was improved, as shown in Fig. 5.35b–d, f. Similar conclusions are also found in [43], which indicates that the yield load or bond behavior of seismic-damaged reinforcing bars in RC beam-column joints was significantly delayed or improved after repairing with BFRP sheets from the load-strain loops. The corroded longitudinal reinforcements in column of specimen RBC-3 slipped in the pull stage from the early stage, resulting in larger strains than those of retrofitted specimens with the corrosion rates of 3% and 9%, as shown in Fig. 5.35d–f.

Comparing with unretrofitted specimens, the bond behavior or strains of retrofitted corroded stirrups were improved or decreased, as shown in Fig. 5.36b–d. The BFRP sheets exhibited a good performance in increasing the bond behavior and decreasing the strains of corroded stirrups in beam.

The load-strain loops of stirrups in column are shown in Fig. 5.37. BFRP sheets exhibited a better influence on the improvement of bond behavior and strain reduction of the corroded stirrups in column outside the core area, than those in the core area, as shown in Fig. 5.37b–d.

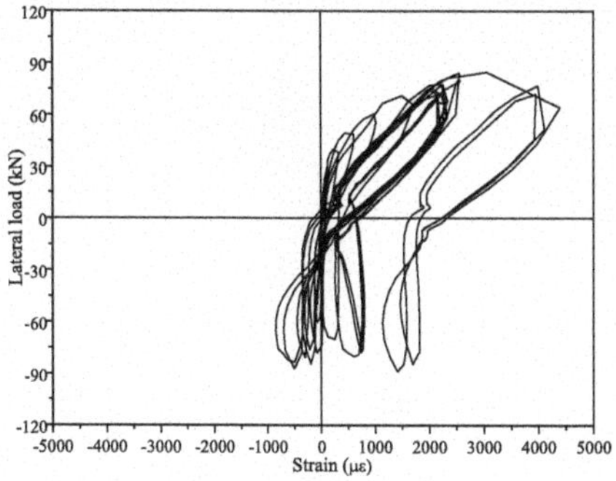


(a) BL21, BC-0, $\rho_w = 0$

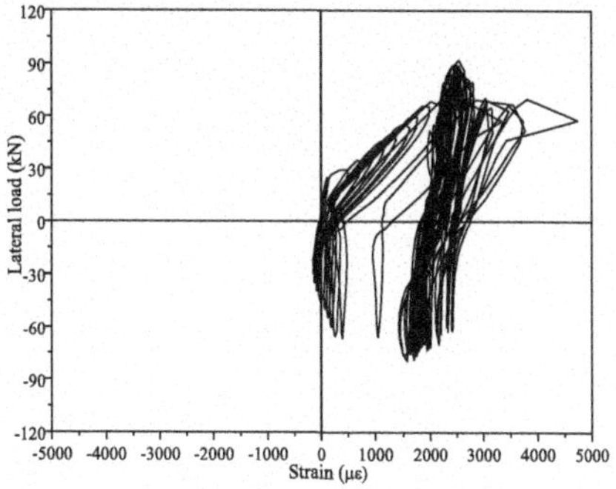


(b) BL33, BC-1, $\rho_w = 3\%$

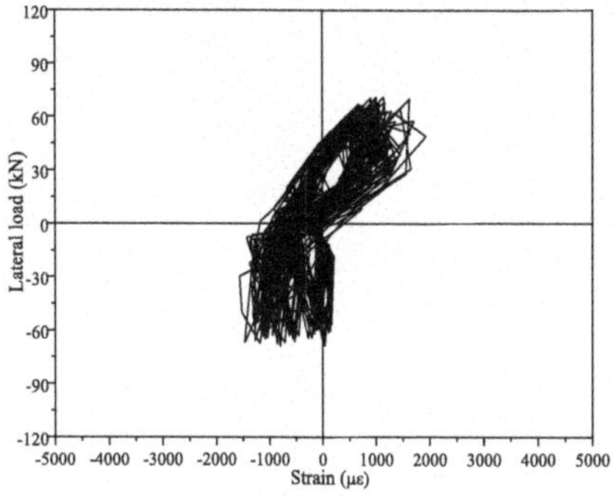


(c) BL33, BC-3, $\rho_w = 15\%$

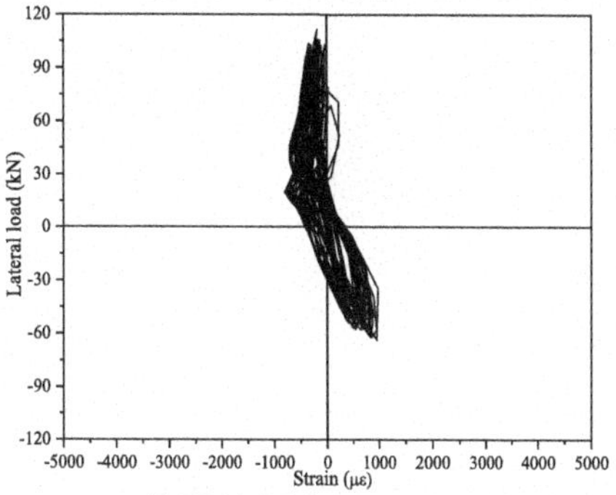


(d) BL33, RBC-1, $\rho_w = 3\%$

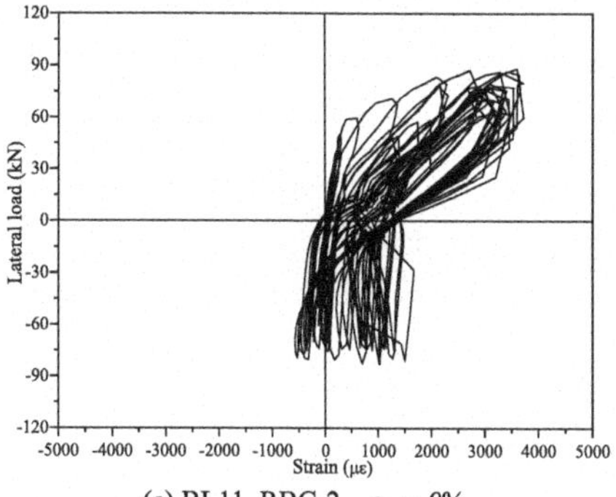


(e) BL11, RBC-2, $\rho_w = 9\%$

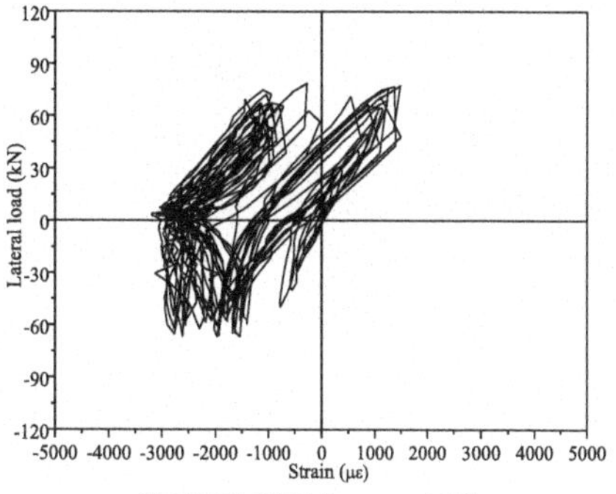


(f) BL33, RBC-3, $\rho_w = 15\%$

**Fig. 5.34** Load-strain loops of longitudinal reinforcements in beam

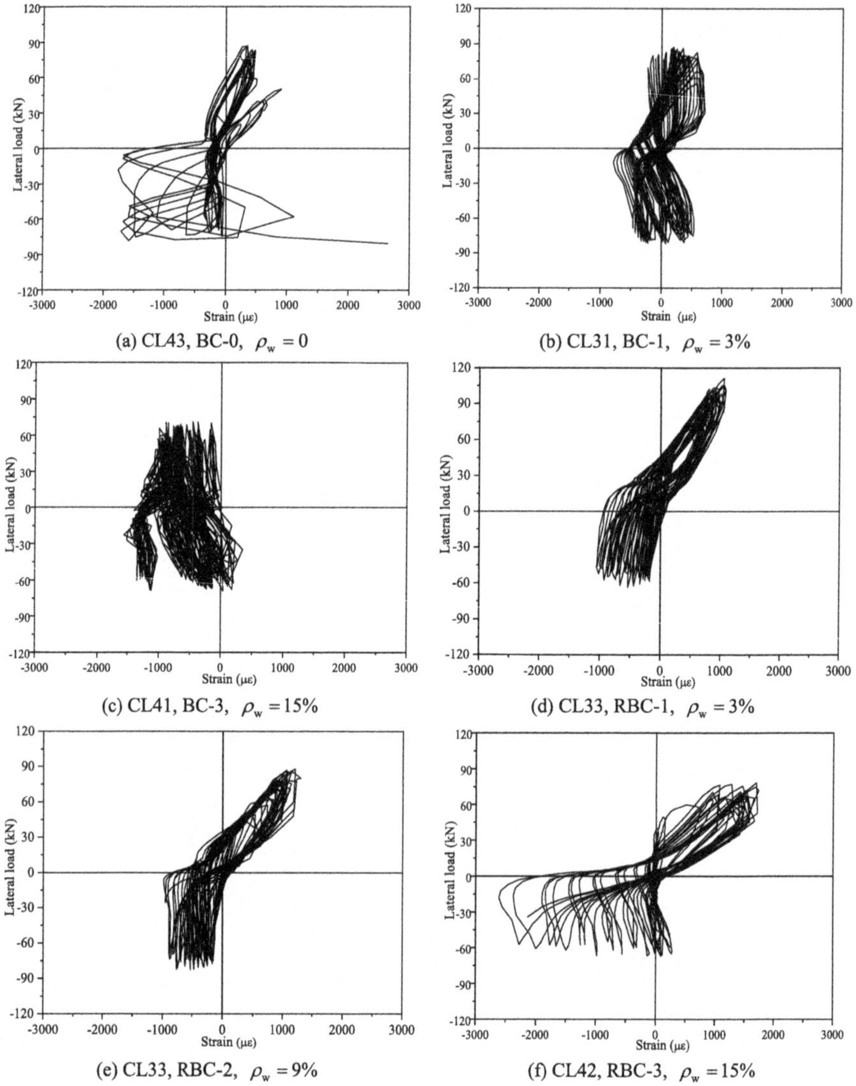


**Fig. 5.35** Load-strain loops of longitudinal reinforcements in column

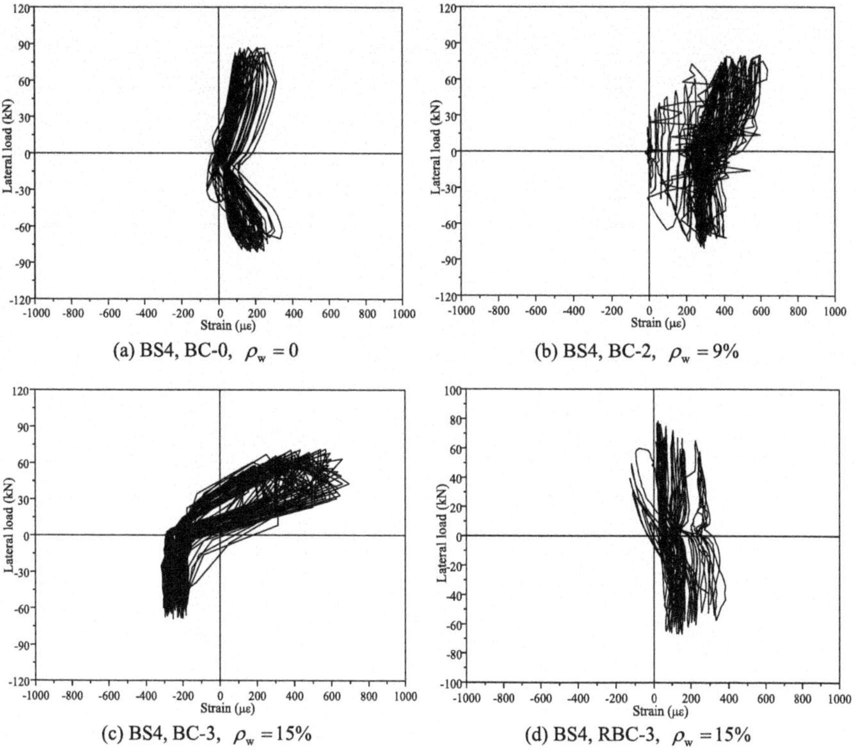


**Fig. 5.36** Load-strain loops of stirrups in beam

# 5.3 Retrofitting of Seismic-Damaged Beam-Column Joints Under Different Corrosion Rates with BFRP Sheets

## *5.3.1 Experimental Program*

An experimental investigation on six RC interior beam-column joints was conducted. Although investigation on the seismic performance of corroded beam-column joints with the corrosion rate of 9% and the retrofitted joints with BFRP sheets with the same rate has been conducted in [12], investigation on the seismic performance of corroded beam-column joints under different earthquake damages after repairing with BFRP sheets is still limited. To verify the retrofitting effectiveness of BFRP sheets on damaged and corroded RC members and comprehend the seismic performance of corroded beam-column joints under different earthquake damages after retrofitting with BFRP sheets, the corrosion rate of 9% was determined and four beam-column joints with the corrosion rates of 0 and 9% were firstly tested under the cyclic loadings in certain displacement angles to simulate the earthquake damages. After that, these damaged specimens were retrofitted with externally bonded BFRP

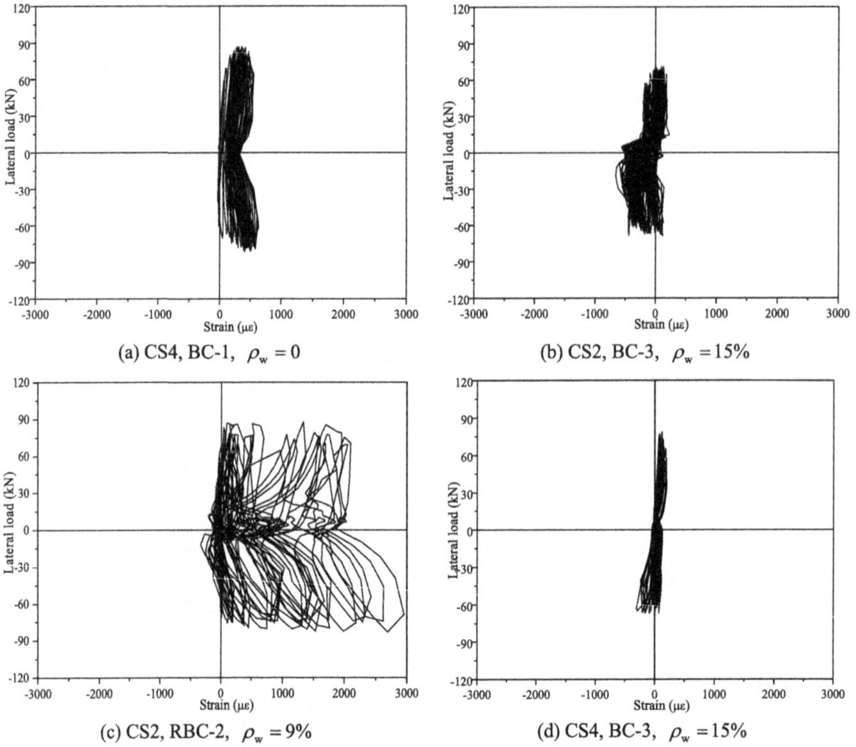


**Fig. 5.37** Load-strain loops of stirrups in column

sheets. Test results were discussed and compared to those of benchmark specimen and uncorroded specimen retrofitted with BFRP sheets.

Detailed descriptions of all tested specimens are described in Table 5.12. Benchmark specimen BC0 was directly tested without any retrofitting, while specimen RBC1 with an affix of "R" was uncorroded joint retrofitted with BFRP sheets. Specimens ERBC1, ERBC-2, ERBC-3, and ERBC4 with an affix of "E" were pre-damaged and then retrofitted with BFRP sheets, among which ERBC2-ERBC4 with the corrosion rate of 9% were conducted with the accelerating corrosion.

Besides, another group of experimental study on five RC interior beam-column joints was conducted.

For evaluating the seismic rehabilitation of corroded joints retrofitted with BFRP sheets under the earthquake damage, three corroded specimens with the corrosion rates of 0, 3%, and 9% were firstly tested under the cyclic loading in a certain displacement angle to assume the earthquake damage. Subsequently, these damaged specimens were retrofitted by sticking the BFRP sheets on the surface. Test results were contrasted to these of uncorroded specimen retrofitted with BFRP sheets and reference specimen. Table 5.12 depicts the details of tested specimens. Reference specimen BC0 was tested without any retrofitting, while specimen RBC1 with a

**Table 5.12** Details of tested specimens

| Specimen | Corrosion rates (%) | Damage level | *n* | Concrete strength grade | Retrofitting method |
|---|---|---|---|---|---|
| BC0 | 0 | – | 0.2 | C30 | – |
| RBC1 | 0 | – | 0.2 | C30 | BFRP |
| ERBC1 | 0 | Moderately damaged | 0.2 | C30 | BFRP |
| ERBC2 | 9 | Moderately damaged | 0.2 | C30 | BFRP |
| ERBC3 | 9 | Severely damaged | 0.2 | C30 | BFRP |
| ERBC4 | 9 | Destructively damaged | 0.2 | C30 | BFRP |

*Note n* is the axial load ratio of the column

prefix of "R" was an uncorroded joint retrofitted with BFRP sheets. Seismic-damaged specimens SRBC1, SRBC-2, and SRBC3 with corrosion rates of 0, 3%, and 9%, which had a prefix of "S," were retrofitted with BFRP sheets after the pre-damage. In this group, specimens BC0, RBC1, ERBC1, and ERBC2 were also discussed, and ERBC1-ERBC2 were designated as SRBC1 and SRBC3, respectively.

(1) Specimens Design and Materials

Ordinary concrete utilized in the experimental test was offered by the commercial concrete, the components proportions of which are shown in Table 5.1. The dimension and reinforcement form of specimens BC0, RBC1, ERBC1, ERBC-2, and RBC3 are same as the descriptions shown in Fig. 5.1. The average values of yielding strength of ribbed and plain bars were 335 and 235 MPa, respectively. On the basis of Chinese Standard GB/T 50081-2019 [17], three specimens with the dimensions of 150 mm × 150 mm × 150 mm or 150 mm × 150 mm × 300 mm were tested to calculate the cubic compressive strength $f_{cu}$, or elastic modulus $E_c$ of cast-in-place concrete, respectively. The average values of $f_{cu}$ and $E_c$ were 39.3 and 3.24 × $10^4$ MPa, respectively.

(2) Accelerating Corrosion Test

The setup of accelerated corrosion process of specimens with the corrosion rates of 3%, 9%, and 15% was as same as that shown in Fig. 5.2. The system was composed of a direct current supply, wires, stainless-steel pipes, and 5.0% NaCl solution. Before turning on the power, the specimens have to be immersed in a sink with the NaCl solution for at least 7 days [5, 13]. The stainless-steel pipes deposited in the NaCl solution and "U" type embedded anchors connecting to reinforcing bars in the specimen were set as the cathode and anode, respectively. These reinforcing bars included beam, column longitudinal reinforcements, and the transversal reinforcements were bound by the iron wire, and integrally corroded under the current. Similar arrangements of accelerating corrosion had also been applied on the corroded RC beams [20, 48, 49], columns [50–54], walls [5, 55], and beam-column joints [12, 56–59]. An electric current flowed out from the positive pole of the direct current supply over concrete permeated by NaCl solution, then to the stainless-steel pipes, finally returned to the negative pole of the direct current supply. During the test, the reinforcing bars were oxidized due to the loss of electronics, contributing to the corrosion [60, 61].

Faraday's law is commonly applied for determining the theoretical loss of reinforcing bar mass by Eq. (5.1) [61, 62].

Although the thickness distribution of corrosion products on the surfaces of reinforcing bars was non-uniform along the section direction, the corrosion along the length of reinforcing bars could be considered as uniform within a certain corrosion rate, as reported in [63–67]. Therefore, the average corrosion rates could be referred as the corrosion levels of corroded reinforcing bars. Besides, evaluating the mass loss of all the sections of reinforcing bars could also be utilized by the original mass of reinforcing bars. Several methods reported in [20, 21, 68, 69] were proposed to evaluate the mass loss of corroded reinforcing bars, which indicated that the mass of uncorroded reinforcing bars could be calculated by the product of average weight per unit length and the actual length of the reinforcing bars or the volume and density of reinforcing bar.

A modified method for calculating the mass loss based on the product of the corrosion rate and original mass of uncorroded reinforcing bars, which were calculated by the product of volume and density of reinforcing bar, was proposed by Eq. (5.11). That method could also be utilized for evaluating the mass loss of corroded reinforcing bars in beam-column joints and shear walls, as reported in [5, 12].

$$\Delta m = \rho m_0 = \frac{\pi}{4}\sum d_i^2 L_i \rho_d \rho \tag{5.11}$$

where $m_0$ is the original mass of uncorroded reinforcing bars, in g; $d_i$ is the diameter of reinforcing bars, in mm; $L_i$ is the cumulative connecting length of $d_i$- diameter bars in concrete, in mm, which was taken into account of the contact length of reinforcing bars to the concrete under the corrosion process. Similar calculating method of the length of corroded reinforcing bars can also be found in [70–72], which conducted tests on the bond behavior between corroded reinforcing bars and concrete, and $\rho_d$ is the density of reinforcing bar, in g/mm$^3$.

The duration of corrosion $t$ related to the target corrosion rate and current intensity is derived by Eqs. (5.1) and (5.11) as follows:

$$t = \frac{m_0 \rho Z F}{M I} = \frac{\pi \sum d_i^2 L_i \rho \rho_d Z F}{4 M I} \tag{5.12}$$

where the corrosion rate of reinforcing bars could be evaluated by the duration $t$ and current intensity $I$. For specimens SRBC2 and SRBC3 with the target corrosion rates of 3% and 9%, the duration time of which were 38.9 and 58.3 d when the reinforcing bars were impressed with an adjustable current intensity of 1.5 and 3 A, respectively.

After the cyclic loading test, the corrosion rates of corroded reinforcing bars were calculated by the ratio of the weight of reinforcing bars before and after testing. Firstly, corroded reinforcing bars were removed from the specimens with a certain length and the rust on the surfaces of reinforcing bars was removed or cleaned out by sandpaper or cotton balls with alcohol, respectively. Secondly, the corrosion products of reinforcing bars are washed out by the 12% hydrochloric acid solution and water [57]. Thirdly, the weight of reinforcing bars was recorded after the reinforcing bars were dried out by the drying box, as depicted in Fig. 5.38a, b. And diameters of the 6-mm closed plain bars, as well as the 14-mm and 16-mm ribbed bars of all tested specimens were also checked out and determined by testing the reinforcing bars within the deviations of ± 0.3, ± 0.4, and ± 0.4 mm, respectively, according to the Chinese Standard GB/T 1499.1–2017 [73] and GB/T 1499.2-2018 [74]. Besides, the corrosion rates of reinforcing bars were also checked out by a method published on patent [75], as depicted in Fig. 5.38c.

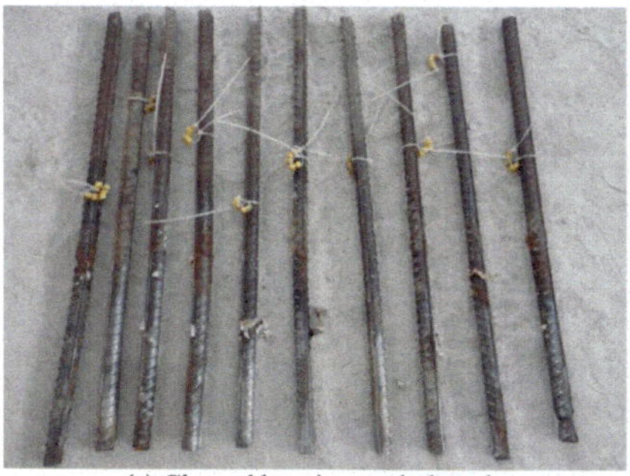
(a) Cleaned bars in certain length

(b) Drying box

(c) Checking out the corrosion rates by the method published on patent

**Fig. 5.38** Test on the actual corrosion rates of corroded reinforcing bars after the cyclic loading test

The average corrosion rates of corroded reinforcing bars with the diameters of 6, 14, and 16 mm were 5.38%, 2.74%, and 3.54% or 12.38%, 8.95%, and 8.54% for specimens SRBC2 or SRBC3, respectively.

(3) Pre-damaged Schemes

For simulating the moderately, severely, and destructively damaged levels of earthquake, the pre-damages of specimens were determined by interlayer displacement angles of 1/90, 1/90, 1/45, and 1/28 after the accelerating corrosion test, which were about 18, 18, 36, and 58 mm, respectively. Detailed loading history is depicted in Fig. 5.4. Tested specimens were dominated by the force before yielding at the first stage and the displacement after yielding at the second stage, respectively, until the pre-damage displacement was obtained as reported in [35].

Tested specimens were governed by the force before yielding at the first stage and the displacement after yielding at the second stage, respectively, until the target pre-damage displacements were obtained as proposed in [35]. For quantifying the damage extent of tested specimens after the pre-damages, stiffness damage index $(DI)_K$ was calculated by Eq. (5.13), as proposed in [76]:

$$(DI)_K = 1 - (K_{\text{final}}/K_{\text{initial}}) \tag{5.13}$$

where $K_{\text{initial}}$ is the initial slope of the base shear-top defection relationship resulting from the pushover analysis of the frame before subjecting it to the earthquake ground motion; and $K_{\text{final}}$ is the initial slope of the same relationship but after subjecting the frame to the earthquake time history.

$K_{\text{initial}}$ and $K_{\text{final}}$ were taken as the initial stiffness of benchmark specimen BC0 and the finial stiffness of specimens ERBC1, ERBC-2, ERBC-3, and ERBC4 after the pre-damages, respectively. The stiffness damage index $(DI)_K$ of specimens ERBC1-ERBC4 was 61.2%, 63.2%, 77.6%, and 87.2%, respectively.

(4) Retrofitting Programs

Considering several methods that have been reported in previous studies, [11, 23, 24] detailed retrofitting schemes are proposed for improving the seismic performance of corroded or damaged specimens to experience the ductile performance [10, 77], as described in Fig. 5.27. By the proposed retrofitting scheme, the seismic performance such as the load-bearing capacity, stiffness, ductility, and cumulative energy dissipation of corroded and earthquake-damaged joints was hoped to be recovered to that of benchmark specimen without reinforcement corrosion and earthquake damage.

Wrapping RC members with FRP sheets have been confirmed to be a sufficient arrangement [6, 10, 77]. BFRP sheets were applied for retrofitting the damaged specimens. Nonetheless, some preparations needed to be accomplished before attaching the BFRP sheets. Firstly, the shattered concrete and dust on the surface of the beam-column joints should be cleared away. Secondly, cracked concretes are retrofitted with the epoxy cement mortar [78], and extra curing time for at least one day is needed after these specimens being retrofitted [79]. When the epoxy cement mortar was calcified, the surfaces of specimens were sanded and a 20 mm radius was polished around the corners of the column and beam. Thirdly, gluing the BFRP sheets by assorted and periodically checked coatings was brushed on the surface of FRP sheet. Lastly, other attentions were paid to guarantee the efficiency of these coatings, and the epoxy resins were needed to be cared for several days.

Results reported in [42] indicate that the additional shear capacity provided by the FRP sheets $V_j$ can be calculated by Eq. (5.14) as follows:

$$V_j = \psi_f n_f f_f t_f h_f \tag{5.14}$$

where $\psi_f$ is the efficiency coefficient of FRP and calculated as 0.85 $(1.2 - 0.1n)$. $n$ is the layers of FRP sheets on the single side of concrete and taken as 1, when the layers of FRP sheets are larger than or equal to 6, the value of $n$ is taken as 6; $n_f$ is the layers of FRP sheets; $f_f$ is the tensile strength of FRP, in MPa; $t_f$ is the thickness of one layer of FRP sheet, in mm; and $h_f$ is the width of FRP sheet along the direction of column section height in joint core and taken as 300 of the beam height, in mm.

According to Eq. (5.14), the additional shear capacities provided by one-three layers of FRP sheets were 156.1, 141.9, and 127.7 kN, respectively. For comparing the seismic performance of earthquake-damaged and corroded joints to that of corroded joints or corroded joints retrofitted with BFRP sheets as the same corrosion rate and verifying the effectiveness of BFRP sheets, the retrofitting scheme was chosen and determined for comparing the seismic performance of earthquake-damaged and corroded joints retrofitted with one layer of BFRP sheets to that of corroded joints retrofitted with the same method as reported in [12]. BFRP sheet with the thickness of 0.121 mm was utilized for retrofitting the beam-column joints. The ultimate average strain, the tensile elastic modulus, and ultimate stress of BFRP sheets obtained by the tests on the basis of Chinese Standard GB/T 3354 [80] were 2.2%, 105 GPa, and 2300 MPa, respectively. And the tested results of elastic modulus, ultimate strain, and ultimate tensile strength of epoxy resin were 2600 MPa, 3.5%, and 45.8 MPa, respectively.

(5) Test Device and Loading History

The layout of test setup included the loading system and instrumentation was same as that shown in Fig. 5.3. And the loading history and detailed arrangement and preparations on the strain gauges of tested specimens were also same as that shown in Figs. 5.4 and 5.5.

### 5.3.2 *Crack Pattern and Failure Modes*

The final failure of benchmark specimen BC0, which was directly tested to failure without any retrofitting, is described in Fig. 5.39a. When the specimen was controlled by the force, micro-flexural cracks were noticed at the lateral load of 40 kN. These cracks began to widen with the test proceeded, and cross-inclined cracks could be noticed on the core area at the lateral load of 60 kN. When the specimen was controlled by the displacement, the previous cracks progressed and widened with a maximum width of 0.75 mm at the displacement of 20 mm (1.18% drift). Simultaneously, penetrated cracks occurred on the beams. The number of cracks on the surface of tested beam-column joints increased with the increasing displacement. The crushing and spalling of the concrete in the core area within a specific region were observed, while the maximum width of cracks was 2.0 mm at the displacement of 52 mm (3.06% drift). A considerable spalling of concrete in the core area and exposing of reinforcing bars were observed, whereas the load-bearing capacity of the specimen BC0 started to decline at the displacement of 68 mm (4.00% drift). The test was ceased when the load-bearing capacity of the specimen BC0 decreased by 15% of the peak load.

The final failure of specimen RBC1, which was retrofitted with BFRP sheets and directly tested to failure without any damage, is described in Fig. 5.39b. A slightly cracking sound was heard at the lateral load of 40 kN, which possibly attribute to the fracture of hardened resins. These sounds became loud and constant with the test proceeded, while no obvious cracks on the surface of concrete were noticed at the force-control stage. When the tested joint was controlled by displacement, BFRP sheets at the corners of specimen RBC1 fractured and bulged at the displacement of 44 (2.59% drift) and 60 mm (3.53% drift), respectively. It could be due to the crushing of concrete at the core area, the entirely bulging and delaminating of BFRP sheets from the surface occurred at the displacement of 76 mm (4.47% drift), as well as the load-bearing capacity decreased. The test was finished at the displacement of 100 mm (5.88% drift). No obvious fractures of the column or beams of the specimen RBC1 were noticed during the test, illustrating that better integration and seismic performance of the beam-column joint would be accomplished by retrofitting with BFRP sheets, as reported in [6].

Final failure modes of specimens ERBC1-ERBC4, which were retrofitted with BFRP sheets and tested to failure after being pre-damaged, are described in Fig. 5.39c–f. Similar failure modes of these retrofitted specimens could be found, experiencing the process of the BFRP sheets delaminating to local stirrups bucking and the core concrete crushing. When the specimens were controlled by the force, continuously and longer cracking sounds were heard that possibly attributed to the cracking of resins. These cracks went on progressing expeditiously with the sounds hearing constantly and vehemently as the load increased. When the tested joints were controlled by the displacement, BFRP sheets of specimens ERBC1-ERBC4 began to fracture at the corners of core area at the displacement of 44 (2.59% drift), 36 (2.12% drift), 24 (1.41% drift), and 24 mm (1.41% drift), respectively. An obvious bulging of BFRP sheets was soon observed attributed to the crushing of concrete at the core area at the displacement of 76 (4.47% drift), 68 (4.00% drift), 66 (3.88% drift), and 64 mm (3.76% drift), respectively. Afterward, the load-bearing capacities of specimens ERBC1-ERBC4 decreased obviously with the delaminating of BFRP sheets from the surfaces. Tests were ceased when the load-bearing capacities of the

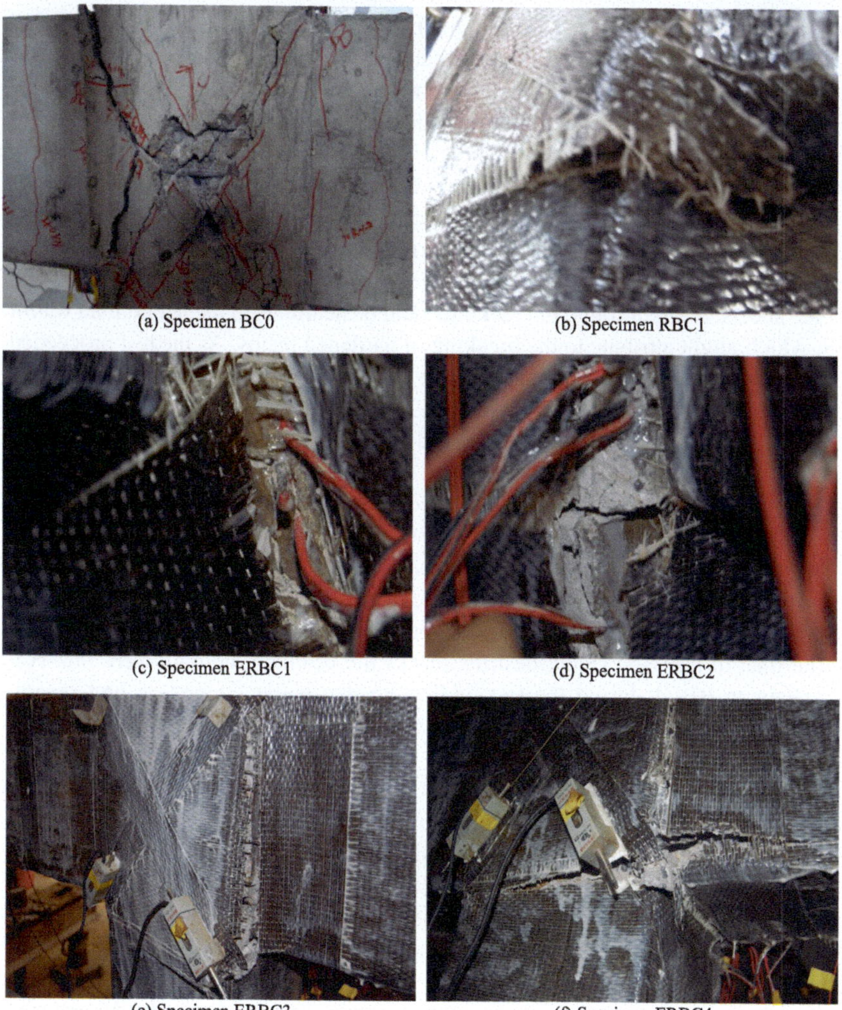

(a) Specimen BC0
(b) Specimen RBC1
(c) Specimen ERBC1
(d) Specimen ERBC2
(e) Specimen ERBC3
(f) Specimen ERBC4

**Fig. 5.39** Final failure modes of all test specimens

specimens ERBC1-ERBC4 decreased by 15% at the displacement of 100 (5.88% drift), 112 (6.59% drift), 98 (5.76% drift), and 96 mm (5.65% drift), respectively.

Despite the fact that considerable bulging of the BFRP sheets and spalling of the crushed concrete were observed in the test, the load-bearing capacities and displacements of damaged corroded joints significantly increased. In the push and pull stage, the concrete at the core area of the joint suffers alternate stress, contributing to cracking and spalling, and the prematurely slipping of corroded stirrups in column at the early stage [12]. That contributed to the premature buckling of corroded stirrups in the core area. After the corroded specimens being retrofitted with BFRP sheets, the concrete in core area was confined by the BFRP that contributed to the spalling and cracking of the concrete, and the bond behavior between the concrete and stirrups improved. The buckling of longitudinal reinforcements was not observed, but can be observed in other researches on the corroded joints. It could be due to the effective confinement of stirrups and BFRP sheets. The BFRP sheets supported in confining the core area were demonstrated to be an effective method to delay the concrete spalling and crushing, as well as the stirrups buckling, which was significant for retrofitting the ductility and seismic performances of specimens ERBC1-ERBC4.

Specimen RBC1 was not pre-damaged but tested to failure after retrofitting with BFRP sheets. A slight sound of cracking could be heard when the lateral load was 40 kN, which might attribute to the breaking of hardened resins. As the test proceeded, the sounds were loud and continuous. However, no evident cracks of the concrete at the surface were observed in the first stage. In the displacement-control stage, BFRP sheets fractured at the corners and bulged at the core area when the displacement was 44 (2.59% drift) and 60 mm (3.53% drift), respectively. Due to the crushing of concrete at the core area, the BFRP sheets were entirely bulged and delaminated from the surface when the displacement was 76 mm (4.47% drift), as well as the load-bearing capacity decreased. Test was terminated when the displacement was 100 mm (5.88% drift), and shearing failure mode could be found in Fig. 5.40d. The X-typed strip and sheets of BFRP in core area fractured with the bulging of concrete in the final mode. No evident destructions of the beams or column were observed in specimen RBC1 during the test, indicating that better integration and seismic behavior of the retrofitted joint could be obtained by applying with BFRP sheets, as reported in [6].

Specimens SRBC1, SRBC2, and SRBC3 were the seismic-damaged joints retrofitted by BFRP sheets with corrosion rates of 0, 3%, and 9%, respectively, and retested to failure. These retrofitted specimens exhibited the generally similar shearing failure modes, which experienced the development of failure from BFRP sheets delaminating to local buckling of stirrups and crushing of the core concrete. In the load-control stage, cracking sounds were heard at the lateral load of 60, 40, and 40 kN momently, longer, and continuously for specimens SRBC1-SRBC3, respectively. These sounds were attributed to the cracking resins. As the lateral load increased, these cracks propagated rapidly as the sounds became continuous and loud. In the displacement-control stage, BFRP sheets started to firstly fracture at the corners of the joint with a limited spalling of concrete when the displacement was 44 (2.59% drift), 36 (2.12% drift), and 36 mm (2.12% drift) for specimens SRBC1-SRBC3, respectively. The bulging of BFRP sheets could be soon observed due to the crushing of the concrete in the core area when the displacement was 76 (4.24% drift), 68 (4.00% drift), and 68 mm (4.00% drift), respectively. Subsequently, the load-bearing capacity of these specimens deteriorated evidently. Tests were terminated when the load-bearing capacity dropped by 15% when the displacement was 100 (5.88% drift), 112 (6.59% drift), and 112 mm (6.59% drift), respectively, and the final failure was shown in Fig. 5.40c–e. Although serious bulging of the FRP sheets and spalling of the crushed concrete in core area were found in the test, the load-bearing capacity and

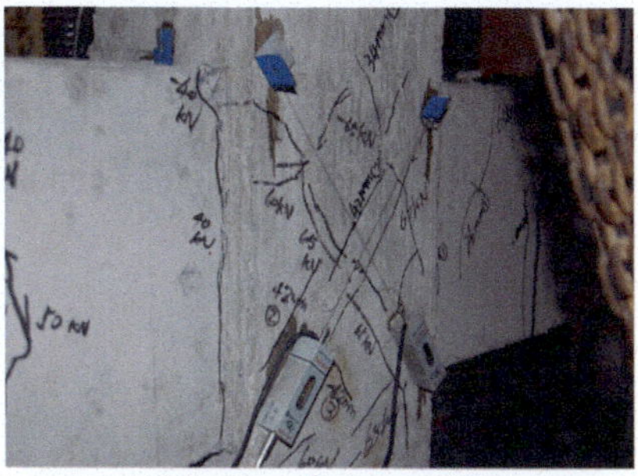

(a) Core area of specimen BC0

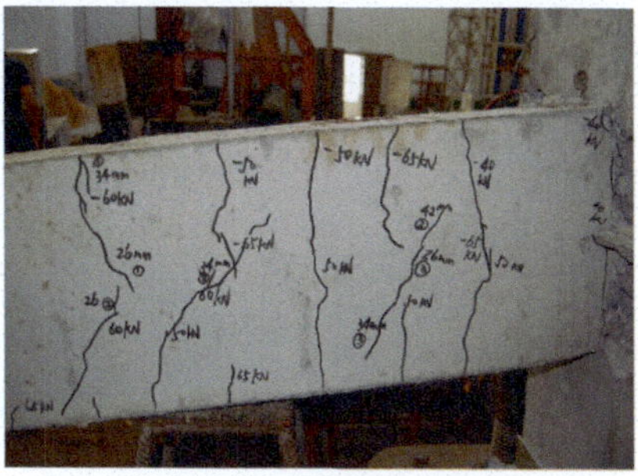

(b) Right beam of specimen BC0

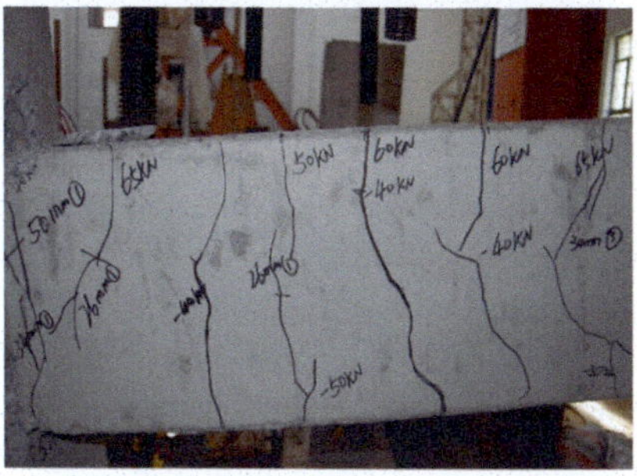

(c) Left beam of specimen BC0

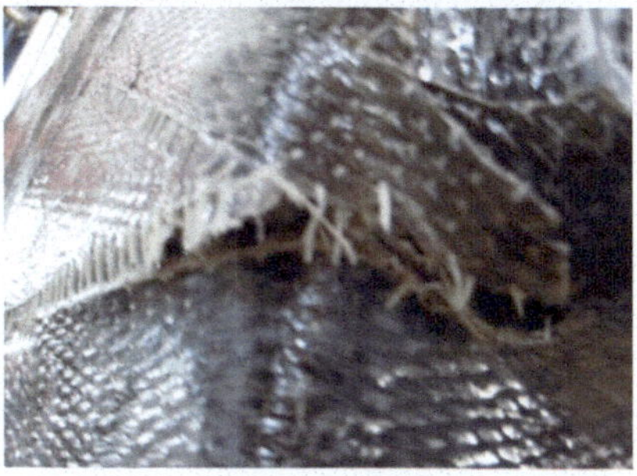

(d) Specimen RBC1

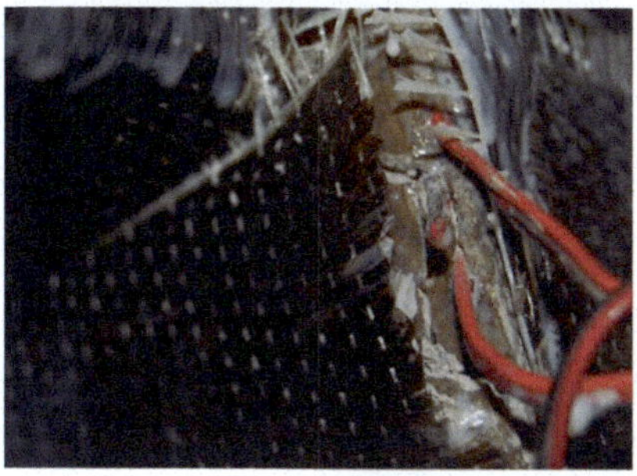

(e) Specimen SRBC1

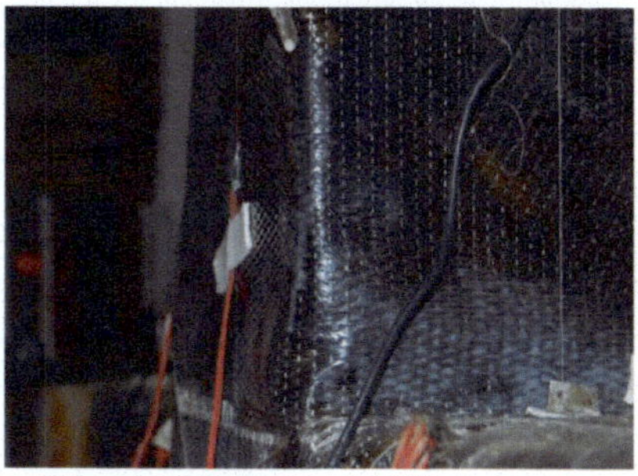

(f) Specimen SRBC2

(g) Specimen SRBC3

**Fig. 5.40** Final failure modes of all tested specimens

displacement of damaged corroded joints were significantly improved. The confinement of core area supported by the BFRP sheets was demonstrated to be an effectual method to delay the concrete spalling and crushing as well as the stirrups buckling, which was significant for retrofitting the ductility and seismic behavior of specimens SRBC1-SRBC3.

### 5.3.3 *Hysteretic Response and Skeleton Curves*

Figure 5.41 describes the lateral load–displacement of tested specimens. Peak load and displacement of specimens can be obtained from the hysteretic loops [6], the encircled regions of which were also utilized to evaluate the energy dissipation capacity of specimens under the cyclic loadings, as reported in [29, 81]. The peak load of specimen RBC1 was considerably larger than that of benchmark specimen BC0, while the displacement also increased, as described in Fig. 5.41a and b.

The superiority of higher ductility for specimen RBC1 was limited in a sense. Besides, the hysteretic loops of specimens ERBC1, ERBC2, ERBC3, and ERBC4 exhibited the obvious pinching phenomena in Fig. 5.41c–f, which indicated that the seismic performance of beam-column joints was considerably affected by the earthquake damage and the affecting degrees increased as the increasing earthquake damages. Especially the hysteretic loop of specimen ERBC4 under the destructive damage was smaller than those of other specimens. After retrofitting with BFRP sheets, the maximum load-bearing capacities or displacements of specimens ERBC1, ERBC2, ERBC3, and ERBC4 increased or were recovered to those of benchmark specimen BC0. Thus, the hysteretic responses of earthquake-damaged and corroded beam-column joints could be improved after retrofitting with BFRP sheets.

The skeleton curves utilized for evaluating the load-bearing capacity and ultimate displacement of all tested specimens are described in Fig. 5.42 and Table 5.13. As described in Fig. 5.42a, the skeleton curve of specimen BC0 was encircled by that of specimen RBC1, the maximum load-bearing capacity of which increased by 7.6% comparing to that of specimen BC0, demonstrating that the displacement and load-bearing capacity of beam-column joints considerably increased by retrofitting with BFRP sheets. The load-bearing capacities of specimens ERBC1-SRBC4 were lower than that of specimen BC0 when the lateral displacement was less than 30 mm at the early stage, which might attribute to the cracking of concrete and yielding of reinforcing bars under the earthquake damage. However, the skeleton curve of benchmark specimen BC0 was enveloped by those of specimens ERBC1-ERBC4 at the later stage due to the confining BFRP sheets on the surfaces of specimens. The maximum load-bearing capacities or ultimate displacements of specimens ERBC1-ERBC4 increased by 1.5%, – 4.8%, – 7.5%, and – 8.6% or 14.9%, 8.9%, 12.0%, and 19.4% contrasting with that of specimen BC0, respectively (Table 5.13).

The load-bearing capacities of earthquake-damaged specimens ERBC2 and ERBC3 with certain corrosion rate of 9% would not be recovered to that of specimen BC0 after retrofitting with BFRP sheets, while the ultimate displacements were

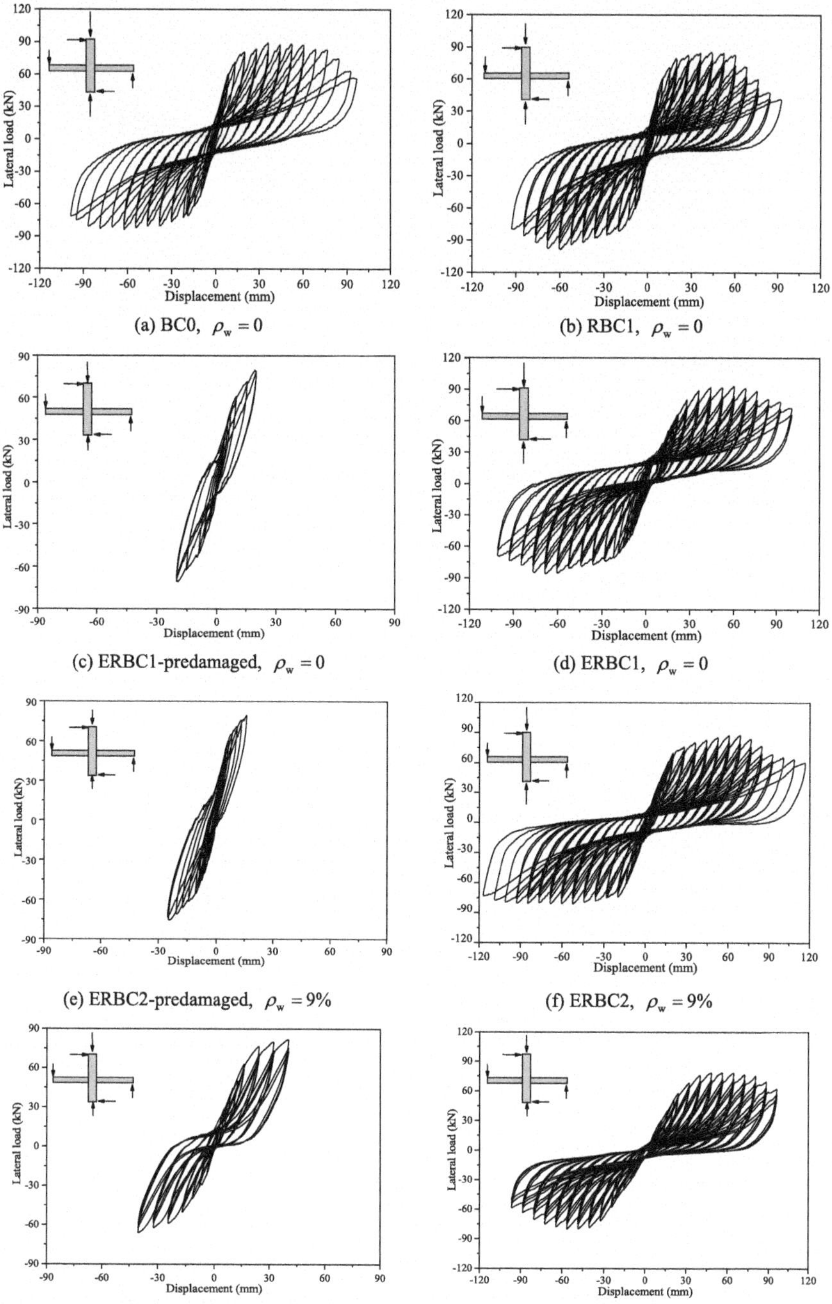


(a) BC0, $\rho_w = 0$ (b) RBC1, $\rho_w = 0$

(c) ERBC1-predamaged, $\rho_w = 0$ (d) ERBC1, $\rho_w = 0$

(e) ERBC2-predamaged, $\rho_w = 9\%$ (f) ERBC2, $\rho_w = 9\%$

(g) ERBC3-predamaged, $\rho_w = 9\%$ (h) ERBC3, $\rho_w = 9\%$

**Fig. 5.41** Hysteretic curves of specimens

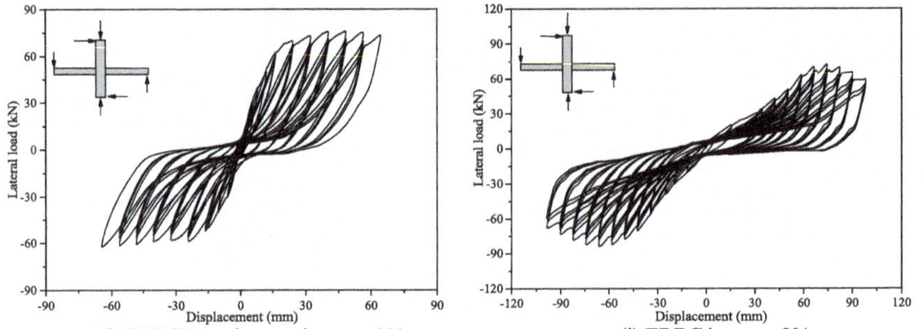


(i) ERBC4-predamaged, $\rho_w = 9\%$ (j) ERBC4, $\rho_w = 9\%$

**Fig. 5.41** (continued)

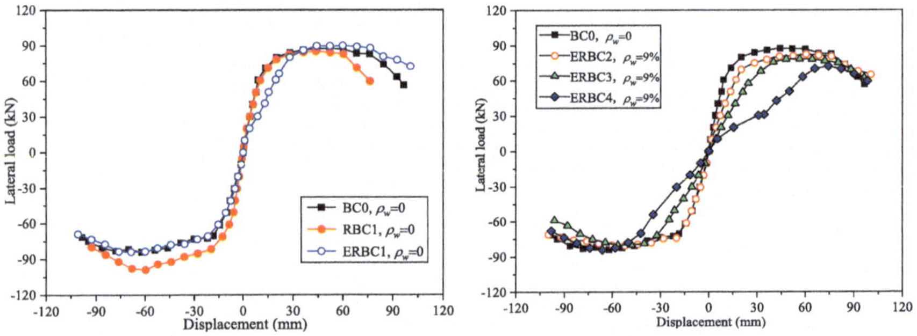


**Fig. 5.42** Skeleton curves of hysteretic responses for all test specimens

**Table 5.13** Test results of corroded specimens with the corrosion rate of 9% in different damages

| Specimen | $P_m$(kN) | $\Delta_y$(mm) | $\Delta_u$(mm) | $\mu$ | $\theta_u$ |
|---|---|---|---|---|---|
| BC0 | 85.6 | 14.8 | 82.3 | 5.6 | 1/21 |
| RBC1 | 92.1 | 15.7 | 86.8 | 5.5 | 1/20 |
| ERBC1 | 86.9 | 15.3 | 94.6 | 6.2 | 1/18 |
| ERBC2 | 81.5 | 14.0 | 89.6 | 6.4 | 1/19 |
| ERBC3 | 79.2 | 24.5 | 92.2 | 3.8 | 1/21 |
| ERBC4 | 78.2 | 45.6 | 98.3 | 2.2 | 1/17 |

*Note* $P_m$ is max load-bearing capacity; $\Delta_y$ is yielding displacement; $\Delta_u$ is ultimate displacement; $\mu$ is the displacement ductility ratio; and $\theta_u$ is the drift

effectually retrofitted to that of specimen BC0. That might attribute to the passive restraint effect provided by the BFRP sheets. Furthermore, the increasing degrees of ultimate displacements of specimens ERBC2, ERBC3, and ERBC4 increased as the degrees of earthquake damage increased, indicating that the retrofitting effectiveness was considerably affected by the confinement of BFRP sheets.

**Table 5.14** Results of corroded specimens with the corrosion rate of 0, 3%, and 9% in the same damages

| Specimen | $P_m$(kN) | $\Delta_y$(mm) | $\Delta_u$(mm) | $\mu$ | $\theta_u$ |
|---|---|---|---|---|---|
| BC0 | 85.6 | 14.8 | 82.3 | 5.56 | 1/21 |
| RBC1 | 92.1 | 15.7 | 86.8 | 5.53 | 1/20 |
| SRBC1 | 86.9 | 15.3 | 94.6 | 6.18 | 1/18 |
| SRBC2 | 85.0 | 14.2 | 84.3 | 5.94 | 1/20 |
| SRBC3 | 81.5 | 14.0 | 89.6 | 6.40 | 1/19 |

*Note* $P_m$ is max load-bearing capacity; $\Delta_y$ is yielding displacement; $\Delta_u$ is ultimate displacement; $\mu$ is the displacement ductility ratio; and $\theta_u$ is the drift

The load–displacement responses of all tested specimens are depicted in Fig. 5.43. Maximum load-bearing capacity, yielding, and ultimate displacement can be obtained from the hysteretic loops [79], the encircled regions of which were also applied for calculating the energy dissipation capacity of specimens under the cyclic loadings, as reported in [81]. A symmetrically seismic performance of reference specimen BC0 could be found in Fig. 5.43a, which might attribute to the cracking pattern and the different development of cracks on the beams or columns as depicted in Fig. 5.40a–c. That was a common phenomenon, and similar performances of tested beam-column joints can also be found in [40, 41, 78, 82]. The materials and performance of tested specimens were uniform and symmetrical before cracking, however, those of specimens were non-uniform and asymmetric after the concrete cracked. The load-bearing capacity and seismic performance of advanced cracked beams decreased and were lower than those of succeeding cracked beams, contributing to the unsymmetrical seismic performance of beam-column joint. Microcracks could be observed at the left beam when the lateral load was 40 kN in the push stage, as depicted in Fig. 5.40b. Initial damage was firstly found in left beam, contributing to the more cracks and weaker hysteretic responses of left beam than those of right beam in later loading stage, as depict in Fig. 5.40a and c. Therefore, the seismic performance of specimen BC0 was unsymmetrical.

Besides, the maximum load-bearing capacity of specimen RBC1 was larger than that of reference specimen BC0, while the displacement was limitedly increased, as depicted in Fig. 5.43a and b. The asymmetric behaviors of RBC1 might attribute that: (1) different cracking patterns of the beams; (2) the asymmetry of loading device; (3) the distinct rehabilitation of BFRP applied on the surface of specimen. For specimen RBC1, the advantages of a higher ductility for RC members provided by BFRP sheets were limited to some extent. Furthermore, obvious pinching phenomena in the hysteretic loops of specimens SRBC1, SRBC2, and SRBC3 could be observed in Fig. 5.43c–e and were not fuller than that of specimen BC0. However, after retrofitting with BFRP sheets, the maximum load-bearing capacity and displacement of specimens SRBC1-SRBC3 were recovered to those of specimen BC0. Therefore, BFRP sheet could be applied for improving and retrofitting the hysteretic responses of the seismic-damaged corroded joints in the cyclic loading.

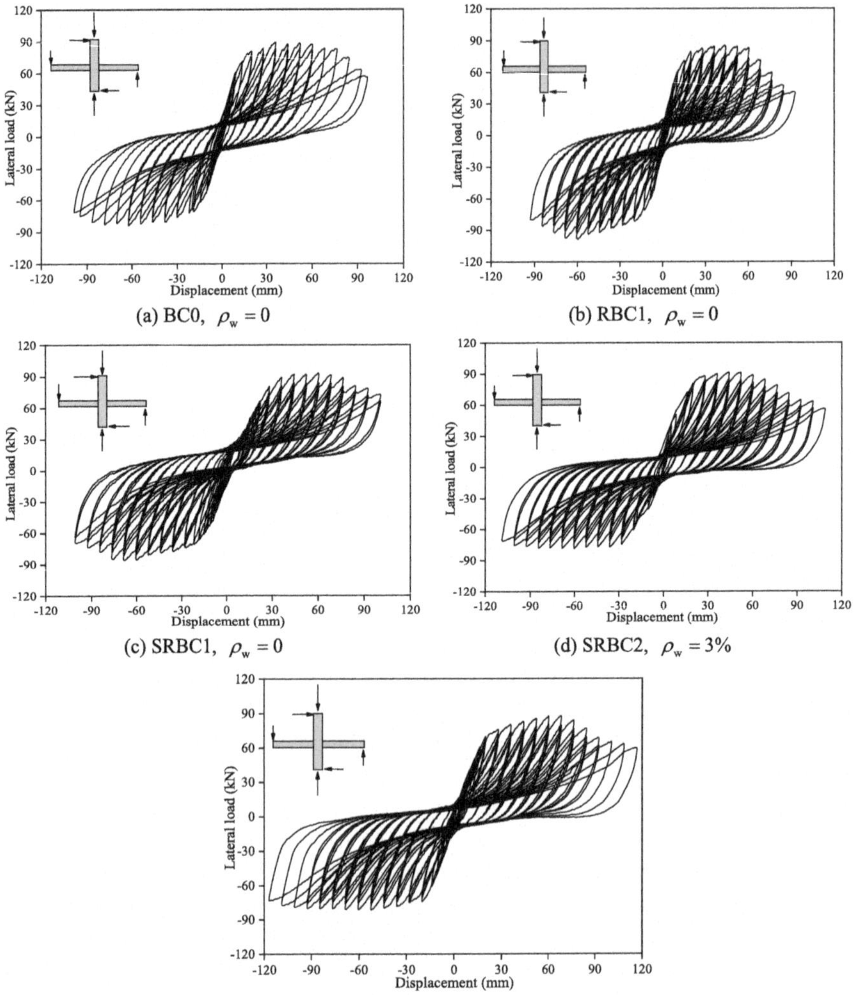


(a) BC0, $\rho_w = 0$

(b) RBC1, $\rho_w = 0$

(c) SRBC1, $\rho_w = 0$

(d) SRBC2, $\rho_w = 3\%$

(f) SRBC3, $\rho_w = 9\%$

**Fig. 5.43** Hysteretic curves of specimens

For evaluating the load-bearing capacity and ultimate displacement, the envelope curve and test results of all tested specimens are depicted in Fig. 5.44 and Table 5.15. The skeleton curves of specimens BC0 and RBC1 were not symmetric, while the skeleton curves of specimens SRBC1, SRBC2, and SRBC3 exhibited similar and relatively symmetric behaviors. Besides, the skeleton curve of specimen BC0 was enveloped by that of specimen RBC1 only in the pull stage, the maximum load-bearing capacity of which increased by 7.6% comparing to that of specimen BC0. Although the load-bearing capacities of specimens SRBC1, SRBC2, and SRBC3 were slightly lower than that of specimen BC0 when the lateral displacement was

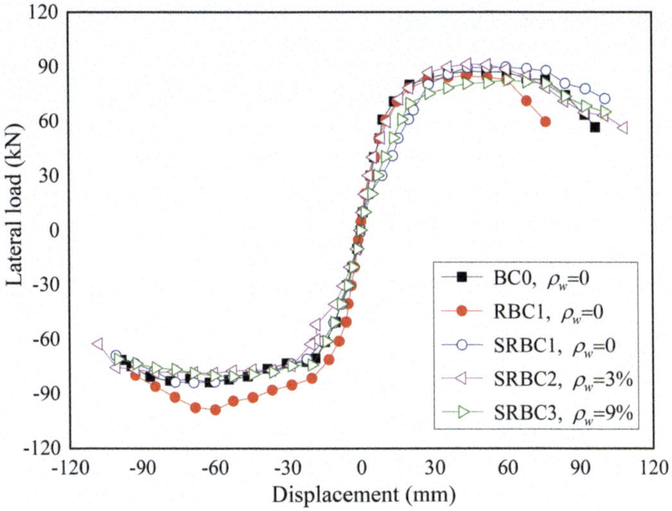


**Fig. 5.44** Envelope curves of hysteretic responses for all tested specimens

less than 30 mm at the early stage, the envelope curve of which could envelop that of specimen BC0 at the middle and later stages. The maximum load-bearing capacities of specimens SRBC1, SRBC2, and SRBC3 with corrosion rates of 0, 3%, and 9% increased by 1.5% or decreased by 0.7% and 4.8% contrasting with that of specimen BC0, respectively, while the ultimate displacements of those enhanced by 14.9%, 2.4%, and 8.9%.

The load-bearing capacity of seismic-damaged corroded specimens SRBC2 and SRBC3 with corrosion rates of 3% and 9% could not be recovered to that of reference specimen BC0 after retrofitting with BFRP sheets, while the ultimate displacements and energy dissipation were effectively retrofitted to those of specimen BC0 that might attribute to the passive restraint effect provided by the BFRP sheets. Besides, the improving degrees of ultimate displacements of those specimens were lower than that of specimen SRBC1, which indicated that the retrofitting effectiveness was also influenced by the reinforcement corrosion.

**Table 5.15** Experimental and theoretical results of shear strength of the reference specimen BC0

| Experimental results (kN) | Theoretical results (kN) | | | | Ratio of theoretical to experimental results | | | |
|---|---|---|---|---|---|---|---|---|
| $V_{et,0}$ | $V_j$[89] | $V_{jhd}$[90] | $V_n$[91] | $V_m$[92] | $V_j/V_{et,0}$ | $V_{jhd}/V_{et,0}$ | $V_n/V_{et,0}$ | $V_m/V_{et,0}$ |
| 223.27 | 233.69 | 344.70 | 415.91 | 361.90 | 1.047 | 1.544 | 1.863 | 1.621 |

### 5.3.4 Displacement Ductility

Ductility is considered as an important parameter to determine the deformation capacity when the RC member suffers the plastic deformation without apparent deterioration of load-bearing capacity under the earthquake [83, 84]. The displacement ductility ratio $\mu$ and the drift $\theta_u$ were generally applied for evaluating the ductility of specimens, and calculated by the Eqs. (5.15) and (5.16).

$$\mu = \frac{\Delta_u}{\Delta_y} \tag{5.15}$$

$$\theta_u = \frac{\Delta_u}{h_w} \tag{5.16}$$

where $h_w$ and $\Delta_u$ are the quantified height of column and ultimate displacement, respectively, in mm. A few methods reported in researches [35–37] have been proposed for determining the yielding point of RC members, among which the "energy equivalent method" is generally used [5, 35]. Based on the previous method, the yielding displacement $\Delta_y$ was defined, in mm, and $\Delta_u$ was the ultimate displacement of specimen with the 85% of the maximum load-bearing capacity $P_{\max}$. Results of all tested specimens are described in Table 5.14.

The yielding displacements of specimens RBC1, ERBC1, ERBC2, ERBC3, and ERBC4 increased by 6.1%, 3.4%, – 5.4%, 65.5%, and 208.1%, respectively, comparing to that of benchmark specimen BC0. The yielding displacements of corroded specimens ERBC3 and ERBC4 under the severe and destructive damages increased significantly after retrofitting with BFRP sheets, which were higher than those of uncorroded specimens RBC1 and ERBC1. However, the yielding displacement of corroded specimen ERBC1 under the moderate damage would not be recovered to that of benchmark specimen BC0 as supposed.

The displacement ductility ratios or drifts of specimens RBC1, ERBC1, ERBC2, ERBC3, and ERBC4 increased by – 1.8%, 10.7%, 14.3%, – 32.1%, and – 60.7% or 5%, 16.7%, 10.5%, 0, and 23.5% comparing to those of benchmark specimen BC0, respectively. The displacement ductility ratios of specimens ERBC1 and ERBC2 could be recovered to that of benchmark specimen BC0, while those of specimens RBC1, ERBC3, and ERBC4 could not. That might attribute to the considerable increase of yielding displacements of specimens ERBC3-ERBC4, contributing to the ultimate rotations of which decreased. Similar conclusions can also be found in research [35], which indicates that the ultimate rotations of specimens decreased with the increasing damaging degrees. The displacement ductility ratios of specimens ERBC3 and ERBC4 were 3.8 and 2.2, respectively, and were both greater than 2, which coincided with the minimum demand of Seismic Design Criteria for the ductile members [85]. Therefore, the test results were reasonable and the ductility of earthquake-damaged and corroded beam-column joints could be utilized for the temporary retrofitting or urgent application under the severe and destructive damages.

Besides, the drifts of specimens ERBC1-ERBC4 under the earthquake damage would be recovered to that of benchmark specimen BC0 after retrofitting with BFRP sheets.

The yielding displacements of specimens RBC1, SRBC1, SRBC2, and SRBC3 increased by 6.1%, 3.4%, – 4.1%, and – 5.4%, respectively, comparing to that of reference specimen BC0. The yield displacement of damaged specimen SRBC1 increased after retrofitting with BFRP sheets comparing to that of specimen BC0, however, was lower than that of undamaged specimen RBC1. The yield displacements of specimens SRBC2 and SRBC3 were lower than that of specimen BC0 and decreased as the corrosion rates increased. Besides, the drift or displacement ductility ratios of specimens RBC1, SRBC1, SRBC2, and SRBC3 increased by 5.0%, 16.7%, 5.0%, and 10.5% or – 0.5%, 11.2%, 6.8%, and 15.1% comparing to those of specimen BC0, respectively. This indicated that utilizing BFRP sheet can rehabilitate the ductility of corroded or uncorroded specimens after the pre-damage to recover to that of reference specimen.

### 5.3.5 Stiffness Degradation

Stiffness is commonly characterized as the ability to endure the deformations of the RC structure or members [86, 87]. Secant stiffness, considering as the slope of the straight line connecting the peak points at each cycle in the stage of push and pull, is utilized for evaluating the stiffness degradation of specimens [36, 87] and calculated by Eq. (5.9). The secant stiffness of specimens BC0, ERBC2, ERBC3, and ERBC4 are described in Fig. 5.45. Initial stiffness of specimens ERBC2-ERBC4 was 9.7, 3.0, and 2.0 kN/mm, which increased by 18.3%, – 63.4%, and – 75.6% comparing to that of benchmark specimen BC0 with 8.2 kN/mm, respectively.

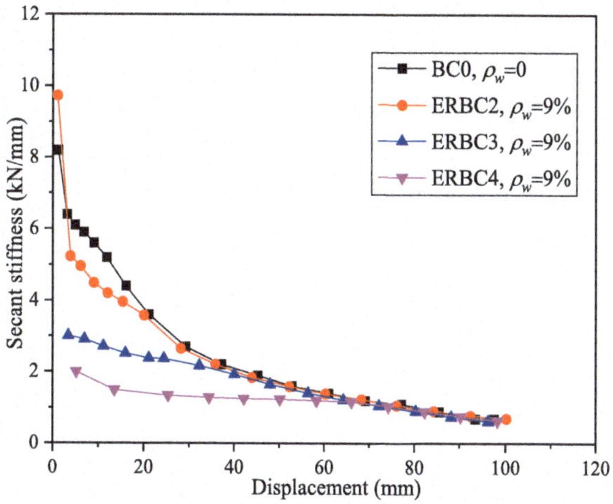


**Fig. 5.45** Secant stiffness degradation of specimens BC0, ERBC2, ERBC3, and ERBC4

Initial stiffness of specimen ERBC2 with the corrosion rate of 9% under the moderate damage could be retrofitted to that of benchmark specimen BC0, while specimens with the same corrosion rate under the severe and destructive damages could not. That might attribute to the residual cracks of concrete and yielding of corroded reinforcing bars, resulting in considerable damages on the beam-column joints and the decrease in the initial stiffness.

Notably, it could be due to the effective confinements on the beam-column joints, the stiffness degradations of specimens ERBC3 and ERBC4 were lower than those of other specimens when the lateral displacement was less than 40 mm. Subsequently, the stiffness of tested specimens gradually exhibited a similar degradation rate and decreased as the increasing displacement. The final stiffness of specimens ERBC2, ERBC3, and ERBC4 was 0.7, 0.6, and 0.6 kN/mm, which decreased by 0, 14.3%, and 14.3% comparing to that of benchmark specimen BC0 with 0.7 kN/mm, respectively. The stiffness degradation of specimens ERBC2-ERBC4 was considerably improved, which might attribute to the confinement of BFRP sheets on the surfaces of beam-column joints. Retrofitting the stiffness and reducing the stiffness degradation of earthquake-damaged corroded beam-column joints with BFRP sheets were reasonable and effective.

The secant stiffness of tested specimens is depicted in Fig. 5.46. Initial stiffness of specimens RBC1, SRBC1, SRBC2, and SRBC3 was 8.9, 9.8, 9.0, and 6.6 kN/mm, which increased by 8.5%, 19.5%, 9.8%, and − 19.5% comparing to that of reference specimen BC0 with 8.2 kN/mm, respectively. Retrofitting specimens with FRP sheets can increase the initial stiffness contrasted with specimen BC0 [36], however, the initial stiffness of specimen SRBC3 with a corrosion rate of 9% was not increased as expected, probably attributed to the severe reinforcement corrosion that resulted in the damage to the initial stiffness. As the test continued, continuous degradation of all tested specimens could be found. Notably, the stiffness degradation of specimen RBC1 exhibited a lower rate than those of other specimens when the lateral displacement was less than 30 mm, indicating that the stiffness was clearly influenced by the reinforcement corrosion at the early stage. Subsequently, the stiffness of all tested specimens gradually exhibited a similar degradation rate and decreased as the increasing displacement. The final stiffness of specimens RBC1, SRBC1, SRBC2, and SRBC3 was 0.7, 0.7, 0.7, and 0.7 kN/mm, which increased by 16.7%, 16.7%, 16.7%, and 16.7% contrasting with that of specimen BC0 with 0.6 kN/mm, respectively. Stiffness degradations of retrofitted specimens RBC1, SRBC1, SRBC2 and SRBC3 were significantly improved, which might attribute to the constrainment of the BFRP sheets on the joints. Retrofitting the stiffness of seismic-damaged corroded specimens with corrosion rates of 0, 3%, and 9% by BFRP sheets was reasonable and effective.

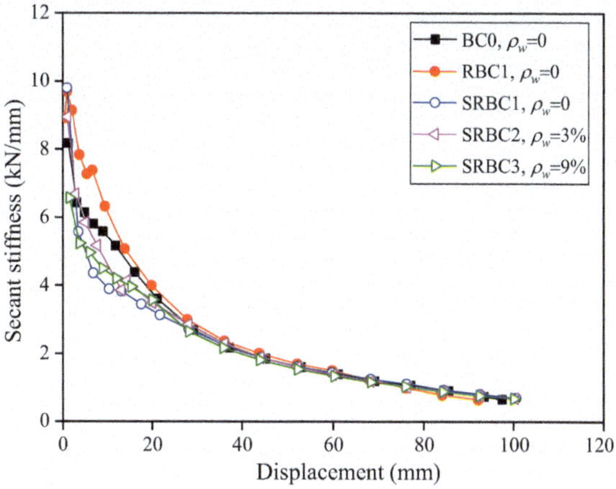


**Fig. 5.46** Secant stiffness degradation of specimens BC0, RBC1, SRBC1, SRBC2, and SRBC3

## 5.3.6 *Energy Dissipation Capacity*

Energy dissipation of specimen is generally estimated by the encircled region of hysteretic loop in each cycle, while the cumulative energy dissipation is obtained through accumulating the energy dissipation of entire loops over the test [77, 88]. Figure 5.47 describes the cumulative energy dissipation of tested specimens. Better energy dissipated capacity of benchmark specimen BC0 than those of retrofitted specimens ERBC2-ERBC4 could be observed during the test. That might attribute to the pre-damages and reinforcement corrosion of beam-column joints, resulting in the brittle failure modes and pinching phenomena visible.

Additionally, the total cumulative energy dissipations of specimens ERBC2, ERBC3, and ERBC4 were 34.3, 32.3, and 36.7 kN·m, which increased by 14.3%,

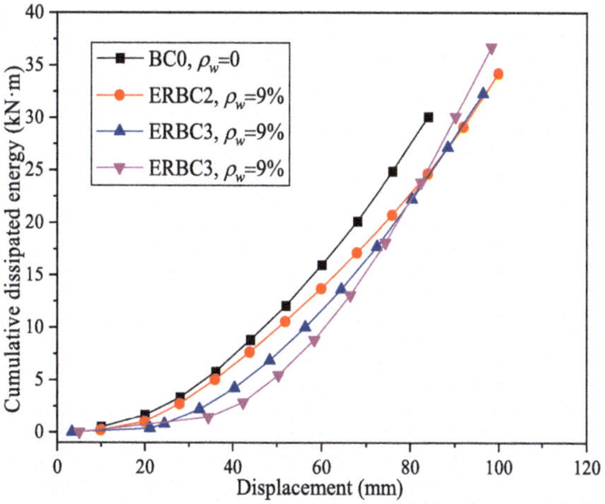


**Fig. 5.47** Cumulative dissipated energy of specimens BC0, ERBC2, ERBC3, and ERBC4

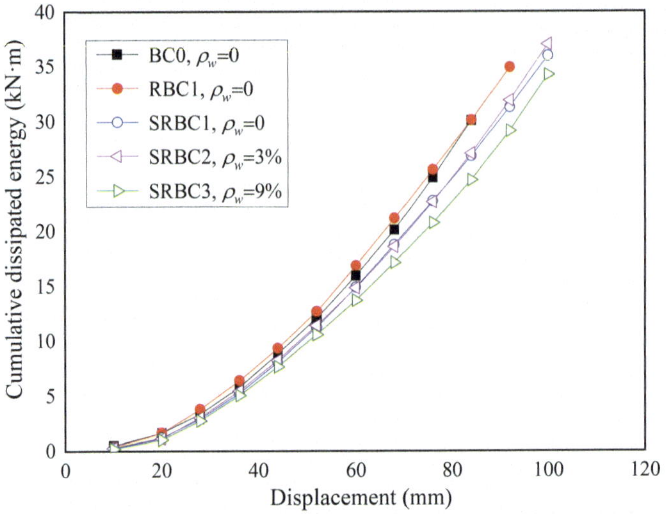


**Fig. 5.48** Cumulative dissipated energy of specimens BC0, RBC1, SRBC1, SRBC2, and SRBC3

7.7%, and 22.3% comparing to that of benchmark specimen BC0 with 30.0 kN · m, respectively. The cumulative energy dissipations of earthquake-damaged corroded beam-column joints considerably increased and were recovered to that of benchmark specimen by retrofitting with BFRP sheets.

Figure 5.48 depicts the cumulative dissipated energy of all tested specimens. Both specimens BC0 and RBC1 exhibited better behaviors on the energy dissipated capacity than those of specimens SRBC1, SRBC2, and SRBC3, which might attribute to the pre-damage and reinforcement corrosion of joints that resulted in the brittle failure modes and the pinching phenomena visible in the test. The enhancement of stiffness, load-bearing capacity, and ductility of retrofitted specimens were accountable for the increase of energy dissipation. Additionally, the overall cumulative energy dissipation of specimens RBC1, SRBC1, SRBC2, and SRBC3 was 34.9, 36.0, 37.0, and 34.3 kN·m, which increased by 16.3%, 20.0%, 23.3%, and 14.3% comparing to that of reference specimen BC0 with 30.0 kN·m, respectively. The energy dissipation capacities of seismic-damaged corroded joints with corrosion rates of 0, 3%, and 9% were adequately recovered to that of reference specimen by retrofitting with BFRP sheets.

### *5.3.7 Analysis of Shear Strength*

(1) Shear Strength of the Reference Joint

Based on the Chinese Standard GB 50010-2010 [89], the shear strength of core area of the joint, $V_j$, is calculated by Eq. (5.17):

$$V_j = \frac{1}{\gamma_{RE}}\left(1.1\eta_j f_t b_j h_j + 0.05\eta_j N\frac{b_j}{b_c} + f_{yv} A_{svj}\frac{h_{b0} - a_s'}{s}\right) \tag{5.17}$$

where $\gamma_{RE}$ is the seismic coefficient of shear strength ($\gamma_{RE} = 0.85$); $\eta_j$ is the restraint influencing coefficient of the beam on the joint ($\eta_j = 1.0$) [35]; $b_c$ and $h_j$ are the sufficient width and height of column, in mm, respectively; $b_j$ is the sufficient width of the core area, in mm; $h_{b0}$ is the sufficient height of the beam, in mm; $a_s\prime$ is the distance from the center point of the longitudinal reinforcements in beams to the surface of the concrete, in mm; $f_{yv}$, $A_{svj}$, and $s$ are the yielding strength, in N, cross-sectional area, in mm$^2$, and spacing values, in mm of the stirrups in the column, respectively; $f_t$ is the tensile strength of the concrete, in MPa; and $N$ is the axial load applied on the top of the column, in N.

Based on the Eurocode 8 [90], the shear strength of core area of the joint, $V_{jhd}$, can be calculated by Eqs. (5.18) and (5.19).

$$V_{jhd} = \eta f_{ck}\sqrt{1 - \frac{v_d}{\eta}}b_j h_{jc} \tag{5.18}$$

$$\eta = 0.6(1 - f_{ck}/250) \tag{5.19}$$

where $f_{ck}$ is the compressive strength of the concrete, in MPa; $b_j$ is the effective joint width, in mm and taken as the minimum of $b_c$ and $b_w + 0.5h_c$, where $b_c$, $b_w$, and $h_c$ are the cross-sectional dimension of column, cross-sectional dimension of column, and thickness of confined parts of a wall section, or width of the web of a beam, respectively; $v_d$ is the normalized axial force in the column above the joint and taken as 0.2; and $h_{jc}$ is the distance between extreme layers of column reinforcement, in mm.

According to the ACI 352R [91], the shear strength of the joint $V_n$, can be calculated by Eq. (5.20).

$$V_n = 0.083\gamma\sqrt{f_c'}b_j h_c \tag{5.20}$$

where $\gamma$ is the coefficient depended on the type of connection. The core joints were effectively confined on two opposite vertical faces and the coefficient $\gamma$ is taken as 15 (Type 2 connection); $f_c'$ is specified compressive strength of concrete in the connection, in MPa; $b_j$ is the effective joint width, in mm, and taken as the minimum of $\frac{b_b+b_c}{2}$, $b_b + \sum\frac{mh_c}{2}$, and $b_c$. The term $b_b$, $b_c$, and $m$ are the width of the longitudinal beam, in mm, width of column transverse to the direction of shear, in mm, and slope to define the effective width of joint transverse to the direction of shear, respectively; and $h_c$ is the depth of the column in the direction of joint shear being considered, in mm.

Considering the CSA code [92], the shear strength of the joint, $V_m$, can be calculated by Eq. (5.21).

$$V_m = 1.6\lambda\phi_c\sqrt{f_c'}A_j \tag{5.21}$$

where $\lambda$ is the factor to account for low-density concrete ($\lambda$ is taken as 1.0 for normal density concrete); $\phi_c$ is the resistance factor for concrete ($\phi_c = 0.65$); $f'_c$ is the strength of the concrete in the joint; and $A_j$ is the minimum cross-sectional area within a joint in a plane parallel to the axis of the reinforcement generating the shear in the joint, and equal to the gross area of the column section, in mm$^2$.

Results of the shear strength of the core area of reference specimen calculated by the Chinese Standard GB 50010-2010 [89], Eurocode 8 [90], ACI 352R [91], and CSA code [92] are depicted in Table 5.17. The deviation between the theoretical and experimental results calculated on the basis of Chinese Standard GB 50010-2010 was 4.7%, which was considerably less than those calculated on the basis of Eurocode 8, ACI 352R, and CSA code. That might attribute to the overestimated shear strength, indicating that calculating the theoretical shear strength of core area of tested joints based on the Chinese Standard GB 50010-2010 was more feasible and reasonable.

Besides, models for evaluating the shear strength of corroded and damaged joints were still needed. Considering several calculating methods proposed in [35, 42], the experimental and theoretical results of all tested specimens were depicted in Fig. 5.51. The maximum deviations of the experimental and theoretical results of tested specimens proposed in GB 50010-2010 [89], Xu [35], and Cheng [42] were 17.9%, 15.9%, and 8.0%, respectively, which were both larger than those of Cheng [42]. Although the deviations of the experimental and theoretical results of beam-column joints proposed in Cheng [42] were small, the calculating methods only considered the influence of seismic damage and FRP retrofitting on the beam-column joints. For evaluating the shear strength of core area of corroded beam-column joints under the seismic damages retrofitted with FRP sheets, the existing methods needed to be modified.

Therefore, a modified method for evaluating the theoretical shear strength of core area of beam-column joints considering the influence of reinforcement corrosion, seismic damage, and FRP rehabilitation needed to be proposed.

(2) Shear Strength of the Retrofitted Joints

The shear strength of the unretrofitted joint core obtained from the test $V_{er}$ could be calculated by the equations as follows according to Eqs. (5.22) and (5.23) proposed in [93, 94]:

$$V_{er} = T_l + T_r - V_c \tag{5.22}$$

$$T_l = f_y A_s \tag{5.23}$$

where $T_l$ and $T_r$ are the tensile strength of the upper and lower reinforcing bars in the left and right beams, in kN, respectively, and calculated by substituting Eq. (5.23); $V_c$ is the horizontal shear strength of column and equal to the maximum horizontal loads of the tested joints, in kN; $f_y$ and $A_s$ are the yielding strength and residual cross-section areas of the corroded longitudinal reinforcements and calculated by Eqs. (5.40) and (5.41), respectively, according to the equations proposed in [94].

$$f_y = (1 - 0.005\rho) f_{y0} \tag{5.24}$$

$$A_s = (1 - 0.01\rho) A_{s0} \tag{5.25}$$

where $f_y$ and $f_{y0}$ are the yielding strength of the corroded and uncorroded reinforcing bars, in MPa, respectively; $A_s$ and $A_{s0}$ are the residual and initial cross-section area of corroded and uncorroded reinforcing bars, in mm$^2$, respectively; and $\rho$ is the corrosion rate, in %.

The yielding strength of uncorroded reinforcing bars with the 14-mm diameter and 6-mm diameter obtained from the test was 405.23 and 261.11 MPa, respectively. The maximum horizontal loads of corroded joints with the tested corrosion rates of 0, 2.74%, and 8.95% were 85.6, 79.4, and 76.9 kN, respectively, as reported in [12]. Based on the Eqs. (5.38)–(5.41), the tested shear strength of corroded joint cores with the tested corrosion rates of 0, 2.74%, and 8.95% was 288.02, 279.01, and 248.06 kN, respectively. The theoretical shear strength of corroded joint cores with the corrosion rates of 0, 2.74%, and 8.95% obtained by substituting Eqs. (5.24) and (5.25) into Eq. (5.17), among which the value of $\gamma_{RE}$ was taken as 1.0, was 247.36, 239.93, 214.36 kN, respectively. The ratios of theoretical or tested shear strength of corroded joint cores with the corrosion rates of 0, 2.74%, and 8.95% to those of reference joint with the corrosion rate of 0 were 1, 0.989, and 0.964 or 1, 0.969, and 0.861, respectively. The ratios of corresponding tested ratios to theoretical ratios were 1, 0.980, and 0.893, respectively. An influence coefficient of reinforcement corrosion on the load-bearing capacity $\omega_\rho$ was used for evaluating the effect of reinforcement corrosion on the shear strength of joint cores. $\omega_\rho$ was obtained by the regression analysis as depicted in Fig. 5.52 and is shown in Eq. (5.26).

$$\omega_\rho = 1.04 - 0.04e^{\rho/7.1} \tag{5.26}$$

where $\rho$ is the corrosion rate, in %.

The model for predicting the theoretical shear strength of corroded joint core $V_{j,RC}^c$ considering the influence coefficient of reinforcement corrosion on the load-bearing capacity $\omega_\rho$ and corrosion rate $\rho$ was proposed and calculated by Eq. (5.27).

$$\begin{cases} V_{j,RC}^c = (1.1\eta_j f_t b_j h_j + 0.05\eta_j N \dfrac{b_j}{b_c} + f_{yv}^c A_{svj}^c \dfrac{h_{b0} - a_s'}{s})\omega_\rho \\ f_{yv}^c = (1 - 0.005\rho) f_{yv};\quad A_{svj}^c = (1 - 0.01\rho) A_{svj} \end{cases} \tag{5.27}$$

where $f_{yv}^c$ and $f_{yv}$ are the yielding strength of the corroded and uncorroded stirrups, in MPa, respectively; $A_{svj}^c$ and $A_{svj}$ are the residual and initial cross-section area of corroded and uncorroded stirrups, in mm$^2$, respectively.

Based on Eq. (5.27), the theoretical shear strength of corroded joint cores with the corrosion rates of 0, 2.74% and 8.95% was 247.36, 239.93, and 214.36 kN, respectively. The ratios of theoretical shear strength of corroded joint cores with the corrosion rates of 0, 2.74%, and 8.95% to that of reference joint were 1, 0.970, and

0.867, which were almost equal to those of tested results of 1, 0.969, and 0.861, respectively.

The shear strength of retrofitted corroded joint cores $V_j$ was composed of the shear strength of the joint core provided by the concrete and stirrups $V^c_{j,RC}$ and the shear strength provided by the FRP sheet $V_{j,FRP}$, and calculated by Eq. (5.28).

$$V_j = V^c_{j,RC} + V_{j,FRP} \tag{5.28}$$

Besides, the shear strength provided by the FRP sheets $V_{j,FRP}$ in Eq. (5.28) could be calculated by Eqs. (5.29)-(5.31) according to the equation proposed in [42].

$$V_{j,FRP} = \psi_f n_f f_f t_f h_f \tag{5.29}$$

$$\psi_f = 0.85(1.2 - 0.1n) \tag{5.30}$$

$$\psi_{bf} = \psi^0_f(1.2 - 0.1n) \tag{5.31}$$

where $n_f$ is the total layers of FRP sheets; $f_f$ is the tensile strength of FRP, in MPa; $t_f$ is the thickness of one layer of FRP sheet, in mm; $h_f$ is the width of FRP sheet along the direction of column section height in joint core and taken as 300 of the beam height, in mm; $\psi_f$ is the efficiency coefficient of FRP. Based on the investigations on the corroded joints retrofitted with BFRP sheets reported in [12], the efficiency coefficient of BFRP sheets was defined as $\psi_{bf}$ and calculated by Eq. (5.31); $\psi^0_f$ is the coefficient; and $n$ is the layers of FRP sheets on the single side of concrete and taken as 1, when the layers of FRP sheets are larger than or equal to 6, the value of $n$ is taken as 6.

The tested result of shear strength of retrofitted joint cores $V_{jh,r}$ could be calculated by Eqs. (5.32) and (5.33) according to the equations proposed in [95].

$$V_{jh,r} = T_l + T_r + 2T_{FRP} - V_c \tag{5.32}$$

$$T_{FRP} = f_f A_f \tag{5.33}$$

where $A_f$ is the cross-section area of the L-wrapped FRP sheets applied on the beams, in mm$^2$.

The maximum horizontal loads of corroded joints with the corrosion rates of 0, 2.74%, and 8.95% retrofitted with BFRP sheets were 92.1, 88.1, and 83.1 kN, respectively, as reported in [12]. The tested results of shear strength of the corroded joints cores or corroded joints cores retrofitted with BFRP sheets calculated by Eqs. (5.27) and (5.32) and (5.33) were 288.02, 279.01, and 248.06 kN or 392.84, 381.63, and 353.18 kN, respectively, when the corrosion rates were 0, 2.74%, and 8.95%. The tested results of shear strength provided by BFRP sheets of retrofitted joint cores with the corrosion rates of 0, 2.74%, and 8.95% were 104.82, 102.62, and 105.12

kN, respectively. Based on Eqs. (5.29) and (5.31), the values of $\psi_f^0$ of retrofitted joints with the corrosion rates of 0, 2.74%, and 8.95% were 0.571, 0.559, and 0.573, the deviations between which and the average value of 0.568 were 0.5%, -1.6%, and 0.9%, respectively. Therefore, the value of $\psi_f^0$ was taken as 0.568.

Based on Eq. (5.29), the theoretical or tested results of shear strength provided by BFRP sheets of retrofitted joint cores with the corrosion rates of 0, 2.74%, and 8.95% were 104.33 kN or 104.82, 102.62, and 105.12 kN, respectively. The deviations between theoretical and tested shear strength of retrofitted joint cores with the corrosion rates were 0, 2.74%, and 8.95% were − 0.4%, 1.6%, and − 0.7%, respectively.

The model for predicting the theoretical shear strength of the corroded joint core retrofitted with BFRP sheets $V_j^{cf}$ considering the shear strength provided by BFRP sheets and corrosion rate was proposed and calculated by Eq. (5.34).

$$\begin{cases} V_j^{cf} = (1.1\eta_j f_t b_j h_j + 0.05\eta_j N\dfrac{b_j}{b_c} + f_{yv}^c A_{svj}^c \dfrac{h_{b0} - a_s'}{s})\omega_\rho + \psi_{bf} n_f f_f t_f h_f \\ \psi_{bf} = 0.568(1.2 - 0.1n) \end{cases} \tag{5.34}$$

(3) Influence of the Reinforcement Corrosion and Material Damage on the Strength

The beam-column joints were damaged by the seismic damage, which also resulted in the deterioration of strength or performance of the material and structure. Therefore, the strength reducing factor $\alpha_F$ was utilized to evaluate the seismic damages on the shear strength of corroded joint cores retrofitted with BFRP sheets $V_u^{cfd}$ as Eq. (5.35) and calculated by Eq. (5.36) as follows according to the equations proposed in [14].

$$V_u^{cfd} = (1.1\eta_j f_t b_j h_j + 0.05\eta_j N\frac{b_j}{b_c} + f_{yv}' A_{svj}' \frac{h_{b0} - a_s'}{s})\omega_\rho + \psi_{bf} n_f f_f t_f h_f \tag{5.35}$$

$$\alpha_F = 1 + \beta_1 D + \beta_2 D^2 \tag{5.36}$$

$$f_{yv}' = \alpha_F f_{yv} \tag{5.37}$$

$$f_y' = \alpha_F f_y \tag{5.38}$$

where $f_{yv}'$ and $f_y'$ are the residual yielding strength of stirrups and longitudinal reinforcements after the seismic damage, in MPa; $f_{yv}$ and $f_y$ are the residual yielding strength of stirrups and longitudinal reinforcements before the seismic damage, in MPa, and calculated by Eq. (5.24); $D$ is the damaging index and taken as 0.5 for the moderate damage, as reported in [96]; $\beta_1$ and $\beta_2$ are the coefficients, which can be calculated by the equations as follows according to Eqs. (5.39) and (5.40) [96].

$$\beta_1 = 0.127 - 0.000586 f'_{yv} + 0.229\left(\frac{f'_{yv}}{1000}\right)^2 + 0.00143 f_c \tag{5.39}$$

$$\beta_2 = -1.013 + 0.585n - 1.762n^2 + \frac{10.959}{f_c} \tag{5.40}$$

where $n$ is the axial load ratio; and $f_c$ is the axial compressive strength of concrete, in MPa.

The value of $\alpha_F$ calculated by Eqs. (5.36)–(5.40) was 0.8577. The theoretical shear strength of seismic-damaged corroded joint cores retrofitted with BFRP sheets with the corrosion rates of 0, 2.75%, and 8.95% were 341.95, 335.09, and 311.07 kN, respectively. The ratios of theoretical to tested shear strength of specimens SRBC1-SRBC3 of 344.88, 333.73, and 308.54 kN calculated by Eqs. (5.25)–(5.26) and (5.29)–(5.30) were 0.992, 1.004, and 1.008, respectively.

Besides, anther influence coefficient of seismic damages on the load-bearing capacity $\omega_d^m$ could also be utilized to evaluate the moderately seismic damages on the shear strength of corroded joint cores retrofitted with BFRP sheets $V_j^{cfd}$ based on Eq. (5.34) and calculated by Eq. (5.41).

$$V_j^{cfd} = (1.1\eta_j f_t b_j h_j + 0.05\eta_j N\frac{b_j}{b_c} + f_{yv}^c A_{svj}^c \frac{h_{b0} - a'_s}{s})\omega_p \omega_d^m + \psi_{bf} n_f f_f t_f h_f \tag{5.41}$$

The maximum horizontal loads of damaged joints SRBC1, SRBC2, and SRBC3 depicted in Table 5.14 or undamaged joints reported in [12] with the same corrosion rates were 86.9, 85.0, and 81.5 kN or 92.1, 88.1, and 83.1 kN, respectively. Based on Eqs. (5.32) and (5.33), the tested results of shear strength of the seismic-damaged joints cores or undamaged corroded joint cores retrofitted with BFRP sheets were 344.88, 333.73, and 308.54 kN or 392.84, 381.63, and 353.18 kN. The ratios of the theoretical shear strength of undamaged corroded joint cores to that of damaged corroded joint cores calculated by Eq. (5.41) were considered to be equal to those between tested strength. Therefore, the calculated values of $\omega_d^m$ of tested joints with the corrosion rates of 0, 2.75%, and 8.95% were 0.826, 0.820, and 0.812, respectively, the deviation between which and average value of 0.819 were 0.8%, 0.1%, and −0.9%. Therefore, the value of $\omega_d^m$ was taken as 0.819. The ratios of theoretical strength of damaged joints cores to undamaged joint cores with the same corrosion rates of 0, 2.74%, and 8.95% were 87.3%, 87.4%, and 87.8%, which were almost equal to those of tested strength of 87.8%, 87.4%, and 87.4%, respectively. Besides, the theoretical shear strength of seismic-damaged corroded joint cores retrofitted with BFRP sheets with the corrosion rates of 0, 2.75%, and 8.95% were 306.92, 300.83, and 279.89 kN, respectively. The ratios of theoretical to tested shear strength of specimens SRBC1, SRBC2, and SRBC3 of 344.88, 333.73, and 308.54 kN calculated by Eqs. (5.25) and (5.26) were 0.890, 0.904, and 0.907, respectively. Ratios obtained from the proposed model Eq. (5.41) had a similar result as those obtained from Eq. (5.35) proposed in

[14], indicating that Eq. (5.41) could also be utilized to evaluate the shear strength of seismic damaged and corroded joint cores retrofitted with BFRP sheets.

Consequently, the model for predicting theoretical shear strength of seismic-damaged corroded joint cores retrofitted with BFRP sheets $V_j^{cfd}$ considering the influence coefficient of reinforcement corrosion on the load-bearing capacity $\omega_\rho$, the shear strength provided by the BFRP sheets, and influence coefficient of seismic damages on the load-bearing capacity $\omega_d^m$ was proposed and evaluated by Eq. (5.42).

$$\begin{cases} V_j^{cfd} = (1.1\eta_j f_t b_j h_j + 0.05\eta_j N\frac{b_j}{b_c} + f_{yv}^c A_{svj}^c \frac{h_{b0} - a_s'}{s})\omega_\rho \omega_d^m + \psi_{bf} n_f f_f t_f h_f \\ \omega_\rho = 1.04 - 0.04e^{\rho/7.1} \\ f_{yv}^c = (1 - 0.005\rho) f_{yv} \\ A_{svj}^c = (1 - 0.01\rho) A_{svj} \\ \psi_{bf} = 0.568(1.2 - 0.1n) \\ \omega_d^m = 0.819 \end{cases} \tag{5.42}$$

(4) Discussion of the Experimental and Theoretical Results

A comparison of tested and theoretical results of tested specimens is depicted in Table 5.18 and Fig. 5.53 The deviations between theoretical and tested results of shear strength of specimens BC0, RBC1, SRBC1, SRBC2, and SRBC3 based on the model were – 14.7%, – 10.5%, – 10.8%, – 9.7%, and – 9.6%, respectively. The deviations of tested results were small, and a good agreement between the theoretical and tested results could be found. However, the calculated results of retrofitted specimens were slightly less than tested results, which might attribute to the underestimated support of BFRP sheets.

Besides, Eq. (5.42) could also be utilized for evaluating the theoretical shear strength of specimens reported in [6] and [97], which were seismic-damaged and retrofitted with BFRP and CFRP, respectively. The deviations between theoretical and tested results of shear strength of these specimens reported in [6] and Chang et al. [97] were in the arranges from 43.1% to – 20.9%, and 3.2% to – 8.9%, respectively, indicating that a relatively good agreement between the theoretical and tested results could be found. Similar to the important coefficients of components to analyze the robust reported in [98, 99], another important parameter Γ was proposed for evaluating the reliability of the comparison of theoretical and tested results by Eq. (5.43):

$$\Gamma = \left| \frac{V_j - V_{er}}{V_j} \right| \times 100\% \tag{5.43}$$

where $V_j$, and $V_{er}$ are the theoretical and tested results of shear strength of the joint and calculated by Eqs. (5.42) and (5.22)–(5.23) and (5.28)–(5.29), in kN, respectively.

The values of Γ of tested specimens BC0, RBC1, SRBC1, SRBC2, and SRBC3 were 16.4%, 11.7%, 12.1%, 10.8%, and 10.6%, respectively, which were small and indicated that the comparison of the theoretical and tested results of corroded and damaged specimens was robust. Besides, the values of Γ of tested specimens reported in [6] or Chang et al. [97] were 30.1%, 25.2%, 25.1%, 26.4%, and 10.6% or 6.6%, 0.4%, 3.1%, 9.7%, 4.9%, 2.4%, and 0.2%, respectively. That indicated that the comparison of the theoretical and tested results of tested specimens based on the model was relatively robust.

Utilizing the proposed model Eq. (5.42) for evaluating the theoretical shear strength of core area of the RC beam-column joints considering the influence of reinforcement corrosion, seismic damage, and FRP rehabilitation was feasible and reasonable in a sense.

On the basis of Chinese Standard GB 50011-2010 [89], the model for predicting the theoretical shear strength of corroded joint core retrofitted with BFRP sheets $V_j^{cf}$ considering the shear strength provided by BFRP sheets and corrosion rate was calculated by Eq. (5.44).

$$\begin{cases} V_j^{cf} = \left(1.1\eta_j f_t b_j h_j + 0.05\eta_j N\dfrac{b_j}{b_c} + f_{yv}^c A_{svj}^c \dfrac{h_{b0} - a_s'}{s}\right)\omega_\rho + \psi_{bf} n_f f_f t_f h_f \\ \omega_\rho = 1.04 - 0.04e^{\rho/7.1};\ \psi_{bf} = 0.568(1.2 - 0.1n) \\ f_{yv}^c = (1 - 0.005\rho) f_{yv} \\ A_{svj}^c = (1 - 0.01\rho) A_{svj} \end{cases} \tag{5.44}$$

where $\eta_j$ is the restraint influencing coefficient of the beam on the joint ($\eta_j = 1.0$); $b_c$ and $h_j$ are the sufficient width and height of column, in mm, respectively; $b_j$ is the effective width of the core area, in mm; $h_{b0}$ is the effective height of the beam, in mm; $a_s\prime$ is the distance from the center point of the longitudinal reinforcements in beams to the surface of the concrete, in mm; $f_t$ is the tensile strength of the concrete, in MPa; $N$ is the axial load applied on the top of the column, in N; $s$ is the spacing values of the stirrups in the column, in mm; $n_f$ is the total layers of FRP sheets; $f_f$ is the tensile strength of FRP, in MPa; $t_f$ is the thickness of one layer of FRP sheet, in mm; $h_f$ is the width of FRP sheet along the direction of column section height in joint core and taken as 300 of the beam height, in mm; $\psi_{bf}$ is the efficiency coefficient of BFRP; $n$ is the layers of FRP sheets on the single side of concrete and taken as 1, when the layers of FRP sheets are larger than or equal to 6, the value of $n$ is taken as 6; $f_{yv}^c$ and $f_{yv}$ are the yielding strength of corroded and uncorroded stirrups, in MPa, respectively; $A_{svj}^c$ and $A_{svj}$ are the residual and initial cross-section area of corroded and uncorroded stirrups, in mm$^2$, respectively; $\omega_\rho$ is the influence coefficient of reinforcement corrosion on the load-bearing capacity; and $\rho$ is the corrosion rate, in %.

The influence coefficient of earthquake damages on the load-bearing capacity $\omega_d^m$ has been proposed and taken as 0.819 for moderate damages, $\omega_d^m(\theta)$ was proposed for predicating the different earthquake damages on the shear strength of corroded joint cores retrofitted with BFRP sheets $V_j^{cfd}$ based on the equation proposed in research [14] and calculated by Eq. (5.45):

$$V_j^{cfd} = (1.1\eta_j f_t b_j h_j + 0.05\eta_j N \frac{b_j}{b_c} + f_{yv}^c A_{svj}^c \frac{h_{b0} - a_s'}{s})\omega_\rho \omega_d^m(\theta) + \psi_{bf} n_f f_f t_f h_f \tag{5.45}$$

where $\theta$ is the pre-damage drift, and taken as 0, 1/90, 1/45, and 1/28 for the undamaged, moderately, severely, and destructively damaged levels of earthquake, respectively.

The decreasing degrees of theoretical shear strength of specimens ERBC2, ERBC3, and ERBC4 calculated by Eq. (5.45) to that of specimen RBC3 were supposed to be equal to those of tested results. Therefore, the calculated values of $\omega_d^m(\theta)$ of tested specimens RBC3 and ERBC2-ERBC4 with the corrosion rate of 8.95% were 1, 0.812, 0.297, and 0.160, respectively. $\omega_d^m(\theta)$ could be obtained by the regression analysis as depicted in Fig. 5.49 and Eq. (5.62):

$$\omega_d^m(\theta) = \frac{1.08}{1 + (\theta/0.02)^{2.72}} - 0.08 \tag{5.46}$$

The values of $\omega_d^m(\theta)$ obtained from Eq. (5.31) were 1, 0.819, 0.384, and 0.105 when the pre-damage drift was 0, 1/90, 1/45, and 1/28, respectively. Based on Eq. (5.46), the theoretical shear strength of specimens RBC3, ERBC2, ERBC3, and ERBC4 retrofitted with BFRP sheets with the actual corrosion rates of 8.95% were 318.69, 279.89, 186.64, and 126.84 kN, respectively. The decreasing degrees of theoretical shear strength of damaged specimens ERBC2, ERBC3, and ERBC4 to that of undamaged specimen RBC3 were 12.2%, 41.4%, and 60.2%, which were almost equal to those of tested results of 12.6%, 47.3%, and 56.5%, respectively.

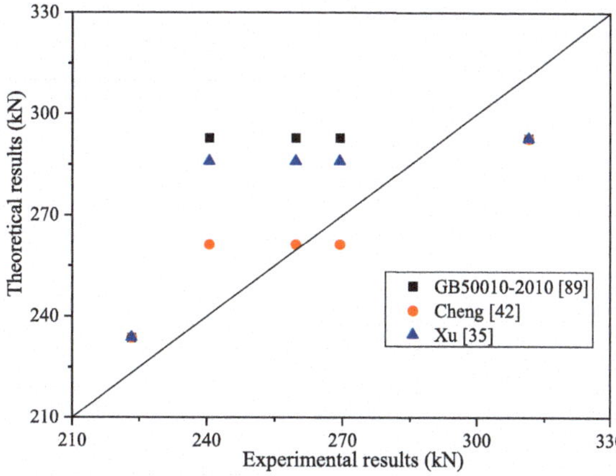


**Fig. 5.49** Comparison of experimental and theoretical results based on the models

That indicated that Eq. (5.45) could be applied for evaluating the shear strength of earthquake-damaged and corroded joint cores retrofitted with BFRP sheets.

Thus, the model for predicting the shear strength of earthquake-damaged corroded beam-column joint cores retrofitted with BFRP sheets $V_j^{cfd}$ supposing the influence of reinforcement corrosion on the load-bearing capacity $\omega_\rho$, the shear strength provided by the BFRP sheets, and influence coefficient of earthquake damages on the load-bearing capacity $\omega_d^m(\theta)$ was modified by Eq. (5.47).

$$\begin{cases} V_j^{cfd} = \left(1.1\eta_j f_t b_j h_j + 0.05\eta_j N\dfrac{b_j}{b_c} + f_{yv}^c A_{svj}^c \dfrac{h_{b0} - a_s'}{s}\right)\omega_\rho \omega_d^m(\theta) + \psi_{bf} n_f f_f t_f h_f \\ \omega_\rho = 1.04 - 0.04e^{\rho/7.1} \\ \psi_{bf} = 0.568(1.2 - 0.1n) \\ f_{yv}^c = (1 - 0.005\rho) f_{yv} \\ A_{svj}^c = (1 - 0.01\rho) A_{svj} \\ \omega_d^m(\theta) = \dfrac{1.08}{1 + (\theta/0.02)^{2.72}} - 0.08 \end{cases} \tag{5.47}$$

Table 5.16 and Fig. 5.50 describe the comparison of theoretical and tested results of all specimens. The deviations between theoretical and tested results of shear strength of specimens BC0, RCB1, ERBC1, ERBC2, ERBC3, and ERBC4 based on proposed model were – 14.1%, – 10.5%, – 10.8%, – 9.3%, 0.3%, and – 17.4%, respectively. The deviations of tested specimens were small, and relatively good agreement between theoretical and tested results could be found. Furthermore, the deviations between theoretical and tested results of shear strength of specimens in [6, 97] were 43.1%, – 52.0%, – 17.3%, – 18.1%, and – 9.6% or – 0.6%, 6.7%, 9.8%, – 3.6%, – 0.1%, 3.5%, and 6.0%, respectively.

The deviations were relatively small, and good agreement between the theoretical and tested results of tested specimens could be found in a sense, indicating that the proposed models could also be utilized to evaluate the moderately and severely damaged uncorroded beam-column joints retrofitted with CFRP or BFRP sheets.

Besides, the theoretical results of retrofitted specimens were slightly less than those of tested results, which might attribute to the underestimated shear strength provided by BFRP sheets.

**Table 5.16** Comparison of tested and theoretical results of tested specimens

| Specimen | Reference | Corrosion rates (%) | Damage level | Types of FRP | $V_j$(kN) | $V_{er}$(kN) | Ratio of theoretical to tested results |
|---|---|---|---|---|---|---|---|
| BC0 | This chapter | 0 | – | – | 247.36 | 288.02 | 0.859 |
| RBC1 | | 0 | – | One layer of BFRP | 351.69 | 392.84 | 0.895 |
| SRBC1 | | 0 | Moderately damaged | One layer of BFRP | 307.73 | 344.88 | 0.892 |
| SRBC2 | | 3 | Moderately damaged | One layer of BFRP | 301.22 | 333.73 | 0.903 |
| SRBC3 | | 9 | Moderately damaged | One layer of BFRP | 278.86 | 308.54 | 0.904 |
| J0 | [6] | 0 | – | – | 188.36 | 131.64 | 1.431 |
| J1 | | 0 | Severely damaged | Two layers of BFRP | 284.23 | 355.72 | 0.799 |
| J2 | | 0 | Moderately damaged | Two layers of BFRP | 284.04 | 355.22 | 0.800 |
| J3 | | 0 | Moderately damaged | Two layers of BFRP | 284.04 | 358.92 | 0.791 |
| J4 | | 0 | Moderately damaged | Two layers of BFRP | 318.13 | 351.92 | 0.904 |
| SJ1 | [97] | 0 | Damaged | | 361.23 | 384.90 | 0.939 |
| SJ3 | | 0 | Damaged | Two layers of CFRP | 368.64 | 367.16 | 1.004 |
| SJ4 | | 0 | Damaged | Two layers of CFRP | 372.34 | 360.75 | 1.032 |
| SJ5 | | 0 | Damaged | Two layers of CFRP | 361.23 | 396.43 | 0.911 |
| SJ6 | | 0 | Damaged | Two layers of CFRP | 372.34 | 390.42 | 0.954 |
| SJ7 | | 0 | Damaged | Two layers of CFRP | 372.34 | 381.37 | 0.976 |
| SJ8 | | 0 | Damaged | Two layers of CFRP | 368.64 | 369.34 | 0.998 |

*Note* $V_j$ is the theoretical values of the shear strength calculated by Eq. (5.42); $V_{er}$ is the tested values of the shear strength calculated by Eqs. (5.22), (5.23), (5.28) and (5.29)

**Table 5.17** Comparison of theoretical and tested results of test specimens

| Specimen | Reference | Corrosion rates (%) | Damage level | Types of FRP | $V_j$(kN) | $V_{er}$(kN) | $V_j/V_{er}$ |
|---|---|---|---|---|---|---|---|
| BC0 | This Chapter | 0 | – | – | 247.36 | 288.02 | 0.859 |
| RBC1 | | 0 | – | One layer of BFRP | 351.69 | 392.84 | 0.895 |
| ERBC1 | | 0 | Moderately damaged | One layer of BFRP | 307.73 | 344.88 | 0.892 |
| ERBC2 | | 9 | Moderately damaged | One layer of BFRP | 279.89 | 308.54 | 0.907 |
| ERBC3 | | 9 | Severely damaged | One layer of BFRP | 186.64 | 186.05 | 1.003 |
| ERBC4 | | 9 | Destructively damaged | One layer of BFRP | 126.84 | 153.55 | 0.826 |
| J0 | [6] | 0 | – | – | 188.36 | 131.64 | 1.431 |
| J1 | | 0 | Severely damaged | Two layers of BFRP | 149.55 | 311.67 | 0.480 |
| J2 | | 0 | Moderately damaged | Two layers of BFRP | 284.04 | 343.29 | 0.827 |
| J3 | | 0 | Moderately damaged | Two layers of BFRP | 284.04 | 346.99 | 0.819 |
| J4 | | 0 | Moderately damaged | Two layers of BFRP | 318.13 | 351.92 | 0.904 |
| SJ1 | [97] | 0 | Damaged | – | 361.23 | 363.34 | 0.994 |
| SJ3 | | 0 | Damaged | Two layers of CFRP | 368.64 | 345.60 | 1.067 |

(continued)

**Table 5.17** (continued)

| Specimen | Reference | Corrosion rates (%) | Damage level | Types of FRP | $V_j$(kN) | $V_{er}$(kN) | $V_j/V_{er}$ |
|---|---|---|---|---|---|---|---|
| SJ4 | | 0 | Damaged | Two layers of CFRP | 372.34 | 339.19 | 1.098 |
| SJ5 | | 0 | Damaged | Two layers of CFRP | 361.23 | 374.87 | 0.964 |
| SJ6 | | 0 | Damaged | Two layers of CFRP | 368.64 | 368.86 | 0.999 |
| SJ7 | | 0 | Damaged | Two layers of CFRP | 372.34 | 359.81 | 1.035 |
| SJ8 | | 0 | Damaged | Two layers of CFRP | 368.64 | 347.78 | 1.060 |

*Note* $V_j$ is the theoretical values of the shear strength calculated by Eq. (5.47); and $V_{er}$ is the tested values of the shear strength calculated by Eqs. (5.26)–(5.29)

**Table 5.18** Test results of all specimens

| Specimen | $P_u$(kN) | $\Delta_y$(mm) | $\Delta_u$(mm) | $\mu$ | $\theta_u$ |
|---|---|---|---|---|---|
| BC0 | 78.10 | 22.77 | 94.47 | 3.82 | 1/16 |
| SBC1 | 85.95 | 25.39 | 102.13 | 4.02 | 1/15 |
| SBC2 | 82.05 | 28.34 | 112.39 | 3.97 | 1/14 |
| SBC3 | 86.22 | 23.43 | 93.30 | 3.98 | 1/16 |
| SBC4 | 79.18 | 27.05 | 105.32 | 3.89 | 1/14 |
| SBC5 | 83.98 | 26.53 | 102.13 | 3.85 | 1/15 |

*Note* $P_u$ is ultimate load-bearing capacity; $\Delta_y$ is yielding displacement; $\Delta_u$ is ultimate displacement; $\mu$ is the displacement ductility ratio; and $\theta_u$ is the ultimate rotation

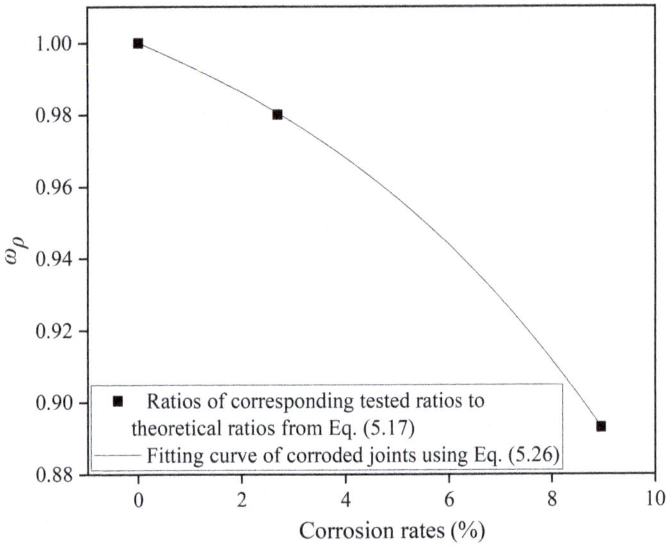


**Fig. 5.50** Relationship between the influence coefficient of $\omega_\rho$ and corrosion rate

**Fig. 5.51** Comparison of tested and theoretical results based on the proposed model

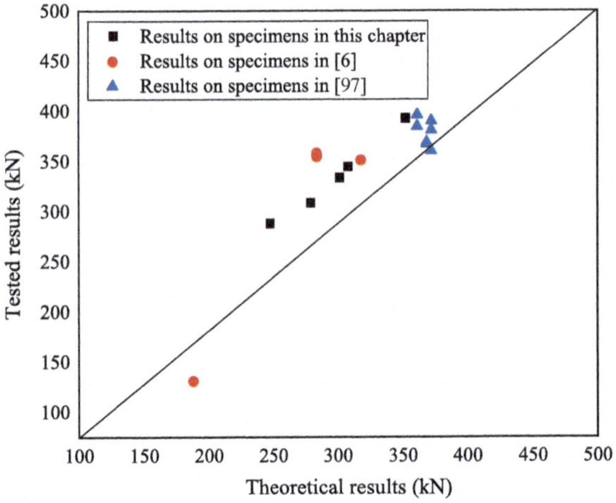

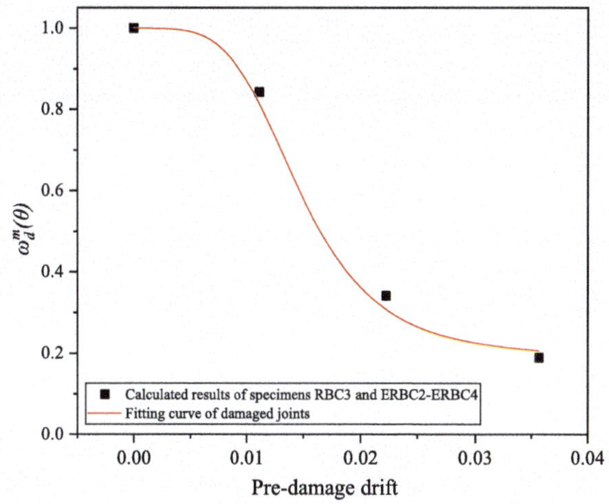


**Fig. 5.52** Relationship between the influence coefficient $\omega_d^m$ and pre-damage drift

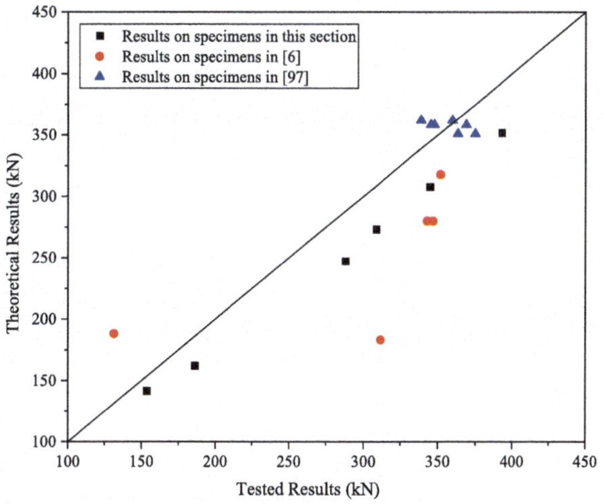


**Fig. 5.53** Comparison of theoretical results based on the proposed model and tested results

# 5.4 Seismic Performance of Beam-Column Joints Retrofitted with BFRP Sheets in Different Methods

## *5.4.1 Experimental Program*

(1) Specimens and Materials

Six 1/2-scale RC interior beam-column joints designed as specimens BC0, SBC1, SBC2, SBC3, SBC4, and SBC5 [100] were tested in accordance with the Chinese Standard GB 50010-2010 [89], respectively. According to Chinese Standard GB/T 50081-2019 [17], the cubic compressive strength $f_{cu}$, cylindrical axial compressive strength $f_c'$ characterized as 0.79 $f_{cu}$ [101], and elastic modulus $E_c$ of concrete were 39.3, 31.0, and 3.24 × $10^4$ MPa, respectively. Dimensions and reinforcement of specimens are same as the descriptions depicted in Fig. 5.1. The average values of yielding strength of ribbed and plain reinforcements and stirrups were 335 and 235 MPa, respectively.

A thickness of 0.121 mm BFRP sheets were applied for the test specimens, and the average values of tensile elastic modulus, ultimate strain, and ultimate tensile strength were 1.05 × $10^5$ MPa, 2.2%, and 2.30 × $10^3$ MPa, respectively.

(2) Strengthening Scheme

Specimens SBC1-SBC5 were strengthened in five various methods with BFRP sheets, as depicted in Fig. 5.56a–e, respectively.

Specimen SBC1 was strengthened for improving the load-bearing capacity of the beams, as reported in [40, 102]. Three layers of 100 mm width BFRP sheets were wrapped around the beams at the surface on left and right sides of the core area, and 50 mm width BFRP sheets spaced at 100 mm were wrapped around the beams with the same layers as above. The end regions of column above and below the joint area were wrapped in transverse direction with three layers of 300 mm width BFRP sheets, as depicted in Fig. 5.54a.

Considering the practical effect of RC slab on the beams, specimen SBC2 was strengthened to investigate the increasing levels of whole beams and column, as reported in [6]. Two layers of BFRP sheets with 800 mm and 525 mm width were wrapped along the axis of the upper and lower column, respectively. Meanwhile, two layers of U-wrapped BFRP sheets along the axis of the left and right beams were applied to the surface, the ends of which had a distance of 60 mm to the top surface of beams. In accordance with a 1/2-scale beam-column joint, the distance of 60 mm was considered as the practical thickness of 120 mm RC slab. Finally, one-layer BFRP strip with 50 mm width was designed as the anchors and stuck at the end of U-wrapped BFRP sheets, as depicted in Fig. 5.54b.

Considering the practical effect of RC slab on the beams as specimen SBC2, the specimen SBC3 was strengthened within partial lengths. Three layers of 200 mm width L-typed BFRP sheets, 300 mm length along the column and 500 mm length along the beam, were stuck at each corner of the joint. Other three layers of BFRP sheets with 300 mm width were wrapped around the upper and lower column, as depicted in Fig. 5.54c.

In specimens SBC4 and SBC5, the joints were strengthened by utilizing the BFRP sheets with two layers in the plastic region, as depicted in Fig. 5.54d–e. L-typed BFRP sheets with 200 mm width were arranged at four corners of the joints. These sheets were attached exactly extending with a height of 300 mm from the beam surfaces, and a length of 500 mm from the column surfaces. Then, straight BFRP sheets in 250 mm or 300 mm width were arranged along the column or beam axis with a length of 300 mm or 500 mm on the front and back sides, respectively. Meanwhile, the upper and lower column ends near to the core areas were wrapped with a height of 300 mm BFRP sheets. The difference between specimens SBC4 and SBC5 was

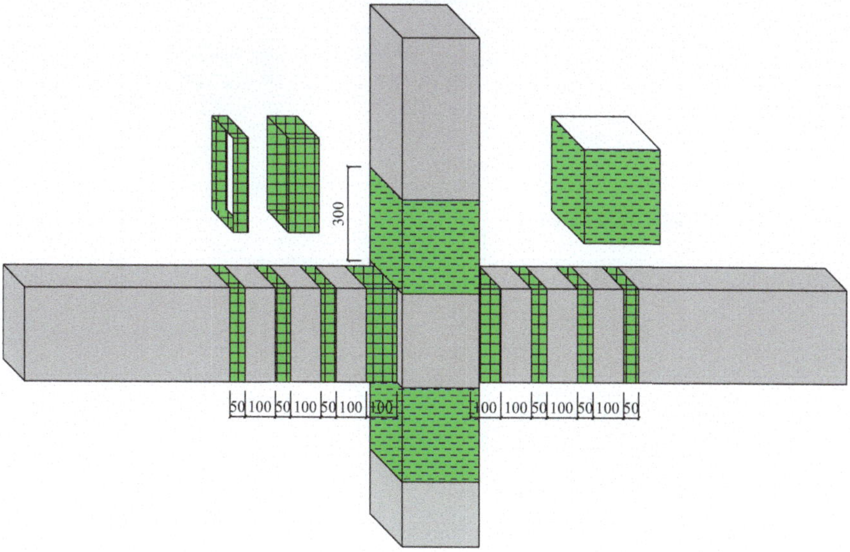


(a) Specime SBC1

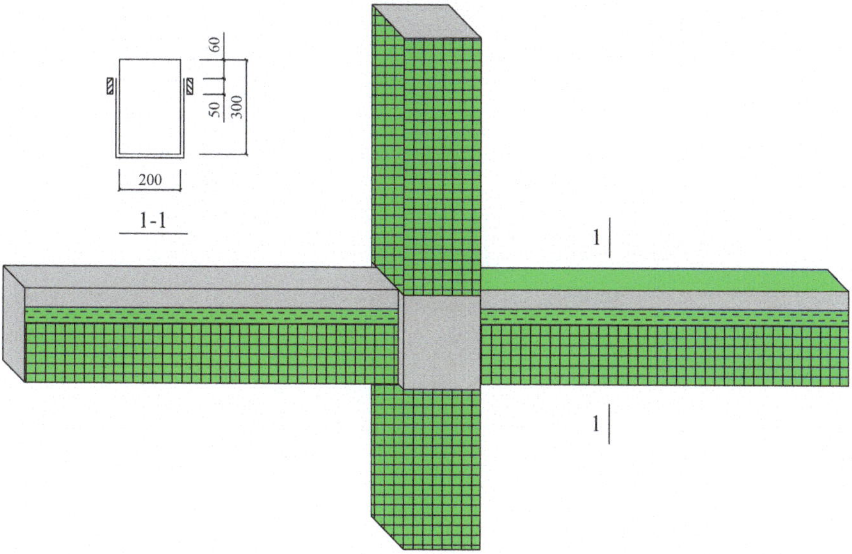


(b) Specime SBC2

**Fig. 5.54** Details of strengthening schemes of all specimens (all units in mm)

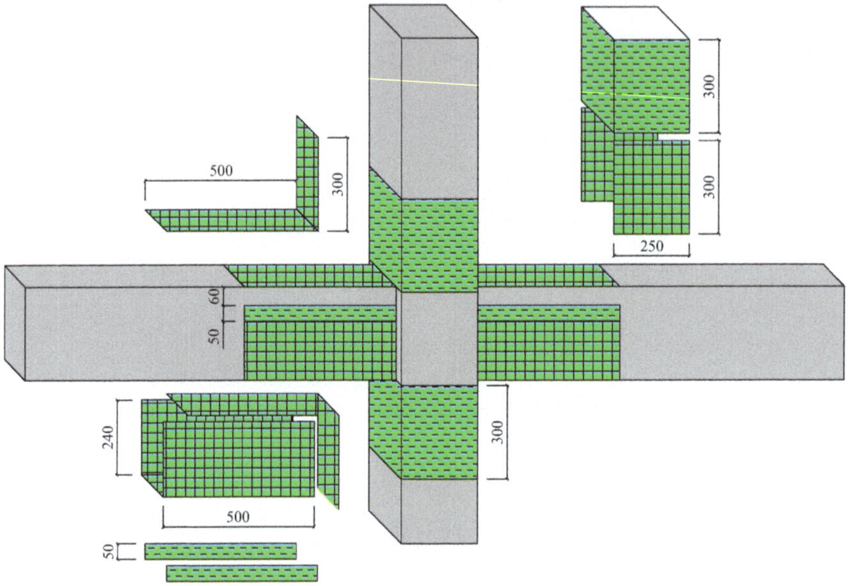


(c) Specime SBC3

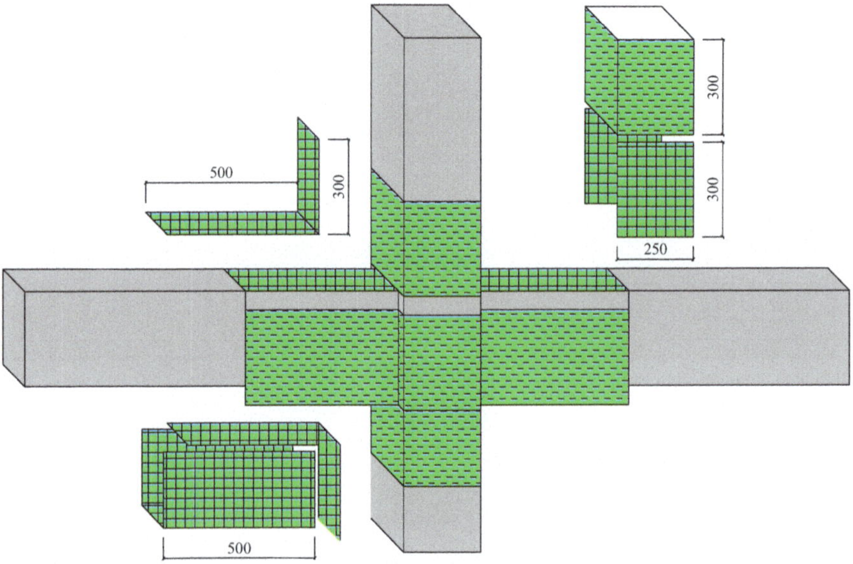


(d) Specime SBC4

**Fig. 5.54** (continued)

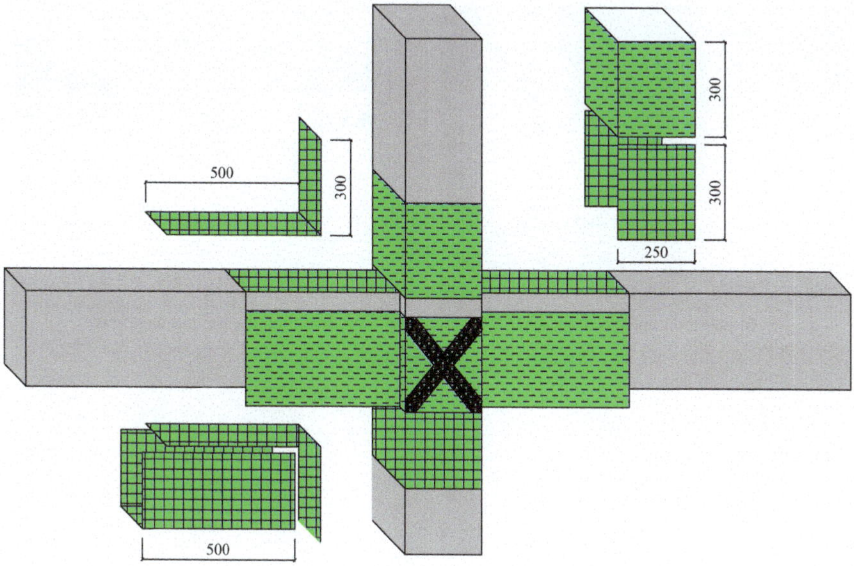


(e) Specime SdBC5

**Fig. 5.54** (continued)

that the later had X-typed BFRP strips strengthening the core areas to improve the load-bearing capacity on the front and back sides. Each strip was arranged through the diagonals of joint on two sides of the core area in 50 mm width, as depicted in Fig. 5.54d.

Before applying the BFRP sheets, several preparations were needed for all strengthened specimens, as depicted in Fig. 5.55. Firstly, the surface of the specimen was ground smoothly and formed with a 20 mm radius of the corners at the strengthening regions. Then, undercoat utilizing the mixed epoxy resins was applied after the dust on the surfaces was removed. Cropped BFRP sheets were applied on the specimen with careful installations before the resins cooled down. Another layer of epoxy resin was utilized as the overcoat followed by the sheets were completely installed on the surface. Lastly, the strengthened specimens were placed and cured for at least seven days before the test.

(3) Test Setup and Loading History

The layout of test setup included the loading system and instrumentation was the same as that shown in Fig. 5.3. And the loading history of tested specimens was also the same as that shown in Fig. 5.4.

(4) Instrumentation

For recording the deformations of specimens, LVDTs were arranged at the top of column and beams, as well as the corners of joint core area. Additionally, two LVDTs

(a) Grinding the specimen surface

(b) Removing the dust on the surface

(c) Applying with the resin undercoat

(d) Installing the BFRP sheets

(e) Applying with the resin overcoat

(f) Placing the strengthened specimens for careful curing

**Fig. 5.55** Detailed preparation procedures for the strengthened specimens

were also arranged at the surface of the specimens to measure the shear deformations under the cyclic load [103], as depicted in Fig. 5.4a.

For evaluating the performance of reinforcements and stirrups in the loading process, strain gauges were arranged with various locations for specimens SBC1, SBC2, SBC3, SBC4, and SBC5, as depicted in Fig. 5.56. A total number of forty strain gauges were arranged on the beam longitudinal reinforcements, column longitudinal reinforcements, beam stirrups, and column stirrups. Meanwhile, three groups of strain gauges were divided and arranged to evaluate the performance of concrete and BFRP on specimens BC0, SBC1, SBC2, SBC3, SBC4, and SBC5, as depicted in

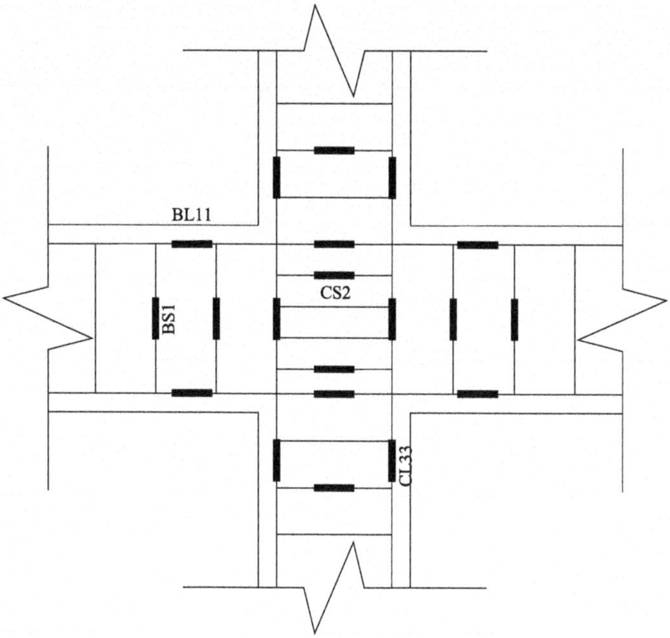


**Fig. 5.56** Layout of strain gauges on reinforcing steels

Fig. 5.57. The first group of six strain gauges, which were designed as SC1-SC6 in specimen BC0, and SF1-SF6 in specimens SBC1, SBC2, SBC3, SBC4, and SBC5, located at a distance of 200 mm adjacent to the core area. The second group of four strain gauges in test specimens, among which only two strain gauges were arranged in specimen SBC2 and SBC3, located at a distance of 500 mm adjacent to the core area, as depicted in Fig. 5.57. Additionally, the third group of strain gauges was arranged for various purposes. As depicted in Fig. 5.57a, e and f, four strain gauges designed as SC(F)7-SC(F)10 were arranged with 45 angles at the surface of concrete and BFRP sheets, respectively. Two strain gauges were arranged in vertical directions at one side of the beam surface in specimen BC0, and other four strain gauges were arranged in vertical directions in specimens SBC2 and SBC3, as depicted in Fig. 5.57a, c and d.

### 5.4.2 *Crack Pattern and Failure Modes*

(1) Reference Specimen BC0

The failure mode of specimen BC0 is depicted in Fig. 5.58a, and the specimen BC0 exhibited a comparative fragility and failed with a shear mode. The flexural cracks adjacent to the core area were firstly observed at the external surface of the beams

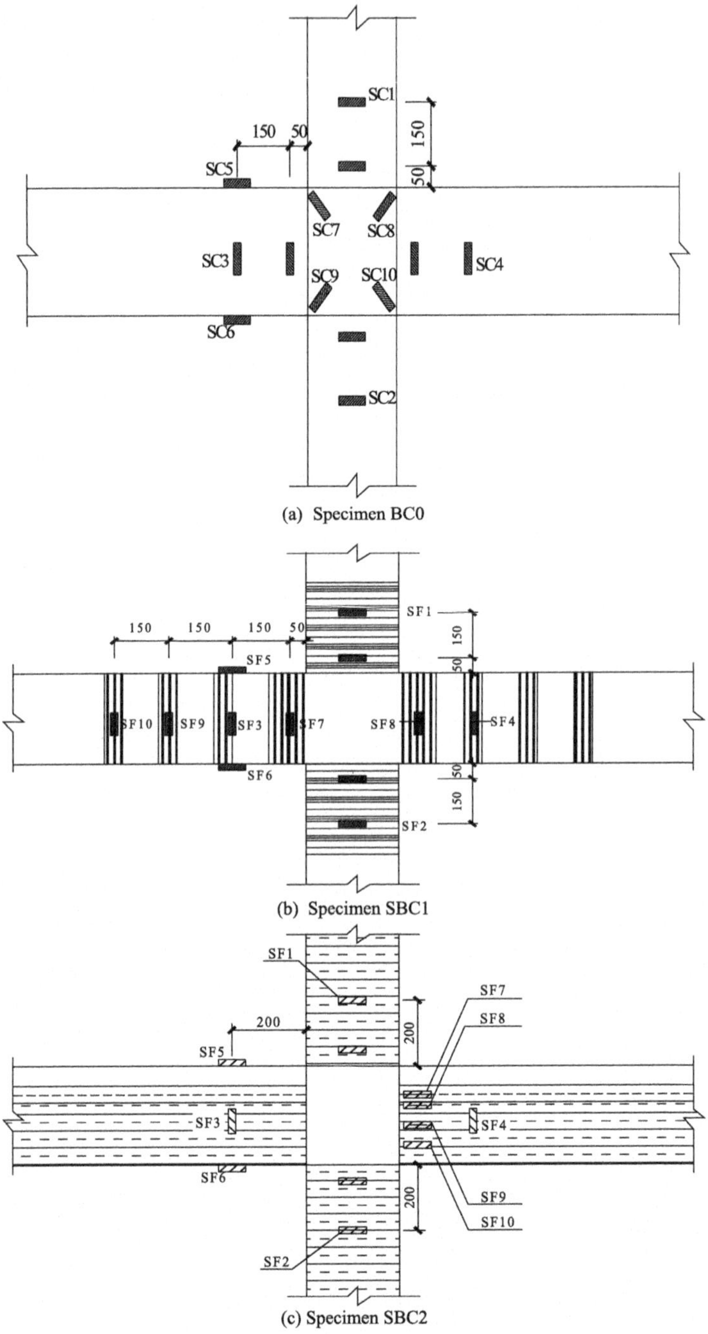


**Fig. 5.57** Layout of strain gauges on the surface of BFRP sheets

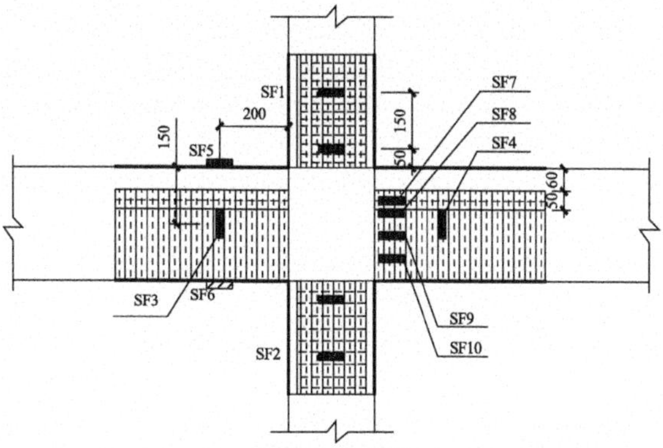


(d) Specimen SBC3

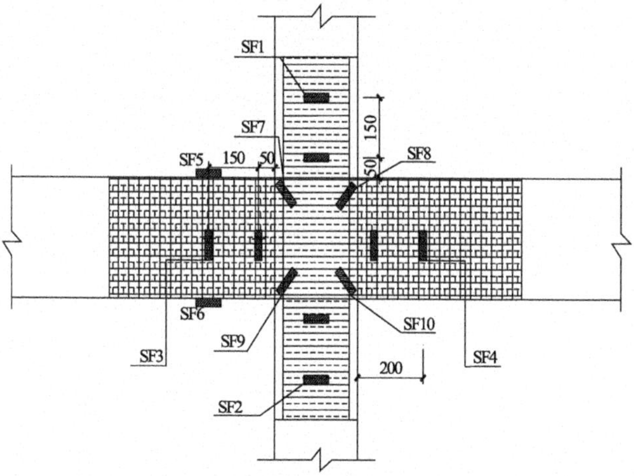


(e) Specimen SBC4

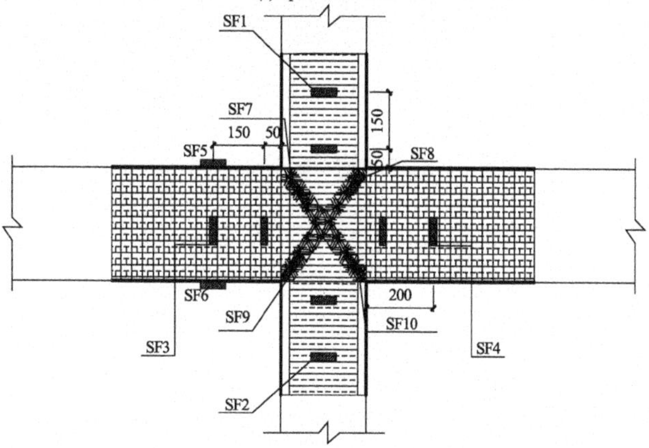


(f) Specimen SBC5

**Fig. 5.57** (continued)

with the lateral load of 40.0 kN, and then developed to the top surface. Cracks continuously occurred with the increase of the lateral loads. At the end of force-control stage, diagonal cracks were observed at the lateral load of 60.0 kN in the joint core area. These diagonal cracks extended to the corners of the joint area, and the visible crossing diagonal cracks were formed at the lateral load of -60.0 kN in the pull direction.

As the test proceeded, the diagonal cracks widen, whose max width reached 2.5 mm, as well as the concrete spalled from the beams at the displacement of 50.0 (3.28% drift) mm. A significant spalling of the concrete in core area and the exposure

(a) Specimen BC0

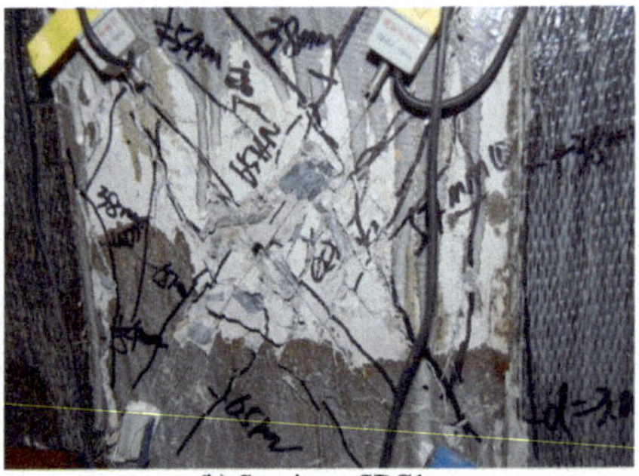
(b) Specimen SBC1

(c) Specimen SBC2

(d) Specimen SBC3

(e) Specimen SBC4

(f) Specimen SBC5

**Fig. 5.58** Final failure mode of all test specimens

of stirrups occurred at the displacement of 74.0 (4.85% drift) mm, which resulted in the decreasing of the load-bearing capacity.

(2) Strengthened Specimen SBC1, SBC2, SBC3, SBC4, and SBC5

The final failure of specimen SBC1 is depicted in Fig. 5.58b. Micro-flexural cracks adjacent to the core area could be observed at the beams with a distance of 300 mm at the lateral load of 40.0 kN. Locations of these cracks were removed from the ends of the beams by utilizing the BFRP strips with 100 mm width. Diagonal cracks could be found in the core area at the lateral load of 65.0 kN. When the lateral displacement was 54.0 (3.54% drift) mm, the core concrete cracked rapidly under the cyclic load with the appearance of vertical cracks at the corners of core area. Large amounts of cracks and the exposure of stirrups could be found in the core area, which made the significant decrease of the load-bearing capacity at the lateral displacement of 78.0 (5.11% drift) mm.

The final failure of specimen SBC2 is depicted in Fig. 5.58c. Due to the utilization of BFRP sheets wrapped on the surface of whole beams, the micro-flexural cracks adjacent to the core area were observed at the beams when the lateral load was 40.0 kN. Meanwhile, two transverse cracks occurred at the corners of the core area, which were attributed to the cracking of hardened resins. When the lateral load was 60.0 kN, diagonal cracks were observed at the joint core area, as well as the flexural cracks at the beam were perforated. Diagonal X-typed cracks occurred and the core concrete spalled at the lateral displacement of 50.0 (3.28% drift) and 74.0 (4.85% drift) mm, respectively, while the load-bearing capacity of specimen had no obvious deterioration. No evident destructions of beams or column were found in specimen SBC2, indicating that a better integrity of the strengthened joints could be obtained by strengthening with whole length of BFRP sheets than that reported in [6].

The final failure of specimen SBC3 is depicted in Fig. 5.58d. Similar tendency of the concrete cracking in specimen SBC3 could be observed at the first stage as the specimen SBC2. The utilized length of BFRP sheet in specimen SBC3 was shorter than that of specimen SBC2, which contributed to the severe cracks occurring at the beams. No evident debonding of BFRP sheets at the end of beams was observed in specimens SBC2 and SBC3, indicating that the BFRP strip utilized as the anchors was an effective way, as reported in [104].

As depicted in Fig. 5.58e–f, specimens SBC4 and SBC5 had a similar tendency of the concrete cracking, which indicated that the cracks occurred at the surface of BFRP sheet on the beams at the lateral load of 60.0 kN. Due to the core concrete spalled, the surface of the BFRP sheets on specimens SBC4 and SBC5 was bulged and transverse fractured when the lateral displacement was 64.0 (4.20% drift) and 70.0 (4.59% drift) mm, respectively. When the lateral displacement was 80.0 (5.25% drift) mm, obvious abruption of BFRP sheet in core area and the sheet delaminated from the concrete surface. However, the X-typed BFRP strips on specimen SBC5 were fractured at the corners of joints, and the core concrete were severely crushed. Meanwhile, the L-typed sheets on the beams were delaminated from the concrete surface at the lateral displacement of 86.0 (5.64% drift) mm.

### 5.4.3 Hysteretic Response and Load–Displacement Curves

A comparison of load–displacement hysteretic loops between reference and strengthened specimens is depicted in Fig. 5.59. The encircled region of hysteretic loop, which was proven to be a basic method, can be utilized for evaluating the energy dissipation of specimen in the cyclic load, as reported in [29]. Hysteretic loop of specimen BC0 exhibited an early pinching phenomenon and short yielding platform, which was attributed to the stiffness degradation and concrete spalling, as depicted in Fig. 5.59a. Comparatively, the hysteretic loops of specimens SBC1, SBC2, SBC3, SBC4, and SBC5 exhibited full and fusiform curves, indicating that the seismic performance of beam-column joints was significantly increased after strengthening with BFRP sheets. Additionally, specimens SBC2, SBC3, SBC4, and SBC5 exhibited longer yielding platforms and lower stiffness degradation than those of reference specimen BC0.

The envelope curve, which was obtained from the load–displacement hysteretic loop and consisted of the peak load and corresponding displacement during the test in the first circle, is depicted in Fig. 5.60. Linear curves could be observed for reference and strengthened specimens at the few circles, indicating that these RC beam-column joints were remaining in the elastic stage. Due to the reinforcing bars yielding, the surface concrete of specimens soon cracked and the confinement of FRP sheets came into effect.

As depicted in Table 5.18, the ultimate load-bearing capacity of specimens BC0, SBC1, SBC2, SBC3, SBC4, and SBC5 was 78.1, 86.0, 82.1, 86.2, 79.2, and 84.0 kN, which increased by 10.1%, 5.1%, 10.4%, 1.4%, and 7.6% contrasting with that of reference specimen BC0, respectively. The envelope curve of reference specimen BC0 was encircled by that of strengthened specimens, indicating that the load-bearing capacity and displacements of specimens SBC1-SBC5 were larger than those of specimen BC0. However, envelope curves of specimens SBC3, SBC4, and SBC5 were not symmetrical in the process of pull and push as expect, which were possibly attributed to the asymmetry of the test setup. Besides, specimen SBC2 exhibited the largest ultimate displacement among these strengthened specimens, indicating that strengthening the column and beams with whole length of BFRP sheets was a relatively effective method.

### 5.4.4 Ductility

As depicted in Table 5.18, contrasting with the reference specimen BC0, the yielding or ultimate displacements of specimens SBC1, SBC2, SBC3, SBC4, and SBC5 significantly increased by 11.5%, 24.5%, 2.9%, 18.8%, and 16.5% or 8.1%, 19.0%, − 1.2%, 11.5%, and 8.1%, respectively. The yielding and ultimate displacements of beam-column joint were improved by strengthening with BFRP sheets, however,

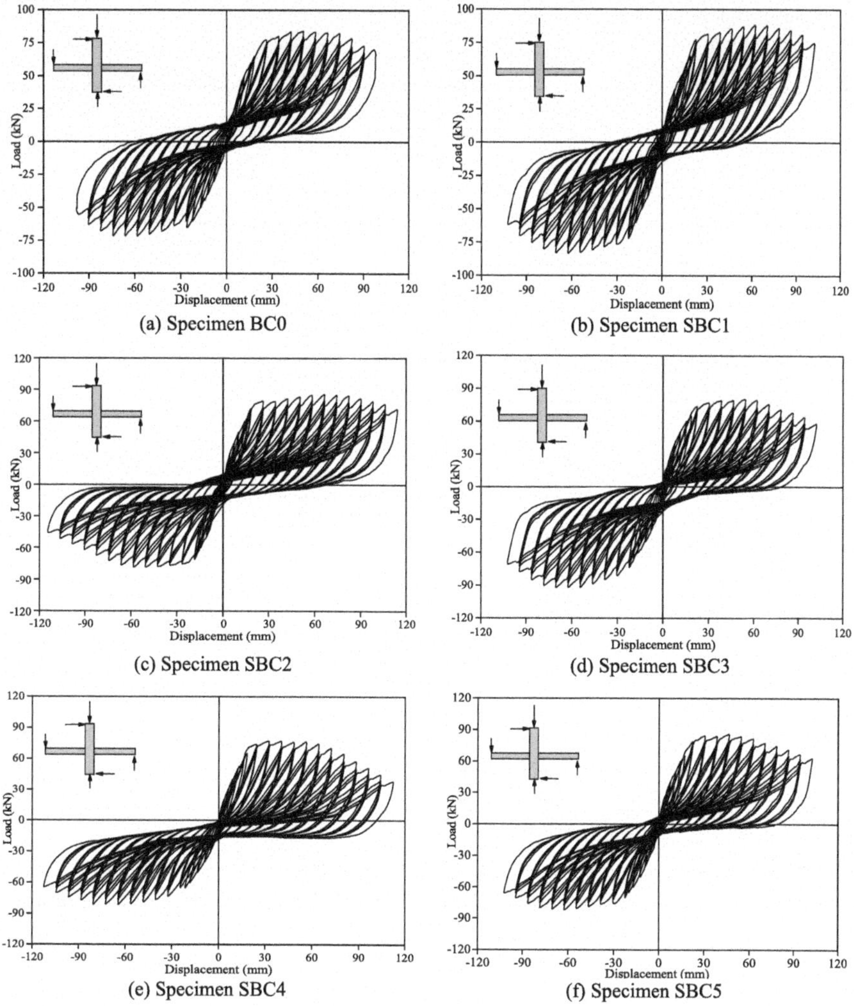


**Fig. 5.59** Load–displacement hysteretic responses of all test specimens

the ultimate displacement of specimen SBC3 was not increased as expected. Additionally, the displacement ductility ratios or ultimate rotation of specimens SBC1, SBC2, SBC3, SBC4, and SBC5 increased by 5.2%, 3.9%, 4.2%, 1.8%, and 0.8% or 6.7%, 14.3%, 0, 14.3%, and 6.7% contrasting with those of reference specimen BC0, respectively. No evident increase in the ultimate rotation of specimen SBC3 was found contrasted with that of reference specimen BC0. Generally, ductility ratios and ultimate rotation of beam-column joints increased after strengthening with BFRP sheets.

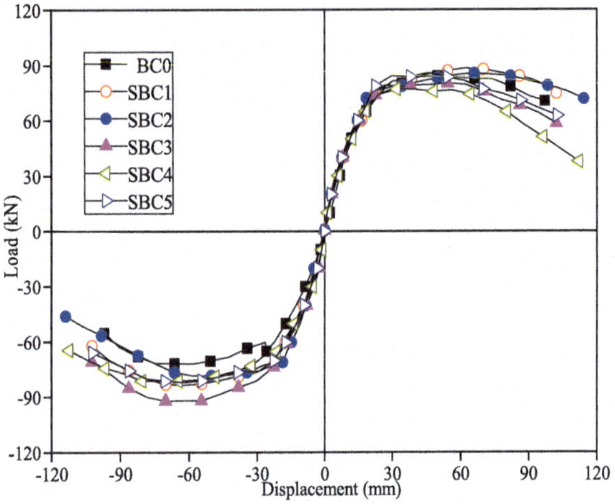


**Fig. 5.60** Envelope curves of hysteretic loops for reference and strengthened specimens

## 5.4.5 *Stiffness Degradation*

The secant stiffness is considered as the slope of the peak point of the first circle in the process of cyclic load [36], and can be calculated by Eq. (5.9) [5, 35, 83, 105].

A comparison of the secant stiffness of reference and strengthened specimens is depicted in Fig. 5.61. The initial stiffness of specimens BC0, SBC1, SBC2, SBC3, SBC4, and SBC5 was 4.7, 5.4, 5.1, 8.3, 8.1, and 8.2 kN/mm, which increased by 14.9%, 8.5%, 76.6%, 72.3%, and 74.5% contrasting with that of reference specimen BC0, respectively, as depicted in Table 5.19. The initial stiffness of strengthened specimens significantly increased, among which specimens SBC3-SBC5 had larger initial stiffness than those of other strengthened specimens may attribute to the confinement of BFRP sheets in the core area. As the test proceeded, the stiffness of all test specimens decreased as the displacement increased, while the strengthened specimens exhibited a lower stiffness degradation than that of reference specimen. No evident changes among these strengthened specimens SBC1-SBC5 in stiffness degradation could be found, as reported in [78]. The final stiffness of specimens BC0, SBC1, SBC2, SBC3, SBC4, and SBC5 was 0.6, 0.8, 0.7, 0.7, 0.7, and 0.7 kN/mm, which increased by 33.3%, 16.7%, 16.7%, 16.7%, and 16.7% contrasting with that of reference specimen BC0, respectively. The initial and final stiffness both significantly increased after strengthening with BFRP sheets, which were beneficial for RC structures suffering the sudden loss of stiffness under the earthquake as reported in [79, 83].

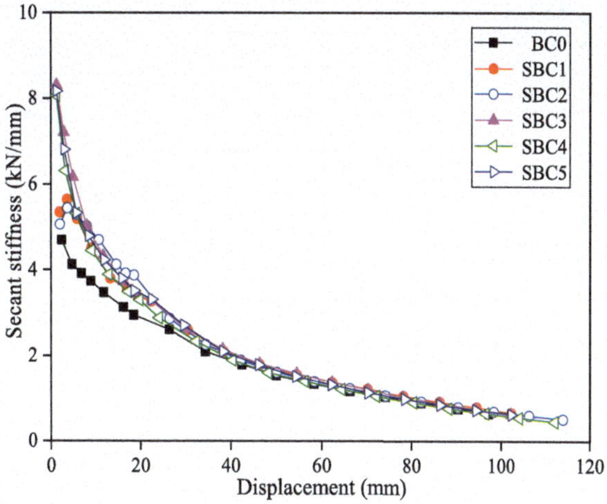


**Fig. 5.61** Secant stiffness degradation of reference and strengthened specimens

**Table 5.19** Test results of secant stiffness and energy dissipation

| Specimen | Layers of BFRP sheets | $K_0$ (kN/mm) | Ratio of strengthened specimens to specimen BC0 (%) | $K$ (kN/mm) | Ratio of strengthened specimens to specimen BC0 (%) | $Q$ (kN·m) | Ratio of strengthened specimens to specimen BC0 (%) |
|---|---|---|---|---|---|---|---|
| BC0 | – | 4.7 | 100 | 0.6 | 100 | 70.3 | 100 |
| SBC1 | 3 | 5.4 | 114.9 | 0.8 | 133.3 | 90.6 | 122.8 |
| SBC2 | 2 | 5.1 | 108.5 | 0.7 | 116.7 | 105.1 | 149.5 |
| SBC3 | 3 | 8.3 | 176.6 | 0.7 | 116.7 | 87.2 | 124.0 |
| SBC4 | 2 | 8.1 | 172.3 | 0.7 | 116.7 | 77.1 | 109.7 |
| SBC5 | 2 | 8.2 | 174.5 | 0.7 | 116.7 | 75.6 | 107.5 |

*Note* $K_0$ is the initial secant stiffness; $K$ is the final secant stiffness; and $Q$ is the cumulative dissipated energy

## 5.4.6 Energy Dissipation Capacity

A comparison of the calculated equivalent viscous damping ratios of all test specimens at the first circle is depicted in Fig. 5.62. The initial $\xi_{eq}$ of specimens BC0, SBC1, SBC2, SBC3, SBC4, and SBC5 was 4.8%, 9.9%, 8.3%, 10.3%, 9.0%, and 9.1%, which significantly increased by 106.3%, 72.9%, 114.6%, 87.5%, and 89.6% contrasting with that of reference specimen BC0, respectively, especially the specimens SBC1 and SBC3 strengthened with three layers of BFRP sheets. The initial equivalent viscous damping ratios and the dissipated energy capacity could be increased by the hardened BFRP sheets and epoxy resin on the surface of beam-column joints. After the resins cracked, equivalent viscous damping ratios of specimens SBC1, SBC2, SBC3, SBC4, and SBC5 decreased, then all test specimens

Fig. 5.62 Equivalent viscous damping ratio $\xi_{eq}$ of test specimens

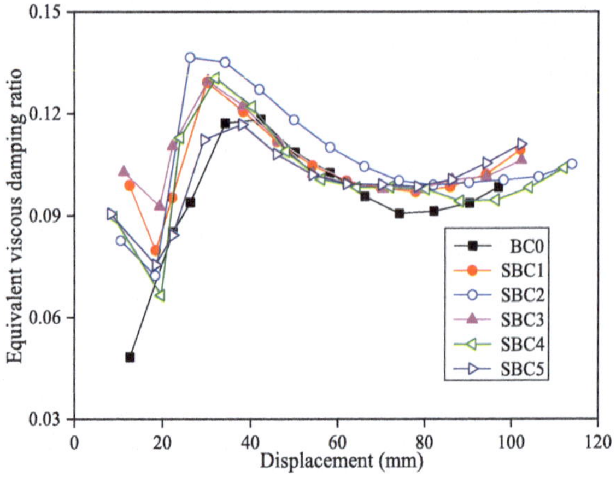


increased and finally decreased with the increase of displacement. The final $\xi_{eq}$ of specimens BC0, SBC1, SBC2, SBC3, SBC4, and SBC5 was 9.8%, 11.0%, 10.5%, 10.6%, 10.4%, and 11.1%, which increased by 12.2%, 7.1%, 8.2%, 6.1%, and 13.3% contrasting with that of the reference specimen BC0, respectively.

The cumulative dissipated energy of all test specimens is depicted in Fig. 5.63, values of which increased with the increase of displacement. The total cumulative energy dissipation of specimens BC0, SBC1, SBC2, SBC3, SBC4, and SBC5 was 70.3, 90.6, 105.1, 87.2, 77.1, and 75.6 kN · m, which increased by 28.9%, 49.5%, 24.0%, 9.7%, and 7.5% contrasting with that of the reference specimen BC0, respectively, as depicted in Table 5.21. Strengthening the beam-column joints with three layers of BFRP sheets indicated that the cumulative energy dissipation was significantly increased than that of specimens strengthened with two layers, as well as the specimen strengthened with two layers of BFRP sheets in whole lengths of beams and columns. Thus, utilizing the BFRP for strengthening the beam-column joints and increasing the layers of BFRP sheets could significantly increase the energy dissipation capacity of RC members under the earthquake.

## 5.4.7 *Shear Deformation of Core Area*

The shear strain of the core area is utilized to evaluate the shear deformations and considered as an important investigation of the beam-column joint [106, 107]. The shear strains were utilized to contrast the effect of the strengthening schemes of BFRP sheets on the interior RC beam-column joints.

Two LVDTs designed as D1 and D2 were installed on the surface of the specimens to measure the deformations, as depicted in Fig. 5.3a. The average shear strains of test specimens are calculated by Eq. (5.48) [106]:

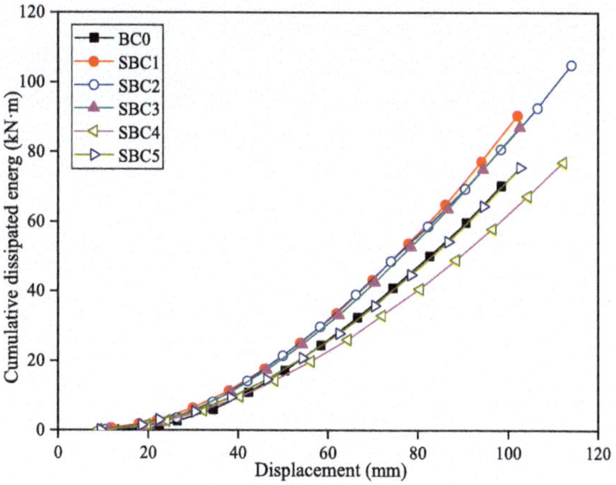


**Fig. 5.63** Cumulative dissipated energy of test specimens

$$\gamma = \frac{\sqrt{a^2 + b^2}}{2ab}(\delta_1 - \delta_2) \tag{5.48}$$

where $\gamma$ is the shear strain in the core area; $a$ and $b$ are the dimensions of the core area, as depicted in Fig. 5.64, in mm; $\delta_1$ and $\delta_2$ are the measured displacements of the LVDTs, in mm, respectively.

Figure 5.65 depicts the relationship between the lateral load and shear strain, while the shear strains of specimen SBC2 were lost for some reason. The strain curves of test specimens were different with each other. The shear strain of specimens BC0 was 0.0055 at the peak load, then the strain increased with the increase of lateral load. After that, the concrete in the core area crushed soon and LVDTs fell. The shear strains of specimens SBC1, SBC3, SBC4, and SBC5 were 0.0045, 0.0051, 0.0027, and 0.0014

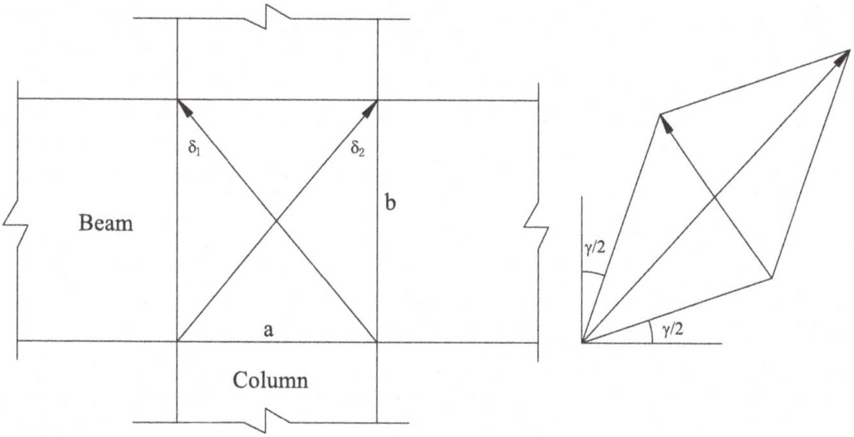


**Fig. 5.64** Deformations of the core area

at the peak load, which decreased by 18.2%, 7.3%, 50.9%, and 74.5% contrasting with that of the reference specimen BC0, respectively. As depicted in Fig. 5.65b–c, the load-bearing capacity and stiffness of the beams and column increased after strengthening with strips, wrapped sheets, and L-typed sheets in specimens SBC1 and SBC3, however, which might contribute to more shear deformation concentrating in the core area. With BFRP sheets stuck on the core area, the shear deformations of which in specimens SBC4 and SBC5 were constrained, as depicted in Fig. 5.65d–e. Test results indicated that the proposed strengthening scheme applied in specimens SBC4 was feasible and cost-efficient. Furthermore, the shear strains of specimen SBC5 were relatively smaller than those of other strengthened specimens, which might attribute to the confinement of X-typed BFRP strips stuck on the core area. The shear strains and deformations of the core area could be effectively reduced by additional BFRP sheets or strips with X-typed.

### 5.4.8 Strains of Reinforcements, Stirrups, and BFRP Sheets

(1) Strains of Reinforcements and Stirrups

Partial strains of longitudinal reinforcements, stirrups in beams and column are depicted in Figs. 5.66, 5.67, 5.68 and 5.69. The arrangement of strain gauges is depicted in Fig. 5.56, and the strains of reinforcements and stirrups were obtained by the first cycle of each lateral displacement.

The strain-lateral displacement of longitudinal reinforcements BL11 in beams is depicted in Fig. 5.66, and exhibits an obvious tension of the reinforcing bars. Strains of the longitudinal reinforcements remained linear and were not high in the initial loading stage, which indicated that the reinforcing bars were in elastic stage and not yielded yet. The curves of specimens SBC1, SBC2, SBC3, SBC4, and SBC5 were identical with that of reference specimen BC0, indicating that the strains of reinforcing bars were limitedly restrained by the FRP sheets. As the test proceeding, the strains increased as the displacement increased. After the reinforcing bars yielded, the strains of which were large under the increasing displacements. In the later loading stage, strains of longitudinal reinforcements in specimens SBC1-SBC5 were significantly lower than those of specimen BC0 when the concrete cracked. This is due to the collaborative work of FRP sheets and reinforcing bars under the lateral load, making the reinforcing bars experience less load and lower strain than before [61]. As depicted in Fig. 5.66, the max strains of specimens BC0, SBC1, SBC2, SBC3, SBC4, and SBC5 were 3638, 3419, 1566, 2857, 3349, and 2877 $\mu\varepsilon$, which decreased by 6.0%, 57.0%, 21.5%, 7.9%, and 20.9% contrasting with the reference specimen BC0, respectively. Besides, the strains of longitudinal reinforcements in specimen SBC2 were obviously lower than those of reference specimen and other strengthened specimens, which may attribute to the whole length of FRP sheets experiencing a larger load and the continuous restraint of the FRP sheets on the longitudinal reinforcements.

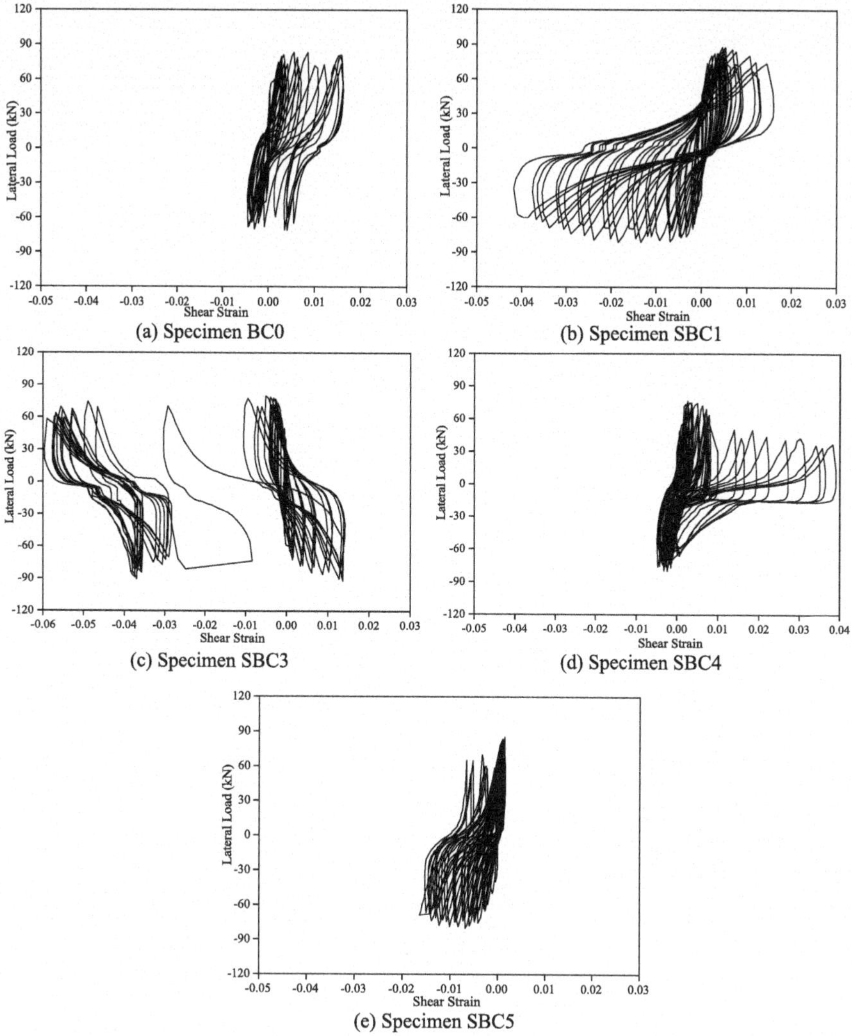


**Fig. 5.65** Relationship of the lateral load versus shear strain

The strain-lateral displacement of stirrup BS1 in beams is depicted in Fig. 5.67 and exhibits an obvious tension of the reinforcing bars. Tensile strains of the stirrups remained linear and were nearly symmetry in the pull and push stage of the initial loading stage. The curves of specimens SBC1, SBC2, SBC3, SBC4, and SBC5 were identical with those of reference specimen BC0. As the test proceeding, the strains increased as the displacement increased. In the later loading stage, strains of stirrups in specimens SBC1, SBC2, SBC3, SBC4, and SBC5 were lower than those of specimen BC0 when the concrete cracked. As depicted in Fig. 5.67, the maximum strains of specimens BC0, SBC1, SBC2, SBC3, SBC4, and SBC5 were 487, 487,

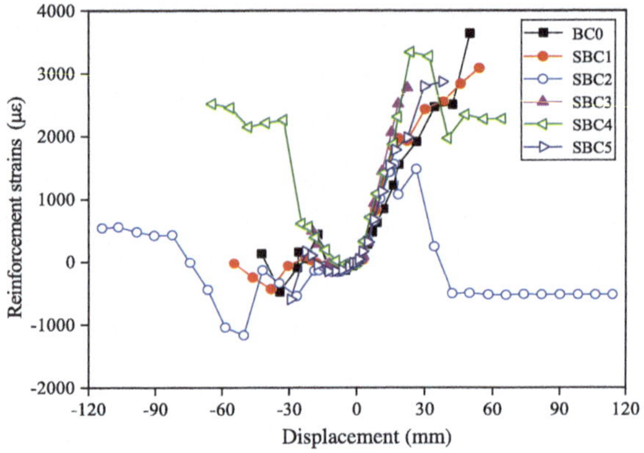


**Fig. 5.66** Strains of longitudinal reinforcements in beams of test specimens

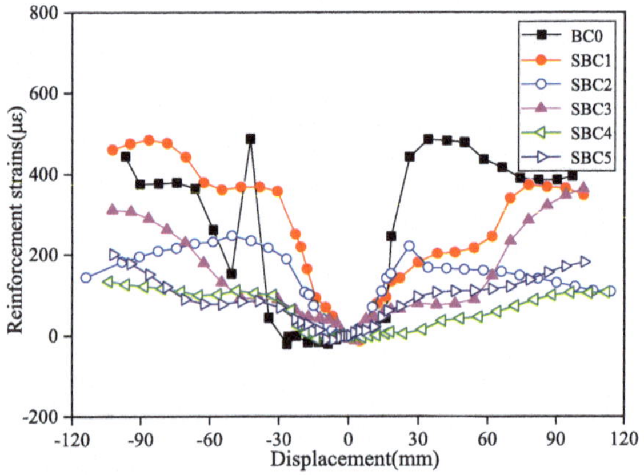


**Fig. 5.67** Strains of stirrup in beams of test specimens

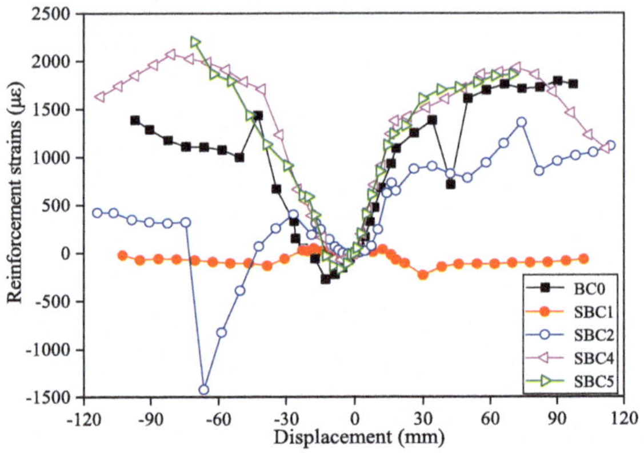


**Fig. 5.68** Strains of longitudinal reinforcements in columns of test specimens

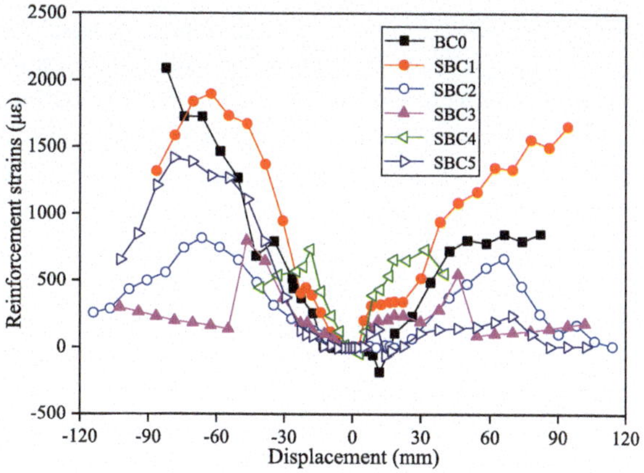


**Fig. 5.69** Strains of stirrup in columns of test specimens

250, 365, 135, and 203 με, which decreased by 0%, 48.7%, 25.1%, 72.3%, and 58.3%, respectively, contrasting with those of reference specimen BC0. Similarly, the strains of stirrups in specimen SBC2 were notably low. Additionally, strains of stirrups in all test specimens were no more than 500 με, which were considerably less than those of longitudinal reinforcements in specimens, indicating that stirrups played an important role in the shear resistance, and no shear failure occurred in the test. In coincidence with the design concept of strong shear weak bending, stirrups with less strains contribute to the formation of bending plastic hinge at the beam end [108].

The strain-lateral displacement of longitudinal reinforcements CL33 in columns is depicted in Fig. 5.68, among which the strains of longitudinal reinforcements in columns of specimen SBC3 was lost or damaged. Strains of longitudinal reinforcements in columns were relatively less than those of which in beams of test specimens, indicating that the reinforcing bars in columns had a good performance on experiencing the axial load and lateral load with FRP sheets under the constraint of transverse FRP sheets. Consequently, the yielding of longitudinal reinforcements in beams was prior to that in columns, which was in coincidence with the failure mode of bending of the beams and no cracks occurring on the columns. Besides, strains of longitudinal reinforcements in columns of specimens SBC1 and SBC2 were obviously less than those of specimens BC0, SBC4, and SBC5, indicating that the strains of reinforcing steels were influenced by the layers and strengthening schemes of BFRP sheets.

(2) Strains of Reinforcements and Stirrups

Figure 5.69 depicts the strain-lateral displacement of stirrup CS2 in columns of all test specimens. Transverse deformations of columns were mainly undertaken by stirrups experiencing a higher value of strains, however, the strains of FRP sheets were relatively lower indicating that no evident effects were found in constraining the columns at the initial stage before the concrete cracked as reported in [109]. The max strains of specimens BC0, SBC1, SBC2, SBC3, SBC4, and SBC5 were 2090,

1900, 821, 802, 737, and 1419 με, which decreased by 9.1%, 60.7%, 61.6%, 64.7%, and 32.1%, respectively, contrasting with that of reference specimen BC0. Strains of strengthened specimens SBC1, SBC2, SBC3, SBC4, and SBC5 were significantly lower than those of reference specimen BC0. As the test proceeding, strains of the stirrups increased as the displacement increased. The transverse deformations of columns increased, and the concrete was constrained by the stirrups and transverse wrapping FRP sheets at the later loading stage.

Figure 5.70 depicts the strains of concrete surface SC1 of specimen BC0, and FRP sheets SF1 at the corresponding position of specimens SBC1, SBC2, SBC3, SBC4, and SBC5. The strains of test specimens exhibited significant differences at the pull and push loading stages. Contrasting with the concrete strains of specimen BC0, FRP sheets strains of specimens SBC1, SBC2, SBC3, SBC4, and SBC5 exhibited relatively lower values at the initial stage and indicated that the FPR sheets had a limited effect on the constraint of concrete transverse deformations, which were mainly constrained by the stirrups as depicted in Fig. 5.70. The increasing strains of FRP sheets may attribute to the constraint of concrete with stirrups at the later loading stage [109]. Maximum strains of specimens BC0, SBC1, SBC2, SBC3, SBC4, and SBC5 were 260, 37, 37, 31, 107, and 34 με, which decreased by 85.8%, 85.8%, 88.1%, 58.8%, and 86.9%, respectively, contrasting with that of reference specimen BC0. The measured FRP sheets strains in the test were considerably less than the ultimate average strain of 2.2%, which may be attribute to the crushed concrete of core area and deteriorated capacity of the joint, indicating that the RC structures or members had sufficient ductility margin under the earthquake. Furthermore, no specific anchorage was utilized for the FRP sheets in the beam-column joint, and the strength of FPR could predictably not be fully achieved. Similar conclusions of FRP strengthening specimens are also reported in [79, 110].

Figure 5.71 depicts the strains of concrete surface at the top and bottom of the beam SC5-SC6 of specimen BC0, and FRP sheets SF5-SF6 of specimens SBC1, SBC2, SBC3, SBC4, and SBC5 at the corresponding position. However, for some reason, the

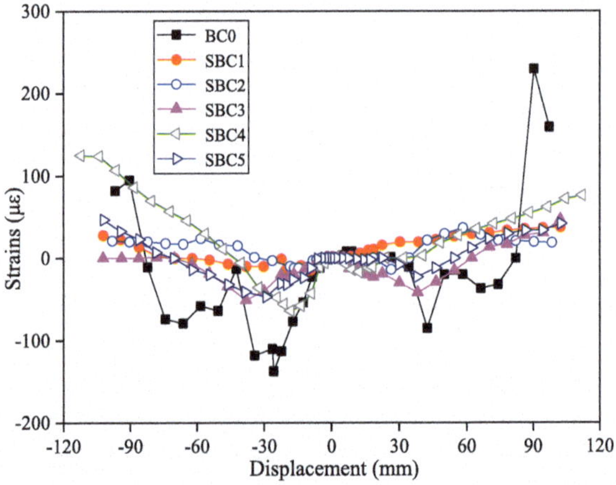


**Fig. 5.70** Strains of concrete or FRP sheets on the columns of test specimens

FRP sheets strains of specimen SBC2 were lost or damaged. As depicted in Fig. 5.71a, strains of SC5 and SC6 were significantly different and asymmetric indicating that the top and bottom surface of concrete at the same cross-section experienced an uneven loading, where compressive strain could be observed. While the FRP sheets strains were relatively large, contributing to constraining the concrete transverse deformations of concrete at the later loading stage, as depicted in Fig. 5.71b–e. Besides, the strains of beams in specimen BC0 decreased with a short descent segment after the max value. However, the FRP strains of strengthened specimens exhibit a long descent segment indicating that the beam-column joints have a good performance of plastic deformation after strengthening with BFRP sheets, which is beneficial for RC structures or members under the earthquake [5, 83].

## 5.5 Summary

Investigations on the seismic performance of corroded RC beam-column joints, corroded joints retrofitted with BFRPs, earthquake-damaged and corrosion-damaged joints retrofitted with BFRPs, uncorroded joints strengthened with BFRPs in different methods were conducted. The models for the seismic performance of joints in consideration of earthquake damage, corrosion rate, and FRP confinement were also proposed. The major results obtained from this chapter were summarized as follows:

(1) The reinforcement corrosion affected the seismic performance of RC beam-column joints. The reinforcement corrosion rate increased and load-bearing capacity of corroded RC beam-column joints decreased with the decrease of half-cell potentials. The spalling and crushing of concrete, as well as the buckling of the corroded stirrups occurred in the corroded joints. The load-bearing capacity and cumulative energy dissipation of corroded joints decreased with the increase of corrosion rate. The models for the load-bearing capacity and reinforcement corrosion rate of RC beam-column joints were proposed in consideration of half-cell potentials.

(2) The seismic performance of corroded RC beam-column joints improved by retrofitting with BFRP sheets. The application of BFRP sheets to confine the corroded RC beam-column joints improved the bond behavior between reinforcing bars and concrete. The ductile failure mode of the corroded RC beam-column joints retrofitted with BFRP sheets was observed.

(3) The seismic performance of earthquake-damaged and corroded RC beam-column joints could be retrofitted with BFRP sheets. The spalling and crushing of concrete, as well as the buckling of stirrups of damaged joints were delayed by retrofitting with BFRP sheets. The initial or final stiffness, ultimate rotations, displacement ductility ratios, cumulative energy dissipation of damaged joints improved by retrofitting with BFRP sheets. The models for the shear

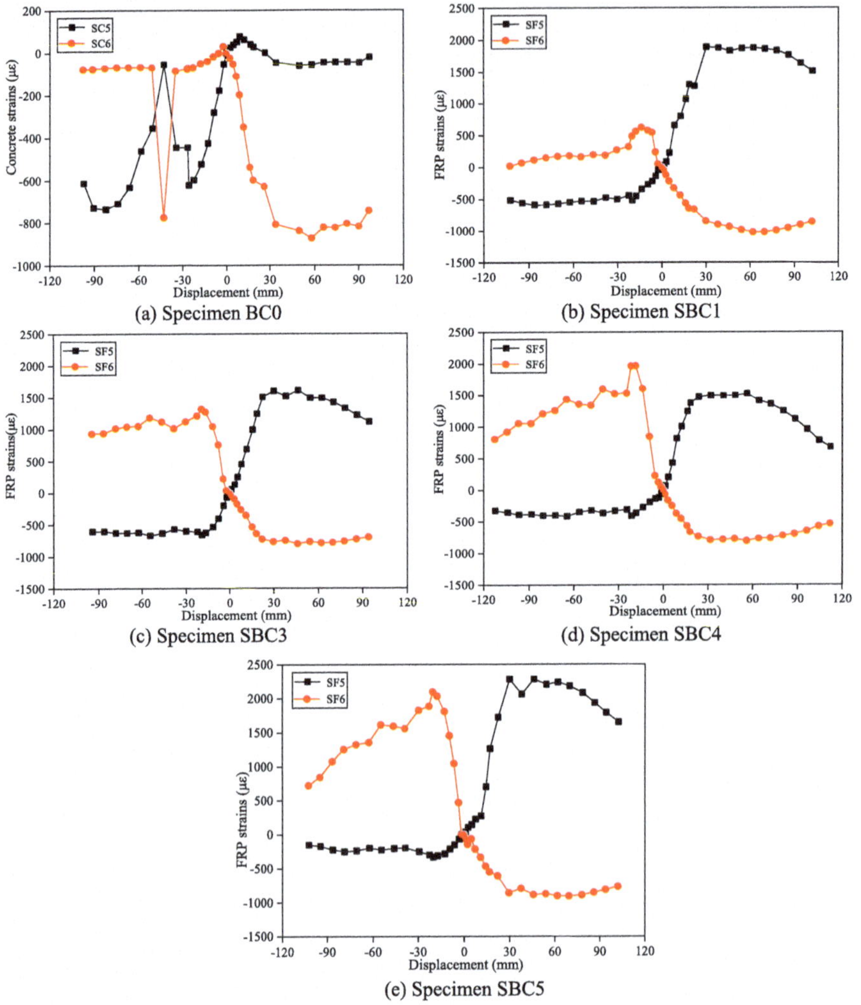


**Fig. 5.71** Strains of concrete and FRP sheets on the beams

strength of joint core area of earthquake-damaged and corroded RC beam-column joints were proposed in consideration of earthquake damage, corrosion rate, and retrofitting with FRP sheets.

(4) The seismic performance of RC beam-column joints could be improved by strengthening with BFRP sheets in five types. The hysteretic loops of strengthened RC beam-column joints exhibited full and fusiform curves. The initial or final secant stiffness, displacement ductility ratios, ultimate rotation, load-bearing capacities, the final equivalent viscous damping ratios, and the total cumulative energy dissipation of RC beam-column joints improved by strengthening with BFRP sheets compared to those corresponding values of joints

without strengthening. The shear deformations of the core area of RC beam-column joints were effectively reduced by the confinement of BFRP sheets. The strengthening techniques for increasing the seismic performance of joints were developed in consideration of the influence of slabs, layers, regions, and types of FRP sheets.

## References


1. Kanchanadevi A, Ramanjaneyulu K (2018) Effect of corrosion damage on seismic behaviour of existing reinforced concrete beam-column sub-assemblages. Eng Struct 174:601–617
2. Zhang XC, Li B (2020) Damage characteristics and assessment of corroded RC beam-column joint under cyclic loading based on acoustic emission monitoring. Eng Struct 205:110090
3. Shen DJ, Deng SC, Zhang JY et al (2015) Behavior of reinforced concrete box beam with initial cracks repaired with basalt fiber-reinforced polymer sheet. J Reinf Plast Compos 34(18):1540–1554
4. Shen DJ, Zeng X, Zhang JY et al (2019) Behavior of RC box beam strengthened with basalt FRP using end anchorage with grooving. J Compos Mater 53(23):3307–3324
5. Shen DJ, Yang Q, Huang CB et al (2019) Tests on seismic performance of corroded reinforced concrete shear walls repaired with basalt fiber-reinforced polymers. Constr Build Mater 209:508–521
6. Lu ZD, Lei SU, Jiangtao YU (2011) Experimental study on the seismic behaviour of strengthened concrete column-beam joints by simulated earthquake. Procedia Eng 14:1871–1878
7. Sasmal S, Khatri CP, Karusala R (2015) Numerical simulation of performance of near-surface mounted FRP-upgraded beam–column sub-assemblages under cyclic loading. Struct Infrastruct Eng 11(8):1012–1027
8. Karayannis CG, Sirkelis GM (2008) Strengthening and rehabilitation of RC beam-column joints using carbon-FRP jacketing and epoxy resin injection. Earthquake Eng Struct Dynam 37(5):769–790
9. Said A, Nehdi M (2008) Rehabilitation of RC frame joints using local steel bracing. Struct Infrastruct Eng 4(6):431–447
10. El-Amoury T, Ghobarah A (2002) Seismic rehabilitation of beam–column joint using GFRP sheets. Eng Struct 24(11):1397–1407
11. Realfonzo R, Napoli A, Pinilla JGR (2014) Cyclic behavior of RC beam-column joints strengthened with FRP systems. Constr Build Mater 54:282–297
12. Shen DJ, Li M, Liu C et al (2021) Seismic performance of corroded reinforced concrete beam-column joints repaired with BFRP sheets. Constr Build Mater 307:124731
13. Li M, Shen DJ, Yang J et al (2013) Relationship between half-cell potentials and load carrying capacity of corroded reinforced concrete beam-column joints. Appl Mech Mater 351–352:939–944
14. Li M, Shen DJ, Yang Q et al (2022) Rehabilitation of seismic-damaged reinforced concrete beam-column joints with different corrosion rates using basalt fiber-reinforced polymer sheets. Compos Struct 289:115397
15. Shen DJ, Li M, Yang Q et al (2022) Seismic performance of earthquake-damaged corroded reinforced concrete beam-column joints retrofitted with basalt fiber-reinforced polymer sheets. Struct Infrastruct Eng 1–17
16. Shen DJ, Li M, Kang JC et al (2021) Experimental studies on the seismic behavior of reinforced concrete beam-column joints strengthened with basalt fiber-reinforced polymer sheets. Constr Build Mater 287:122901

17. Ministry of Housing and Urban-Rural Development of the People's Republic of China (2019) Standard for test methods of concrete physical and mechanical properties: GB/T 50081-2019. China Architecture and Building Press, Beijing (in Chinese)
18. Poursaee A, Hansson CM (2009) Potential pitfalls in assessing chloride-induced corrosion of steel in concrete. Cem Concr Res 39(5):391–400
19. Yuan Y, Ji Y, Shah SP (2007) Comparison of two accelerated corrosion techniques for concrete structures. ACI Struct J 104(3):344–347
20. Hou L, Zhou B, Guo S et al (2019) Corrosion behavior and flexural performance of reinforced concrete/ultrahigh toughness cementitious composite (RC/UHTCC) beams under sustained loading and shrinkage cracking. Constr Build Mater 198:278–287
21. Mak MWT, Desnerck P, Lees JM (2019) Corrosion-induced cracking and bond strength in reinforced concrete. Constr Build Mater 208:228–241
22. Li Q, Jin X, Yan D et al (2021) Study of wiring method on accelerated corrosion of steel bars in concrete. Constr Build Mater 269:121286
23. Li B, Qian K (2011) Seismic behavior of reinforced concrete interior beam-wide column joints repaired using FRP. J Compos Constr 15(3):327–338
24. Mosallam A, Allam K, Salama M (2019) Analytical and numerical modeling of RC beam-column joints retrofitted with FRP laminates and hybrid composite connectors. Compos Struct 214:486–503
25. Lam ES, Li B, Xue Z et al (2019) Experimental studies on reinforced concrete interior beam-column joints strengthened by unsymmetrical chamfers. Eng Struct 191:575–582
26. Khan MI, Al-Osta MA, Ahmad S et al (2018) Seismic behavior of beam-column joints strengthened with ultra-high performance fiber reinforced concrete. Compos Struct 200:103–119
27. Ministry of Housing and Urban-Rural Development of the People's Republic of China (2015) Specification for seismic test of buildings: JGJ/T 101-2015. China Architecture and Building Press, Beijing (in Chinese)
28. Hoffmann K (1996) Practical hints for the installation of strain gages. Hottinger Baldwin Messtechnik GmbH, p 56
29. Ouyang LJ, Gao WY, Zhen B et al (2017) Seismic retrofit of square reinforced concrete columns using basalt and carbon fiber-reinforced polymer sheets: a comparative study. Compos Struct 162:294–307
30. Gunes O, Lau D, Tuakta C et al (2013) Ductility of FRP–concrete systems: Investigations at different length scales. Constr Build Mater 49:915–925
31. Lam SSE, Wu B, Wong YL et al (2003) Drift capacity of rectangular reinforced concrete columns with low lateral confinement and high-axial load. J Struct Eng 129(6):733–742
32. Pour-Ghaz M, Isgor OB, Ghods P (2009) Quantitative interpretation of half-cell potential measurements in concrete structures. J Mater Civ Eng 21(9):467–475
33. ASTM C876-09 (2009) Standard test method for half-cell potentials of uncoated reinforcing steel in concrete. American Society for Testing and Materials, West Conshohocken, PA
34. Antonopoulos CP, Triantafillou TC (2003) Experimental investigation of FRP-strengthened RC beam-column joints. J Compos Constr 7(1):39–49
35. Xu CX, Peng S, Liu XQ et al (2020) Analysis of the seismic behavior of CFRP-strengthened seismic-damaged composite steel-concrete frame joints. Journal of Building Engineering 28:101057
36. Roy B, Laskar AI (2018) Cyclic performance of beam-column subassemblies with construction joint in column retrofitted with GFRP. Structures 14:290–300
37. Faleschini F, Gonzalez-Libreros J, Zanini MA et al (2019) Repair of severely-damaged RC exterior beam-column joints with FRP and FRCM composites. Compos Struct 207:352–363
38. Ilia E, Mostofinejad D (2019) Seismic retrofit of reinforced concrete strong beam–weak column joints using EBROG method combined with CFRP anchorage system. Eng Struct 194:300–319
39. Sola E, Ožbolt J, Balabanić G et al (2019) Experimental and numerical study of accelerated corrosion of steel reinforcement in concrete: Transport of corrosion products. Cem Concr Res 120:119–131

40. Mostofinejad D, Hajrasouliha M (2019) 3D beam–column corner joints retrofitted with X-shaped FRP sheets attached via the EBROG technique. Eng Struct 183:987–998
41. Liu GY (2011) Experimental study on seismic behavior of corroded reinforced concrete beam joints. Central South University, Hunan (in Chinese)
42. Cheng L (2008) Investigation of BFRP-reinforced pre-damaged concrete beam-column joints under reversed cyclic loading. Tongji University, Shang Hai (in Chinese)
43. Su L, Lu Z, Zhang K et al (2010) Experimental study on BFRP-reinforced pre-damaged concrete column-beam joints by simulated earthquake. J Southeast Univ (Nat Sci Ed) 40(3):559–564 (in Chinese)
44. Lin HW, Zhao YX, Ožbolt J et al (2017) The bond behavior between concrete and corroded steel bar under repeated loading. Eng Struct 140:390–405
45. Liu XF, Li Y (2018) Experimental study of seismic behavior of partially corrosion-damaged reinforced concrete columns strengthened with FRP composites with large deformability. Constr Build Mater 191:1071–1081
46. Lei L, Zheng S, Ning Z (2013) Performance-based comprehensive evaluation on the seismic behavior of the corroded RC column. Chin Civil Eng J 46(S2):50–56 (in Chinese)
47. Li Q, Niu D, Xiao Q et al (2018) Experimental study on seismic behaviors of concrete columns confined by corroded stirrups and lateral strength prediction. Constr Build Mater 162:704–713
48. Ou YC, Nguyen ND (2016) Influences of location of reinforcement corrosion on seismic performance of corroded reinforced concrete beams. Eng Struct 126:210–223
49. Ye Z, Zhang W, Gu X (2018) Deterioration of shear behavior of corroded reinforced concrete beams. Eng Struct 168:708–720
50. Rajput AS, Sharma UK (2018) Corroded reinforced concrete columns under simulated seismic loading. Eng Struct 171:453–463
51. Firouzi A, Abdolhosseini M, Ayazian R (2020) Service life prediction of corrosion-affected reinforced concrete columns based on time-dependent reliability analysis. Eng Fail Anal 117:104944
52. Radhi MS, Gorgis IN, Hassan MS (2021) Analytical analysis of design-oriented models for forecasting the performance of CFRP-confined corrosion-affected concrete columns. Constr Build Mater 313:125491
53. Choe G, Shinohara Y, Kim G et al (2020) Concrete corrosion cracking and transverse bar strain behavior in a reinforced concrete column under simulated marine conditions. Appl Sci 10(5):1794
54. Jia J, Zhao L, Wu S et al (2020) Experimental investigation on the seismic performance of low-level corroded and retrofitted reinforced concrete bridge columns with CFRP fabric. Eng Struct 209:110225
55. Kwapisz M, Vorwagner A, Rebhan M et al (2018) Investigations on existing concrete cantilever walls subjected to reinforcement corrosion. In: High tech concrete: where technology and engineering meet: proceedings of the 2017 fib symposium, held in Maastricht, The Netherlands, June 12–14, 2017. Springer International Publishing, pp 2083–2091
56. Yang J, Guo T, Chai S (2020) Experimental and numerical investigation on seismic behaviours of beam- column joints of precast prestressed concrete frame under given corrosion levels. Structures 27:1209–1221
57. Yang J, Guo T, Chai S et al (2019) Cyclic tests on beam-column joints of SCOPE system with different key slot lengths and corrosion levels. Eng Struct 201:109406
58. Luo XY, Cheng JF, Xiang P et al (2020) Seismic behavior of corroded reinforced concrete column joints under low-cyclic repeated loading. Archives Civil Mech Eng 20:1–20
59. Li XG, Jun MJ, Zhou YM, Shi PP et al (2018) Experimental study on seismic performance of corroded RC frame joints after fire. J Build Struct 39(03):84–92
60. Dong JF, Zhao YX, Wang K et al (2017) Crack propagation and flexural behaviour of RC beams under simultaneous sustained loading and steel corrosion. Constr Build Mater 151:208–219
61. Zhang HF, Wu J, Jin FY et al (2019) Effect of corroded stirrups on shear behavior of reinforced recycled aggregate concrete beams strengthened with carbon fiber-reinforced polymer. Compos B Eng 161:357–368

62. Maaddawy TA, Soudki KA (2003) Effectiveness of impressed current technique to simulate corrosion of steel reinforcement in concrete. J Mater Civ Eng 15(1):41–47
63. Chen JW, Fu CQ, Ye HL et al (2020) Corrosion of steel embedded in mortar and concrete under different electrolytic accelerated corrosion methods. Constr Build Mater 241:117971
64. Liu JH, Luo XY, Xiao Y (2021) Non-uniform corrosion and distribution models of residual cross-sectional areas of rebars. J Huazhong Univ Sci Technol (Nat Sci Ed) 49(11):83–88 (in Chinese)
65. Jin NG, He JH, Fu CQ et al (2020) Study on experimental method and morphology of accelerated non-uniform corrosion of steel bars. J Zhejiang Univ (Eng Ed) 54:483–490 (in Chinese)
66. Al-Hammoud R, Soudki K, Topper TH (2010) Bond analysis of corroded reinforced concrete beams under monotonic and fatigue loads. Cement Concr Compos 32(3):194–203
67. Shen DJ, Wu SX (2004) Experimental study on the properties and models of corroded reinforcement in concrete exposured to the atmospheric environment. China Concr Cement Prod 46–50 (in Chinese)
68. Dong BQ, Fang GH, Liu YQ et al (2017) Monitoring reinforcement corrosion and corrosion-induced cracking by X-ray microcomputed tomography method. Cem Concr Res 100:311–321
69. Zheng AH, Li S, Zhang DF et al (2021) Shear strengthening of RC beams with corrosion-damaged stirrups using FRP grid-reinforced ECC matrix composites. Compos Struct 272:114229
70. Alhawat M, Ashour A (2019) Bond strength between corroded steel reinforcement and recycled aggregate concrete. Structures 19:369–385
71. Fu CQ, Fang DM, Ye HL et al (2021) Bond degradation of non-uniformly corroded steel rebars in concrete. Eng Struct 226:111392
72. Shang H, Chai X (2021) Bond behavior between corroded steel bar and concrete under reciprocating loading history of beam type specimens. Eng Struct 247:113112
73. China Iron and Steel Association (2017) Steel for the reinforcement of concrete—Part 1: hot rolled plain bars: GB/T 1499.1-2017. Standards Press of China (in Chinese)
74. China Iron and Steel Association (2018) Steel for the reinforcement of concrete—Part 2: hot rolled ribbed bars: GB/T 1499.2-2018. Standards Press of China (in Chinese)
75. Shen DJ, Zhang JY, Cui ZH et al (2012) Device and method for measuring corrosion rate of reinforcement along length direction. China, ZL 201210303757.9 (in Chinese)
76. Ghobarah A, Abou-Elfath H, Biddah A (1999) Response-based damage assessment of structures. Earthquake Eng Struct Dynam 28(1):79–104
77. Zgür Yurdakul Ö, Avşar Ö (2015) Structural repairing of damaged reinforced concrete beam-column assemblies with CFRPs. Struct Eng Mech 54(3):521–543
78. Obaidat YT, Abu-Farsakh GAFR, Ashteyat AM (2019) Retrofitting of partially damaged reinforced concrete beam-column joints using various plate-configurations of CFRP under cyclic loading. Constr Build Mater 198:313–322
79. Li B, Qian K, Tran CTN (2013) Retrofitting earthquake-damaged RC structural walls with openings by externally bonded FRP strips and sheets. J Compos Constr 17(2):259–270
80. General Administration of Quality Supervision, Inspection and Quarantine of People's Republic of China (1999) Test method for tensile properties of oriented fiber reinforced plastics (GB/T 3354-1999). Standards Press of China, Beijing (in Chinese)
81. Yuan F, Chen MC, Pan JL (2019) Experimental study on seismic behaviours of hybrid FRP–steel-reinforced ECC–concrete composite columns. Compos B Eng 176:107272
82. Wang GL, Dai JG, Bai YL (2019) Seismic retrofit of exterior RC beam-column joints with bonded CFRP reinforcement: An experimental study. Compos Struct 224:111018
83. Shen DJ, Yang Q, Jiao Y et al (2017) Experimental investigations on reinforced concrete shear walls strengthened with basalt fiber-reinforced polymers under cyclic load. Constr Build Mater 136:217–229
84. Attari N, Youcef YS, Amziane S (2019) Seismic performance of reinforced concrete beam–column joint strengthening by frp sheets. Structures 20:353–364

85. Caltrans SDC (2019) Seismic design criteria version 2.0. State of California, Department of Transportation, Sacramento, CA
86. Khorramian K, Sadeghian P (2019) Performance of high-modulus near-surface-mounted FRP laminates for strengthening of concrete columns. Compos B Eng 164:90–102
87. Hou W, Xu SL, Ji DS et al (2019) Seismic performance of steel plate reinforced high toughness concrete coupling beams with different steel plate ratios. Compos B Eng 159:199–210
88. Sun FJ, Pang SH, Zhang ZW et al (2020) Retrofitting seismically damaged steel sections encased concrete composite walls using externally bonded CFRP strips. Compos Struct 236:111927
89. Ministry of Housing and Urban-Rural Development of the People's Republic of China (2010) Code for seismic design of buildings: GB 50010-2010. China Architecture and Building Press, Beijing (in Chinese)
90. European Committee for Standardization (2005) Design of structures for earthquake resistance-part 1: general rules, seismic actions and rules for buildings: EN 1998-1. Brussels
91. ACI (2002) Recommendations for design of beam-column connections in monolithic reinforced concrete structures: 352R-02. American Concrete Institute, Farmington Hills
92. Canadian Standards Association (2004) Design of concrete structures. Canadian Standards Association, Mississauga, Ont
93. Zhang XC, Li B (2018) Shear-strength capacity assessment of corroded reinforced concrete beam-column joints. J Perform Constr Facil 32(5):04018067
94. Du YG, Clark LA, Chan AHC (2005) Residual capacity of corroded bars. Mag Concr Res 57(3):135–147
95. Lee WT, Chiou YJ, Shih MH (2010) Reinforced concrete beam–column joint strengthened with carbon fiber reinforced polymer. Compos Struct 92(1):48–60
96. Hadi MNS, Tran TM (2014) Retrofitting nonseismically detailed exterior beam–column joints using concrete covers together with CFRP jacket. Constr Build Mater 63:161–173
97. Chang ZF, Yang ZY, Li SJ (2016) Experimental study on CFRP-strengthened damaged concrete column-beam joints. J Huazhong Univ Sci Technol (Nat Sci Ed) 44:65–69,75 (in Chinese)
98. Choi JH, Chang DK (2009) Prevention of progressive collapse for building structures to member disappearance by accidental actions. J Loss Prev Process Ind 22(6):1016–1019
99. Liang H, Long L (2012) A quantification method of structural robustness. Eng Mech 29(8):213–219
100. LI M (2014) Study on seismic performance of reinforced concrete beam-column joints retrofitted with basalt fiber reinforced polymer. Hohai University, Nanjing (in Chinese)
101. Chen JF, Teng JG (2001) Anchorage strength models for FRP and steel plates bonded to concrete. J Struct Eng 127(7):784–791
102. Youssef MA, Alam MS, Nehdi M (2008) Experimental investigation on the seismic behavior of beam-column joints reinforced with superelastic shape memory alloys. J Earthquake Eng 12(7):1205–1222
103. Yang J (2012) Study on Seismic Performance and Retrofitting Measures of Corroded Reinforced Concrete Beam-Column Joints. Hohai University, Nanjing (in Chinese)
104. Wang ZH, Du XL, Zhan JD et al (2009) Experimental study on the seismic behavior of beam-column joints strengthened with AFRP sheets. J Beijing Univ Technol 35(01):30–35 (in Chinese)
105. Xu G, Li A (2019) Seismic performance of a new type precast concrete sandwich wall based on experimental and numerical investigation. Soil Dyn Earthq Eng 122:116–131
106. Le-Trung K, Lee K, Lee J et al (2010) Experimental study of RC beam–column joints strengthened using CFRP composites. Compos B Eng 41(1):76–85
107. Sasmal S, Voggu S (2019) Performance evaluation of differently designed beam-column joints under cyclic shear-torsion loading. Eng Struct 197:109447
108. Congxiao W, Weishan L, Yun Z (2015) Experimental study on seismic behaviors of new energy-dissipative prefabricated concrete frame structure joints. Chin Civil Eng J 48(9):23–30

109. Moradi E, Naderpour H, Kheyroddin A (2020) An experimental approach for shear strengthening of RC beams using a proposed technique by embedded through-section FRP sheets. Compos Struct 238:111988
110. Torabian A, Isufi B, Mostofinejad D et al (2020) Flexural strengthening of flat slabs with FRP composites using EBR and EBROG methods. Eng Struct 211:110483

# Chapter 6
# Retrofitting of Corrosion-Damaged Reinforced Concrete Shear Walls with BFRP Sheets

RC shear walls are commonly used in high-rise buildings to resist lateral loads such as winds and earthquakes [1]. Their benefits such as high load-bearing capacity and high energy dissipation make them an attractive alternative in the construction of high-rise buildings particularly in areas of seismic activity. However, one of the main problems limiting their practical application is stiffness degradation during an earthquake event. As a result, strengthening approaches for RC shear walls including the use of thin steel sheets or FRP [2–7], and the use of steel/FRP hybrid system [8] has increasingly developed. It has been proved that bonding of FRP sheets on RC shear walls is the most cost-effective alternative in strengthened buildings in a very short time [9]. The use of thin steel sheets indeed improves the bending and vibration characteristics of walls; however, FRP materials improve even more as result of better flexibility. Moreover, the use of FRP for strengthening existing buildings has been drawn attention, due to the low weight-strength ratio, short installation periods, corrosion resistance, and low intervention on the structures [10, 11]. The load-bearing capacity and durability of structures can also be improved effectively.

The seismic performance of RC structures significantly decreased after the reinforcing bars corroded. However, the assessment and predicting models of the seismic performance of corroded RC shear walls were comparatively limited. Existing researches mainly concentrate on the modeling of corroded RC beams [12], columns [13, 14], and beam-column joints [15], and several models for predicting the corroded members are also proposed. The models based on ASCE 41-13 taking into account the effect of corrosion of reinforcing bars on the load-bearing capacities of transverse reinforcement and concrete of corroded columns have been proposed in [16]. In these methods, a good coincidence of the tested and calculated results of load-bearing capacities of corroded RC short columns utilizing the residual cross-sectional areas of reinforcing bars after corrosion was found. Besides, a two-dimensional numerical model for investigating the essential influences on the load-bearing capacities of corroded joints was also recommended in [17]. The modified model based on the Strut-and-Tie model were utilized for predicting the load-bearing capacity of shear



D. Shen, *Seismic Performance of Corroded Reinforced Concrete Structures Retrofitted with FRP*, https://doi.org/10.1007/978-981-99-7984-4_6

walls strengthened or retrofitted with BFRP, as reported in [17, 18]. Although efforts for accurately modeling the corrosion of reinforcing bars on the load-bearing capacities of columns and beam-column joints have been made, the methods for predicting the load-bearing capacities of corroded shear walls were yet limited. Therefore, analytical model considering the influence of corrosion of reinforcing bars on the shear strength or FE modeling for simulating the deterioration of corroded shear walls are needed and significant.

Firstly, the corrosion-damaged shear walls were accomplished by the accelerating corrosion test and tested under the cyclic loadings. The seismic performances of corrosion-damaged shear walls were investigated, as well as the analytical and FE models for predicting the corroded walls were proposed [19]. Secondly, the corrosion-damaged beam-column joints were retrofitted with BFRP sheets, and the seismic performances of these walls retrofitted with BFRP sheets were discussed and compared to those of uncorroded joints. Lastly, the seismic performance of uncorroded shear walls retrofitted with BFRP sheets in different methods considering the method of fan anchor, X-shaped, and lateral BFRP strip were investigated. Besides, models for predicting the shear strength of shear walls were proposed in [17, 18].

## 6.1 Seismic Performance of Corrosion-Damaged Shear Walls

### *6.1.1 Experimental Program*

(1) Description of Wall Specimens

The experimental program presents in this work involves four specimens (denoted as SHW-1, SHW-2, SHW-3 and SHW-4). These specimens are of the cantilever type with a stiff beam at the top and a heavily reinforced foundation block at the base. The specimens have rectangular 120 mm × 1000 mm cross-sections and height 1600 mm. The boundary elements are rectangular and of a cross-section of 225 mm × 120 mm. These boundary elements are reinforced with six reinforcing bars of 10-mm diameter. Confinement in the boundary elements is provided by 8-mm diameter closed stirrups spaced at 150 mm. Figure 6.1 shows the details of the wall specimens. Concrete strength is 30 MPa. Average values of measured yielding strengths of 10-mm and 8-mm diameter bars are 335 and 235 MPa, respectively.

(2) Accelerated Corrosion Method

An accelerated corrosion method was used to accelerate the corrosion of the reinforcing bars. The specimens were fully immersed in an aqueous solution of 5.0% sodium chloride by weight for seven days before being subjected to the accelerated corrosion procedure. Before and during this procedure, the temperature of the water over each concrete specimen was held constant at 25 ± 2 °C. The specimens are shown in Fig. 6.2.

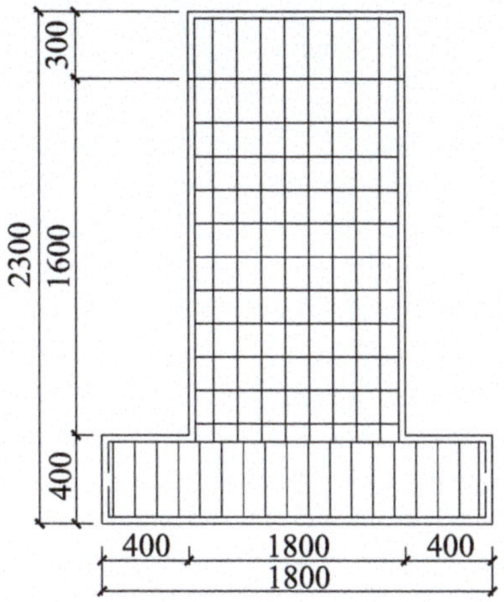


**Fig. 6.1** Geometry and reinforcement details of all specimens (all units in mm)

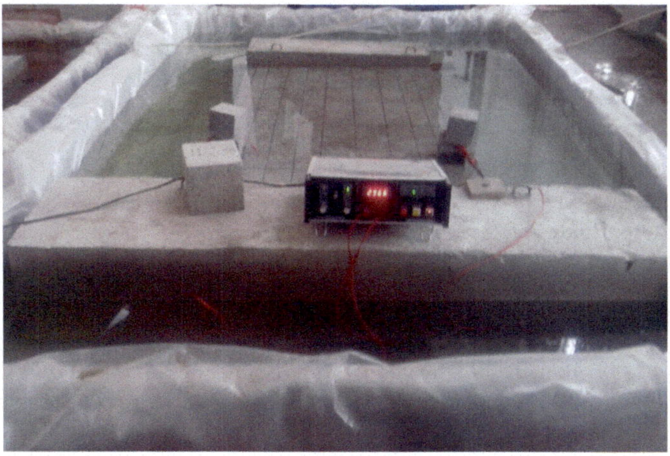

**Fig. 6.2** Accelerated corrosion test of shear wall specimens

An adjustable direct current supply providing 24 V constant potentials at 0–3A was used for the accelerated corrosion process. The direct current was applied to the main reinforcing bar embedded in the concrete, using the reinforcing bar as the anode and a copper plate as the cathode. The current was passed from the reinforcing bars to the copper plate placed inside of 5.0% salt solution (NaCl). The setup for the accelerated corrosion process is shown in Fig. 6.2.

The designed mass loss $\Delta m$ of the reinforcing bars due to corrosion was calculated according to Faraday's law using Eq. (6.1).

$$\Delta m = \frac{MIt}{ZF} \tag{6.1}$$

where $t$ is the duration of exposure; $I$ is the average magnitude of electrical current; $M$ is the molar mass for iron (56 g/mol for iron); $F$ is the Faraday's constant and taken as 96,500 A·s; and $Z$ is the ionic charge (2 for Fe).

The quantity of charge applied for any given electrolysis is given by the product of time and current. For the corrosion process, for each mol of iron that oxidizes, 2 mol of electrons are given out, consuming a charge of 2 × 96,487 C. The mass loss is then calculated by multiplying the applied charge by the molar mass and dividing by the charge needed per mole. This technique was successfully adopted by several researchers to induce corrosion in RC elements [20].

In order to attain the corrosion rate 3%, 9%, and 15%, the Eqs. (6.2) and (6.3) can be used to evaluate the total time that the current was applied on the reinforcing bars.

$$\Delta m = \frac{1}{4}\pi D^2 L \gamma \eta \tag{6.2}$$

$$t = \frac{\pi D^2 L \gamma \eta F}{4MI} = 1.35 \times 10^4 \frac{L\eta}{I} \tag{6.3}$$

where $D$ is the diameter of the reinforcing bars, in mm; $L$ is the total length of reinforcing bars, in mm; $\gamma$ is the density of the iron and equal to $7.86 \times 10^{-3}$ g/mol; $\eta$ is the corrosion rate, in %.

The duration time $t$ of exposure can be evaluated by Eq. (6.3) according to Eqs. (6.1) and (6.2). The time $t$ is shown in Table 6.1.

(3) Arrangement of Measuring Half-Cell Potentials

The half-cell potentials technique can be used to locate corroding reinforcing bars without disrupting the concrete cover to the reinforcing steel. A copper–copper sulfate half-cell 200 concrete corrosion tester is shown in Fig. 6.3. It consists of a rigid tube or container composed of a dielectric material that is nonreactive with copper or copper sulfate, a porous wooden or plastic plug that remains wet by capillary action, and a copper rod that is immersed within the tube in a saturated solution of copper sulfate. The solution shall be prepared with reagent grade copper sulfate crystals dissolved in distilled. The solution may be considered saturated when an

**Table 6.1** Duration time of duration

| Specimens | Current intensity (A) | Corrosion rate (%) | Date (days) |
|---|---|---|---|
| SHW-0 | – | – | – |
| SHW-1 | 1.0 | 3 | 18 |
| SHW-2 | 3.0 | 9 | 18 |
| SHW-3 | 3.0 | 15 | 30 |

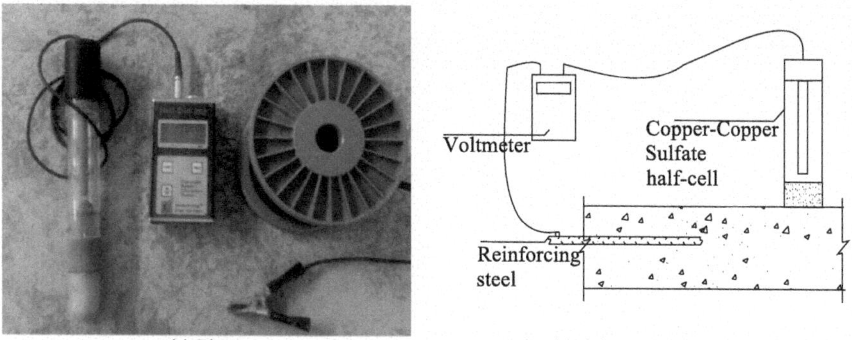


(a) Photo (b) Schema of the half-cell

**Fig. 6.3** Half-cell 200 concrete corrosion tester

excess of crystals lies at the bottom of the solution. Before testing the specimen, pre-wetting on the concrete surface should be conducted [21].

The lateral surface of the specimens SHW-2, SHW-3, and SHW-4 was divided into 24 meshes as illustrated in Fig. 6.4.

(4) Experimental Setup

Figure 6.5 shows the test setup, including loading system, and instrumentation. The wall base was fixed to the strong floor of the laboratory by high strength steel bolts. All specimens had stiff RC stubs at top and bottom for fixing with loading frame. All specimens were subjected to a constant axial load of 325 kN throughout the test and a number of cycles of quasi-static cyclic loads at the head beam which was created by a horizontal hydraulic actuator. Magnitudes of the lateral loads were measured by a load cell (500 kN push ↔ 500 kN pull capacity). Large databases of wall tests [22] indicate that the vast majority of the tests have been subjected to axial load ratios less than 0.1. To conveniently compare the test results with the future test results, the axial load ratio of 0.09 was adopted for all specimens.

The horizontal loading history was divided into two parts at both forward and backward cycles, as shown in Fig. 6.6 and reported in Liao et al. [23]. The force-control stage was performed at the load levels increasing from $\pm 0.25P_y$ to $\pm 0.75P_y$ by $\pm 0.125P_y$. Only one cycle was imposed at each load level in this stage. During the displacement control stage, two cycles were imposed at the incremental displacement levels increasing from $\pm \Delta_y$ to $\pm 2\Delta_y$ by $\pm 0.125\Delta_y$. The tests were stopped when the lateral loads of specimens had dropped by 15% from their ultimate lateral loads. The top displacement and the lateral load were monitored during the test. Cracks were marked on the surfaces of specimens at the peak load of each half cycle. The test data is collected by the computerized data acquisition system where the horizontal load–displacement curve has been drawn and the yielding of reinforcement and pre-damaged displacement can be determined. LVDTs were used for displacement measurements and their arrangement is shown in Fig. 6.5.

Fig. 6.4 Measuring points of half-cell potentials

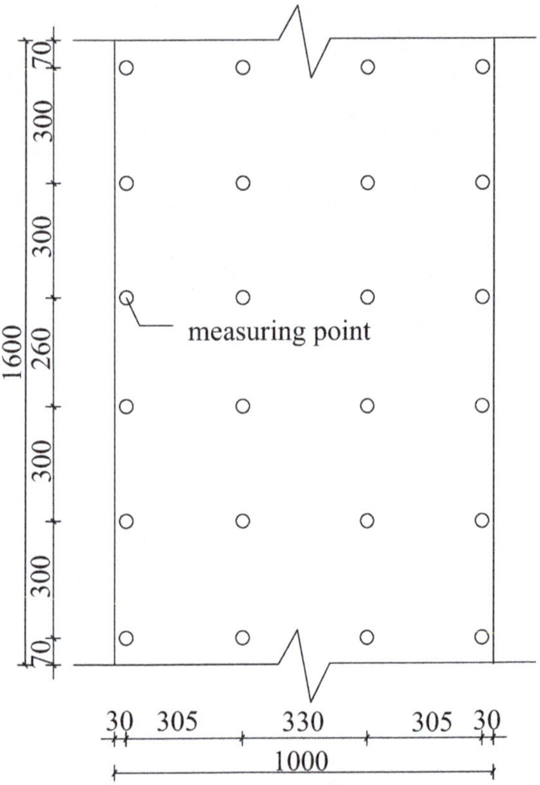


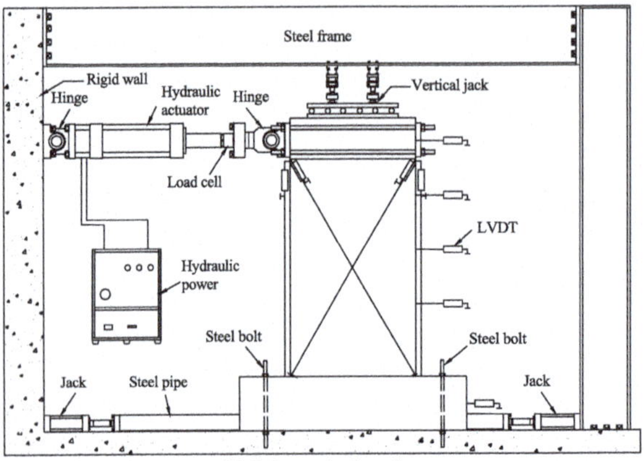


(a) Schematic drawing

(b) Photo

**Fig. 6.5** Test setup layout

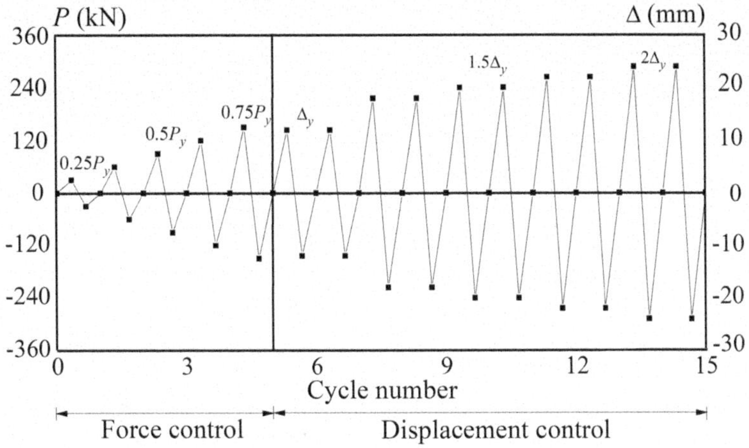


**Fig. 6.6** Horizontal loading history

### *6.1.2 Relationship Between Half-Cell Potentials, Corrosion Rate, and Load-Bearing Capacity*

(1) Relationship Between Half-cell Potentials and Corrosion Probability

The half-cell potentials can be used to evaluate the corrosion probability according to American standard ASTM C876-91 [24]. If potentials over an area are more positive than − 200 mV, there is a greater than 90% probability that no reinforcement corrosion occurrs in that area at the time of measurement. If potentials over an area are in the range of − 350 mV to − 200 mV, corrosion activity of the reinforcing bar in that area is uncertain. If potentials over an area are more negative than − 350 mV, there is a greater than 90% probability that reinforcement corrosion occurrs in that area at the time of measurement. Before electrification, the potentials of all shear walls were in the range of − 200 mV to − 350 mV. It illustrates that corrosion activity of the reinforcing bar in that area was uncertain. After electrification, the potentials of SHW-2 and SHW-3 were in the range of − 200 mV to − 350 mV. It illustrates that corrosion activity of the reinforcing bar in that area was still uncertain. However, the potentials of SHW-4 were more negative than − 350 mV. It illustrates that there was a greater than 90% probability that reinforcement corrosion of SHW-4 occurrs.

(2) Relationship Between Half-cell Potentials and Corrosion Rate

Before electrification, the average value of half-cell potentials of specimen SHW-2 was − 272 mV. After 1A current was applied on the specimen SHW-2 for 18 days, the half-cell potentials average value of specimen SHW-2 was − 322 mV. The half-cell potentials decreased at 50 mV and at the degree of 18.4%. The corrosion rate of SHW-2 reached 3%.

Before electrification, the half-cell potentials average value of specimen SHW-3 was − 263 mV. After 3A current was applied on the specimen SHW-3 for 18 days,

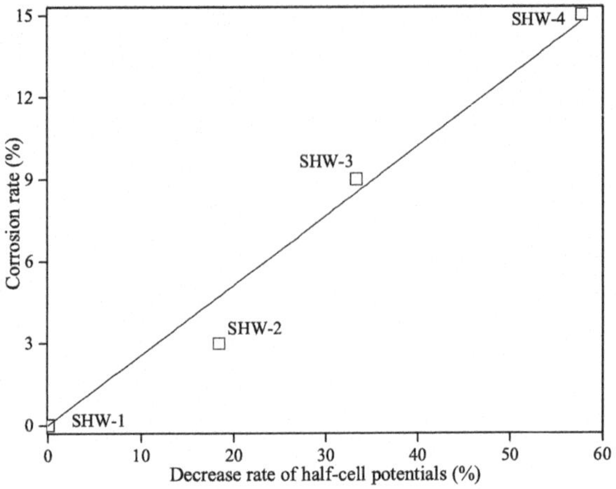


**Fig. 6.7** Relationship between corrosion rate and decrease of half-cell potential of specimens

the half-cell potentials average value of specimen SHW-3 was − 350 mV. The half-cell potentials decreased at 87 mV and at the degree of 33.3%. The corrosion rate of SHW-3 reached 9%.

Before electrification, the half-cell potentials average value of specimen SHW-4 was − 272 mV. After 3A current was applied on the specimen SHW-4 for 30 days, the half-cell potentials average value of specimen SHW-4 was − 429 mV. The half-cell potentials decreased at 157 mV and at the degree of 57.7%. The corrosion rate of SHW-4 reached 15%.

Generally, the corrosion rate of walls increased with the decrease of half-cell potentials. Figure 6.7 represents that there is very significantly linear relationship between them. The trend line is given as Eq. (6.4).

$$\rho_w = 0.255\beta \tag{6.4}$$

where $\beta$ is decrease rate of half-cell potentials in %; and $\rho_w$ is corrosion rate, in %.

(3) Relationship Between Half-cell Potentials and Load-bearing Capacity

After electrification, load-bearing capacity of walls decreased with the half-cell potentials decreasing. The load-bearing capacity of corrosion-damaged specimen SHW-2, SHW-3, and SHW-4 was lower at the rate of 9.0%, 15.5%, and 20.6% than that of normal specimen SHW-1, respectively. When the corrosion rate of walls was in range of 0 to 15%, Fig. 6.8 represents that there is nonlinear relationship between corrosion rate and the ratio of load-bearing capacity of corroded RC walls and normal wall. The relation of them is given as Eq. (6.5).

$$R(\rho) = 0.00084\rho_w^2 - 0.024\rho_w + 1 \tag{6.5}$$

where $R(\rho)$ is the ratio of load-bearing capacity of corroded RC walls and normal wall; and $\rho_w$ is corrosion rate, in %.

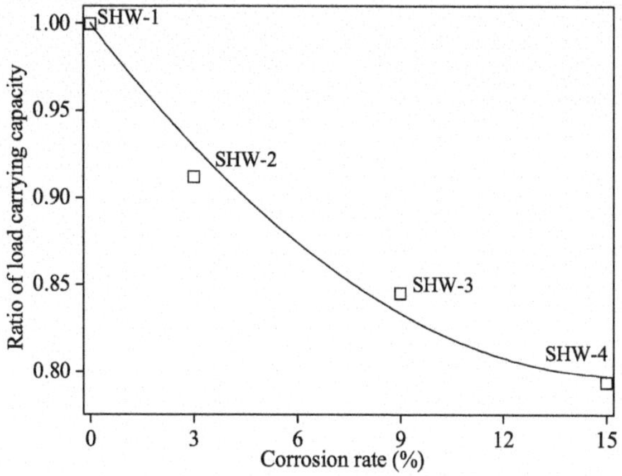


**Fig. 6.8** Relationship between the ratio of load-bearing capacity and load-bearing capacity rate of specimens

When the corrosion rate of walls was in range of 0–15%, Fig. 6.9 represents that there is very significantly linear relationship between the half-cell potentials and load-bearing capacity. The trend line is given as Eq. (6.6).

$$R(\rho) = -0.0004\rho_w + 1 \tag{6.6}$$

where $R(\rho)$ is the ratio of load-bearing capacity of corroded RC walls and normal wall; and $\rho_w$ is decrease rate of half-cell potentials, in %.

(4) Relationship Between Decrease Rate of the Half-Cell Potentials and Load-bearing Capacity

Decrease rate of loading-bearing capacity of walls increased with decrease rate of half-cell potentials. When the decrease rate of half-cell potentials of specimens SHW-2, SHW-3, and SHW-4 reached 18.4%, 33.3%, and 57.7%, the load-bearing capacity

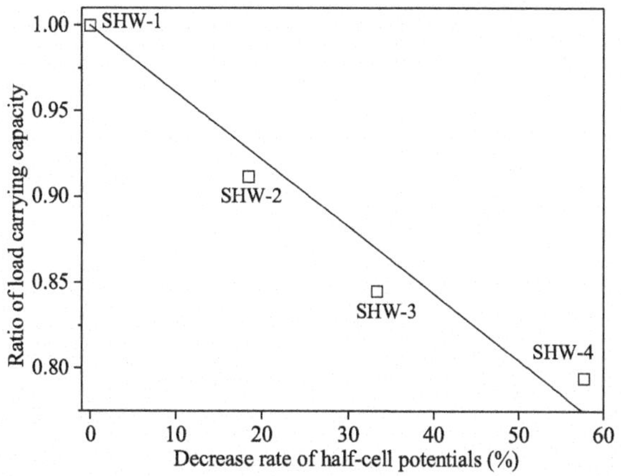


**Fig. 6.9** Relationship the ratio of load-bearing capacity and decrease rate of the half-cell potentials

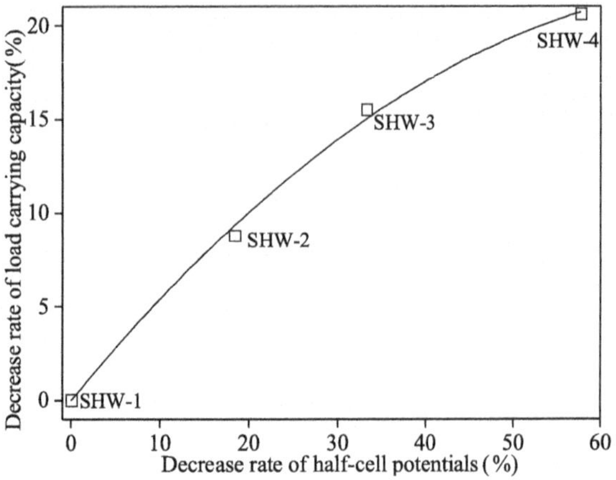


**Fig. 6.10** Relationship between decrease rate of the half-cell potentials and load-bearing capacity

was lower at the rate of 9.0%, 15.5%, and 20.6% than that of normal specimen SHW-1, respectively.

When the corrosion rate of walls was in range of 0–15%, Fig. 6.10 represents that there is nonlinear relationship between the decrease rate of half-cell potentials and decrease rate of loading-bearing capacity. In conclusion, the Eq. (6.7) is given below.

$$\gamma = -0.004\beta^2 + 0.58\beta \tag{6.7}$$

where $\beta$ is decrease rate of half-cell potentials, in %; and $\gamma$ is the decrease rate of load-bearing capacity.

### 6.1.3 *Crack Pattern*

The failure modes of corroded specimens SHW1, SHW2, and SHW3 were similar to that of benchmark specimen SHW0, experiencing the developments from the cracking of concrete to limited buckling of the longitudinal reinforcement at the base of wall; however, the crack initiation and propagation were dependent on the corrosion rate.

In force control stage, cracking of the concrete at the base of the wall could be first observed when the lateral load was 110.4, 100.8, 85.0, and 76.9 kN for specimens SHW0, SHW1, SHW2, and SHW3, respectively. Adjacent to the base of the walls in tension, horizontal flexural cracks were observed and then the initial obliquely cracks developed along the height of the walls.

Cracks of concrete at the overhead areas of the walls were observed and the horizontal cracks occurred at the boundary elements simultaneously under the condition that the lateral load increased. Figure 6.11 shows that initial cracks prolonged to the inner parts of the walls and formed the diagonal shear cracks at the end of the loading stage.

In the stage controlled by displacement, the horizontally flexural cracks near the bottom developed to the center of the wall. Meanwhile, diagonal shear cracks could be observed along almost the whole height of the wall. However, diagonal cracks of specimens with higher corrosion rate were more intensive and had a larger width, as shown in Fig. 6.11. At the base of the wall, the concrete cover spalled and the reinforcing bars exposed when the horizontal displacement was 32.0 (1.82% drift), 28.1 (1.61% drift), 27.5 (1.57% drift), and 24.2 mm (1.38% drift) for specimens SHW0, SHW1, SHW2, and SHW3, respectively. In a short time, the specimens failed instantly with limited buckling of longitudinal reinforcements after the concrete spalled. The spalling area of the concrete cover was larger and the concrete crushed

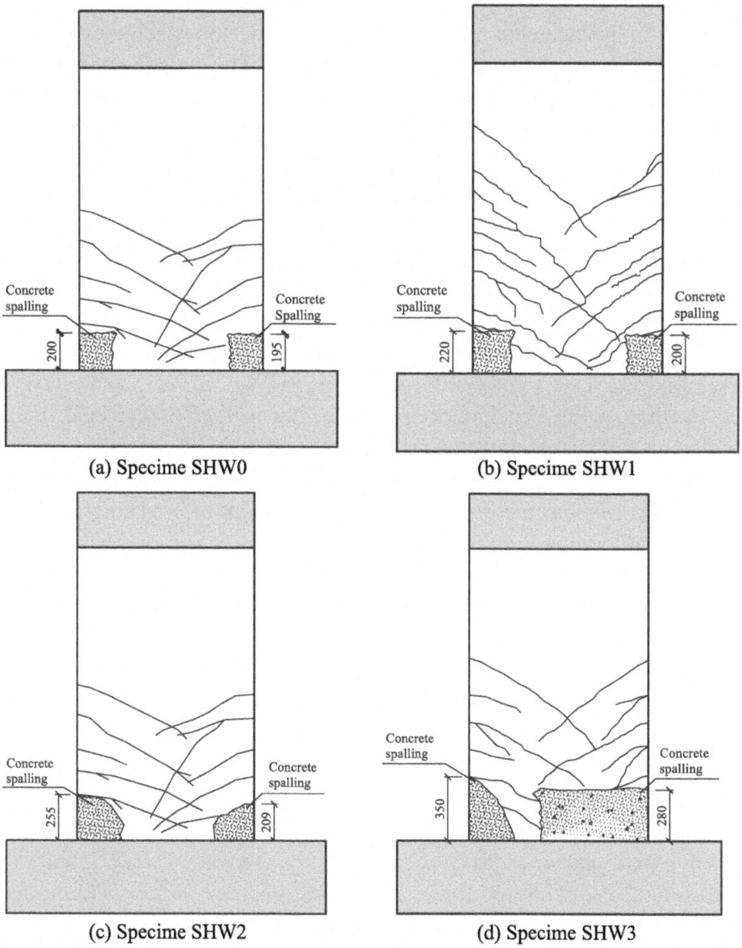


**Fig. 6.11** Crack pattern and concrete spalling area of specimens with different corrosion rates (all units in mm)

more severely when the corrosion rate increased. Due to premature cracks along the reinforcing bars caused by corrosion products, corroded specimens failed earlier than the benchmark specimen. A progressive development of shearing cracks along with the deterioration of load-bearing capacity of the benchmark wall could be observed after reaching the peak lateral load. However, the severely corroded walls failed in a brittle manner with the main diagonal cracks from the wall base to the head beam suddenly appearing when reaching the peak lateral load, and the rapid decrease of shear strength were observed in the post-peak stage. Brittle post-peak behaviors could be found in corroded RC shear walls, especially the walls with high corrosion rates.

### *6.1.4 Hysteretic Response and Skeleton Curves*

The lateral load–displacement hysteretic loops of all tested specimens are shown in Fig. 6.12, and the internal region of these loops could also be utilized for estimating the energy dissipation capacity. The hysteretic curve of specimen SHW0 was roughly spindle-shaped, while apparent pinching phenomena were observed in the hysteretic curves of specimens SHW1, SHW2, and SHW3 and increased as the corrosion rate increased. Higher corrosion rate led to further decreases in ductility and energy dissipation capacity.

The enveloped curve obtained from the hysteretic loop could also be utilized for defining the ductility or stiffness. Figure 6.12 also shows the enveloped curves of all tested specimens from the averaged values of absolute maximum loads or corresponding displacements in cyclic loadings for per cycle of hysteretic loop. The enveloped curves of corroded specimens SHW1, SHW2, and SHW3 were enclosed by that of the benchmark specimen due to their lower peak loads and ultimate displacements, demonstrating that the load-bearing capacities of corroded walls considerably decreased after reinforcing bars corroded.

Ordinary load–displacement curve of tested specimen suffered horizontally cyclic loading and axial loading is shown in Fig. 6.13.

The yielding strength $P_y$ and corresponding displacement $\Delta_y$ were determined on the basis of "energy equivalent method", as reported in [17]; $P_u$ is the ultimate strength and defined as 85% of the peak lateral load $P_{\max}$; $\Delta_u$ is the ultimate displacement of corresponding $P_u$. The enveloped curves of specimens SHW1, SHW2, and SHW3 declined rapidly after reaching the peak without obvious yielding platforms, demonstrating that the specimens failed in brittle patterns due to reinforcement corrosion. Table 6.2 shows the test results of all tested specimens.

The yielding or ultimate strength of specimens SHW1, SHW2, and SHW3 decreased by 0.9%, 8.5%, and 15.1% or 4.0%, 11.0%, and 16.4% compared with that of benchmark specimen SHW0, respectively. Specimen SHW1 with a slight corrosion rate of 3% exhibited lower decreasing degrees in the yielding and ultimate strength than those of specimens SHW2 and SHW3 with severe corrosion rates of 9% and 15%. Similar conclusion could also be found in the comparison of the yielding displacement. Moreover, the ultimate displacements of specimens

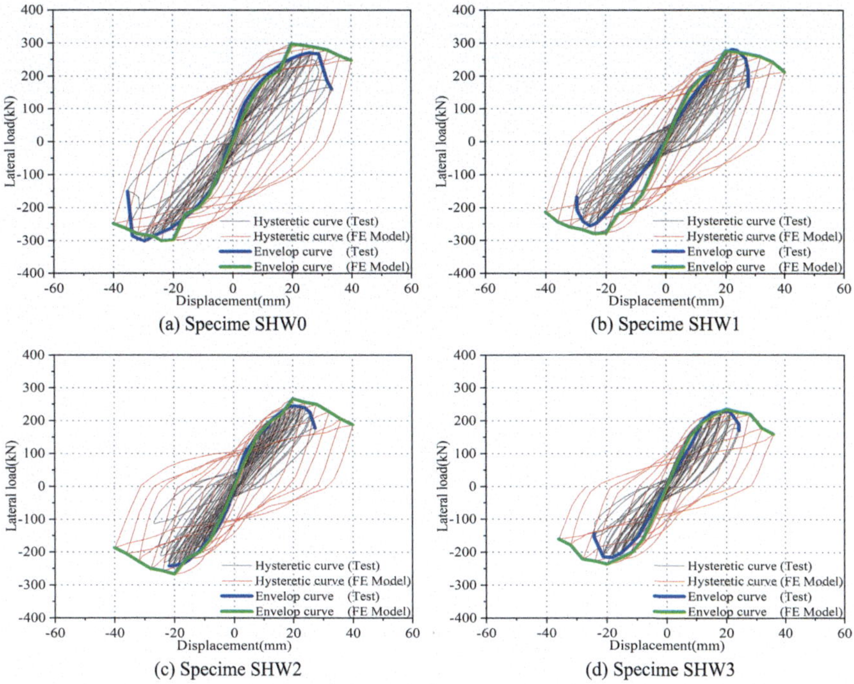


**Fig. 6.12** Load–displacement hysteretic loops of specimens

**Fig. 6.13** Definition of yielding displacement and lateral load

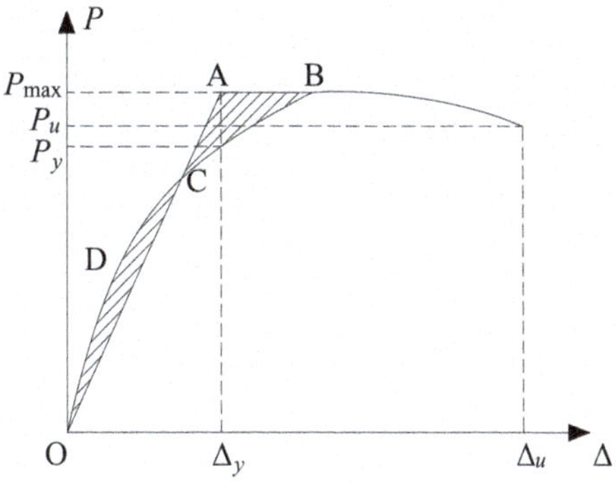


SHW1, SHW2, and SHW3 decreased by 12.8%, 24.3%, and 27.2% compared with that of benchmark specimen SHW0, respectively. The corrosion of reinforcing bars exhibited a significant influence on the load-bearing capacities and displacements of corroded shear walls, especially for those with severe corrosion rates of 9% and 15%, which contributed to the decreasing degrees of 8.5%–16.4% and 10.1%–27.2%, respectively. A coincident conclusion on the ultimate strength and displacements of

**Table 6.2** Test results of strengths and displacements of corroded specimens

| Specimen | Corrosion rate (%) | $P_y$ (kN) | Decrease ratio compared to SHW0 (%) | $P_u$ (kN) | Decrease ratio compared to SHW0 (%) | $\Delta_y$ (mm) | Decrease ratio compared to SHW0 (%) | $\Delta_u$ (mm) | Decrease ratio compared to SHW0 (%) |
|---|---|---|---|---|---|---|---|---|---|
| SHW0 | 0 | 225.3 | 0 | 272.7 | 0 | 16.8 | 0 | 31.3 | 0 |
| SHW1 | 3 | 223.2 | 0.9 | 261.8 | 4.0 | 16.0 | 4.8 | 27.3 | 12.8 |
| SHW2 | 9 | 206.2 | 8.5 | 242.7 | 11.0 | 15.1 | 10.1 | 23.7 | 24.3 |
| SHW3 | 15 | 191.3 | 15.1 | 228.0 | 16.4 | 14.6 | 13.1 | 22.8 | 27.2 |

*Note* $P_y$ is yielding strength; $P_u$ is ultimate strength; $\Delta_y$ is yielding displacement; and $\Delta_u$ is ultimate displacement severely corroded RC shear walls, exhibiting the decreasing degrees of 6.0%–22.0% and 8.1%–32.0%, could be found in [25].

### 6.1.5 Stiffness Degradation

The stiffness is ordinarily described as the capability to undergo the deformations of RC members or structures under the earthquake [26]. The secant stiffness is ordinarily applied for estimating the stiffness degradation of tested specimens at per loading cycle and calculated by Eq. (6.8) as follows [27, 28].

$$K_i = \frac{|+F_i| + |-F_i|}{|+X_i| + |-X_i|} \tag{6.8}$$

where $K_i$, $F_i$, and $X_i$ are secant stiffness, peak load, and displacement corresponding to the maximum load in the $i$th cycle in kN/mm, kN, mm, respectively.

Figure 6.14 shows the secant stiffness of tested specimens. The initial stiffness of specimens SHW1, SHW2, and SHW3 was 12.5, 11.3, and 10.9 kN/mm, which decreased by 6.7%, 15.7%, and 18.7% compared with that of benchmark specimen SHW0 of 13.4 kN/mm, respectively. The initial stiffness of corroded shear walls significantly decreased over 15% when the corrosion rate was larger than 9%. When the cyclic numbers were 5 or 17, the stiffness of specimens SHW1, SHW2, and SHW3 was 9.0, 7.9, and 7.4 kN/mm or 3.2, 3.5, and 3.1 kN/mm, which decreased by 15.9%, 26.2%, and 30.8% or 11.1%, 2.8%, and 13.9% compared with those of benchmark specimen SHW0 of 10.7 kN/mm or 3.6 kN/mm, respectively. The stiffness of corroded specimens degraded rapidly compared with that of the benchmark specimen. For lightly corroded specimen SHW1 with corrosion rate of 3%, minor strength and stiffness degradation occurred when compared with that of the benchmark specimen. While for severely corroded specimens SHW2 and SHW3 with corrosion rates of 9% and 15%, considerable stiffness degradations could be observed in the later loading stages of the curves.

### 6.1.6 Ductility and Energy Dissipation Capacity

Generally, two parameters, i.e., displacement ductility ratio and drift, are utilized for evaluating the elastic–plastic deformation of RC members [29]. The displacement ductility ratio is characterized as $\mu = \Delta_u/\Delta_y$, where $\Delta_y$ and $\Delta_u$ are yielding and ultimate displacement in mm, respectively. The drift $\theta_u$ is characterized as the ratio of ultimate displacement $\Delta_u$ to the height $h$ in mm, which is calculated from the wall base to the center of the loading point. Test results of ductility and drift of all tested specimens are shown in Table 6.3. The displacement ductility ratio or drift of corroded specimens SHW1, SHW2, and SHW3 decreased by 8.1%, 15.6%, and 16.1% or

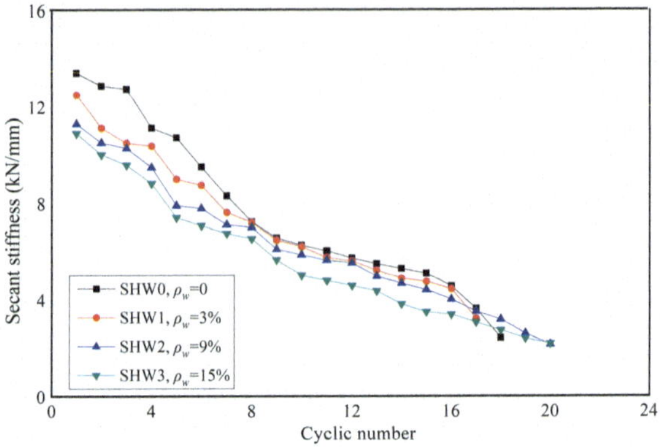


**Fig. 6.14** Secant stiffness degradation of all tested specimens

13.6%, 22.7%, and 27.1% compared with those of benchmark specimen SHW0, respectively. Ductility of severely corroded specimens SHW2-SHW3 decreased by over 15% compared with that of benchmark specimen SHW0, which was larger than that of corroded specimens SHW1. A coincident conclusion on the ductility ratio of severely corroded RC shear wall with the corrosion rate of 16.56%, exhibiting the decreasing degrees of 23.0%, could also be found in [30]. That indicated that the ductility of RC walls significantly decreased after the reinforcing bars corroded, contributing to the decrease of deformations and brittle behaviors of RC shear walls.

Results of the cumulative dissipated energy $Q$ of all tested specimens at the $i$th cycle are shown in Fig. 6.15. When the cyclic numbers were 8 or 16, the cumulative dissipated energy of specimens SHW1, SHW2, and SHW3 was 5.5, 4.2, and 3.7 kN · mm or 32.5, 20.7, and 16.9 kN ·mm, which decreased by – 37.5%, – 5.0%, and 7.5% or – 18.2%, 24.7%, and 38.5% compared with those of benchmark specimen SHW0 of 4.0 kN · mm or 27.5 kN · mm, respectively. The cumulative dissipated energy of specimen SHW1 was larger than that of benchmark specimen with the corrosion rate of 3%, which might attribute to corrosion products filling the space of concrete and contribute to the increase of dissipated energy capacity. However, the completely cumulative dissipated energy of specimens SHW1, SHW2, and SHW3 was 36.4,

**Table 6.3** Test results of ductility, drift, and energy dissipation of corroded specimens

| Specimen | Corrosion rate (%) | $\mu$ | Decrease ratio compared to SHW0 (%) | $\theta_u$ | Decrease ratio compared to SHW0 (%) | $Q$(kN mm) | Decrease ratio compared to SHW0 (%) |
|---|---|---|---|---|---|---|---|
| SHW0 | 0 | 1.86 | 0 | 1/51 | 0 | 41.6 | 0 |
| SHW1 | 3 | 1.71 | 8.1 | 1/59 | 13.6 | 36.4 | 12.5 |
| SHW2 | 9 | 1.57 | 15.6 | 1/66 | 22.7 | 32.1 | 22.8 |
| SHW3 | 15 | 1.56 | 16.1 | 1/70 | 27.1 | 27.1 | 34.9 |

*Note* $\mu$ is displacement ductility ratio; $\theta_u$ is drift; and $Q$ is cumulative dissipated energy

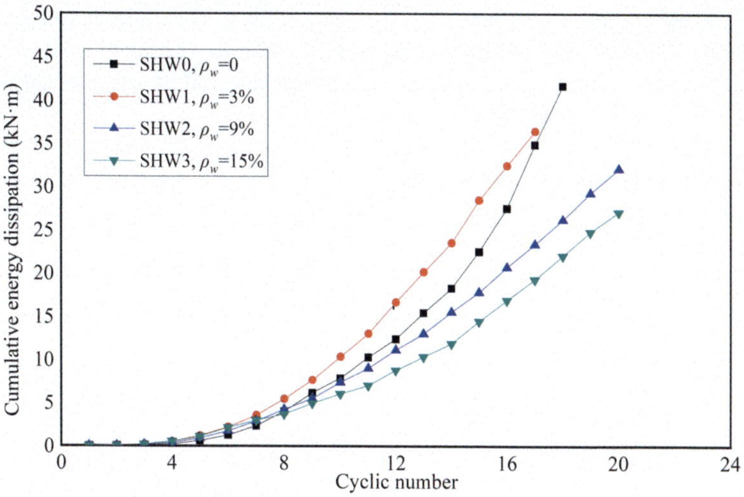


**Fig. 6.15** Cumulative dissipated energy of all specimens at each loading cycle of corroded specimens

32.1, and 27.1 kN · mm, which decreased by 12.5%, 22.8%, and 34.9% compared with that of benchmark specimen SHW0 of 41.6 kN ·mm, respectively. The decrease in completely cumulative dissipated energy of lightly corroded specimen SHW1 was relatively small, while for severely corroded specimens SHW2 and SHW3, the cumulative dissipated energy significantly decreased by over 20% of that of benchmark specimen, as shown in Table 6.3.

### *6.1.7 Analysis of Shear Strength*

Several applicable methods for evaluating the shear strength of RC walls have been recommended in researches, among which the model proposed in ACI 318-14 [31] is generally utilized. The main parameters of tested walls included the compressive concrete strength, cross-sectional dimensions, aspect ratio, and axial load ratio were considered, contributing to the shearing mode of walls. The models based on ACI 318-14 for evaluating the load-bearing capacities of corroded RC shear walls in consideration of the influence of concrete and damaged reinforcing bars by corrosion were modified.

The shear strength of benchmark specimen provided by concrete proposed in ACI 318-14 [31] is given as Eq. (6.9).

$$V_c = 0.27\sqrt{f_c'}t_w d + \frac{Nd}{4l_w} \tag{6.9}$$

where $l_w$ is the length of specimen; $N$ is the vertical load; $f_c'$ is the concrete strength under the condition of compression; $t_w$ is the width of specimen; and $d$ is the distance from the extreme compression fiber to the centroid of longitudinal tensile reinforcement and taken as $0.8l_w$.

The rust formed by the corrosion damages in concrete resulted in the volumetric expansion and increasing splitting stresses. The reduced strength of concrete in this area after cracking and spalling attributed to the increasing splitting stresses is calculated by Eq. (6.10) [9].

$$f_{cc}' = \frac{f_c'}{1 + K\varepsilon_1/\varepsilon_{co}} \tag{6.10}$$

where $K$ is the coefficient correlated with the diameter and roughness of the reinforcing bars, which is taken as 1.0 reported in [32]; $\varepsilon_{co}$ is the strains of the corresponding maximum stress in condition of compression; and $\varepsilon_1$ is the strains of cracked concrete under the condition of tension at the right angles to the direction of the applied compression, and calculated by the Eq. (6.11).

$$\varepsilon_1 = \frac{2\pi n_{\text{bars}}(\upsilon_{rs} - 1)X}{b_0} \tag{6.11}$$

where $n_{\text{bars}}$ is the count of compressive reinforcing bars; $b_0$ is the perimeter values of the cross-section of the wall; $\upsilon_{rs}$ is the ratio of volume of corroded reinforcing bars after expanding to that of benchmark reinforcing bars, which bases on the kind of corrosion products and taken as 2.0 reported in [32]; and $X$ is the depth of the corrosion and taken as $0.5(d_b - d_c)$, as reported in [33].

The diameters as well as mechanical properties decreased after the reinforcing bars corroded [33], contributing to the decrease of yielding strength or cross-sectional areas of transverse reinforcements. Thus, models for evaluating the residual yielding strength or cross-sectional areas of reinforcing bars after corroding were calculated by Eqs. (6.12)–(6.13), as reported in [34]. The yielding strength provided by corroded reinforcing bars $V_{sc}$ could be calculated by Eq. (6.14).

$$f_{ytc} = (1 - 0.005\rho_c)f_{y0} \tag{6.12}$$

$$A_{vc} = (1 - 0.01\rho_c)A_{v0} \tag{6.13}$$

$$V_{sc} = \frac{A_{vc} f_{ytc} d}{s} \tag{6.14}$$

where $A_{v0}$ and $f_{y0}$ are the cross-sectional area and yielding strength of benchmark transverse reinforcements, in mm$^2$ and MPa, respectively; $A_{vc}$ and $f_{ytc}$ are the cross-sectional area and yielding strength of corroded transverse reinforcements, in mm$^2$ and MPa, respectively; $s$ is the spacing of transverse reinforcements, in mm; and $\rho_c$ is the corrosion rate, in %.

Finally, the shear strength of corroded RC walls $V_{nc}$ was calculated by Eq. (6.15).

**Table 6.4** Comparison of shear strength of all tested specimens

| Specimen | Test results of ultimate shear strength (kN) | Calculated shear strength based on ACI 318-14 (kN) | Deviations on ACI 318-14 and test results (%) | Simulation results of shear strength (kN) | Deviations on simulation and test results (%) |
|---|---|---|---|---|---|
| SHW0 | 272.7 | 272.3 | − 0.2 | 296.9 | 8.9 |
| SHW1 | 261.8 | 266.4 | 1.8 | 274.9 | 5.0 |
| SHW2 | 242.7 | 255.6 | 5.3 | 259.6 | 7.0 |
| SHW3 | 228.0 | 245.1 | 7.5 | 232.3 | 1.9 |

$$\begin{cases} V_{nc} = 0.27\sqrt{f'_{cc}}t_w d + \frac{Nd}{4l_w} + \frac{A_{vc} f_{ytc} d}{s} \\ f'_{cc} = \frac{f'_c}{1+K\varepsilon_1/\varepsilon_{co}} \\ f_{ytc} = (1 - 0.005\rho_c) f_{y0} \\ A_{vc} = (1 - 0.01\rho_c) A_{v0} \end{cases} \quad (6.15)$$

Table 6.4 shows the comparison of shear strength of all tested specimens. The deviations between the results of calculated shear strength based on ACI 318-14 of specimens SHW0, SHW1, SHW2, and SHW3 and those of tested shear strength were − 0.2%, 1.8%, 5.3%, and 7.5%, respectively. The deviations of specimens SHW0, SHW1, SHW2, and SHW3 increased as the corrosion rate increased, while the maximum deviation was less than 10%.

### *6.1.8 Simulation Analysis on Shear Wall*

An analytical model by ABAQUS for predicting the nonlinear behavior of corroded shear walls was proposed. The model was based on the constitutive models for concrete and reinforcing bars and clarified the influence of corrosion of reinforcing bars on the seismic performance of the shear wall.

(1) Constitutive Model for Concrete and Bars

The nonlinear behavior of concrete in FE analysis is generally evaluated by concrete damaged plasticity (CDP) model based on Chinese Standard GB 50010-2010 [35]. The stress, inelastic strain, and damage factor should be determined depending on the relationship between the strain and stress, as reported in [35]. The compressive stress–strain relationship utilized in the model could be expressed.

$$\sigma_c = \begin{cases} \frac{\rho_c n}{n-1+x^n} E_c \varepsilon & x \le 1 \\ \frac{\rho_c}{\alpha_c (x-1)^2 + x} E_c \varepsilon & x > 1 \end{cases} \quad (6.16)$$

$$\rho_c = \frac{f_{c,r}}{E_c \varepsilon_{c,r}} \quad (6.17)$$

$$n = \frac{E_c \varepsilon_{c,r}}{E_c \varepsilon_{c,r} - f_{c,r}} \tag{6.18}$$

$$x = \frac{\varepsilon}{\varepsilon_{c,r}} \tag{6.19}$$

where $\sigma_c$ is the stress of concrete in the condition of compression, in MPa; $\alpha_c$ is the parameters in the downgrade period of stress–strain curve of concrete in the condition of uniaxial compression, and acquired by [35]; $f_{c,r}$ is the characteristic values of concrete in the condition of uniaxial compression, in MPa; $\varepsilon_{c,r}$ is the strains of the corresponding $f_{c,r}$ and obtained by [35]; and $E_c$ is the elastic modulus of concrete in the condition of uniaxial compression, in MPa.

The tensile stress–strain relationship of CDP model could be expressed.

$$\sigma_t = \begin{cases} \rho_t(1.2 - 0.2x^5)E_c\varepsilon & x \le 1 \\ \frac{\rho_t}{\alpha_t(x-1)^{1.7}+x}E_c\varepsilon & x > 1 \end{cases} \tag{6.20}$$

$$\rho_t = \frac{f_{t,r}}{E_c \varepsilon_{t,r}} \tag{6.21}$$

$$x = \frac{\varepsilon}{\varepsilon_{t,r}} \tag{6.22}$$

where $\sigma_t$ is the stress of concrete in the condition of tension, in MPa; $f_{t,r}$ is the characteristic values of concrete in the condition of uniaxial tension, in MPa; $\alpha_t$ is the parameters in the downgrade period of stress–strain curve of concrete in the condition of uniaxial tension, and acquired by [35]; and $\varepsilon_{c,r}$ is the strains of the corresponding $f_{t,r}$ and obtained by [35].

The damage factors $D_t$ and $D_c$ were calculated by the inelastic curves of the stress–strain correlation of concrete under the tension and compression, respectively, while $D_t$ was not considered in the early stage of tension with the straight line. Besides, other recommended parameters that needed to be considered in the FE analysis of ABAQUS [15] were summarized in Table 6.5.

For better simulating the mechanical properties of reinforcing bars, the trilinear model [35] shown in Fig. 6.16 was utilized.

**Table 6.5** Parameters of the CDP model

| Parameter | $\xi$ | $\eta$ | $\lambda$ | $K$ | $\mu$ |
|---|---|---|---|---|---|
| Value | 30 | 0.1 | 1.16 | 0.6667 | 0.001 |

*Note* $\xi$ is the Expansive angle; $\eta$ is the eccentricity; $\lambda$ is the ratio of the biaxial ultimate compressive strength to the uniaxial ultimate compressive; $K$ is the ratio of secondary stress invariants on the tensile to compressive meridian planes; and $\mu$ is the viscosity parameter

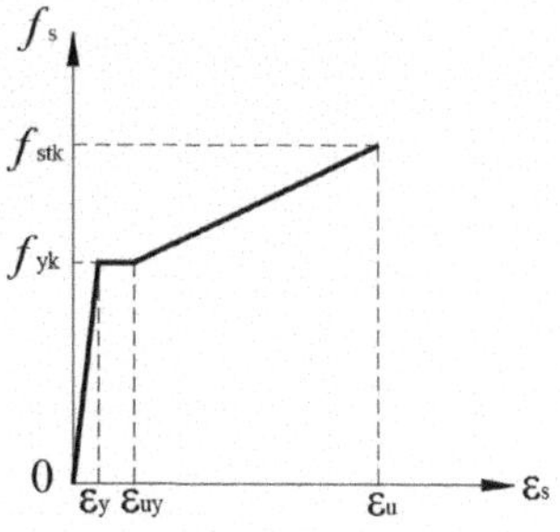


**Fig. 6.16** Constitutive model of tensile reinforcing bars

$$\sigma_s = \begin{cases} E_s \varepsilon_s & \varepsilon_s \le \varepsilon_y \\ f_{yk} & \varepsilon_y < \varepsilon_s \le \varepsilon_{uy} \\ f_{yk} + k(\varepsilon_s - \varepsilon_{uy}) & \varepsilon_{uy} < \varepsilon_s \le \varepsilon_u \\ 0 & \varepsilon_s > \varepsilon_u \end{cases} \tag{6.23}$$

where $\sigma_s$ is the tensile stress of reinforcing bars in MPa; $E_s$ is the elastic modulus of reinforcing bars, in MPa; $f_{yk}$ and $f_{stk}$ are the standard values of yielding and ultimate strength, in MPa, respectively; $\varepsilon_s$ is the strains; $\varepsilon_y$ is the strains of the corresponding yielding strength and equal to $f_{yk}/E_s$; $\varepsilon_{uy}$ is the initial strains of the hardening reinforcing bars; $\varepsilon_u$ is the ultimate strains; and $k$ is the slope of reinforcing bars at the hardening stage and equal to $(f_{stk} - f_{yk})/(\varepsilon_u - \varepsilon_y)$.

Besides, another simplified model for evaluating the yielding and ultimate strength of reinforcing bars after corroding could also be calculated by Eqs. (6.24)–(6.25) as proposed in [36].

$$f_{yc} = \frac{1 - 1.23\rho_c}{1 - \rho_c} f_{y0} \tag{6.24}$$

$$f_{uc} = \frac{1 - 1.245\rho_c}{1 - \rho_c} f_{u0} \tag{6.25}$$

where $f_{yc}$ and $f_{y0}$ are the yielding strength of corroded and benchmark reinforcing bars, in MPa, respectively; $f_{uc}$ and $f_{u0}$ are the ultimate strength of corroded and benchmark reinforcing bars, in MPa, respectively; and $\rho_c$ is the corrosion rate, in %.

(2) Element Type

The eight-node linear solid elements and truss elements of three-dimensional FE model were utilized in simulating the concrete and reinforcing bars as shown in Fig. 6.17, respectively. Cartesian sections are utilized for simulating the bond behavior between the longitudinal reinforcements and concrete [37], while the bar-slip behavior between reinforcing bars and concrete in transverse direction was not considered. That indicated that no bond slip of reinforcing bars in transverse direction was observed in the process of cyclic loadings. The CEB-TIP model [38] was utilized

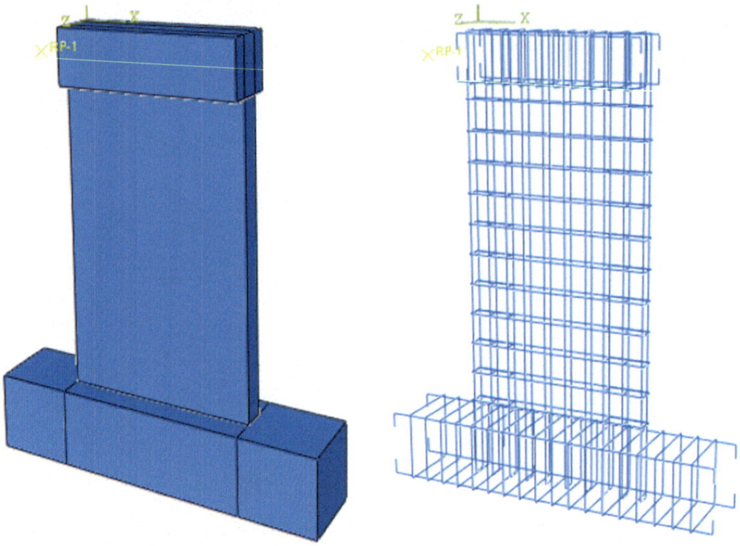

(a) Concrete (b) Bars

**Fig. 6.17** FE modeling

for simulating the correlation between the bond stress and slip of uncorroded reinforcing bars. Besides, a decreasing coefficient for evaluating the bond stress between corroded reinforcing bars and concrete was calculated and shown in Eq. (6.26), as proposed in [39].

$$\beta_c = \begin{cases} 1 + 0.5625\rho_c - 0.3375\rho_c^2 + 0.05565\rho_c^3 - 0.003\rho_c^4 & \rho_c \leq 7\% \\ 2.0786\rho_c^{-1.0369} & \rho_c > 7\% \end{cases} \tag{6.26}$$

where $\rho_c$ is the corrosion rate, in %.

Furthermore, boundary conditions of this model and that of cyclic test were consistent. The out-plane translational motion of the head beam was restrained and the in-plane translational motion of head beam was released, suffering the combined loads of axial and horizontal loads. Plane loadings were adopted for simulating the axial loads of 325 kN on the head beam. Meanwhile, the translational motions on the *X*, *Y*, and *Z* directions of wall base were fixed. The loading history consisted of displacement control, and the mesh size of the automatic meshing technique was 50 mm ×50 mm ×50 mm, as reported in [40].

(3) Comparison of Tested and FE Model Results

The comparison of hysteretic curves on the test and FE model results of corroded walls SHW0, SHW1, SHW2, and SHW3 was shown in Fig. 6.12. Generally, the numerical simulation results coincided with tested results when the corrosion rate was less than 15%. Better agreements could be found in the curves of specimens

SHW0, SHW1, and SHW2 than those of specimen SHW3, indicating that seriously corroded shear walls deteriorated more unstably in the process of pull and push stage. Besides, the results of simulation on ultimate shear strength of specimens SHW0, SHW1, SHW2, and SHW3 were 296.9, 274.9, 259.6, and 232.3 kN, which were 8.9%, 5.0%, 7.0%, and 1.9% larger than those of tested shear strength of 272.7, 261.8, 242.7, and 228.0 kN shown in Table 6.4, respectively. The deviations of all specimens were small, indicating that the simulation results well coincided with those of tested results and FE model could be made use of predicting the shear strength of corroded shear walls as well. However, the regions of hysteretic curves of specimens SHW0, SHW1, SHW2, and SHW3 obtained from the FE model were larger than those obtained from tested results. The proposed FE model considering the decrease of yielding strength, ultimate strength, and bond stress of corroded reinforcing bars could be utilized for predicting the shear strength of corroded shear walls in a sense.

Figure 6.18 shows the equivalent plastic strain (PEEQ) of concrete of all tested specimens. The accumulative plastic strain of all components of tested specimens at each location in the whole loading process could be observed by the PEEQ. Similar PEEQ characteristics of specimen SHW0, SHW1, and SHW2 with corrosion rates of 0–9% were found in Fig. 6.18a–c, while serious bulging and crushing of concrete of specimen SHW3 with the corrosion rate of 15% was found in Fig. 6.18d. The plastic strains of the base of corroded wall panels were larger than those of other regions in the initial loading stage, while the areas of plastic deformations expanded to the upper part of the wall panels as the loads increased. At the end of the loading stage, the maximum strains of concrete at the base of the wall panels reached the ultimate strain of concrete in compression, indicating that the concrete crushed at this moment. That coincided with the cracking patterns of all tested specimens. The PEEQ of specimens SHW0, SHW1, SHW2, and SHW3 increased as the corrosion rate increased. By comparing the hysteretic curves and PEEQ of numerical simulation to those of test results of specimens SHW0, SHW1, SHW2, and SHW3, the proposed FE model could be made use of evaluating the experimental results and seismic performance of corroded shear walls.

## 6.2 Seismic Performance of Corrosion-Damaged Shear Walls Retrofitted with BFRP Sheets

### *6.2.1 Experimental Program*

(1) Specimen Details and Materials

Four RC shear wall specimens designated as SHW0, SHW1, SHW2, and SHW3 with the expected corrosion rate of 0, 3%, 9%, and 15% were tested. The mix proportions of concrete are shown in Table 6.6. The cubic compressive strength of concrete $f_{cu}$,

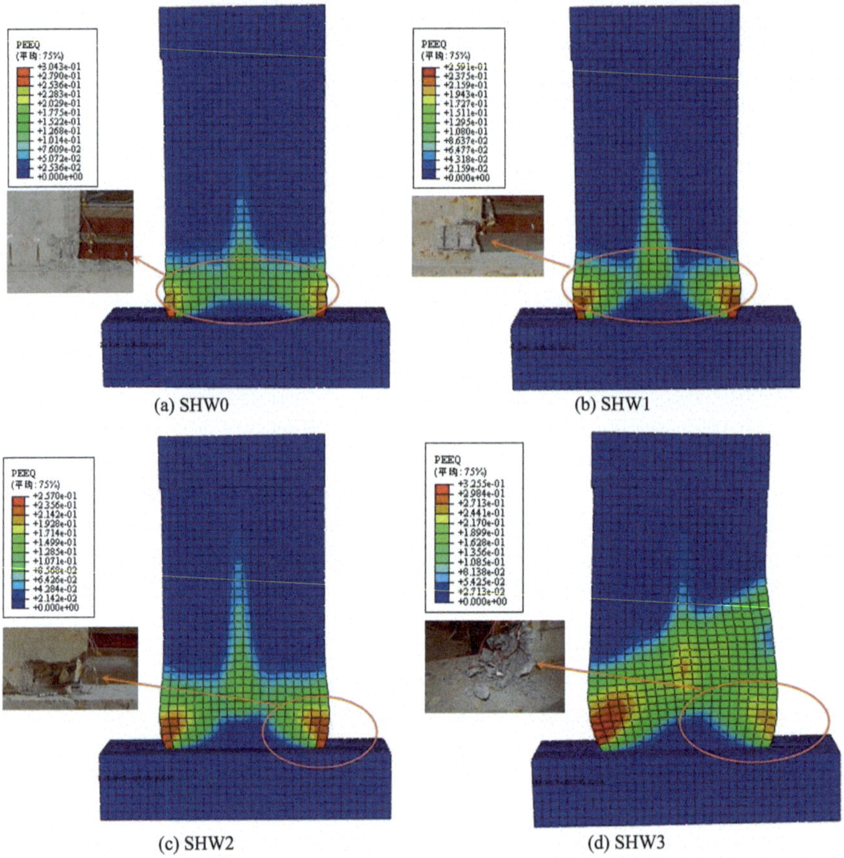


**Fig. 6.18** Equivalent plastic strain of concrete

the cylindrical axial compressive strength of concrete $f_c'$ defined as 0.79 $f_{cu}$ [41], and elastic modulus of concrete $E_c$ was 39.3, 31.0, and 3.24 × $10^4$ MPa, respectively.

Geometric dimensions and reinforcement details of all specimens were identical, as shown in Fig. 6.1. Average values of measured yielding strengths of 10, 8, and 20 mm diameter reinforcing bars were 335, 235, and 335 MPa, respectively.

(2) Accelerated Corrosion Process

**Table 6.6** Mix proportions of concrete (kg/m$^3$)

| Cement | Sand | Water | Coarse aggregates | Fly ash | Ground granulated blast furnace slag | Superplasticizer |
|---|---|---|---|---|---|---|
| 212 | 721 | 168 | 1123 | 70 | 70 | 4.22 |

Specimens SHW1, SHW2, and SHW3 were subjected to accelerated electrolytic corrosion, as reported in [42]. As shown in Fig. 6.2, corrosion-damaged specimens SHW1, SHW2, and SHW3 was also conducted by the setup for accelerating the corrosion process.

For specimens SHW1, SHW2, and SHW3 with the expected corrosion rates of 3%, 9%, and 15%, the conduction time was 6, 9, and 30 days when the current intensity of 3A was impressed on the reinforcing bars, respectively.

(3) Retrofitting of Specimens

Following the accelerated corrosion process of specimens SHW1, SHW2, and SHW3, the specimens were retrofitted. The retrofitted specimens were referred to with a prefix of "*R*" to the specimen names. For example, specimen RSHW1 had the same corrosion rate with specimen SHW1.

The retrofitting scheme for corroded RC shear walls is shown in Fig. 6.19, as reported in [43, 44]. ACI 440 [45] has proven that completely wrapping the FRP system around the section on all four sides is the most effective retrofitting scheme. Before the application of BFRP sheets, careful preparation of the walls was conducted with grinding of several areas to achieve a completely smooth surface and founding of the corners (at a radius of about 20 mm). Then one layer of a closed sheet was applied around the circumference of the specimens along the full height of 1600 mm, consisting of one layer of 0.121 mm thick BFRP sheet with the fibers oriented in the direction perpendicular to the longitudinal axis of the specimen. The average value of BFRP tensile elastic modulus, ultimate average strain, and ultimate tensile strength was $1.05 \times 10^5$ MPa, 2.2%, and $2.30 \times 10^3$ MPa, respectively. This easy-to-apply retrofitting scheme attempted to provide effective confinement with BFRP sheets as well as recover the degraded strength and ductility due to reinforcement corrosion to avoid brittle failure and progressive collapse of RC structures during an earthquake event.

(4) Test Setup and Procedure

The layout of test setup included the loading system and instrumentation in this chapter was the same as that shown in Fig. 6.5. And the loading history was also same as that shown in Fig. 6.6.

### 6.2.2 Crack Patterns and Failure Modes

The failure modes of all retrofitted specimens were generally similar and all retrofitted specimens experienced a progression of failure from BFRP debonding to local buckling of the longitudinal reinforcement and bottom concrete crushing.

In the force-control stage, the concrete at the interface of the BFRP sheets and foundation first cracked at the lateral load of 210.0, 200.0, and 176.8 kN for specimens RSHW1, RSHW2, and RSHW3, respectively. These first cracks were debonding

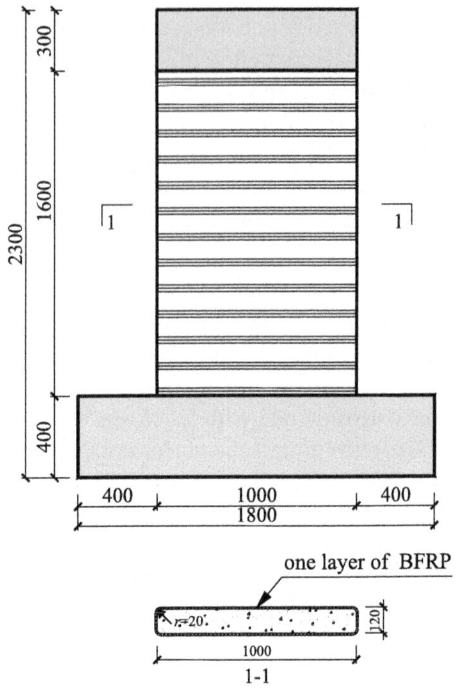


**Fig. 6.19** Retrofitting scheme for corroded specimens with different corrosion rates (all units in mm)

cracks along the length of the walls, as shown in Fig. 6.20a, which were followed by horizontal slight rupture of BFRP sheets near the bottom of the walls in the boundary elements. When the lateral load increased, the sheets in the upper regions of the walls gradually ruptured, as shown in Fig. 6.20b, and the tiny sound could be heard. At the end of the force-control phase, the debonding cracks extended to the full length of the interface of the BFRP sheets and foundation.

In the displacement-control stage, the bottom concrete in the boundary elements expanded outward under the axial load in combination with the lateral cyclic load, as shown in Fig. 6.20c. The BFRP sheets in the boundary elements were completely ruptured and the concrete spalled due to the sudden loss of the confinement at the lateral displacement of 20.8 (1.19% drift), 20.2 (1.15% drift), and 18.2 mm (1.04% drift) for specimens RSHW1, RSHW2, and RSHW3, respectively. The concrete cover at the bottom of the wall rapidly spalled due to the loss of the FRP confinement and degrading bond between concrete cover and the corroded reinforcing bar, and then the reinforcing bar exposed at the lateral displacement of 28.1 (1.61% drift), 27.5 (1.57% drift), and 24.5 mm (1.40% drift) for specimens RSHW1, RSHW2, and RSHW3, respectively. Soon after the spalling of concrete cover, specimens failed immediately with local buckling of longitudinal reinforcement, as shown in Fig. 6.20d. The confinement to the compressed concrete in the plastic hinge regions of the retrofitted specimens provided by the BFRP sheets was effective, denoted by the delaying of the concrete spalling and crushing as well as longitudinal reinforcement buckling.

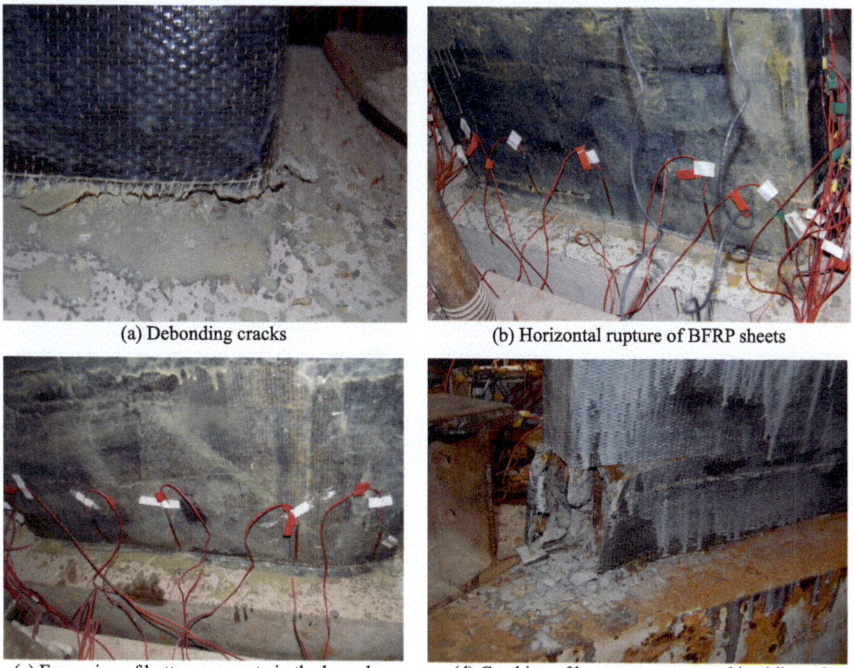

(a) Debonding cracks
(b) Horizontal rupture of BFRP sheets
(c) Expansion of bottom concrete in the boundary elements
(d) Crushing of bottom concrete and buckling of longitudinal reinforcement

**Fig. 6.20** Observed failure of retrofitted specimens

However, severe cracks along the interface of the BFRP sheets and foundation were observed in comparison with the original specimen, this was mainly associated with the gradual absence of FRP confinement in the loading process, which was introduced to increase the shear strength and ductility of the corroded specimens.

### *6.2.3 Hysteretic Response and Load–Displacement Curves*

Figure 6.21 shows hysteretic loops of all specimens, which is a key index of seismic performance. The internal area of hysteretic loops can be used as an estimation of the energy dissipation capacity. The hysteretic loop of original specimen was approximately spindle-shaped with no evident pinching phenomenon, while the hysteretic behavior of the retrofitted specimens was generally better than that of the corresponding corroded specimens but similar to or slightly poorer than that of the original specimen. The BFRP sheets should ideally improve the hysteretic behavior of the retrofitted specimens, although actual confinement offered by BFRP sheets was inefficient in elongated rectangular cross-sections.

Envelope curves of hysteretic loops can be used to define ductility and strength degradation. Figure 6.22 shows the envelopes of hysteretic loops of the specimens,

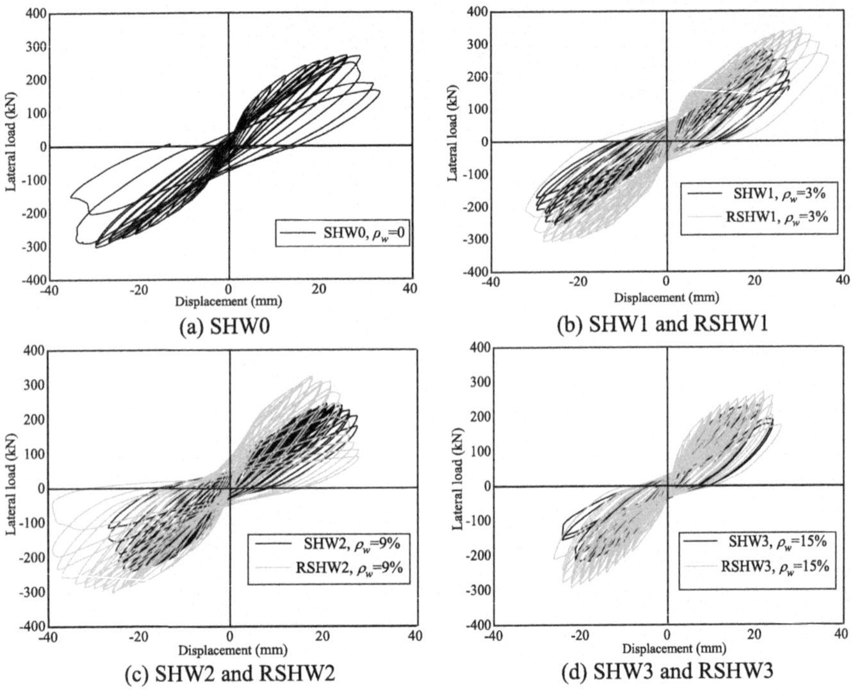


**Fig. 6.21** Load–displacement hysteretic loops of all specimens

which were obtained from averaging the maximum absolute values in positive and negative directions for each cycle of hysteretic loops. There was a very short yielding plateau followed by a gradual hardening up to peak strength, which was followed by a gradual softening after peak strength. The envelope curve of the original specimen was enclosed by that of the retrofitted specimens, which showed that the strength of the retrofitted RC shear walls efficiently recovered.

The load–displacement curve of the specimen subjected to a cyclic load in combination with an axial load is shown in Fig. 6.13, where $P_y$ is the yielding load and $\Delta_y$ is the yielding displacement which are determined based on the assumption that the area of hatched region $S_{OCD}$ equals to that of $S_{ABC}$; $P_u$ is the ultimate load, which is defined as 85% of the peak lateral load $P_{\max}$ and $\Delta_u$ is the ultimate displacement. The test results of strengths of all specimens are shown in Table 6.7. The lightly corroded specimen with a corrosion rate of 3% retained over 95% of its original load-bearing capacity and displacement, which indicated that reinforcement corrosion had a minor effect on the load–displacement curve. However, severely corroded specimens with corrosion rates of 9% and 15% exhibited a significant decrease in both load-bearing capacity and displacement, retaining less than 80% of their original levels. It was possible for all corroded specimens to recover almost to their original load-bearing capacity, but it was only possible for the lightly corroded specimen to recover to its original displacement.

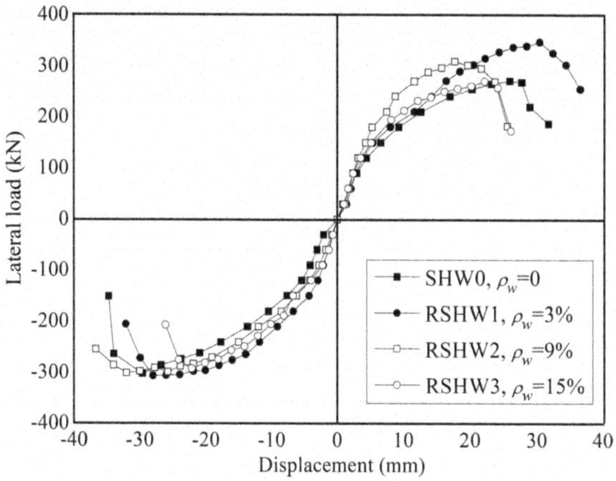


**Fig. 6.22** Envelopes of hysteretic loops of original and retrofitted specimens

### 6.2.4 *Stiffness Degradation*

The stiffness of retrofitted specimens was estimated based on the secant stiffness by Eq. (6.8) at each loading cycle, as shown in Fig. 6.23.

As shown in Fig. 6.23, the stiffness of the retrofitted specimens degraded at a similar rate when compared with that of original specimen. As shown in Table 6.8, the lightly corroded specimen with a corrosion rate of 3% retained over 90% of its initial and final secant stiffness, which indicated that reinforcement corrosion had a limited effect on the stiffness degradation. However, severely corroded specimens with corrosion rates of 9% and 15% exhibited a significant decrease in stiffness characteristics, retaining around 80% of the original level. It was possible for corroded specimens with corrosion rates of 3% and 9% to recover almost to their original initial or final secant stiffness, but it was not possible for the severely corroded with a corrosion rate of 15% to recover to its original initial or final secant stiffness as expected.

### 6.2.5 *Ductility and Energy Dissipation Capacity*

The elastic–plastic deformation is generally quantified by displacement ductility ratios and energy dissipation capacity. The displacement ductility ratio and the drift are defined as reported in Sect. 6.1.6 $\mu = \Delta_u/\Delta_y$, and the drift can be defined as $\theta_u = \Delta_u/h_w$.

The displacement ductility ratio of all specimens was less than 2.0. Results reported in [46] show that the displacement ductility ratio is less than 2.0 when the axial load ratio is less than 0.1. Therefore, results were reasonable. As shown in Table 6.9, the lightly corroded specimen with a corrosion rate of 3% retained over

**Table 6.7** Test results of strengths and displacements of corroded and retrofitted specimens

| Specimen | Corrosion rate (%) | $P_y$ (kN) | Ratio of corroded or retrofitted specimens to SHW0 (%) | $P_u$ (kN) | Ratio of corroded or retrofitted specimens to SHW0 (%) | $\Delta_y$ (mm) | Ratio of corroded or retrofitted specimens to SHW0 (%) | $\Delta_u$ (mm) | Ratio of corroded or retrofitted specimens to SHW0 (%) |
|---|---|---|---|---|---|---|---|---|---|
| SHW0 | 0 | 225.3 | 100 | 272.7 | 100 | 16.8 | 100 | 31.3 | 100 |
| SHW1 | 3 | 223.2 | 99.1 | 261.8 | 96.0 | 16.0 | 95.2 | 27.3 | 87.2 |
| SHW2 | 9 | 206.2 | 91.5 | 242.7 | 89.0 | 15.1 | 89.9 | 23.7 | 75.7 |
| SHW3 | 15 | 191.3 | 84.9 | 228.0 | 83.6 | 14.6 | 86.9 | 22.8 | 72.8 |
| RSHW1 | 3 | 288.3 | 128.0 | 326.7 | 119.8 | 18.0 | 107.1 | 32.1 | 102.6 |
| RSHW2 | 9 | 266.4 | 118.2 | 305.3 | 112.0 | 16.1 | 95.8 | 30.1 | 96.2 |
| RSHW3 | 15 | 243.3 | 108.0 | 281.6 | 103.3 | 15.9 | 94.6 | 26.6 | 85.0 |

*Note* $P_y$ is yielding strength; $P_u$ is ultimate strength; $\Delta_y$ is yielding displacement; and $\Delta_u$ is ultimate displacement

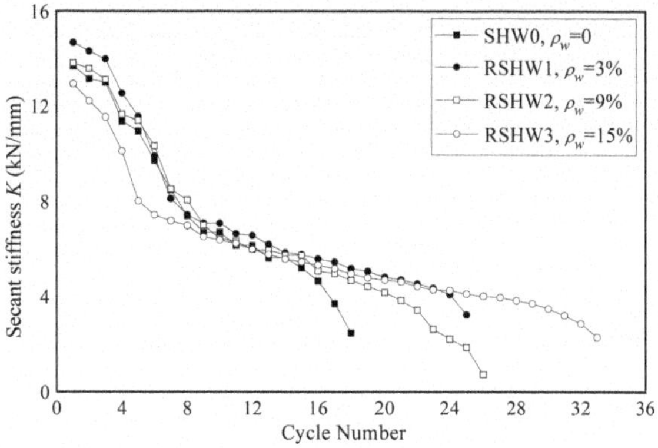


**Fig. 6.23** Secant stiffness degradation of original and retrofitted specimens

**Table 6.8** Test results of stiffness

| Specimen | Corrosion rate (%) | $K_0$ (kN/mm) | Ratio of corroded or retrofitted specimens to SHW0 | $K$ (kN/mm) | Ratio of corroded or retrofitted specimens to SHW0 (%) |
|---|---|---|---|---|---|
| SHW0 | 0 | 13.7 | 100 | 2.5 | 100 |
| SHW1 | 3 | 12.5 | 91.2 | 2.4 | 96.0 |
| SHW3 | 9 | 11.3 | 82.5 | 2.2 | 88.0 |
| SHW3 | 15 | 10.9 | 79.6 | 2.2 | 88.0 |
| RSHW1 | 3 | 14.7 | 107.3 | 3.2 | 128.0 |
| RSHW2 | 9 | 13.8 | 100.7 | 2.7 | 108.0 |
| RSHW3 | 15 | 13.0 | 94.9 | 2.3 | 92.0 |

*Note* $K_0$ is initial secant stiffness; and $K$ is final secant stiffness

90% of its displacement ductility ratio, which indicated that reinforcement corrosion had a limited effect on the ductility. However, severely corroded specimens with corrosion rates of 9% and 15% exhibited a significant decrease in ductility, retaining around 80% of the original level. It was possible for corroded specimens with corrosion rates of 3% and 9% to recover almost to their original displacement ductility ratio and drift, but it was not possible for the severely corroded specimen with a corrosion rate of 15% to recover to its original displacement ductility ratio and drift as expected, which mainly attributed to the following reasons: (a) the localization of elastoplastic deformation at the critical flexural cracks led to buckling of bottom reinforcements; (b) the linear elastic property of BFRP which was not capable of contributing to ductility as reinforcement was; and (c) the inefficiency of BFRP sheets to provide adequate confinement in elongated cross-sections.

Cumulative dissipated energy $Q$ was utilized to evaluate the energy dissipation capacity, as shown in Eq. (6.27).

**Table 6.9** Test results of ductility, drift, and energy dissipation of corroded and retrofitted specimens

| Specimen | Corrosion rate (%) | $\mu$ | Ratio of corroded or retrofitted specimens to SHW0 (%) | $\theta_u$ | Ratio of corroded or retrofitted specimens to SHW0 (%) | $Q$(kN · m) | Ratio of corroded or retrofitted specimens to SHW0 (%) |
|---|---|---|---|---|---|---|---|
| SHW0 | 0 | 1.86 | 100 | 1/51 | 100 | 41.6 | 100 |
| SHW1 | 3 | 1.71 | 91.9 | 1/59 | 86.4 | 36.4 | 87.5 |
| SHW2 | 9 | 1.57 | 84.4 | 1/66 | 77.3 | 32.1 | 77.2 |
| SHW3 | 15 | 1.56 | 83.4 | 1/70 | 72.9 | 27.1 | 65.1 |
| RSHW1 | 3 | 1.78 | 95.7 | 1/50 | 102.0 | 71.9 | 172.8 |
| RSHW2 | 9 | 1.87 | 100.1 | 1/53 | 96.2 | 65.7 | 157.9 |
| RSHW3 | 15 | 1.67 | 89.8 | 1/60 | 85.0 | 57.4 | 138.0 |

*Note* $\mu$ is displacement ductility ratio; $\theta_u$ is drift; $h_e$ is equivalent viscous damping coefficient; and $Q$ is cumulative dissipated energy

$$Q = \sum_{i=1}^{N} Q_i \tag{6.27}$$

where $Q_i$ is the energy dissipation for the $i$th cycle calculated by the hatched area $S_{ABCD}$, as shown in Fig. 6.24; and $N$ is the number of cycles until failure.

Furthermore, the cumulative dissipated energy should be divided into two parts, involving dissipated energy due to reinforcing bar $Q_s$ and dissipated energy due to FRP $Q_f$. Assuming that the FRP was linearly elastic until rupture, as well as only FRP in plastic hinge zones was taken into consideration for energy dissipation. Therefore, dissipated energy due to FRP $Q_f$ can be calculated, as shown in Eq. (6.28).

$$Q_f = 4\int_{0}^{\varepsilon_f^*} 2nE_f\varepsilon_f t_f L_p h d\varepsilon_f \tag{6.28}$$

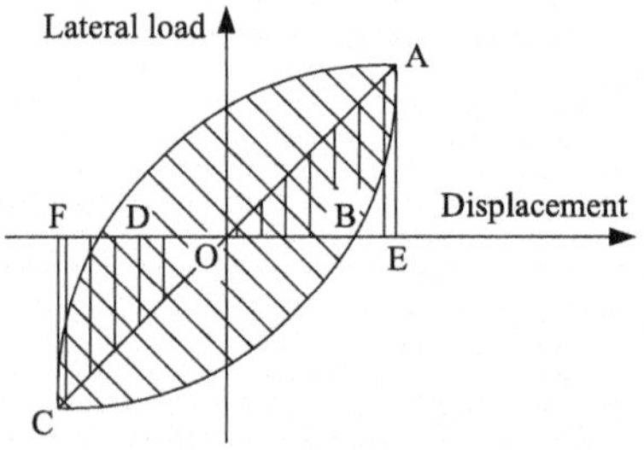


**Fig. 6.24** Definition of energy dissipation for $i$th cycle

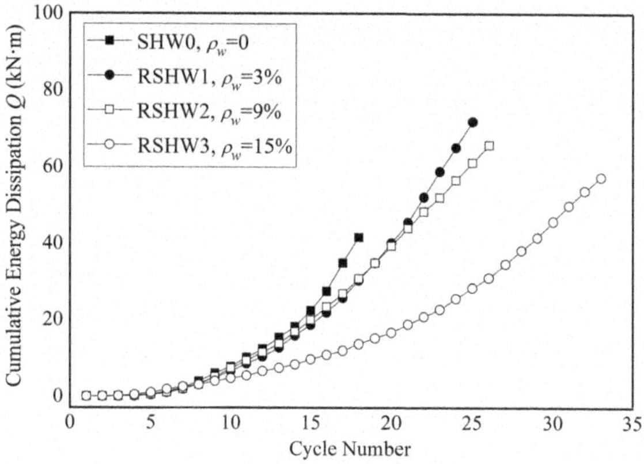


**Fig. 6.25** Cumulative dissipated energy of original and retrofitted specimens at each loading cycle

where $\varepsilon_f$ and $\varepsilon_f^*$ are the strain and final strain of fibers in the direction of the lateral load, respectively; and $L_p$ is the length of plastic hinge zones; and the coefficient 4 means that the FRP fibers in the range of plastic hinge zones totally experienced tension and restoration four times during each loading cycle before reaching the final strain.

The calculated results of cumulative dissipated energy $Q$ of original and retrofitted specimens at each loading cycle are shown in Fig. 6.25. As shown in Table 6.9, the lightly corroded specimen with a corrosion rate of 3% retained less than 90% of its cumulative dissipated energy, which indicated that reinforcement corrosion had an obvious effect on the energy dissipation capacity. However, severely corroded specimens with corrosion rates of 9% and 15% exhibited a significant decrease in energy dissipation capacity, retaining less than 80% of the original level. It was possible for all corroded specimens to fully recover to their original energy dissipation capacity. The dissipated energy due to FRP was 0.16, 0.22, and 0.35 kN·m, which only accounted for 0.2%, 0.3%, and 0.6% of the cumulative dissipated energy. Therefore, FRP did not play a vital role in energy dissipation as a result of debonding failure prior to fibers reaching ultimate tensile strain; however, the dissipated energy of the retrofitted specimens due to reinforcing bar was higher than that of the original specimen. The reason was that there is no outward restraint within the unsupported length of the compressive reinforcing bars between the transverse reinforcements after the concrete cover spalled in the original specimen under the cyclic loads. At the same time, the core concrete surrounded by the longitudinal and transverse reinforcement has a tendency to expand outward, resulting in the buckling of the longitudinal reinforcement. With the FRP confinement, the spalling of the concrete cover and the buckling of the longitudinal reinforcement delayed and occurred after the rupture of BFRP sheets, resulting in higher dissipated energy due to reinforcing bar in retrofitted specimens.

### 6.2.6 Theoretical Modeling

(1) Components of Ultimate Displacement

The ultimate displacement of RC shear walls mainly consists of flexural displacement, shear displacement, and slip displacement [47]. A fiber model for wall section was employed to perform a flexural analysis. The strain in reinforcing bar and concrete shall be assumed directly proportional to the distance from the neutral axis. The constitutive model for FRP-confined concrete proposed in ACI 440 [45] is shown in Fig. 6.26a and Eq. (6.29), respectively.

$$f_c = \begin{cases} E_c\varepsilon_c - \frac{(E_c-E_2)^2}{4f'_c}\varepsilon_c^2 & 0 \le \varepsilon_c \le \varepsilon'_t \\ f'_c + E_2\varepsilon_c & \varepsilon'_t \le \varepsilon_c \le \varepsilon_{ccu} \end{cases} \quad (6.29)$$

where $\varepsilon_c$ is the strain level in concrete; $\varepsilon'_t$ is the transition strain; and $E_2$ is the slope of linear hardening portion of the cumulative model for FRP-confined concrete.

The maximum compressive strain in the FRP-confined concrete $\varepsilon_{ccu}$ is shown in Eq. (6.30).

$$\varepsilon_{ccu} = \varepsilon'_c\left(1.50 + 12\kappa_b\frac{f_l}{f'_c}\left(\frac{\varepsilon_{fe}}{\varepsilon'_c}\right)^{0.45}\right) \quad (6.30)$$

where $\varepsilon'_c$ is the maximum strain of unconfined concrete; $\varepsilon_{fe}$ is the effective strain level in FRP defined as $0.58\,\varepsilon_{fu}$, where $\varepsilon_{fu}$ is the design rupture strain of FRP; $f_l$ is the maximum confining pressure due to FRP sheet defined as $2E_fnt_f\varepsilon_{fe}/D$, where $D$ is the diagonal of the cross-section; and $\kappa_b$ is the shape factor defined as $\frac{A_e}{A_c}\left(\frac{h}{b}\right)^{0.5}$, $A_e$ and $A_c$ are the cross-sectional areas of effectively confined concrete section and concrete in compression member, respectively.

The compressive strength of FRP-confined concrete $f'_{cc}$ is shown in Eq. (6.31), as reported in [48].

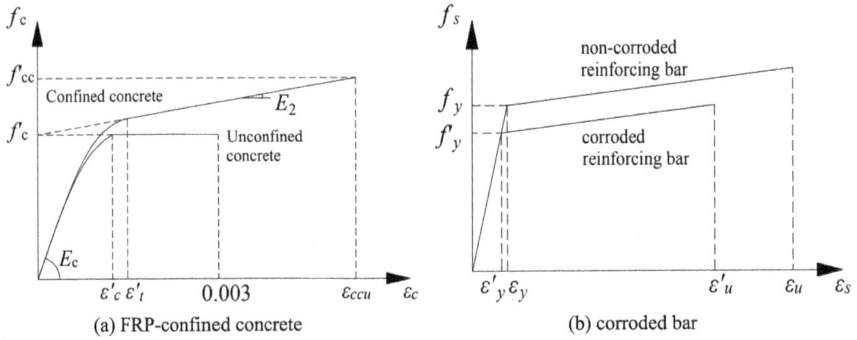


**Fig. 6.26** Constitutive model for sectional analysis

$$\frac{f'_{cc}}{f'_c} = 1 + 3.66\left(\frac{2r}{b}\right)^{0.72}\left(\frac{f_l}{f'_c}\right)^{1.87}\left(\frac{h}{b}\right)^{-2.5} \tag{6.31}$$

where $2r/b$ is the corner radius ratio (corner radius divided by half the width of the wall).

Results reported in [34] show that the shape of the force-extension curve of reinforcing bar does not change substantially for up to corrosion rate of 16%. Therefore, a bilinear constitutive model considering strain hardening for corroded reinforcement was utilized in the analysis, as shown in Fig. 6.26b. The modified yielding stress and ultimate strain of the corroded reinforcing bar are reported in [49], as shown in Eqs. (6.32) and (6.33), respectively.

$$f_y = (1 - 0.005\rho_w)f_{y0} \tag{6.32}$$

$$\varepsilon_u = (1 - 0.05\rho_w)\varepsilon_{u0} \tag{6.33}$$

where $\rho_w$ is the corrosion rate; $f_y$ is the yielding strength of corroded reinforcing bar; $f_{y0}$ is the yielding strength of non-corroded reinforcing bar; $\varepsilon_u$ is the ultimate strain of corroded reinforcing bar; and $\varepsilon_{u0}$ is the ultimate strain of non-corroded reinforcing bar.

The yielding displacement $\Delta_y$ was calculated by integrating the linear curvature distribution along the wall height, as reported in [50] and shown in Fig. 6.27. The plastic displacement $\Delta_p$ was calculated by multiplying $h_w - 0.5L_p$ by the plastic rotation at the base $\theta_p$. The flexural displacement of the wall was obtained as the sum of yield displacement $\Delta_y$ and plastic displacement $\Delta_p$, as shown in Eq. (6.34).

$$\Delta_{flex} = \Delta_y + \Delta_p = \frac{\phi_y h_w^2}{3} + (\phi_u - \phi_y)L_p(h_w - 0.5L_p) \tag{6.34}$$

where $(\phi_u - \phi_y)L_p$ is the plastic rotation $\theta_p$ and is based on the assumption that the plastic curvature is lumped in the center of the equivalent plastic hinge length $L_p$; $\phi_u$ and $\phi_y$ are the ultimate and yielding curvature at the section of the wall base, respectively.

Only one parameter $L_p$ was unknown in Eq. (6.34). Various approaches to estimate the plastic hinge length have been proposed. Results reported in [51] show that the plastic hinge length is assumed as one half of the wall length ($L_p = 0.5L_w$). However, this estimated approach took no account into reinforcement corrosion and FRP confinement, and therefore, it was inappropriate. In addition, the plastic hinge length is defined as the distance between the top and bottom edges of the spalled regions of walls when failure for simplicity, as reported in [52], and the test result of plastic hinge length was 200, 220, 255, and 350 mm or 267, 416, and 471 mm for specimens SHW0, SHW1, SHW2, and SHW3 or RSHW1, RSHW2, and RSHW3, respectively.

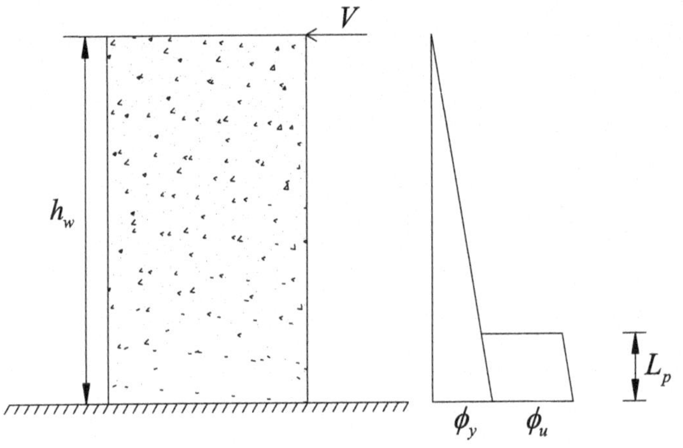


**Fig. 6.27** Flexural component of ultimate displacement using plastic hinge approach

The shear component was estimated with a simplified elastic approach, as shown in Eq. (6.35).

$$\Delta_{\text{shear}} = \frac{1.2 P_{\text{max}} h_w}{G A_g} \tag{6.35}$$

where $P_{\text{max}}$ is the peak lateral load; $A_g$ is the gross section area of the wall; and $G$ is the shear modulus of concrete defined as 40% of modulus of elasticity of concrete.

The slip component was a rigid-body rotation of the walls due to the slip of longitudinal reinforcement in the anchoring concrete. This rotation was additive to the rotation calculated from the flexural analysis. A stepped bond-stress model was utilized for the prediction of the slip component, as reported in [53]. The slip component of ultimate displacement $\Delta_{\text{slip}}$ was obtained by multiplying the slip rotation at the wall base $\theta_s$ by the wall height $h_w$, as shown in Eq. (6.36).

$$\Delta_{\text{slip}} = \theta_s h_w \tag{6.36}$$

For the stepped bond-stress model, it is assumed that the slip rotation will be about the neutral axis. The slip rotation at the wall base $\theta_s$ is shown in Eq. (6.37), as reported in [52].

$$\theta_s = \begin{cases} \frac{\varepsilon_s f_s d_b}{8u_b(d-c)} \\ \frac{d_b}{8u_b(d-c)}(\varepsilon_y f_y + 2(\varepsilon_s + \varepsilon_y)(f_s - f_y)) \end{cases} \tag{6.37}$$

where $u_b$ is the uniform bond stress of $1.0\sqrt{f'_{cC}}$; $d$ is the distance measured from extreme compression fiber to centroid of boundary element; $c$ is the neutral axis depth; $\varepsilon_s$ and $f_s$ are strain and stress of reinforcing bar, respectively; $\varepsilon_y$ and $f_y$ are yielding strain and stress of reinforcing bar, respectively.

As shown in Table 6.10, the proportion of the flexural component in the ultimate displacement for both corroded and retrofitted specimens increased with the increase of the corrosion rate due to larger plastic hinge length in the specimen with higher corrosion rate. The proportion of the shear component in the ultimate displacement for retrofitted specimens was slightly larger than that of corroded specimens due to shear strength enhanced by BFRP sheets. The bond between reinforcing bar and concrete decreased, as the proportion of slip component in the ultimate displacement for the corroded specimens increased with the increase of the corrosion rate. The proportion of slip component in the ultimate displacement for retrofitted specimens was lower than that of corroded specimens, which showed that BFRP sheets provided the effective confinement to reduce the slip between reinforcing bar and concrete.

(2) Effectiveness of FRP Confinement

The provision of FRP confinement to the potential plastic hinge zones can substantially increase both strength and ductility of RC shear walls [54]. Therefore, the study on the effectiveness of FRP confinement was an essential issue to gain a comprehensive understanding of the retrofitting mechanism of corroded RC shear walls using FRP sheets.

The specimen's ratio of the height to the thickness was 8.3 (1000/120), far larger than that of commonly used RC columns, resulting in the non-uniform distribution of confining stress around the perimeter of wall cross-section. The model for confining stress distribution of cross-sections featuring large aspect ratio (over 2) was proposed to study the effectiveness of FRP confinement. The confining stress to the long side of the wall cross-section $\sigma_{rx}$ was still assumed to be a uniform distribution, as shown in Fig. 6.28a. Figure 6.20c shows the typical failure mode of the retrofitted RC shear walls in the plastic hinge zones, where the extreme edge of the cross-section of walls was approximately bordered by a parabolic line (dashed line) along the long side of the wall cross-section as the bottom concrete expanded outward under the axial load in combination with the lateral cyclic load, and BFRP sheets were generally ruptured at the corner (the intersection of long side and short side of the wall cross-section).

According to the elastic theory, the confining stress to the short side of the wall cross-section $\sigma_{ry}$ was assumed to be an imaginary parabolic distribution, varying from a maximum at the corners to a minimum in between, as shown in Fig. 6.28b and reported in [55]. The distribution equation confining stress to the short side of the wall cross-section is shown in Eq. (6.38).

$$\sigma_{ry} = \frac{4\sigma_{ry\max}}{h^2} x^2 \tag{6.38}$$

where $\sigma_{ry\max}$ is the maximum confining stress in the corner to the short side of the wall cross-section; and $x$ is the distance apart from the midpoint.

Considering the equilibrium conditions in two directions of wall cross-section, the confining stress can be expressed by the tensile stress developed in the FRP as

**Table 6.10** Components of ultimate displacements

| Specimen | Corrosion rate (%) | $\Delta_{\text{flex}}$ (mm) | $\Delta_{\text{shear}}$ (mm) | $\Delta_{\text{slip}}$ (mm) | $\Delta_{\text{flex}}/\Delta_u$ (%) | $\Delta_{\text{shear}}/\Delta_u$ (%) | $\Delta_{\text{slip}}/\Delta_u$ (%) |
|---|---|---|---|---|---|---|---|
| SHW0 | 0 | 10.3 | 0.4 | 2.3 | 33.0 | 1.3 | 7.3 |
| SHW1 | 3 | 9.4 | 0.4 | 2.2 | 34.5 | 1.5 | 8.2 |
| SHW2 | 9 | 9.1 | 0.3 | 2.2 | 38.4 | 1.3 | 9.2 |
| SHW3 | 15 | 9.4 | 0.3 | 2.1 | 41.2 | 1.3 | 9.3 |
| RSHW1 | 3 | 9.6 | 0.5 | 2.3 | 30.0 | 1.5 | 7.1 |
| RSHW2 | 9 | 9.5 | 0.4 | 2.2 | 31.6 | 1.5 | 7.5 |
| RSHW3 | 15 | 9.7 | 0.4 | 2.2 | 36.6 | 1.5 | 8.2 |

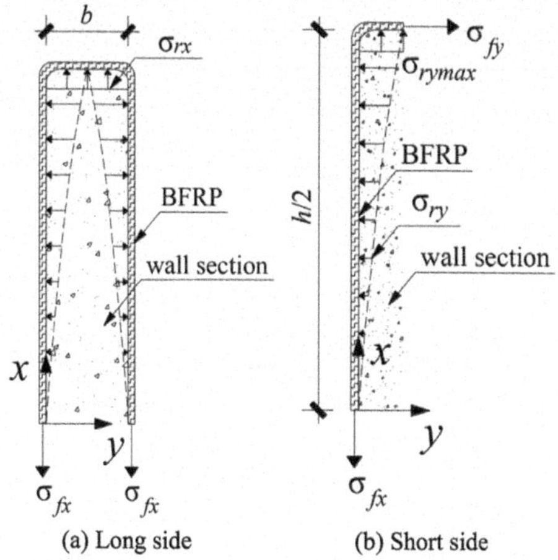


**Fig. 6.28** Assumed distribution of confining stress in two directions

shown in Eqs. (6.39)–(6.40).

$$\sigma_{rx} = \frac{2E_f n t_f \varepsilon_{fx}}{b} \tag{6.39}$$

$$\int_0^{h/2} \sigma_{ry} \mathrm{d}x = E_f n t_f \varepsilon_{fy} \tag{6.40}$$

where $\varepsilon_{fx}$ and $\varepsilon_{fy}$ are the FRP strain along the long side and short side of the wall cross-section, respectively.

After substituting Eq. (6.38) into Eq. (6.40), the maximum confining stress to the corner of the wall cross-section $\sigma_{ry\max}$ can be obtained, as shown in Eq. (6.41).

$$\sigma_{ry\max} = \frac{6E_f n t_f \varepsilon_{fy}}{h} \tag{6.41}$$

Furthermore, the average value of the confining stress to the short side of the wall cross-section $\overline{\sigma}_{ry}$ can be calculated as $2E_f n t_f \varepsilon_{fy}/h$, which accounted for 1/3 of the maximum confining stress in the corner to the short side of the wall cross-section $\sigma_{ry\max}$. The average value of the confining stress to the long side of the wall cross-section $\overline{\sigma}_{rx}$ likewise equaled to $2E_f n t_f \varepsilon_{fx}/b$ as the assumed uniform distribution of confining stress at the long side of the wall cross-section.

The increases in both strength and ductility are highly dependent on the confining stiffness, which is determined as the confining stress developed in the concrete divided by the lateral strain of the concrete. The BFRP sheets wrapped around the circumference of the wall cross-section with fibers in the in-plane direction of the wall, and therefore, the lateral strain of the concrete was evaluated by the compatibility condition in the in-plane direction of the wall, as reported in [54]. As a result, the constant average values of the confining stiffness in two directions are calculated by Eqs. (6.42) and (6.43).

$$E_x = \frac{2E_f n t_f}{b} \tag{6.42}$$

$$E_y = \frac{2E_f n t_f}{h} \tag{6.43}$$

The confining stress to the long side of the wall cross-section was 0.49, 0.53, and 0.57 MPa, when the FRP strain along the long side of the wall cross-section obtained from the test was 2326, 2508, and 2690 με for specimens RSHW1, RSHW2, and RSHW3, respectively. The FRP strain along the short side of the wall cross-section equaled to the ultimate tensile strain (i.e. 0.022) as evidenced by BFRP sheets were generally ruptured at the short side of the wall cross-section, therefore, the average or maximum confining stress to the short side of the wall cross-section remained 0.56 or 1.68 MPa for specimens RSHW1, RSHW2, and RSHW3, respectively. The constant average value of confining stiffness at the long or short side of the wall cross-section was 212 or 25 MPa for specimens RSHW1, RSHW2, and RSHW3, which showed that the major confinement took place at the long side of the wall cross-section as well as FRP confinement to the long side of the wall cross-section primarily contributed to increasing the strength and ductility of the retrofitted RC shear walls.

(3) Strength Considerations

Flexural strength was calculated from sectional analysis based on static equilibrium and compatibility of strains. Ultimate shear strength was calculated from Strut-and-Tie model (STM) proposed by [47]. There were two deteriorating parameters due to reinforcement corrosion-yielding strength and cross-sectional area of reinforcing bar. The residual capacity of corroded reinforcing bar is calculated by Eq. (6.44) [34].

$$A_s = (1 - 0.01\rho_w)A_{s0} \tag{6.44}$$

where $\rho_w$ is the corrosion rate; $A_s$ is the average cross-sectional area of corroded reinforcing bar; and $A_{s0}$ is the initial cross-sectional area of non-corroded reinforcing bar.

Corrosion transforms steel into rust, resulting in the volumetric expansion which can develop the splitting stresses in concrete. The cracking and spalling of concrete induced by these stresses can be modeled by reducing the strength of concrete elements belonging to this region, as reported in [32, 56] and shown in Eqs. (6.10) and (6.11).

STM has been widely used to predict the ultimate shear strength of RC members, inside which complex stress fields were simplified into discrete compressive and tensile force paths. The ultimate shear strength of corroded RC shear walls retrofitted

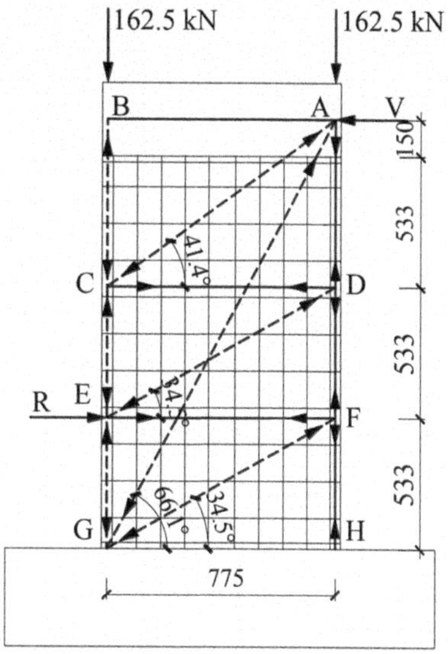


**Fig. 6.29** Strut-and-Tie model for corroded walls retrofitted with BFRP sheets (all units in mm)

with BFRP sheets is predictable based on STM proposed in [47]. The detail of STM for retrofitted specimens is shown in Fig. 6.29.

The transverse reinforcement within the distance of 533 mm, including eight horizontal reinforcing bars, was concentrated in horizontal members CD, and EF, respectively. The six longitudinal reinforcements in boundary elements plus six longitudinal reinforcements in the wall panel were concentrated in vertical members AD, DF, and FH, respectively. The concrete contribution is primarily provided by a direct strut AG (dashed line) from the loading point to the wall base and remains constant after the diagonal cracks occur [57]. It is simply assumed that the strut has a prismatic form with a uniform width of $L_s$ [58]. The strut width can be estimated by calculating the horizontal length of the compression zone at the wall base, as shown in Fig. 6.30a and Eq. (6.45), respectively.

$$L_s = a_s \sin \alpha \tag{6.45}$$

where $a_s$ can be determined using the sectional analysis for the stage when the extreme tensile steel yields [59]; and $\alpha$ is inclined angle of diagonal strut AG, as shown in Fig. 6.30a. For simplicity, $a_s$ can be approximated using the equation reported in [60] for estimating the depth of the flexural compression zone of a column in the elastic stage [59], as shown in Eq. (6.46).

$$a_s = (0.25 + 0.85 \frac{N}{A_w f_c'}) l_w \tag{6.46}$$

where $N$ is the axial load (325.0 kN); and $A_w$ is the net area of the concrete section bounded by the wall thickness $t_w$ and the length of the section in the direction of the

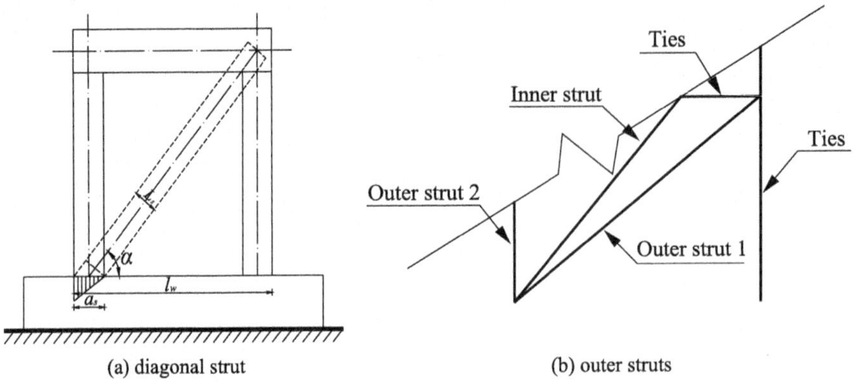


**Fig. 6.30** Descriptions of concrete struts

lateral load $l_w$, as shown in Fig. 6.30a. The cross-sectional area of the inclined strut $A_{\text{strut}}$ is shown in Eq. (6.47).

$$A_{\text{strut}} = L_s \cdot t_w \tag{6.47}$$

The areas of the two outer struts (see Figs. 6.29 and 6.30b), outer strut 1 was an assembly of inclined struts AC, DE, and FG, and outer strut 2 was an assembly of vertical struts BC, CE, and EG, were assumed to be 1/3 and 1/4 the area of the inner strut AG, respectively, as the lateral load is primarily transferred by the inner diagonal strut AG, as reported in [57]. The analysis terminated when all the reinforcing bar ties yield, as reported in [61]. The primary properties of the struts and ties are shown in Tables 6.11 and 6.12, respectively.

According to the observation of the test, BFRP sheets approximately up to 500 mm from the wall base were ruptured. As a result, the horizontal resistance $R$ considering FRP confinement was concentrated at the node $E$ (533 mm above the wall base),

**Table 6.11** Main properties of ties

| Specimen | Ties | Lumped reinforcement | Capacity (kN) | Relative stiffness |
|---|---|---|---|---|
| SHW0 | AD, DF, and FH | 6⏀10 + 6Φ8 | 228.7 | 1.92 |
| | CD and EF | 8⏀8 | 94.6 | 1 |
| SHW1 | AD, DF, and FH | 6⏀10 + 6Φ8 | 218.5 | 1.92 |
| RSHW1 | CD and EF | 8Φ8 | 90.4 | 1 |
| SHW2 | AD, DF, and FH | 6⏀10 + 6Φ8 | 198.8 | 1.92 |
| RSHW2 | CD and EF | 8Φ8 | 82.2 | 1 |
| SHW3 | AD, DF, and FH | 6⏀10 + 6Φ8 | 179.8 | 1.92 |
| RSHW3 | CD and EF | 8Φ8 | 74.4 | 1 |

*Note* ⏀ means 10 mm diameter reinforcing bars with yielding strength of 335 MPa; and Φ means 8 mm diameter reinforcing bars with yielding strength of 235 MPa

**Table 6.12** Main properties of struts

| Specimen | Struts | Angle (°) | Area (mm$^2$) | Relative stiffness |
|---|---|---|---|---|
| SHW0 | AG | 66.1 | 35,575 | 12.80 |
| | AC | 41.4 | 11,858 | 4.27 |
| | DE and FG | 34.5 | 11,858 | 4.27 |
| | BC, CE, and EG | 90.0 | 8894 | 3.20 |
| SHW1<br>RSHW1 | AG | 66.1 | 35,575 | 13.19 |
| | AC | 41.4 | 11,858 | 4.40 |
| | DE and FG | 34.5 | 11,858 | 4.40 |
| | BC, CE, and EG | 90.0 | 8894 | 3.30 |
| SHW2<br>RSHW2 | AG | 66.1 | 35,575 | 14.06 |
| | AC | 41.4 | 11,858 | 4.69 |
| | DE and FG | 34.5 | 11,858 | 4.69 |
| | BC, CE, and EG | 90.0 | 8894 | 3.52 |
| SHW3<br>RSHW3 | AG | 66.1 | 35,575 | 15.06 |
| | AC | 41.4 | 11,858 | 5.02 |
| | DE and FG | 34.5 | 11,858 | 5.02 |
| | BC, CE, and EG | 90.0 | 8894 | 3.76 |

as shown in Fig. 6.29. $R$ was the shear resistance offered by BFRP sheets with fibers in the direction perpendicular to the longitudinal axis of the wall under cyclic load, as well as assumed that the resistance was completely transferred to the wall panel through epoxy without debonding before failure, which could be estimated as reported in [18], as shown in Eq. (6.48).

$$R = 2nt_f\sigma_f d_f(1 + \cot\beta)\sin^2\beta \tag{6.48}$$

where $d_f$ is the useful height of wall section; $\beta$ is the angle between the fiber orientation and the longitudinal axis of the walls; $\sigma_f$ equals to $\varepsilon_f E_f$, which should not exceed $0.75 f_{fu}$, as proposed in ACI 440 [34]; $\varepsilon_f$ is the strain level in FRP, and the value of $\varepsilon_f$ from test for calculating $R$ was 2326, 2508, and 2690 $\mu\varepsilon$ for specimens RSHW1, RSHW2, and RSHW3, respectively; and $f_{fu}$ is the design ultimate tensile strength of FRP.

As shown in Table 6.13, the theoretical results were slightly less than test results, this was mainly because the only contribution of bottom BFRP sheets was considered for simplicity as well as partial sheets had already debonded from the wall surface prior to failure. The contribution of longitudinal reinforcement was considered in calculating the flexural strength, while the STM takes both longitudinal and horizontal bar into consideration for strength calculation. As a result, the shear strength obtained from STM was obviously higher than the corresponding flexural strength based on the sectional analysis. Taking specimens RSHW1 and RSHW3 for example, the average strain predicted by the assumed STM agreed well with the strain history of horizontal reinforcing bars which located 500 mm above the wall base (located vicinity of reinforcing bar tie EF), as shown in Fig. 6.31. Therefore, the ultimate shear strength of retrofitted RC shear wall can be predicted by models based on Li's method in consideration of reinforcement corrosion and FRP confinement.

**Table 6.13** Theoretical flexural and shear strength

| Mock-up | Corrosion rate (%) | $M_R$(kN m) | $V_{MR}$ (kN) | $V_u$ (kN) | $V_{st}$ (kN) | Deviations on STM and test results (%) |
|---|---|---|---|---|---|---|
| SHW0 | 0 | 324.2 | 185.3 | 272.7 | 259.9 | 4.7 |
| SHW1 | 3 | 310.7 | 177.5 | 261.8 | 252.6 | 3.5 |
| SHW2 | 9 | 292.6 | 167.2 | 242.7 | 238.2 | 1.9 |
| SHW3 | 15 | 273.0 | 156.0 | 228.0 | 224.5 | 1.5 |
| RSHW1 | 3 | 310.3 | 177.3 | 326.7 | 276.8 | 15.3 |
| RSHW2 | 9 | 293.2 | 167.5 | 305.3 | 264.4 | 13.4 |
| RSHW3 | 15 | 273.3 | 156.2 | 281.6 | 252.6 | 10.3 |

*Note* $M_R$ is theoretical flexural strength; $V_{MR}$ is theoretical shear strength corresponding to flexural strength; $V_u$ is test results of ultimate shear strength; and $V_{st}$ is theoretical shear strength based on STM, and deviations are calculated as $(V_u - V_{st})/V_u$

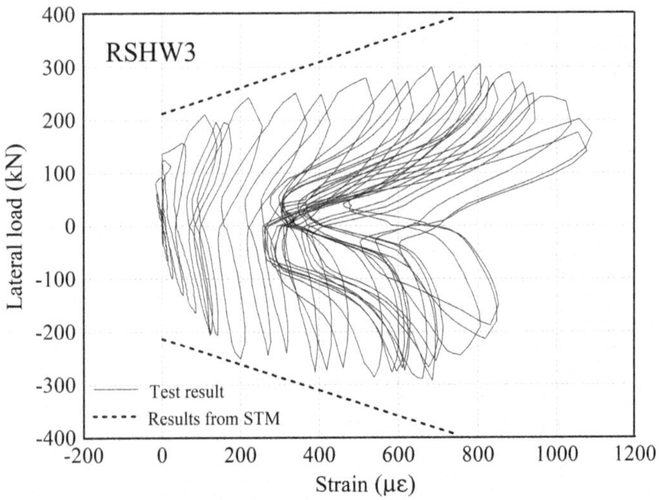


**Fig. 6.31** Strain history of horizontal reinforcement of specimens

## 6.3 Seismic Performance of Shear Walls Retrofitted with BFRP Sheets in Different Methods

### *6.3.1 Experimental Program*

(1) Material Properties

Commercial concrete was used to cast the specimens. The mix proportions of concrete were same as the descriptions shown in Table 6.6. The concrete cubic compressive strength $f_{cu}$, concrete axial compressive strength $f_c$, concrete axial tensile strength $f_t$, compressive elastic modulus $E_c$, and tensile elastic modulus $E_t$ was 39.3, 31.0, 2.33, 3.24 × $10^4$, and 3.29 × $10^4$ MPa, respectively.

The mechanical properties of BFRP strips utilized in this chapter were same as descriptions shown in Sect. 6.2.1. The tested elastic modulus, ultimate strain, and ultimate tensile strength value of the epoxy resin were 2.60 × $10^3$ MPa, 3.5% and 45.8 MPa [62], respectively.

(2) Description of Specimens

Six RC shear wall specimens designated as SHW0, SHW1, SHW2, SHW3, SHW4, and SHW5 were prepared in accordance with the test method recommended by Chinese Standard JGJ3-2010 [63], which represented the part of a shear wall of an existing building.

Geometric dimensions and reinforcing bar details of all specimens were identical, as shown in Fig. 6.1. All specimens were cantilever walls. Each specimen consisted of three subassemblies as follows: (1) the head beam, (2) the wall panel, and (3) the foundation beam. The length, height and thickness of wall were $l$ = 1000 mm, $h$ = 1600 mm, and $t$ = 120 mm, respectively. The aspect ratio of the wall was 1.6, which guaranteed flexural type of failure generally occurred. The web reinforcing bar comprised a double orthogonal grid of 8-mm diameter reinforcing bars vertically and laterally with spacing 110 mm and 150 mm, respectively. The 225 mm × 120 mm boundary elements were reinforced with six reinforcing bars of 10-mm diameter. Confinement in the boundary element was provided by 8-mm diameter closed stirrups spaced at 150 mm (see Fig. 6.32a). Average values of measured yielding strengths of 10-mm diameter reinforcing bars, 8-mm diameter reinforcing bars, and 20-mm diameter reinforcing bars were 335, 235, and 335 MPa, respectively.

Specimen SHW0 was tested without strengthening as the reference specimen, and specimens SHW1, SHW2, SHW3, SHW4, and SHW5 were tested after strengthening with five different configurations of BFRP strips, as shown in Fig. 6.32b–e. BFRP strips were symmetrically applied to both sides of the wall.

The strengthening schemes of specimens SHW1, SHW2, and SHW3 aimed to increase the shear strength, which referred to [64–66]. Two layers of BFRP strips, 200 mm width, were applied through both diagonals of the specimen SHW1 in the form of X-shape. Two layers of 200 mm wide lateral strips, spaced at 350 mm, were used for strengthening of specimen SHW2. Specimen SHW3 was strengthened by

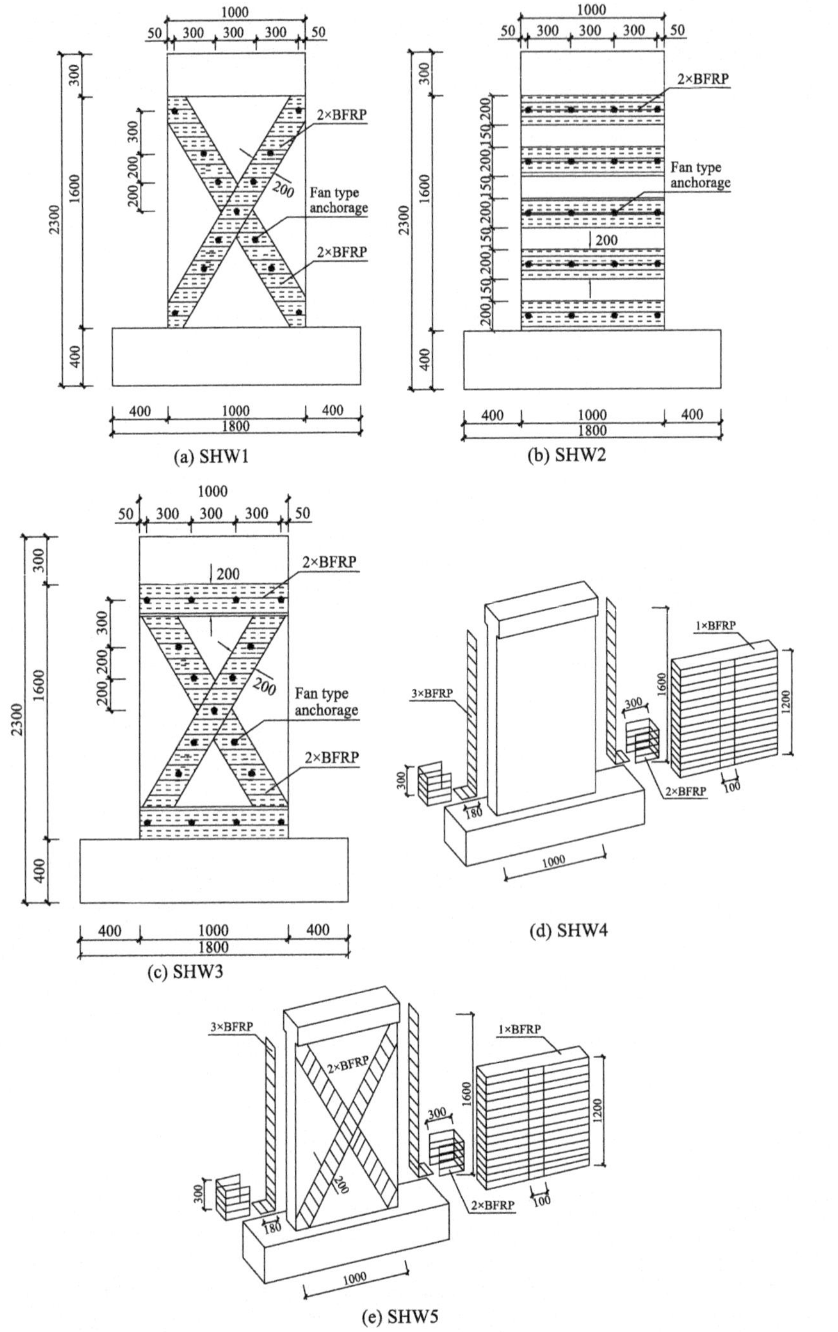


**Fig. 6.32** Details strengthening schemes of specimens SHW1, SHW2, SHW3, SHW4, and SHW5 (all units in mm)

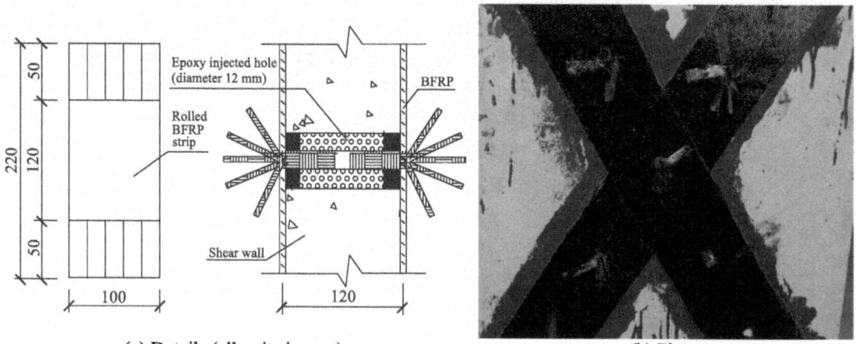


(a) Details (all units in mm) (b) Photo

**Fig. 6.33** Fan type anchor

using combination of X-shaped and lateral BFRP strip. Two layers of 200 mm wide BFRP strips were applied both diagonally and horizontally. In order to enhance the bond behavior between BFRP strips and surface of specimens SHW1, SHW2, and SHW3, the BFRP strips were anchored to concrete with fan type anchors proposed by [64]. The distances between the anchors are shown in Fig. 6.32b–d, and the details and photograph of fan type anchor are shown in Fig. 6.33. The fan type anchors were made from 100 mm wide and 220 mm long BFRP strips by rolling and then 50 mm of them were cut into five pieces and opened in the form of a fan at both ends of the roll. The anchor holes with 12 mm diameters were drilled and the wall surfaces at which BFRP strips will be bonded were roughened for better bonding surface. Surfaces and anchor holes were cleaned by air-blowing and then epoxy was applied to the surfaces with a thickness of 1.5 mm, and injected into anchor holes. After bonding of BFRP strips onto walls, previously prepared fan anchors were passed through the holes by using the wires. The ends of the anchors were opened in the form of the fan (with a diameter of 100 mm) and then were bonded to both sides of the wall to avoid debonding failure of the interface between BFRP strips and concrete.

The strengthening schemes of specimens SHW4 and SHW5 aimed to increase both the flexural and shear strength, which referred to [43, 67]. Strengthening schemes for specimen SHW4 was carried out in three stages as follows: (1) three layers of 120 mm wide, 1780 mm long BFRP strips were bonded to each boundary of the wall in the form of L-shape (180 mm long strips were bonded to the top surface of the foundation beam); (2) two layers of 300 mm wide BFRP strips were bent at 90° at the specimens' boundary elements and then extended for a length of 300 mm; and (3) one layer of 1200 mm wide, 2340 mm long BFRP strips were bonded around the wall from the bottom with a splicing length of 100 mm. Strengthening schemes for specimen SHW5 was mostly same as those of specimen SHW4, except for the addition of two layers of 200 mm wide BFRP strips (using for specimen SHW2) applied through both diagonals of the wall in the form of X-shape just before the third stage.

The application on the BFRP strips to the concrete surface consisted of the following steps [68]: (1) the concrete surface was ground by using a disk grinder to remove any stain and grit; (2) the dust was removed from the surface by using pressurized air; (3) the primer was applied, followed by the resin undercoat; and (4) the BFRP strips was installed, followed by a resin overcoat.

(3) Test Setup and Procedure

The layout of test setup included the loading system and instrumentation in this chapter was same as that shown in Fig. 6.5. And the loading history was also same as that shown in Fig. 6.6.

### *6.3.2 Crack Patterns and Failure Modes*

Reference specimen SHW0 was failed in a flexural manner. The first crack was observed at 54.4 kN lateral load near the base of the wall on the tensile zone. Diagonal cracks along the height of the wall were followed by horizontal flexural cracks. The photograph of the specimen after failure is shown in Fig. 6.34a. The ultimate lateral load of specimen SHW0 was measured 272.7 kN and the lateral drift was 32.0 mm (1.8% drift).

The photograph of specimen SHW1 after failure is shown in Fig. 6.34b. When the lateral drift reached 28.2 mm (1.6% drift), the crushing of concrete cover occurred at the bottom corners of the wall. The measured ultimate lateral load was 300.4 kN and the specimen lost its lateral load-bearing capacity subsequently. Shear cracks were widened rapidly due to debonding of BFRP strips from the wall surface with the increase of lateral drift when fan type anchors damaged. Concrete core area of the wall bottom corners was crushed when the lateral drift reached 34.1 mm (1.9% drift) for the first time.

The photograph of specimen SHW2 after failure is shown in Fig. 6.34c. Concentrated cracks were observed at the both sides of the wall base when the lateral drift reached 32.2 mm (1.8% drift) for the second time, and the corresponding lateral load was 284.3 kN. Plastic hinge was developed at the wall base when the lateral drift reached 38.1 mm (2.1% drift) for the first time. Concrete cover of the wall base was crushed when the lateral drift reached 48.3 mm (2.8% drift) for the third time. BFRP strips were debonded from the wall surface when fan type anchors damaged, and core concrete of the edge of wall base was crushed when the lateral drift reached 54.3 mm (3.1% drift) for the first time. The lateral load-bearing capacity decreased to 0.85 $P_{max}$ when the lateral drift reached 58.3 mm (3.3% drift) and the specimen failed eventually. The destruction of fan type anchors further resulted in BFRP strips rupturing, concrete spalling, and specimen SHW2 failing.

The photograph of specimen SHW3 after failure is shown in Fig. 6.34d. The BFRP strips on the edge of wall were ruptured partly when the lateral drift reached 33.2 mm (1.9% drift) for the second time, and the corresponding lateral load 305.9 kN.

(a) Specimen SHW0 (b) Specimen SHW1

(c) Specimen SHW2 (d) Specimen SHW3

(e) Specimen SHW4 (f) Specimen SHW5

**Fig. 6.34** Failure of tested specimens

The BFRP strips along the diagonals of wall started debonding with the increase of lateral drift. The crushing of concrete cover was observed at the wall base when the lateral drift reached 46.4 mm (2.7% drift) for the third time. The BFRP strips showed more severe debonding when the lateral drift reached 51.4 mm (2.9% drift) for the first time, with spalling of core concrete at the edge of wall base. The lateral load fell to 0.85 $P_{max}$ when the lateral drift reached 54.0 mm (3.0% drift) for the second time. BFRP strips were ruptured and concrete spalled incompletely without destruction of fan type anchors due to combined restriction of X-shaped and lateral strips.

The photograph of specimen SHW4 after failure is shown in Fig. 6.34e. No visible signs of concrete cracks due to the larger area of strengthening. Tiny sound of debonding between concrete and BFRP strips can be heard during the loading. The BFRP strips were ruptured slightly at the wall base when the lateral drift reached

35.6 mm (2.0% drift) for the first time. The BFRP strips were horizontally ruptured at 70 mm above the wall base when the lateral drift reached 41.2 mm (2.4% drift) for the first time.

The photograph of specimen SHW5 after failure is shown in Fig. 6.34f. The failure mode of specimen SHW5 was basically the same as that of specimen SHW4, except for a larger spalling area of concrete and more severe rupture of BFRP strips after failure.

Confinement with FRP for wall type of concrete members with large cross-sectional aspect ratio is usually less effective [48], and the most effective confinement occurs at the corner of the cross-section [69], where BFRP strips were ruptured, as shown in Fig. 6.34c–f.

### *6.3.3 Hysteretic Behavior*

Figure 6.35 compares the load–displacement hysteretic responses of the specimens. Load–displacement hysteretic loop is a critical index of seismic performance, inside which area can be used as a measure of the energy dissipation capacities [64]. As shown in Fig. 6.35, good elasticity was maintained in the beginning and then the significant pinching was observed with the increase of displacement. In general, the pinching in negative direction was much larger than that in the positive direction for all specimens. Additionally, the hysteretic loops of strengthened specimens SHW1, SHW2, SHW3, SHW4, and SHW5 were less pinched compared with that of reference specimen SHW0, i.e., the seismic performance of strengthened specimens was improved to some extent. Specimens SHW4 and SHW5, particularly strengthened with vertical strips at the two edges of wall, of which hysteretic loops were almost symmetric without obvious pinching. This is mainly because of the flexural strengths of them were improved by the vertical strips, which contributed to the energy dissipation characteristics.

The envelope curves of load–displacement hysteretic loops were captured by connecting the peak points of the load–displacement hysteretic loops of each specimen for the first loading cycle, which could be used to evaluate load resistant and deformation capacities, ductility, and strength degradation.

As shown in Fig. 6.36, the envelope curves remained linear for all specimens in the elastic stage, which showed that the strengthened specimens achieve the similar stiffness compared with that of the reference specimen. However, after reaching peak load, the curve of reference specimen decreased the most rapidly without obvious plastic stage. On the contrary, the curves of strengthened specimens had yielding platforms before descending stage. The length of yielding platform varied among the five strengthened specimens, which explicitly showed different ductility capacities. Specimen SHW2 had the longest yielding platform, although its peak load was not the highest. While specimen SHW1 had the shortest yielding platform, it lost load-bearing capacity soon after reaching the peak load. This was mainly due to the debonding of BFRP strips.

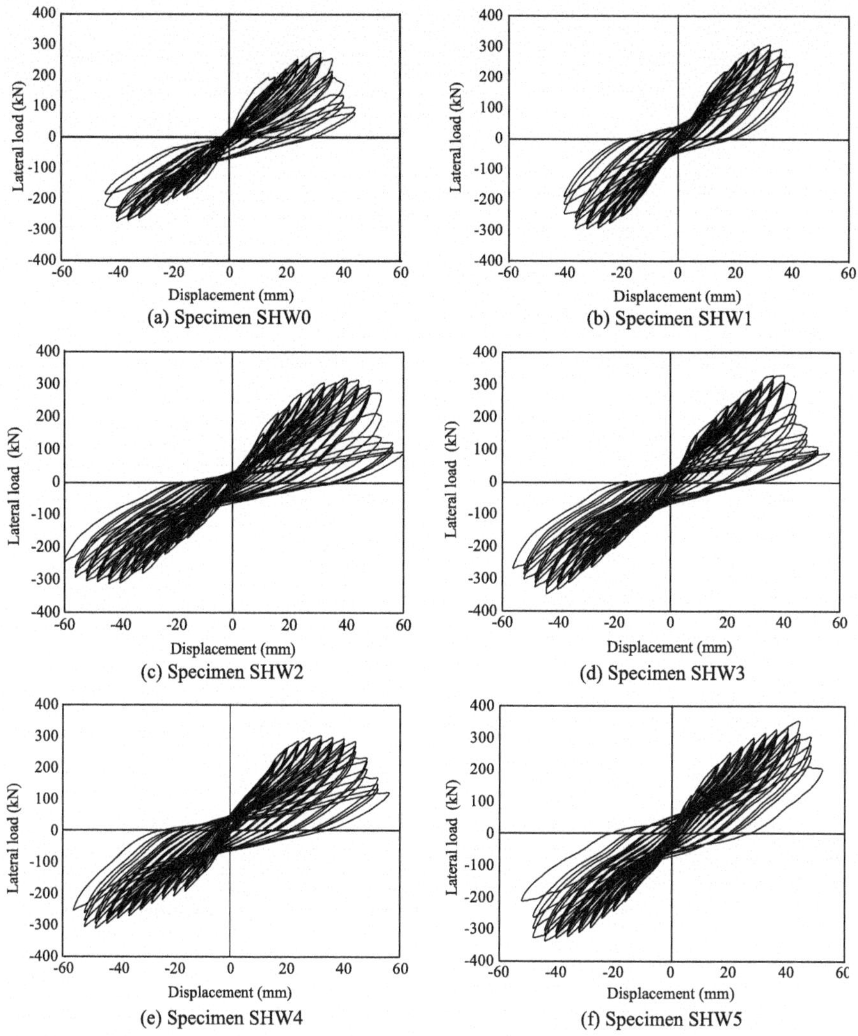


**Fig. 6.35** Load–displacement hysteretic loops of all specimens

## 6.3.4 *Ductility*

Ductility is one of the major issues of concern in achieving the widespread acceptance of FRP-reinforced concrete structures in practice [70]. Four factors, namely yielding displacement $\Delta_y$, ultimate displacement $\Delta_u$, displacement ductility ratio $\mu$, and ultimate rotation $\theta_u$, were taken into consideration to evaluate the ductility behavior. Evaluation of the ductility remained conventional due to the relation between strength and deformation may not have a well-defined yielding point. The yielding displacement $\Delta_y$ corresponding to the load at yielding was evaluated using the method

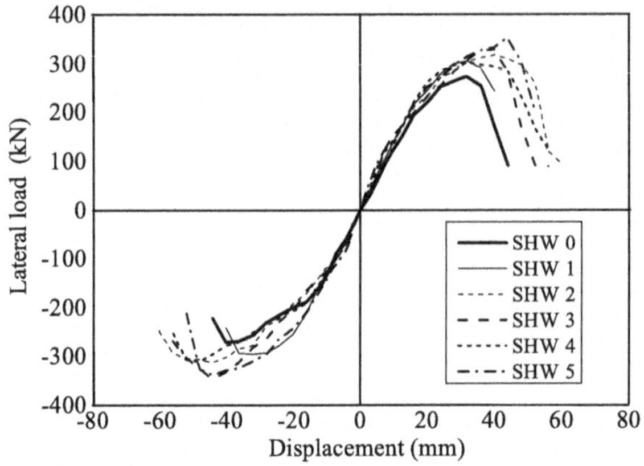


**Fig. 6.36** Envelope curves of hysteretic loops for the first loading cycle

proposed in [71]. The ultimate displacement $\Delta_u$ was defined as displacement corresponding to 85% of peak load on descending stage of envelope curve. Displacement ductility ratio $\mu$ was defined as $\mu = \Delta_u/\Delta_y$, which is proposed in [72, 73]. Ultimate rotation $\theta_u$ was defined as $\theta_u = \Delta_u/h$, where $h$ was actually measured height of wall subassembly of specimen.

As shown in Table 6.14, the yielding or ultimate displacements of specimens SHW0, SHW2-SHW5 were 25.8, 32.1, 33.2, 30.5, and 30.0 mm or 40.5, 55.3, 48.9, 50.8, and 47.4 mm, which were 24.4%, 28.7%, 18.2%, and 16.3% or 36.5%, 20.7%, 25.4%, and 17.0% higher than that of reference specimen SHW0, respectively. The yielding and ultimate displacement of specimen SHW1 was both slightly lower than that of the reference specimen SHW0. The displacement ductility ratios of specimens SHW0, SHW1, and SHW2 and SHW4 were 1.57, 1.71, 1.72, and 1.67, which were 8.9%, 9.6%, and 6.4% higher than that of reference specimen SHW0, respectively. Although the displacement ductility ratio of specimens SHW3 and SHW5 were not clearly increased, the yielding and ultimate displacements were obviously improved compared with those of the reference specimen SHW0. This is unexpected by author, however, results reported in [74] also shows that specimens strengthened both with X-shaped and lateral CFRP strips has the worst ductility capacity. The ultimate rotation of specimens SHW0, SHW2-SHW5 were 1/40, 1/29, 1/33, 1/31, and 1/34, which were 37.9%, 21.2%, 29.0%, and 17.6% higher than that of reference specimen SHW0, respectively. No visible improvement in ultimate rotation of specimen SHW1 was observed compared with that of reference specimen SHW0. In general, the drift capacities of strengthened specimens were higher than that of reference specimen.

**Table 6.14** Summary of test results of displacements

| | Specimen | | | | | |
|---|---|---|---|---|---|---|
| | SHWO | SHW1 | SHW2 | SHW3 | SHW4 | SHW5 |
| $\Delta_y$ (mm) | 25.8 | 23.1 | 32.1 | 33.2 | 30.5 | 30.0 |
| $\Delta_u$ (mm) | 40.5 | 39.6 | 55.3 | 48.9 | 50.8 | 47.4 |
| $\mu = \Delta_u/\Delta_y$ | 1.57 | 1.71 | 1.72 | 1.47 | 1.67 | 1.58 |
| $\theta_u = \Delta_u/h$ | 1/40 | 1/40 | 1/29 | 1/33 | 1/31 | 1/34 |

*Note* $\Delta_y$ is yielding displacement; $\Delta_u$ is ultimate displacement; $\mu$ is displacement ductility ratio; and $\theta_u$ is ultimate rotation

### 6.3.5 Stiffness Characteristic

The actual stiffness values for RC shear walls are well below those corresponding to their elastic properties, even at very low levels of load [43]. The stiffness of the tested walls was evaluated on the basis of secant stiffness by Eq. (6.8), as shown in Fig. 6.37.

The initial or final secant stiffness of specimens SHW0, SHW0, SHW1, SHW2, SHW3, SHW4, and SHW5 were 13.4, 14.9, 15.3, 15.9, 16.5, and 17.3 kN/mm or 2.9, 4.5, 3.2, 3.3, 3.4, and 4.0 kN/mm, which increased by 11.1%, 13.7%, 18.8%, 22.8%, and 29.1% or 55.2%, 10.3%, 13.8%, 17.2%, and 37.9% when compared with that of reference specimen SHW0. For each specimen, the secant stiffness decreased with the increase of cycle number. In general, the strengthened specimens SHW1, SWH2, SHW3, SHW4, and SHW5 not only had higher stiffness but also delayed stiffness degradation compared to the reference specimen SHW0. This property of the strengthened specimens was desirable in an earthquake-like situation. It is observed in the past earthquake that most of the RC structures failed as a result of the sudden loss of stiffness with increasing lateral movement [5]. Therefore, the application on BFRP strips was significantly necessary for the strengthening of structures in quake-prone regions.

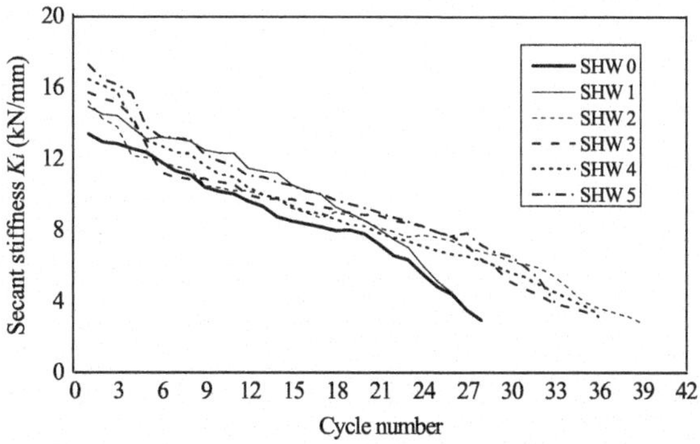


**Fig. 6.37** Comparisons of secant stiffness degradation of all specimens

### 6.3.6 Energy Dissipation Capacity

Failure mechanism of RC members highly depended on the load path and history and largely affects both the ductility and energy dissipation capacity based on the finding of tests. Energy dissipation is a basic structural property of RC elements when subjected to seismic demands [75]. For RC structures designed to reduce damage without collapse due to a seismic event, the input energy can be dissipated through hysteretic responses of RC members, without a significant reduction in strength [76].

The area under the hysteretic loops can be used as a measure of the energy dissipation capacities [43]. Energy dissipations capacities were estimated by calculating the areas inside the hysteretic loops of each specimen and the cumulative energy dissipation capacities were defined as the sum of the area enclosed by all hysteretic loops. Two parameters, namely equivalent viscous damping coefficient $h_e$ [77] and cumulative dissipated energy $Q$ [7], were used to evaluate the energy dissipation capacity of the specimens subjected to cyclic load. Equivalent viscous damping coefficient $h_e$ is determined from the formula, as shown in Eq. (6.49).

$$h_e = \frac{1}{2\pi} \frac{S_{ABCD}}{S_{OAE} + S_{OCF}} \tag{6.49}$$

where $S_{ABCD}$ is energy dissipated in a loading cycle, defined as the area inside the load–displacement hysteretic loop of a specimen at a loading cycle; $S_{OAE}$ and $S_{OCF}$ are energy absorbed by a linear elastic body equivalent to the specimen, defined as the areas inside the triangles *OAE* and *OCF*, as shown in Fig. 6.24. It was noted that a high $h_e$ indicates a large energy dissipation capacity. Cumulative dissipated energy $Q$ was obtained by the accumulation of continuous integration of areas enclosed by every load–displacement hysteretic loop of specimen.

Equivalent viscous damping coefficients of all the specimens at the first loading cycle were calculated, as shown in Fig. 6.38. The equivalent viscous damping coefficients of all the specimens increased and then decreased with the increase of displacement. The initial or final equivalent viscous damping coefficients of specimens SHW0, SHW1, SHW2, SHW3, SHW4, and SHW5 were 2.7%, 2.8%, 3.0%, 3.6%, 4.2%, and 3.9% or 6.0%, 7.9%, 7.6%, 7.5%, 8.3%, and 6.5%, which were 3.7%, 11.1%, 33.3%, 55.5%, and 44.4% or 31.7%, 26.7%, 25.0%, 38.3%, and 8.3% higher than that of reference specimen SHW0, respectively. The equivalent viscous damping coefficients of strengthened specimens effectively increased compared with those of reference specimen.

Figure 6.39 compares the calculated results of cumulative dissipated energy $Q$ of all specimens at each loading cycle. The rate of cumulative dissipated energy of each specimen increased with the increase of cycle number. The final cumulative dissipated energies of specimens SHW0, SHW1, SHW2, SHW3, SHW4, and SHW5 were 56.2, 77.3, 154.7, 131.6, 142.6, and 109.7 kN·m, which increased by 37.5%, 175.3%, 134.2%, 153.7%, and 95.2% when compared with that of reference

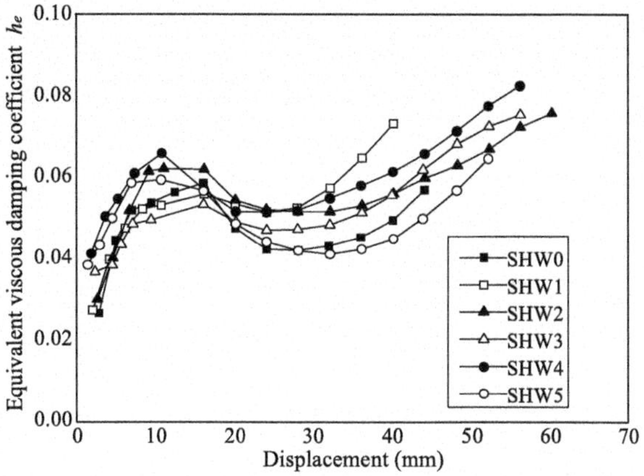


**Fig. 6.38** Equivalent viscous damping coefficient of all specimens

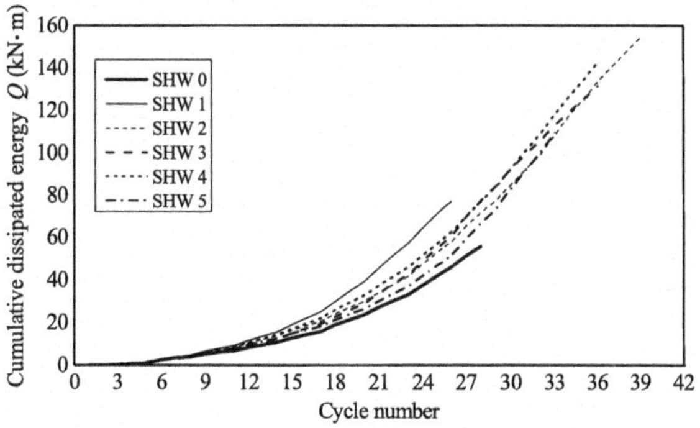


**Fig. 6.39** Cumulative dissipated energy of all specimens at each loading cycle

SHW0, respectively. Energy dissipation capacities of the strengthened specimens were significantly larger than that of the reference specimen.

Therefore, the strengthening with the use of BFRP strips was able to help RC shear walls sustain further inelastic deformations before collapse.

## 6.3.7 Calculation on Load-Bearing Capacity

Table 6.15 shows the theoretical results of load-bearing capacity, involving flexural and shear strengths, estimated before and after the addition of BFRP strips. Flexural strengths $M_{R1}$ were calculated from sectional analysis based on static equilibrium and compatibility of strains. The additional flexural strengths $M_{R2}$ due to BFRP strips at the wall edges for specimens SHW4 and SHW5 were calculated by the tensile stresses developed in the vertical BFRP strips.

Strut-and-Tie models have been used widely for many years to predict the shear strength of RC members, and complex stress fields inside a member are simplified

**Table 6.15** Calculated and measured load-bearing capacity of specimens

| Specimen | Calculated flexural strength due to reinforcing bars $M_{R1}$ (kN · m) | Calculated additional flexural strength due to BFRP strips, $M_{R2}$(kN · m) | Total flexural strength, $M_R$(kN · m) | Calculated shear corresponding to flexural strength $V_{MR}$ (kN) | Calculated shear strength based on STM $V_{st}$ (kN) | Test result of maximum shear strength $V_{meas}$ (kN) | Deviations on theoretical and test results (%) |
|---|---|---|---|---|---|---|---|
| SHW0 | 313.1 | 0 | 313.1 | 178.9 | 259.9 | 272.7 | 4.7 |
| SHW1 | 313.1 | 0 | 313.1 | 178.9 | 271.7 | 300.4 | 9.5 |
| SHW2 | 312.9 | 0 | 312.9 | 178.8 | 305.6 | 314.9 | 3.0 |
| SHW3 | 313.3 | 0 | 313.3 | 179.0 | 309.0 | 337.3 | 8.4 |
| SHW4 | 312.7 | 34.6 | 347.3 | 198.5 | 304.2 | 306.8 | 1.0 |
| SHW5 | 313.6 | 34.6 | 348.2 | 199.0 | 328.2 | 345.4 | 5.0 |

*Note* The deviations are calculated as $(V_{\text{meas}} - V_{\text{st}})/V_{\text{meas}}$

into discrete compressive and tensile force paths. For RC shear wall strengthened with BFRP strips, the shear strength can be calculated on the basis of Strut-and-Tie model proposed by [47].

The concrete contribution is mainly provided by a direct strut AG (dashed line) from the loading point to the base of the wall (see Fig. 6.40) and keeps constant after the diagonal cracks occur [57]. It is assumed that the strut has a prismatic form with a uniform width of $L_s$ for simplicity [58].

The strut width can be estimated with the aid of the horizontal length of the compression zone at the wall base, as shown in Fig. 6.41a and Eq. (6.45), respectively. where $a_s$ can be determined using the sectional analysis for the stage when the extreme tensile steel reaches yielding [59], and $\alpha$ is inclined angle of diagonal strut AG, as shown in Fig. 6.41a. For simplicity, $a_s$ can be approximated using the formula proposed by Park and Paulay [60] for estimating the depth of the flexural compression zone of a column in elastic stage [59], as shown in Eq. (6.46 and 6.47).

The area of the two outer struts, outer strut 1 was assembly of inclined struts AC, DE, and FG, and outer strut 2 was assembly of vertical struts BC, CE, and EG, as shown in Figs. 6.40 and 6.41b, is assumed to be 1/3 and 1/4 the area of the inner strut AG (see Fig. 6.41b), respectively, since the shear force is mainly transferred by the inner diagonal strut AG [57]. The primary properties of the struts and ties are shown in Tables 6.16 and 6.17.

The Strut-and-Tie model applied to calculate the shear strength of all specimens is shown in Fig. 6.40. The transverse reinforcement within the distance of 533 mm, including eight horizontal reinforcing bars, was concentrated in horizontal members CD and EF, respectively. The six longitudinal reinforcements in boundary elements plus six longitudinal reinforcements in the wall panel were concentrated in vertical members AD, DF, and FH, respectively. The analysis terminated when all of the steel tie yields [61]. The details of Strut-and-Tie model strengthened with BFRP strips are shown in Fig. 6.40b–f. Two layers of BFRP strips were applied through both diagonals of specimen SHW1 in the form of X-shape, the diagonal BFRP strips along the strut AG were stretched when the wall subjected to negative load, and diagonal reaction $R_1$ was concentrated at the node $A$. Two layers of lateral strips spaced at 350 mm were used for strengthening of specimen SHW2, only bottom BFRP strips were ruptured when failure, and horizontal reaction $R_2$ was concentrated at the node $E$. Specimen SHW3 was strengthened with combination of X-shaped and lateral BFRP strips, and diagonal reaction $R_1$ and horizontal reaction $R_2$ were concentrated at the node $A$ and $E$, respectively. Specimens SHW4 and SHW5 were somewhere different from specimens SHW1, SHW2, and SHW3, three layers of BFRP strips were bonded along wall edges, and vertical reaction $R_3$ was concentrated at the node A. According to the observation of the experiment, BFRP strips within a distance of approximately 300 mm at the bottom of wall were ruptured, and the horizontal reaction $R_2$ was concentrated at the node E, which was the same with that of specimen SHW2. The diagonal reaction $R_1$ was likewise concentrated at the node A for specimen SHW5, which was the same with that of specimens SHW1 and SHW3. The friction produced by the vertical jack was far less than the shear strength of the specimens, which can be disregarded in the shear strength calculation, as suggested in [3, 43, 45, 47, 64].

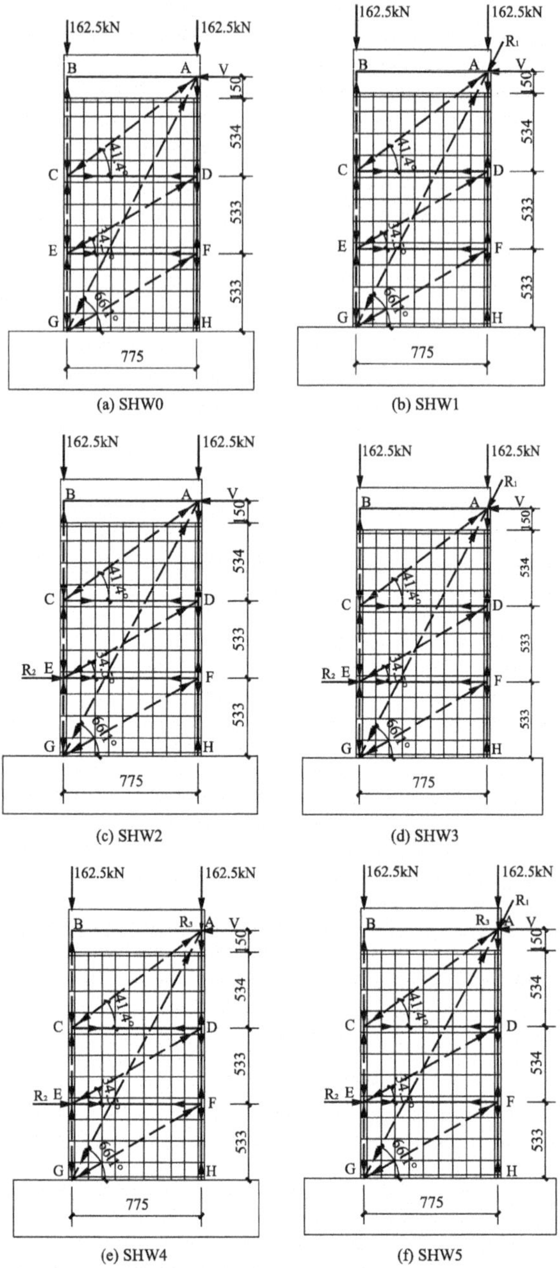


**Fig. 6.40** Detailed Strut-and-Tie model of tested specimens (all units in mm)

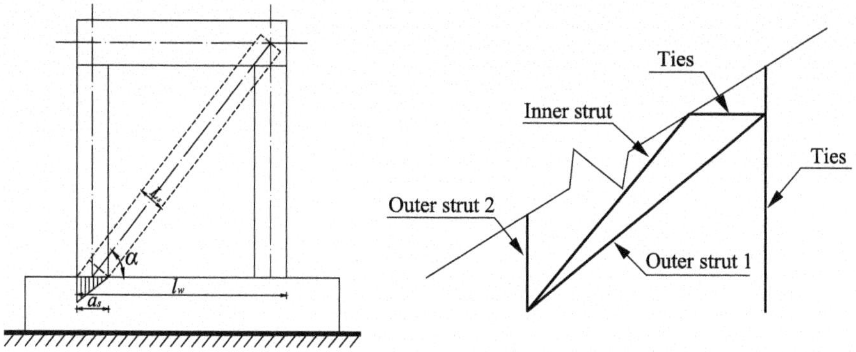


(a) Diagonal strut (b) Outer struts

**Fig. 6.41** Descriptions of concrete struts

**Table 6.16** Main properties of ties of specimens

| Ties | Lumped Reinforcing Bars | Capacity (kN) | Relative stiffness |
|---|---|---|---|
| AD | 6⏀10 + 6Φ8 | 228.7 | 1.92 |
| DF | 6⏀10 + 6Φ8 | 228.7 | 1.92 |
| FH | 6⏀10 + 6Φ8 | 228.7 | 1.92 |
| CD | 8Φ8 | 94.6 | 1 |
| EF | 8Φ8 | 94.6 | 1 |

*Note* ⏀ means 10-mm diameter reinforcing bars with yielding strength of 335 MPa; and Φ means 8-mm diameter reinforcing bars with yielding strength of 235 MPa

**Table 6.17** Main properties of struts of specimens

| Struts | Angle (°) | Area (mm$^2$) | Relative stiffness |
|---|---|---|---|
| AG | 66.1 | 37,121 | 14.23 |
| AC | 41.4 | 12,374 | 4.74 |
| DE | 34.5 | 12,374 | 4.74 |
| FG | 34.5 | 12,374 | 4.74 |
| BC | 90.0 | 9280 | 3.56 |
| CE | 90.0 | 9280 | 3.56 |
| EG | 90.0 | 9280 | 3.56 |

$R_1$, $R_2$, and $R_3$ were the resultants of tensile stresses developed in BFRP strips under cyclic load. It was assumed that the tensile stresses developed in the BFRP strips were transferred into the wall panel through epoxy completely with the use of fan type anchors. Therefore, the resultants equal to the tensile force of FRP, which is proposed by ACI 440 Code [45], as shown in Eqs. (6.50)–(6.52).

$$R_i = 2nt_f\sigma_f d_f(1+\cot\beta)\sin^2\beta,\ i = 1, 2 \tag{6.50}$$

$$R_3 = nt_f d_f \sigma_f \tag{6.51}$$

$$\sigma_f = 0.004 \cdot E_f \leq 0.75 \cdot f_{fu} \tag{6.52}$$

where $n$ is the number of strips; $t_f$ and $d_f$ are their thickness and length, respectively; $\beta$ is the angle between the fiber orientation and the longitudinal axis of the walls; $\sigma_f$ is "effective stress," while the effective strain is limited to 0.4%, as shown in Eq. (6.52); and $f_{fu}$ is design ultimate tensile strength of FRP.

As shown in Table 6.17, the shear strength obtained from theoretical model proposed by Li et al. or experiment were 259.9, 271.7, 305.6, 309.0, 304.2, and 328.2 kN or 272.2, 300.4, 314.9, 337.3, 306.8, and 345.4 kN for specimens SHW0, SHW1, SHW2, SHW3, SHW4, and SHW5, respectively. The deviations on theoretical and test results were 4.7%, 9.5%, 3.0%, 8.4%, 1.0%, and 5.0% for specimens SHW0, SHW1, SHW2, SHW3, SHW4, and SHW5, respectively. The deviations were small. Thus, shear strength of RC shear wall strengthened with BFRP strips can be predicted by models proposed by Li et al. Table 6.17 also shows that the calculated shear strengths of strengthened specimens were all slightly less than test results, arising from only the contribution of BFRP strips near the bottom was considered for simplicity. In addition, the shear strengths of all specimens were greater than the corresponding flexural strengths, which showed that all specimens failed in a flexural manner as expected.

## 6.4 Summary

Investigations on the seismic performance of corroded RC shear walls, corroded walls retrofitted with BFRPs, uncorroded walls strengthened with BFRPs in fan anchor technique were conducted. The models for the seismic performance of walls in consideration of corrosion rate and FRP confinement were also proposed. The major results obtained from the tests were summarized as follows.

(1) The half-cell potentials were used to analyze and evaluate the corrosion rate of reinforcing bars and load-bearing capacity of RC shear walls. The reinforcement corrosion rate increased and load-bearing capacity of RC shear walls decreased with the decrease of half-cell potentials. The relationships between half-cell potentials and corrosion rate or load-bearing capacity were also proposed.

(2) The seismic performance of RC shear walls was influenced by the corrosion of reinforcing bars. The cracking pattern and failure modes of RC shear walls changed because of the reinforcement corrosion. The yielding strength, ultimate strength, ultimate displacement, initial stiffness, displacement ductility ratios,

and drifts of corroded RC shear walls decreased compared with those corresponding values of the uncorroded walls. The model for evaluating the seismic performance of shear walls was proposed in consideration of the corrosion rate of reinforcing bars.

(3) The seismic performance of corroded RC shear walls improved by retrofitting with BFRP sheets. The yielding strength, ultimate shear strength, and cumulative energy dissipation could be recovered to those corresponding values of uncorroded wall when the corrosion rate was lower than 15%. The use of BFRP sheets to confine the corroded RC shear walls improved the bond behavior between steel bars and concrete. And retrofitting with BFRP sheets maintained the flexural type of failure mode of corroded RC shear walls. The model for the ultimate shear strength of corroded RC shear walls retrofitting with BFRP sheets was proposed in consideration of corrosion rate as well as FRP confinement.

(4) The seismic performance of RC shear walls could be improved by strengthening with BFRP sheets. The fan anchor technique was used to improve the bond strength between BFRP sheets and concrete. The ductility and load-bearing capacity could be improved by BFRP sheets with many types especially by lateral sheets. The initial and final secant stiffness, ultimate displacement, ultimate rotation, final equivalent viscous damping coefficients, and cumulative energy dissipation of RC shear walls improved by strengthening with BFRP sheets compared to those of the reference specimen. The model for the shear strength of strengthened RC shear walls was proposed in consideration of strengthening types of BFRP sheets.

## References


1. Farvashany FE, Foster SJ, Rangan BV (2008) Strength and deformation of high-strength concrete shearwalls. ACI Struct J 105(1):21-29
2. Meftah SA, Tounsi A (2007) Lateral stiffness and vibration characteristics of damaged RC coupled shear walls strengthened with thin composite plates. Build Environ 42(10):3596–3605
3. Altin S, Kopraman Y, Baran M (2013) Strengthening of RC walls using externally bonding of steel strips. Eng Struct 49:686–695
4. Christidis KI, Vougioukas E, Trezos KG (2016) Strengthening of non-conforming RC shear walls using different steel configurations. Eng Struct 124:258–268
5. Li B, Qian K, Tran CTN (2013) Retrofitting earthquake-damaged RC structural walls with openings by externally bonded FRP strips and sheets. J Compos Constr 17(2):259–270
6. Dan D (2012) Experimental tests on seismically damaged composite steel concrete walls retrofitted with CFRP composites. Eng Struct 45:338–348
7. Layssi H, Cook WD, Mitchell D (2012) Seismic response and CFRP retrofit of poorly detailed shear walls. J Compos Constr 16(3):332–339
8. Petkune N, Donchev T, Hadavinia H et al (2016) Performance of pristine and retrofitted hybrid steel/fibre reinforced polymer composite shear walls. Constr Build Mater 117:198–208
9. Ehsani MR, Saadatmanesh H (1997) Fiber composites: an economical alternative for retrofitting earthquake-damaged precast-concrete walls. Earthq Spectra 13(2):225–241
10. Wu YF, Zhou YW, He XQ (2010) Performance-based optimal design of compression-yielding FRP-reinforced concrete beams. Compos Struct 93(1):113–123

11. Alcaino P, Santa-Maria H (2008) Experimental response of externally retrofitted masonry walls subjected to shear loading. J Compos Constr 12(5):489–498
12. Biswas RK, Iwanami M, Chijiwa N et al (2020) Effect of non-uniform rebar corrosion on structural performance of RC structures: a numerical and experimental investigation. Constr Build Mater 230:116908
13. Molaioni F, di Carlo F, Rinaldi Z (2021) Modelling strategies for the numerical simulation of the behaviour of corroded RC columns under cyclic loads. Appl Sci 11(20):9761
14. di Carlo F, Meda A, Rinaldi Z (2017) Numerical evaluation of the corrosion influence on the cyclic behaviour of RC columns. Eng Struct 153:264–278
15. Yang J, Guo T, Chai S (2020) Experimental and numerical investigation on seismic behaviours of beam-column joints of precast prestressed concrete frame under given corrosion levels. Structures 27:1209–1221
16. Vu NS, Li B (2018) Seismic performance assessment of corroded reinforced concrete short columns. J Struct Eng 144(4):04018018
17. Shen DJ, Yang Q, Huang CB et al (2019) Tests on seismic performance of corroded reinforced concrete shear walls repaired with basalt fiber-reinforced polymers. Constr Build Mater 209:508–521
18. Shen DJ, Yang Q, Jiao Y et al (2017) Experimental investigations on reinforced concrete shear walls strengthened with basalt fiber-reinforced polymers under cyclic load. Constr Build Mater 136:217–229
19. Li M, Shen DJ, Yang Q et al (2023) Effect of reinforcement corrosion on the seismic performance of reinforced concrete shear walls. Constr Build Mater 377:130977
20. Danish Standards (1980) Pull-out test: DS 2082. Danish Standards Organization, Denmark
21. Pour-Ghaz M, Isgor OB, Ghods P (2009) Quantitative interpretation of half-cell potential measurements in concrete structures. J Mater Civ Eng 21(9):467–475
22. Shegay AV, Motter CJ, Henry RS et al (2015) A database for investigating NZS3101 structural wall provisions. In: Pacific conference on earthquake engineering
23. Liao FY, Han LH, Tao Z (2014) Behaviour of composite joints with concrete encased CFST columns under cyclic loading: experiments. Eng Struct 59:745–764
24. ASTM C (1999) Standard test method for half-cell potentials of uncoated reinforcing steel in concrete. ASTM C 876–991
25. Xin R, Ma P (2021) Experimental investigation on the in-plane seismic performance of damaged masonry walls repaired with grout-injected ferrocement overlay. Constr Build Mater 282:122565
26. Li M, Shen DJ, Yang Q et al (2022) Rehabilitation of seismic-damaged reinforced concrete beam-column joints with different corrosion rates using basalt fiber-reinforced polymer sheets. Compos Struct 289:115397
27. Chen ZY, Zhou Y, Zhong GQ et al (2021) Study on the cyclic shear behavior of damping layer joint for the infilled masonry wall. Constr Build Mater 273:121772
28. Kazem Sharbatdar M, Tajari A (2021) Experimental in-plane seismic strengthening of masonry infilled reinforced concrete frames by engineered cementitious composites (ECC). Constr Build Mater 293:123529
29. Ma Y, Che Y, Gong JX (2012) Behavior of corrosion damaged circular reinforced concrete columns under cyclic loading. Constr Build Mater 29:548–556
30. Zheng SS, Zhou Y, Li QQ et al (2019) Experimental study on aseismic behavior of squat RC shear walls due to chloride ion erosion. Eng Mech 36(12):69–78
31. ACI (2008) Building code requirements for structural concrete: ACI 318-14. American Concrete Institute, Farmington Hills
32. Molina FJ, Alonso C, Andrade C (1993) Cover cracking as a function of rebar corrosion: part 2—numerical model. Mater Struct 26:532–548
33. Vu NS, Yu B, Li B (2016) Prediction of strength and drift capacity of corroded reinforced concrete columns. Constr Build Mater 115:304–318
34. Du YG, Clark LA, Chan AHC (2005) Residual capacity of corroded reinforcing bars. Mag Concr Res 57(3):135–147

35. Ministry of Housing and Urban Rural Development of the People's Republic of China (2010) Standard for code for design of concrete structures: GB 50010-2010. China Architecture and Building Press, Beijing (in Chinese)
36. Zhang WP, Shang DF, Gu XL (2005) Sate-of-the-art of the mechanical properties of corroded steel bars. Ind Constr 1:706–709
37. Joorabchian A, Li Z, Peterman KD (2021) Experimental and numerical investigation of fixed-height cold-formed steel wall assemblies bearing on concrete slabs. Thin-Walled Struct 166:107940
38. Comite Euro-International Du Beton (1993) CEB-FIP model code 1990: design code. Thomas Telford Publishing
39. Xu SX (2003) The models of deterioration and durability evaluation of reinforced concrete structure. Xi'an University of Architecture and Technology, Xi'an. (in Chinese)
40. van Nguyen C, Bui QH, Lambert P (2022) Experimental and numerical evaluation of the structural performance of corroded reinforced concrete beams under different corrosion schemes. Structures 45:2318–2331
41. Chen JF, Teng JG (2001) Anchorage strength models for FRP and steel plates bonded to concrete. J Struct Eng 127(7):784–791
42. Meda A, Mostosi S, Rinaldi Z et al (2014) Experimental evaluation of the corrosion influence on the cyclic behaviour of RC columns. Eng Struct 76:112–123
43. Antoniades KK, Salonikios TN, Kappos AJ (2007) Evaluation of hysteretic response and strength of repaired R/C walls strengthened with FRPs. Eng Struct 29(9):2158–2171
44. Li B, Lim CL (2010) Tests on seismically damaged reinforced concrete structural walls repaired using fiber-reinforced polymers. J Compos Constr 14(5):597–608
45. ACI (2008) Guide for the design and construction of externally bonded FRP systems for strengthening concrete structures: ACI 440.2R-08. American Concrete Institute, Farmington Hills
46. Zhang HM, Zeng S (2014) Experimental study on the seismic performance of shear walls under varied axial compression ratios. Struct Eng 30(5):165–173 (in Chinese)
47. Li B, Pan ZF, Xiang WZ (2015) Experimental evaluation of seismic performance of squat RC structural walls with limited ductility reinforcing details. J Earthquake Eng 19(2):313–331
48. Wu YF, Wei YY (2010) Effect of cross-sectional aspect ratio on the strength of CFRP-confined rectangular concrete columns. Eng Struct 32(1):32–45
49. Du YG, Clark LA, Chan AHC (2005) Effect of corrosion on ductility of reinforcing bars. Mag Concr Res 57(7):407–419
50. Takahashi S, Yoshida K, Ichinose T et al (2013) Flexural drift capacity of reinforced concrete wall with limited confinement. ACI Struct J 110(1):95-104
51. Wallace JW, Orakcal K (2002) ACI 318–99 provisions for seismic design of structural walls. Struct J 99(4):499–508
52. Alarcon C, Hube MA, de la Llera JC (2014) Effect of axial loads in the seismic behavior of reinforced concrete walls with unconfined wall boundaries. Eng Struct 73:13–23
53. Sezen H, Setzler EJ (2008) Reinforcement slip in reinforced concrete columns. ACI Struct J 105(3):280–289
54. Kwan AKH, Dong CX, Ho JCM (2015) Axial and lateral stress–strain model for FRP confined concrete. Eng Struct 99:285–295
55. Maalej M, Tanwongsval S, Paramasivam P (2003) Modelling of rectangular RC columns strengthened with FRP. Cement Concr Compos 25(2):263–276
56. Coronelli D, Gambarova P (2004) Structural assessment of corroded reinforced concrete beams: modeling guidelines. J Struct Eng 130(8):1214–1224
57. Xiang WZ (2007) Seismic performance of structural walls with limited transverse reinforcement. Nanyang Technological University, Singapore
58. Hwang SJ, Fang WH, Lee HJ et al (2001) Analytical model for predicting shear strength of squat walls. J Struct Eng 127(1):43–50
59. Kassem W (2015) Shear strength of squat walls: a strut-and-tie model and closed-form design formula. Eng Struct 84:430–438

60. Park R, Paulay T (1991) Reinforced concrete structures. Wiley
61. Qazi S, Michel L, Ferrier E et al (2015) Strut-and-tie model for a reinforced concrete wall strengthened with carbon fibre-reinforced polymers. Compos Struct 128:87–99
62. Huang CB (2012) Experimental study on seismic behavior and retrofitting measures of multi-aged reinforcement concrete shear walls. Hohai University, Nanjing (in Chinese)
63. Ministry of Housing and Urban Rural Development of the People's Republic of China (2010) Standard for Technical specification for concrete structures of tall building: JGJ3-2010. China Architecture and Building Press, Beijing (in Chinese)
64. Altin S, Anil Ö, Kopraman Y et al (2013) Hysteretic behavior of RC shear walls strengthened with CFRP strips. Compos B Eng 44(1):321–329
65. Yang Q (2018) Seismic analysis of corroded reinforced concrete shear walls strengthened with basalt fiber-reinforced polymers. Hohai University, Nanjing (in Chinese)
66. Cui ZH (2014) Research on seismic performance of reinforcement concrete shear walls retrofitted by basalt fiber reinforced polymers. Hohai University, Nanjing (in Chinese)
67. Antoniades KK, Salonikios TN, Kappos AJ (2005) Tests on seismically damaged reinforced concrete walls repaired and strengthened using fiber-reinforced polymers. J Compos Constr 9(3):236–246
68. Ali-Ahmad M, Subramaniam K, Ghosn M (2006) Experimental investigation and fracture analysis of debonding between concrete and FRP sheets. J Eng Mech 132(9):914–923
69. Song XB, Gu XL, Li YP et al (2013) Mechanical behavior of FRP-strengthened concrete columns subjected to concentric and eccentric compression loading. J Compos Constr 17(3):336–346
70. Zhou YW, Wu YF, Teng JG et al (2009) Ductility analysis of compression-yielding FRP-reinforced composite beams. Cement Concr Compos 31(9):682–691
71. Park R (1989) Evaluation of ductility of structures and structural assemblages from laboratory testing. Bull N Z Soc Earthq Eng 22(3):155–166
72. Zhang HM, Lyu XL, Lu L et al (2007) Influence of boundary element on seismic behavior of reinforced concrete shear walls. Earthq Eng Eng Vibr-Chin Ed 27(1):92–98
73. Todut C, Dan D, Stoian V (2014) Theoretical and experimental study on precast reinforced concrete wall panels subjected to shear force. Eng Struct 80:323–338
74. Wu ZM, Hu CH, Wu YF et al (2011) Application of improved hybrid bonded FRP technique to FRP debonding prevention. Constr Build Mater 25(6):2898–2905
75. QI YS (2015) Study on seismic performance of corroded reinforcement concrete shear walls retrofitted by basalt FRP sheet. Hohai University, Nanjing (in Chinese)
76. Rodrigues H, Varum H, Arêde A et al (2012) A comparative analysis of energy dissipation and equivalent viscous damping of RC columns subjected to uniaxial and biaxial loading. Eng Struct 35:149–164
77. Gong ZG, Lyu XL, Ji SZ (2006) Experimental study on seismic behavior of RC shear walls with different boundary restraints. Struct Eng 22(1):56–61

# Chapter 7
# Bond Behavior Between BFRP Bars and Concrete Under Static and Dynamic Loading

In recent years, the utilization of FRP in new civil engineering structures has gained increasing interest [1–4]. As a novel material, BFRP is developing as a prospective substitute for traditional reinforcing bars in concrete structures [5–8]. BFRP exhibits excellent physical and mechanical properties, such as lightweight [9], high elastic modulus [10–12], high tensile strength [13, 14], and good fatigue resistance [15]. The bond strength between concrete and reinforcing bar is critical to hold serviceability and ultimate limit state within an acceptable limit of civil engineering structures. To enhance the bond performance, adequate anchorage length and sufficient cover thickness are the fundamental requirements that have to be met by bond-slip constitutive law. The uncertainty of sudden load is also the prime concern on the structures as it behaves differently with loading rates. The shocks created by lightening, earthquakes, and volcanoes have been a great worry in civil engineering, especially in high-rise structures. Hence, the response of a higher loading rate has to be considered. The pull-out test elucidates the influence of loading rates on the bond-slip behavior [16]. Extensive research has been undertaken on the pull-out test of steel and FRP bars characterizing the bond stress-slip relation between concrete and FRP bars. However, many efforts have been concentrated only on static behavior, and few work has been done on the cyclic behavior of the bond [17]. The slip of FRP bars sometimes plays an important role when it appears in cyclic loading as excited from seismic loads, especially in cases of beam-column joints and bridges. When medium- or high-rise structures are subjected to severe seismic excitations, the behavior of members in the lower parts of structures is generally controlled by lateral forces, and it causes cracking. These cracks grow larger during the response of the structures to a severe seismic excitation because of cyclic bond deterioration in the joint and in the girder end region [18]. For FRP reinforced elements, there might be a chance of damaging the surrounding concrete and FRP bar surface itself when experiencing a continuous sinusoidal load. Moreover, the continuous deterioration of bond stress also causes fatigue failure. Thus, the behavior of concrete structures that are exposed


D. Shen, *Seismic Performance of Corroded Reinforced Concrete Structures Retrofitted with FRP*, https://doi.org/10.1007/978-981-99-7984-4_7

to dynamic loadings is of great importance in civil engineering. The quick response of the structures is different from that under monotonic loadings.

This chapter investigated the bond performance of BFRP bars in concrete [19]. Experiments were conducted to comprehend the behavior of static, dynamic, and cyclic loadings, aiming at solving the consequences of those regular and irregular loading patterns. Firstly, the influence of BFRP bar diameter and concrete strength on bond strength under static loadings was studied. In addition, the bond mechanism of high strength concrete was also featured [6]. Secondly, the specimens were exposed to dynamic loading and were tested to reveal the bond relationship between concrete and BFRP bars under dynamic loadings [20]. Thirdly, an attempt was made to formulate the bond behavior between BFRP bars and surrounding concrete under cyclic loadings [21].

## 7.1 Bond Stress-Slip Relationship Between BFRP Bars and Concrete Under Static Loading

### *7.1.1 Experimental Program*

The mixture proportions and mechanical properties of concrete are shown in Table 7.1. Ordinary Portland cement (P·II 42.5R) was employed with a Blaine fineness of 375 $m^2$/kg. The fine aggregate was natural river sand with a fineness modulus of 1.93 and maximum size of 1.5 mm. The coarse aggregate used was crushed limestone with maximum particle sizes of 15 mm. The apparent density was 2660 kg/$m^3$. The test of compressive strength of concrete was conducted on 150 mm cubes at the age of 28 days according to Chinese Standard GB/T 50081-2019 [22]. The mechanical properties of the two concrete mixtures are shown in Table 7.1.

The BFRP bars of 10, 14, and 20 mm diameter were used in the test. The surface deformations and characteristics, and BFRP bar diameter are shown in Fig. 7.1. All fibers exhibit the linear stress–strain relationship up to failure load. The mechanical properties of BFRP bars including tensile strength, elastic modulus, and ultimate strain were tested, as shown in Table 7.2.

All specimens for the bond stress-slip relationship test were designed with bond length 5 times the nominal diameter of BFRP bars positioned at the center of the concrete specimens [23, 24], since it was the most popular and well-known method

**Table 7.1** Mixture proportions and mechanical properties of concrete

| Identification of specimen | Mixture proportions (kg/$m^3$) | | | | | Cubic compressive strength (MPa) |
|---|---|---|---|---|---|---|
| | Water | Cement | Sand | Stone | Superplasticizer | |
| C40 | 168 | 420 | 673 | 1176 | 1.26 | 43.6 |
| C70 | 161 | 548 | 591 | 1147 | 3.35 | 71.8 |

**Fig. 7.1** BFRP bars with different diameters

**Table 7.2** Mechanical properties of BFRP bars

| Diameter (mm) | Tensile strength (MPa) | Modulus of elasticity (MPa) | Strain (%) |
|---|---|---|---|
| 10 | 1200 | $5.2 \times 10^4$ | 2.31 |
| 14 | 1000 | $4.8 \times 10^4$ | 2.08 |
| 20 | 900 | $4.5 \times 10^4$ | 2.08 |

that provided a simple means of comparing the relative bond behavior of the different bars, as reported in [25]. Two types of concrete specimens were prepared, including the 150 mm × 150 mm × 150 mm concrete specimens for BFRP bars with a diameter of 10 and 14 mm, as well as the 200 mm × 200 mm × 250 mm concrete specimens for BFRP bars with a diameter of 20 mm, as shown in Fig. 7.2. Two pieces of PVC (4 mm larger than BFRP bar diameter) were provided at each end of the bar inside the cube as a bond breaker. Concrete was put in two layers: first up to the bar level, then the rest of the empty portion. In such a way, it does not disturb the arrangements of wooden mold, BFRP bars, and PVC to avoid the possibility of the formation of voids and irregularities within the cubic or prismatic specimen [26]. Compaction of concrete is done by the vibrator with critical measures ensuring the complete compatibility of the specimens.

Unlike traditional reinforcing bars, FRP bars have very low transverse strength [27]. The tensile properties of the BFRP bars cannot be tested in the same gripping zone as used in the reinforcing bars to avoid the possibility of premature failures [28]. Thus, an anchorage system consisting of a steel pipe filled with an epoxy resin grout was prepared to create the confinement pressure on BFRP bars [29]. A total specimen length of 40 $d$ plus 2 $L$ was provided, as recommended in [30], where $d$ is the nominal diameter of the bar and $L$ is the anchorage length. According to the data provided by the grout manufacturer, epoxy consists of two parts epoxy resins (part A) and a hardener (part B), and the grouts should be allowed to be set for 72 h in laboratory environments. It should ensure that there is no slip during the pull-out test. The arrangement and details of BFRP bars are shown in Fig. 7.2b.

The configurations of twenty specimens are presented in Table 7.3. Each test specimen is identified with a label "CXDY-$n$". The first alphabet C in the specimen

Fig. 7.2 Details of specimens

(a) Photo of mold

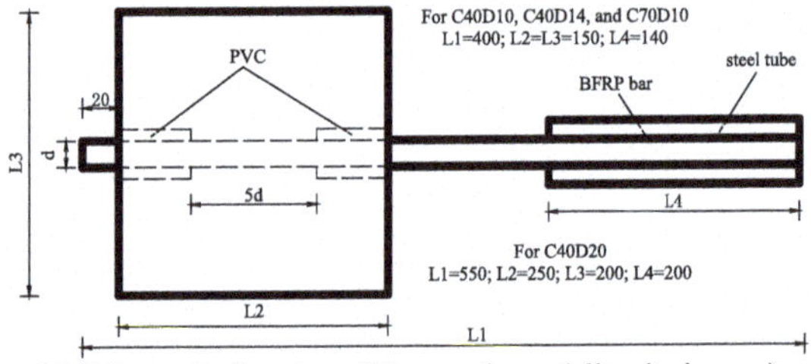


(b) Schematic drawing of the specimen (all units in mm)

name refers to concrete strength (i.e., C = concrete strength grade), X indicates the designed strength of concrete (i.e., 40 and 70 = 40 and 70 MPa cubic compressive strength, respectively), the alphabet D in the specimen name refers to the diameter of BFRP bars, Y indicates the diameter of BFRP bars (i.e., 10 = the diameter of the BFRP bars is 10 mm); and *n* is the progressive test number.

As illustrated in Fig. 7.3, the test was carried out with a servo-hydraulic testing machine (MTS 322). The special gripping system to avoid the possibility of premature failure in the gripping zone of the BFRP rod was selected according to the size of the steel tubes. For all specimens, the initial load was applied at a loading rate of 3.0 kN/min [24]. The pull-out load was, however, applied in displacement control mode to get a post-peak behavior of the bond [25]. A static load was applied on specimens up to the bond failure at a displacement rate of 0.07 mm/s [14]. The load was measured with an electronic loading cell attached to the machine. The bond-slip displacement at the unloaded end of the BFRP bar and concrete displacement of concrete at the rare face of loading are obtained directly from the slip measured by LVDTs, as shown in Fig. 7.3. Load and displacement data were recorded by a DH5922 data logger at a frequency of 10.0 kHz. In the end, all specimens were broken to visualize the bond failure mechanics at the interface of BFRP bars and concrete.

**Table 7.3** Test results of bond strength and slip of all specimens

| Identification of specimen | Bond strength (MPa) | | | | | Free end slip (mm) | | | | |
|---|---|---|---|---|---|---|---|---|---|---|
| | Individual | Average | Results in Eq. (7.5) | Deviations of average value (%) | Deviations of individual value (%) | Individual | Average | Results in Eq. (7.1) | Deviations of average value (%) | Deviations of individual value (%) |
| C40D10-1 | 13.87 | 12.60 | 12.61 | 0.1 | 9.1 | 3.10 | 3.15 | 3.26 | 3.5 | 5.2 |
| C40D10-2 | 11.51 | | | | 9.6 | 3.06 | | | | 6.5 |
| C40D10-3 | 12.74 | | | | 1.0 | 3.47 | | | | 6.1 |
| C40D10-4 | 11.69 | | | | 7.9 | 3.16 | | | | 3.2 |
| C40D10-5 | 13.17 | | | | 4.3 | 2.94 | | | | 10.9 |
| C40D14-1 | 10.88 | 11.58 | 10.21 | 11.8 | 6.2 | 3.97 | 3.90 | 3.74 | 4.1 | 5.8 |
| C40D14-2 | 11.09 | | | | 7.9 | 3.96 | | | | 5.6 |
| C40D14-3 | 11.46 | | | | 10.9 | 3.67 | | | | 1.9 |
| C40D14-4 | 12.60 | | | | 19.0 | 4.16 | | | | 10.1 |
| C40D14-5 | 11.88 | | | | 14.1 | 3.73 | | | | 0.3 |
| C40D20-1 | 8.01 | 8.81 | 9.81 | 11.3 | 22.5 | 4.32 | 4.37 | 4.46 | 2.1 | 3.2 |
| C40D20-2 | 8.71 | | | | 12.6 | 4.02 | | | | 10.9 |
| C40D20-3 | 8.48 | | | | 15.7 | 4.20 | | | | 6.2 |
| C40D20-4 | 9.71 | | | | 1.0 | 4.32 | | | | 3.2 |
| C40D20-5 | 9.12 | | | | 7.6 | 5.00 | | | | 10.8 |
| C70D10-1 | 15.54 | 14.97 | 16.18 | 8.1 | 4.1 | 3.23 | 3.25 | 3.26 | 0.3 | 0.9 |
| C70D10-2 | 16.56 | | | | 2.3 | 3.29 | | | | 0.9 |
| C70D10-3 | 12.84 | | | | 26.0 | 3.32 | | | | 1.8 |
| C70D10-4 | 15.91 | | | | 1.7 | 3.30 | | | | 1.2 |
| C70D10-5 | 14.01 | | | | 15.5 | 3.09 | | | | 5.5 |

Fig. 7.3 Test setup layout

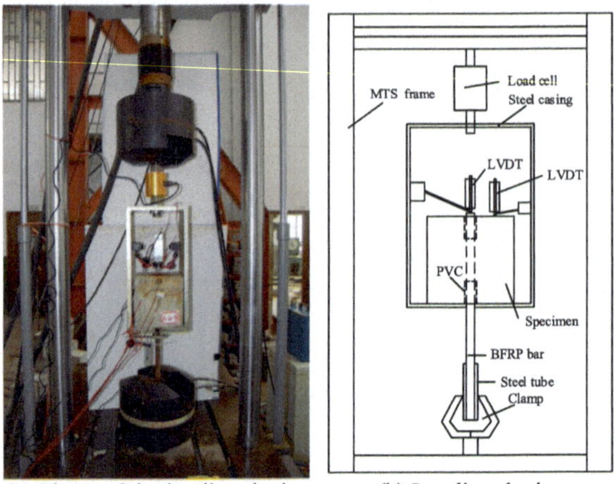


(a) Photo of the loading device (b) Loading device

## 7.1.2 *Failure Modes*

The bond failure modes between concrete and BFRP bars are shown in Fig. 7.4. The nature of the failure mechanism on concrete and BFRP bars was visualized after breaking the tested specimens. The failure between the BFRP bars and concrete was brittle for specimens C40D10, C40D14, and C40D20, as shown in Fig. 7.4a–c. The specimens C40D10, C40D14, and C40D20 with lower compressive strength of concrete showed the crushing of the concrete at bond failure that is accompanied partly by the stripping off of the fiber surface. The close examination of the failure zone revealed that the damages on the FRP surface were limited locally: in some areas, concrete was also stuck on BFRP bars leaving some undamaged parts on the BFRP surface; and in some other parts, the surface of BFRP bars was peeling off before the concrete crushing. For 10 mm FRP bar, the large damages on the FRP surface occurred as compared to the adjacent concrete surface. The stripping off of the surface of FRP bars and concrete crushing seemed to be responsible for bond failures. The white traces that appeared in the FRP surface were due to the crushing of FRP [31]. However, the undamaged part of the BFRP surface was more evident for 14 and 20 mm bars than that for 10 mm bars. The failure was due to the interfacial concrete failures leaving some white patches on the FRP surface. The FRP surface was peeled off or crumbled in some areas but the crushing of concrete was more dominant. It happened because of the less bond stress rather than the smaller diameter of FRP bars.

The failure mode of the bond behavior between the BFRP bars and concrete was influenced by the concrete strength. The failure modes of the bond behavior between the BFRP bars and concrete for specimens C40D10 and C70D10 with the compressive strength of 43.6 and 71.8 MPa were different for the identical diameter of BFRP bars, as shown in Fig. 7.4a, d. The specimen C70D10 with higher compressive strength of concrete showed more ductility in failure as compared to that of the

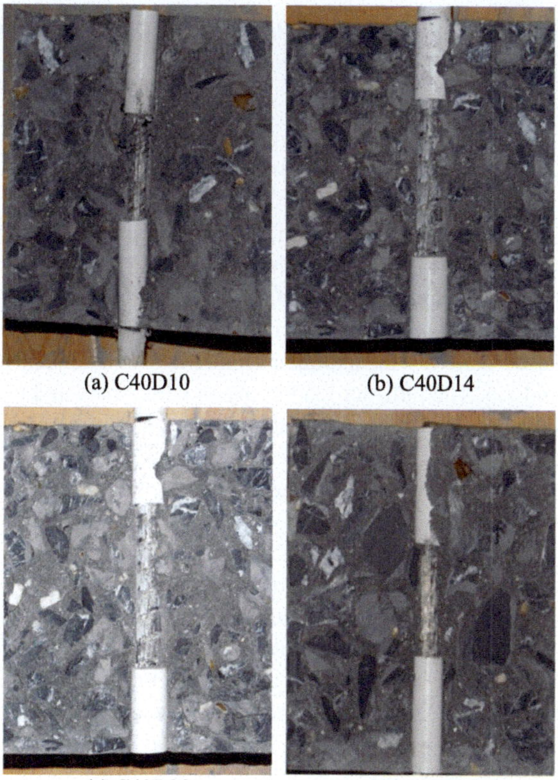

(a) C40D10 (b) C40D14

(c) C40D20 (d) C70D10

**Fig. 7.4** Failure modes for tested specimens

specimen with lower compressive strength. The reason was that the bond failure was completely reflected by inter-laminar shear between BFRP bar layers rather than the crushing of concrete, as reported in [32]. The higher adhesion property of the concrete does not allow separating polymer resin from the circumferential concrete surface because of the weakness in the adhesion property of the FRP composites [33]. This behavior was evidenced by the delamination and detachment of the outer layer of BFRP bars as the outer layer of the strip of BFRP bar (helical wrapping) adhered firmly to the surface of the concrete even after BFRP bar was pulled out, as shown in Fig. 7.4d. Meanwhile, a little concrete crushing was observed, which was the disadvantage of BFRP bars compared with conventional reinforcing bars. This observation indicated that the surface treatment of the FRP bars and the quality of inter-laminar shear governed the pre-peak and post-peak behavior of bond and failure mechanics, especially in high strength concrete.

### 7.1.3 Bond Stress-Slip Relationship Between BFRP Bar and Concrete

The entire bond stress-slip relationship was described by the rise of stress value until its ultimate stress ensuing by softening branch, and then decrease of bond stress to a minimum in succession. Up to the attainment of the peak value, bond stress is complemented by adhesion, friction, deformation, and indentation provided on the surface of FRP bars [34]. The contribution starts to form adhesion property followed by the natural frictional surface of FRP bars [35]. Eventually, all the stress is resisted by the indentation provided on the surface. The bond stress-slip relationships between BFRP bars with the diameter of 10, 14, and 20 mm and concrete are shown in Figs. 7.5 and 7.6. The pull-out load between BFRP bars and concrete increased rapidly with the slight slip value, and later decreased with higher slip value once the maximum pull-out load was reached. The behavior between BFRP bars and concrete could be divided into three phases: micro-slip phase, reinforcing bar pull-out phase, and residual phase [25, 36]. Several researchers have also pointed out that the pull-out load curve for FRP is similar to that for conventional bars in nature. However, the difference in elastic modulus between BFRP bars and conventional reinforcing bars could influence the slip value in identical concrete specimens.

The micro-slip phase encompassed the portion of the curve which was almost linear where the vigorous increase of the stress induced almost no slip, as shown in Fig. 7.6. Although this portion of the curve is often expected to be controlled by adhesion properties of associated material properties, test results showed that the adhesive bond stress decreased with the increase of BFRP bars diameter because large bond stress is needed to mobilize the BFRP bars with small diameter [31]. The stiff slope appeared until the onset of slip for BFRP bars with a small diameter was higher than that with a large diameter.

The concrete pull-out phase is characterized by the nonlinear progression of the curve where an increase in load induces little slip as the maximum pull-out load is approaching [25]. That portion has been represented by a curve after micro-slip phase up to peak stress for specimens with different BFRP bar diameters, as shown in Fig. 7.6. The slip increased abruptly during the pull-out phase followed by the residual phase, and the slip continued increasing with the decreasing load. The resistance to pull-out is provided by two components: the pull-out phase, the interfacial frictional bond between the fiber and polymer matrix; and additional resisting force provided by the mechanical deformations on the fiber surface [37]. In this region, the bond strength of the embedded zone was almost destructed as soon as the load reached its peak value. The pull-out phase appeared after diminishing adhesion properties as the internal crack started to propagate leading to the weaker zone of the concrete around the interface. The crack propagation was largely dependent on the area of the BFRP bars that was in contact with the concrete. However, it was a critical task to distinguish the exact region between the first and second phases. It also showed some discrepancies between the natures of the curves for different diameters of BFRP bars as the stiffness of the curves decreased with the increase of diameter. The thickness

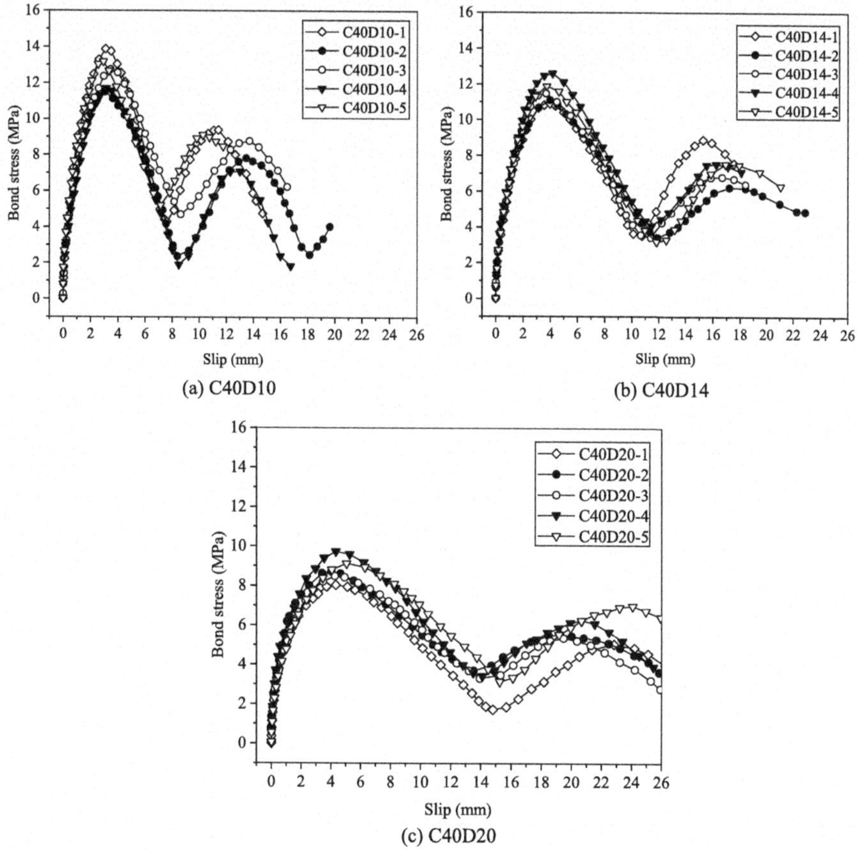


(a) C40D10

(b) C40D14

(c) C40D20

**Fig. 7.5** Relationships between bond stress and slip for different specimens

**Fig. 7.6** Comparisons of relationships between bond stress and slip for specimens with different diameters of BFRP bars

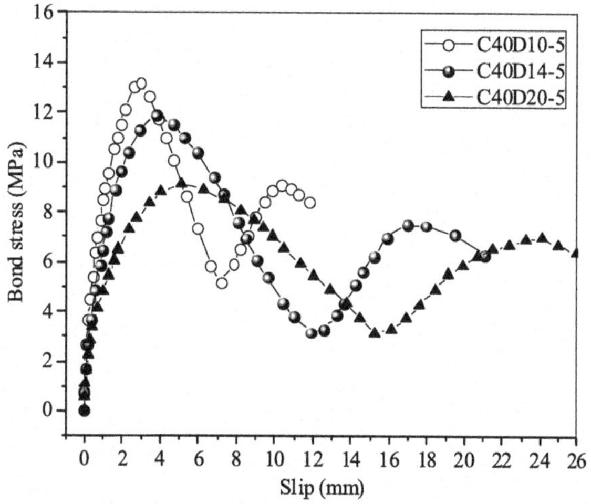

of the cover may also influence the stiffness of the bond; however, the size of the concrete specimen was selected to avoid the splitting failure mode of the cube.

The residual phase represented the rapid reduction of load with a significant increase in slip, as shown in Fig. 7.6, which accounted for the region after the peak value of bond stress. The abrupt increase in slip in that region reflected the severe bond destruction that took place during pulling out. The existing mechanical roughness (helical deformation in this experiment) detaches from the circumferential concrete either by crushing of concrete or displacing or delamination of fiber surface or combined action [23]. Once it debonded, the load decreased drastically until it was inhibited by another neck formed because of uneven surface deformation of the BFRP bar itself or crushed concrete along the embedding length. Moreover, as soon as the helical deformations met another narrow zone, the bond stress started increasing resembling the sinusoidal wave-like curve until it was completely pulled out from the specimen. The reoccurrence of the peak and its interval largely depended on the length of embedment and pitch of the spirally wounded fiber. The previous peak was always higher than the very next peak of the residual curve irrespective of the diameter and concrete strength unless unusual circumstances happened with in the bonding region, as shown in Fig. 7.6. Generally, the maximum bond stress corresponding to the second peak is assumed as 50% of the bond strength regardless the concrete strength and size of FRP bars [38]. The maximum bond stress corresponding to the second peak was about 30%–60% of the bond strength. It showed that the magnitude of the bond stress corresponding to the second peak is largely dependent on the nature of failure mechanics and the amount of damaged bond during the first one. The residual curve suggested that it had lower stiffness (more flat peaks) than previous peaks. It seemed more noticeable in specimens with a higher diameter of the BFRP bars and concrete with higher compressive strength.

### *7.1.4 Bond Strength with Different BFRP Bar Diameters*

Figure 7.5 shows the bond stress-slip relationship between BFRP bars with diameter of 10, 14, and 20 mm and concrete. The comparison of bond strength for the specimens with three different diameters of BFRP bars is shown in Fig. 7.7. The basic nature of the relationship seemed almost the same regardless of BFRP bar diameter. However, the bond strength of the specimens with a smaller diameter of BFRP bar was greater than that with the larger one. The bond strength was 12.60, 11.58, and 8.81 MPa when the diameter of BFRP bars was 10, 14, and 20 mm for specimens C40D10, C40D14, and C40D20, respectively, as shown in Table 7.4 and Fig. 7.7. The bond strength decreased by 8.1% and 30.1% when BFRP bar diameter increased from 10 mm to 14 and 20 mm for specimens C40D10, C40D14, and C40D20, respectively. The results in [23, 31, 39] also show that the bond strength increases with the decrease in FRP bar diameter, and the migration and distribution of stress along the length of the bonded region depend on the bar diameter. Another reason described is Poisson's effect phenomena as in [26, 40]. This effect is generated due to the possible deformations (elongations) of the BFRP bar itself as it consequently affects frictional and mechanical stresses when it undergoes tensile loadings. The reduction of bond stress would be even more pronounced as the increase in the diameter of the bar since the FRP bar possesses the lower stiffness property. The results in [41] also show the bond strength decreases with the increase of FRP bar diameter as it comes from the fabrication process of the specimens. It has also been identified that the bigger diameter bar attracts more water beneath BFRP bar because of the "bleeding action" of the concrete property [31]. This effect is largely noticeable in the specimens with higher water content in concrete. Ultimately, that leads to the formation of voids which possibly reduce the contact surfaces between the BFRP bars and concrete.

The pre-peak and post-peak behavior was also influenced by the diameter of BFRP bar, as shown in Fig. 7.6. The "size effect" concept illustrates the relation between bond stress and BFRP bar diameter. For the specimens with identical material properties and surface treatment, the larger diameter of BFRP bars leads to lower bond strength. It is believed that the stress required to be pulled out is assumed to be concentrated at the circumference of the BFRP bar as soon as it undergoes the load. A heavy amount of energy produced at the interface between the BFRP bar surface

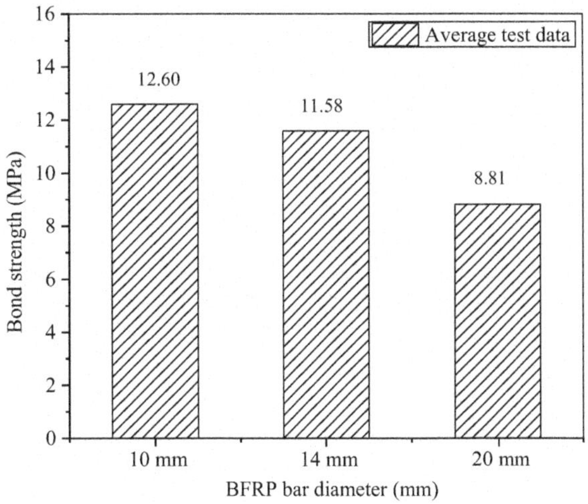


**Fig. 7.7** Comparisons on relationships between bond strength and BFRP bar diameter for specimens

**Table 7.4** Fitting results of bond stress–slip relationship

| Identification of specimen | Diameter (mm) | Parameter $\alpha$ | | | Parameter $p$ | | |
|---|---|---|---|---|---|---|---|
| | | Average value | Results in Eq. (7.8) | Deviations (%) | Average value | Results in Eq. (7.10) | Deviations (%) |
| C40D10 | 10 | 0.411 | 0.417 | 1.5 | 0.404 | 0.402 | 0.5 |
| C40D14 | 14 | 0.375 | 0.377 | 0.5 | 0.343 | 0.343 | 0.0 |
| C40D20 | 20 | 0.308 | 0.317 | 2.9 | 0.256 | 0.254 | 0.8 |

and surrounding concrete was also the reason when pull-out was being undertaken. Finally, that increased the brittleness of the bond behavior. It also suggested that the brittleness of the BFRP bar increased with the decrease in BFRP bar diameter.

### 7.1.5 Effect of BFRP Bar Diameter and Concrete Strength on Slip Corresponding to the Bond Strength

The relationship between the slip corresponding to the bond strength and the diameter of BFRP bars is shown in Fig. 7.8. The average slips corresponding to bond strength were 3.15, 3.90, and 4.37 mm when the diameter of BFRP bars was 10, 14, and 20 mm for specimens C40D10, C40D14, and C40D20, respectively. The slip corresponding to bond strength increased by 23.8% and 38.7% when the diameter of BFRP bars increased from 10 mm to 14 and 20 mm for specimens C40D10, C40D14, and C40D20, respectively. Results in [42] show that the diameter of the bars influences the slip, which depends on the formation and nature of longitudinal cracks in concrete in the bonded zone. Thus, the influence of BFRP bar diameter must be taken into account while determining the slip corresponding to the bond strength. As shown in Fig. 7.8, the relationship between the slip corresponding to the bond strength and the diameter of BFRP bars was linear and could be expressed as Eq. (7.1).

$$s_{max} = s_0 + s_1 d \tag{7.1}$$

where $s_{max}$ and $d$ are the slip corresponding to the bond strength and diameter of BFRP bars, respectively, in mm.

Coefficients $s_0$ and $s_1$ were determined from the regression analysis, and the results were 2.06 and 0.12, respectively. The results of deviations between the free end slip

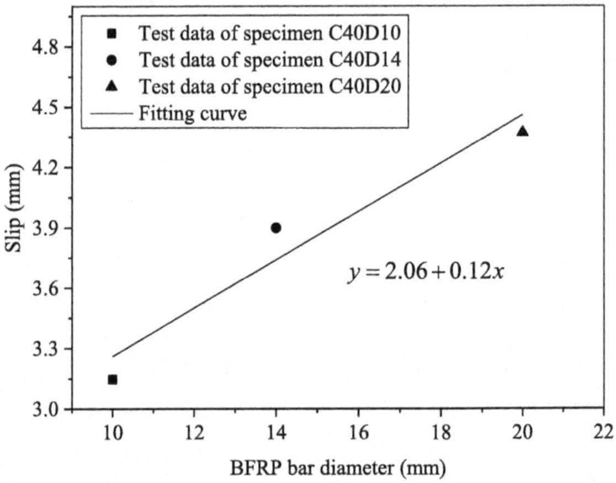


**Fig. 7.8** Relationship between average slip and diameter of BFRP bars of specimens C40D10, C40D14 and C40D20

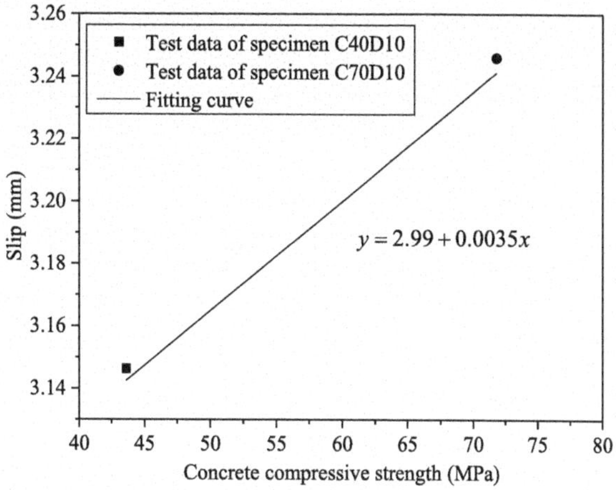


**Fig. 7.9** Relationship between average slip and concrete compressive strength for specimens C40D10 and C70D10

obtained from Eq. (7.1) and the test were small, as shown in Table 7.3. Thus, Eq. (7.1) could be used to predict the free end slip between BFRP bars and concrete.

The relationship between the slip corresponding to the bond strength and concrete strength is shown in Fig. 7.9. The average slips corresponding to the bond strength were 3.15 and 3.25 mm when the concrete compressive strength was 43.6 and 71.8 MPa for specimens C40D10 and C70D10, respectively. The average slips corresponding to the bond strength of the specimens with the 10 mm diameter of BFRP bars increased by 3.17% when the concrete compressive strength increased from 43.6 MPa to 71.8 MPa. Investigation on the bond stress-slip relationship between helically wrapped GFRP bars and concrete showed that the decrease in slip corresponding to bond strength was measured for higher concrete strength [32], and the slip was 4.84 and 4.03 mm for the specimen with the concrete strength of 40 and 70 MPa, respectively. The slip corresponding to the bond strength between BFRP bars and concrete was influenced by concrete strength; however, the degree of impact was small. Thus, the effect of concrete strength on slip corresponding to the bond strength of BFRP bars may be ignored in determining the slip.

### 7.1.6 Bond Strength with Different Concrete Strengths

The bond stress-slip relationship between BFRP bars and concrete for specimen C70D10 with the concrete compressive strength of 71.8 MPa is shown in Fig. 7.10. The comparison of bond stress-slip relationship for specimens C40D10 and C70D10 with different concrete strength is shown in Fig. 7.11. The bond strength was 12.60 and 14.97 MPa when the concrete strength was 43.6 and 71.8 MPa for specimens C40D10 and C70D10, respectively. The bond strength between the BFRP bars with a diameter of 10 mm and the concrete increased by 18.8% when the concrete strength increased by 64.7%. In general, it is believed that higher strength concrete requires a higher amount of cementitious materials, since the use of more cement enhances the compatibility of the concrete unit, and it also improves cohesive properties [39, 43]. Eventually, improvement of the adhesion property reflects its characteristics at the interface between BFRP bars and concrete. It is believed that the chemical bond breaks as the pull-out load exceeds its range, and the natural frictional property of the material comes to resist the load. That led to the eventual higher stress for specimens with higher strength concrete, and the area covered by the curve for specimen C70D10 was also more than that for specimen C40D10. Test results in [44] also show that the bond strength between FRP bars and concrete increased with the increase of concrete strength, and the bond strength increased from 16.64 to 25.21 MPa when concrete compressive strength increased from 29.34 to 47.89 MPa. The nature of the relationships for the specimens with different concrete compressive strength was almost identical, and the flatter curve was obtained for higher strength concrete. This effect continued for the following residual curves. The magnitude of residual peak generally seemed slightly more for the specimens with the concrete compressive strength of 71.8 MPa than that with the concrete compressive strength of 43.6 MPa. However, it may not be always certain because it sometimes depends on the damages caused due to earlier peak stress.

The theoretical models for predicting the bond strength between FRP bars and concrete in consideration of concrete strength, concrete cover depth, the diameter of FRP bars, and the bond length are proposed in [45, 46]. The theoretical model developed in [45] is given in Eq. (7.2).

$$\frac{\tau_{\max}}{\sqrt{f_c}} = 0.03 + 0.14\frac{c}{d} + 9.0\frac{d}{l} \tag{7.2}$$

The theoretical model developed in [46] is given in Eq. (7.3).

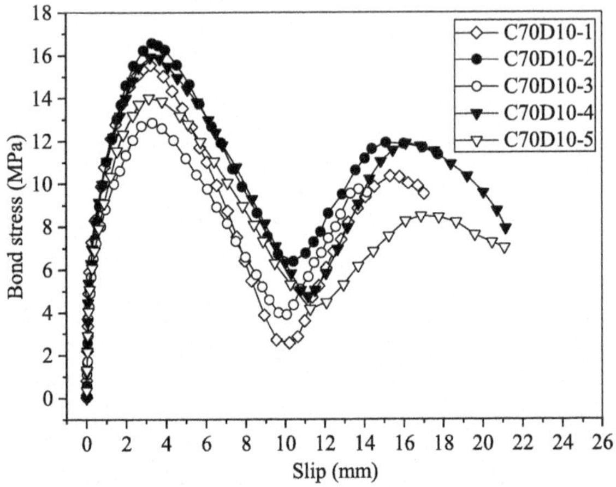


**Fig. 7.10** Bond stress-slip relationship for specimen C70D10

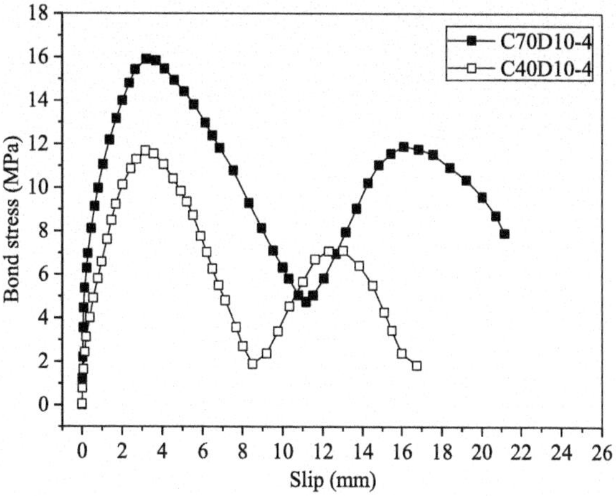


**Fig. 7.11** Comparisons on bond stress-slip relationship for specimens C40D10 and C70D10

$$\frac{\tau_{\max}}{\sqrt{f_c}} = 0.033 + 0.025\frac{c}{d} + 8.3\frac{d}{l} \tag{7.3}$$

where $f_c$, $c$, $d$, and $l$ are the concrete compressive strength, concrete cover depth, diameter of FRP bars, and bonding length, respectively.

The bond strength obtained from test and theoretical models increased with the increase of concrete strength. The bond strength between FRP bars and concrete is proportional to the square root of concrete compressive strength [23, 41, 45, 46, 47]. Thus, relative bond strength was defined as the ratio of bond strength to the square root of concrete compressive strength [36]. The relative bond strength $\tau_R$ can be calculated with Eq. (7.4) [38].

$$\tau_R = \frac{\tau_{\max}}{\sqrt{f_c}} \tag{7.4}$$

Test results showed that the relative bond strength was 1.908, 1.754, and 1.334 when the diameter of the BFRP bars was 10, 14, and 20 mm for specimens C40D10, C40D14, and C40D20, respectively. The ratio of concrete cover depth to diameter of the BFRP bars was 7.00, 4.86, and 4.50 for specimens C40D10, C40D14, and C40D20, respectively. Figure 7.12 shows the relationship between the relative bond strength and ratio of concrete cover depth to diameter of the BFRP bars. As shown in Fig. 7.12, the relationship between the relative bond strength to the ratio of concrete cover depth to diameter of BFRP bars ($c/d$) was linear and could be expressed as Eq. (7.5).

$$\tau_R = \frac{\tau_{\max}}{\sqrt{f_c}} = l + m \cdot \frac{c}{d} \tag{7.5}$$

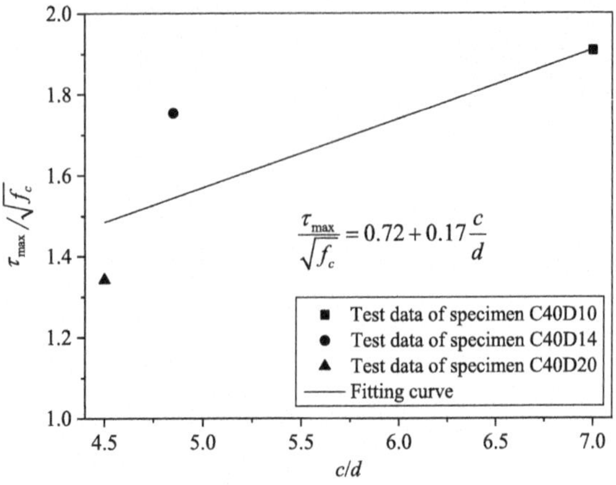


**Fig. 7.12** Relationship between average relative bond strength and *c*/*d* ratio of specimens C40D10, C40D14, and C40D20

where $\tau_R$, $\tau_{max}$, $f_c$, $c$, and $d$ are relative bond strength, bond strength, cubic compressive strength of concrete, concrete cover depth, and diameter of the BFRP bars, respectively.

Coefficient $l$ and $m$ were determined from the regression analysis, and the results were 0.72 and 0.17, respectively. The value of relative bond strength $\tau_R$ from Eq. (7.5) was 1.910, 1.546, and 1.485 when the diameter of the BFRP bars was 10, 14, and 20 mm for specimens C40D10, C40D14, and C40D20, respectively, and the results of the deviations from Eq. (7.5) and the test are 0.1%, 11.8%, and 11.3%. The relative bond strength $\tau_R$ of specimen C70D10 from the test and Eq. (7.5) was 1.767 and 1.910, respectively; and the deviation from Eq. (7.5) and the test was 8.1%, as shown in Table 7.3.

### *7.1.7 Prediction Model for Bond Stress-Slip Relationship*

The bond stress-slip relationship between BFRP bars and concrete was necessary for evaluating the load-bearing capacity of concrete components or structures reinforced with BFRP bars. Some models for bond stress-slip relationship between FRP bars were presented, as reported in [36, 42, 48–50]. The Eligehausen, Popov and Bertero (BPE) model [50], initially more popular and adopted only to describe the bond stress-slip phenomena of reinforcing bar, has been effectively used for FRP bars [44, 51, 52]. This model represented the ascending branch of the bond stress-slip relationship curve, as shown in Eq. (7.6).

$$\frac{\tau}{\tau_{max}} = \left(\frac{s}{s_{max}}\right)^{\alpha}, 0 < s \leq s_{max} \tag{7.6}$$

where $\tau$ and $s$ are bond stress at any point and its corresponding slip value, respectively; $\tau_{max}$ and $s_{max}$ are bond strength and corresponding slip, respectively; $\alpha$ is the unknown parameter to be obtained by using experimental data; and the value of $\alpha$ should be in the range of 0–1 to get meaningful results [49]. The unknown parameter $\alpha$ is obtained by equating the areas ($A_a$) underneath the ascending branch of experimental bond-slip curves [31, 52], as shown in Eq. (7.7).

$$A_a = \int_0^{s_{\max}} \tau_{\max}\left(\frac{s}{s_{\max}}\right)^{\alpha} ds = \frac{\tau_{\max} s_{\max}}{1+\alpha} \tag{7.7}$$

where $A_a$ is the area under the ascending branch of the bond stress-slip curve.

The values of $\alpha$ decreased with the increase of BFRP bar diameter, and the average values were 0.411, 0.375, and 0.308 for the BFRP bars with the diameter of 10, 14, and 20 mm, respectively, as shown in Table 7.4 and Fig. 7.13. The relationship between the parameter $\alpha$ and diameter of BFRP bars can be expressed as Eq. (7.8).

$$\alpha = \alpha_o - \alpha_1 d \tag{7.8}$$

Coefficient $\alpha_0$ and $\alpha_1$ were determined from the regression analysis and the results were 0.517, and 0.01, respectively. The result of the deviations from Eq. (7.8) and the test were 1.5%, 0.5%, and 2.9%, respectively.

After the maximum bond stress of bond stress-slip relationship curve, the bond stress decreased abruptly almost in a linear fashion, which was known as a constant friction branch. The bond stress-slip curve was maintained by the friction component that completed the bond stress-slip relationships. To address the descending branch of the curve, a double branch model has been proposed in [49] as a modification of the BPE model. The descending branch can be expressed as Eq. (7.9).

$$\frac{\tau}{\tau_{\max}} = 1 - p\left(\frac{s}{s_{\max}} - 1\right), s > s_{\max} \tag{7.9}$$

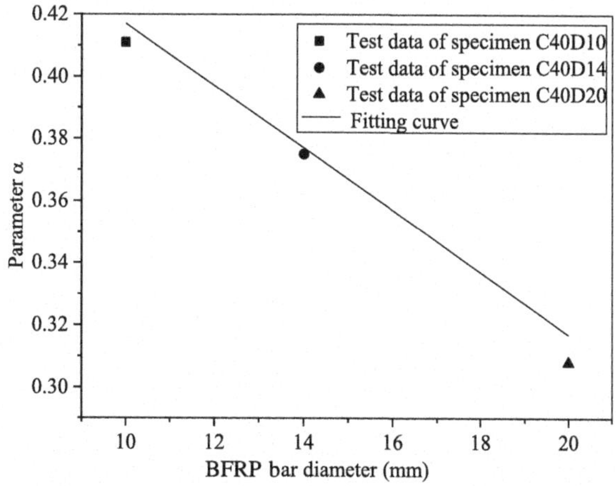


**Fig. 7.13** Relationship between average parameter $\alpha$ and diameter of BFRP bars of specimens C40D10, C40D14, and C40D20

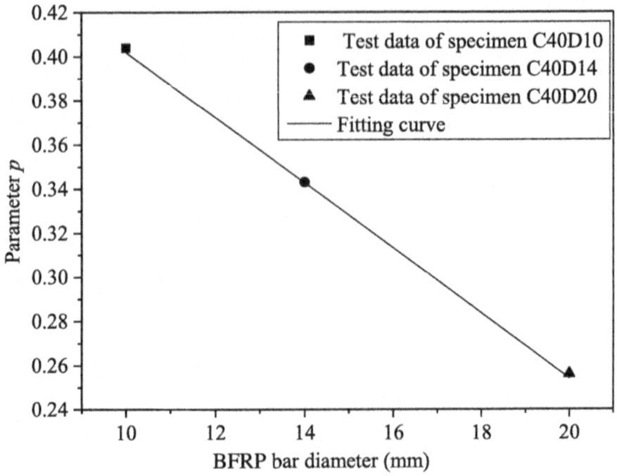


**Fig. 7.14** Relationship between average parameter $p$ and diameter of BFRP bars of specimens C40D10, C40D14, and C40D20

where $p$ is a parameter that denotes the softening and friction component obtained from the regression analysis by balancing the area under softening branch.

The average values of the parameter $p$ for BFRP bars with different diameters and deviations are shown in Table 7.4. The parameter $p$ was 0.404, 0.343, and 0.256 for the BFRP bars with a diameter of 10, 14, and 20 mm, respectively, as shown in Fig. 7.14. The relationship between the parameter $p$ and diameter of BFRP bars was linear and could be expressed as Eq. (7.10).

$$p = p_0 - p_1 d \tag{7.10}$$

Coefficients $p_0$ and $p_1$ were determined from the regression analysis and the results were 0.55, and 0.0148, respectively. The values of parameter $p$ from Eq. (7.10) were 0.402, 0.343, and 0.254 when the diameter of the BFRP bars was 10, 14, and 20 mm, respectively, as shown in Table 7.4. The results of the deviations from Eq. (7.10) and the test are 0.5%, 0, and 0.8%.

Substituting Eqs. (7.6) and (7.9) with Eqs. (7.8) and (7.10), and combining Eqs. (7.1) and (7.5), Eqs. (7.11)–(7.14) was obtained as follows.

$$\frac{\tau}{\tau_{\max}} = \left(\frac{s}{s_{\max}}\right)^{(0.517-0.01d)}, 0 < s \le s_{\max} \tag{7.11}$$

$$\frac{\tau}{\tau_{\max}} = 1 - (0.55 - 0.0148d)\left(\frac{s}{s_{\max}} - 1\right), s > s_{\max} \tag{7.12}$$

$$\tau_{\max} = \left(0.72 + 0.17\frac{c}{d}\right)\sqrt{f_c} \tag{7.13}$$

$$s_{\max} = 2.06 + 0.12d \tag{7.14}$$

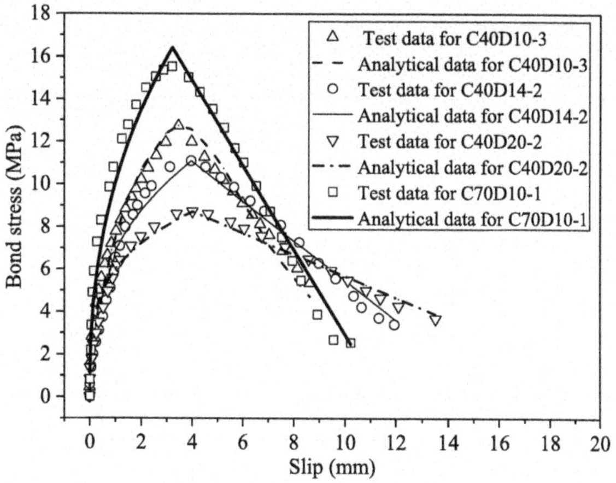


**Fig. 7.15** Fitting results for bond stress-slip relationship

The bond stress-slip relationships for specimens C40D10-3, C40D14-2, C40D20-2, and C70D10-1 obtained from test results and Eqs. (7.11)–(7.14) are shown in Fig. 7.15. The ascending branch for all the specimens demonstrated a high initial stiffness followed by nonlinear characteristics of the curve whereas the descending branch was almost linear. Higher initial stiffness leads to an increase in bond strength without significant slip at the free end of FRP bars [49]. The calculated bond stress-slip relationship obtained from Eqs. (7.11)–(7.14) showed the effect of diameter of BFRP bars on the relationship. The deviations of the bond stress-slip relationships obtained from tests and Eqs. (7.11)–(7.14) were small, as shown in Table 7.3 and Fig. 7.15. Therefore, the calculation method can be used to predict the bond stress-slip relationship between BFRP bars with different diameters and concrete with different compressive strengths. It can be used to evaluate the load-bearing capacity of concrete components or structures reinforced with BFRP bars.

## 7.2 Bond Behavior of Concrete Members Reinforced with BFRP Bars Under Cyclic Loading

### 7.2.1 *Experimental Program*

The mixture proportions and mechanical properties of concrete are shown in Table 7.5. The cylinder compressive strength $f_c'$ could be calculated according to the cubic compressive strength $f_{cu}$ as $f_c' = 0.79 f_{cu}$ [14, 15]. As shown in Table 7.5, the cubic and cylinder compressive strength were 43.60 and 34.44 MPa, respectively. The splitting tensile strength $f_{spl}$ obtained from the test was 3.65 MPa. The relationship between the splitting tensile strength $f_{spl}$ and the cylinder compressive strength $f_c'$ could be expressed as $f_{spl} = 0.59(f_c')^{0.5}$, as reported in [53]. The calculated value

**Table 7.5** Mixture proportions and mechanical properties of concrete

| Mixture proportions ($kg/m^3$) | | | | | Cubic compressive strength (MPa) | Cylinder compressive strength (MPa) |
|---|---|---|---|---|---|---|
| Water | Cement | Fine aggregate | Coarse aggregate | Superplasticizer | | |
| 168 | 420 | 673 | 1176 | 1.26 | 43.60 | 34.44 |

from the model proposed in [53] was 3.46 MPa. The result of the deviation from the test and model was 5.2%. BFRP bars with 10 mm nominal diameter were utilized for all specimens. The fiber volume fraction of BFRP bars was 75%, and BFRP bars had a helical wrapping on the surface.

The configurations of the twenty specimens are shown in Table 7.6. Each pull-out specimen is identified with the label "DXSY-*n*". D means the diameter of BFRP bars (i.e., D = diameter of BFRP bars), *X* means the diameter of BFRP bars (i.e., 10 = diameter of BFRP bars is 10 mm), *S* means the strain rate (i.e., *S* = strain rate), *Y* means the designed strain rate (i.e., 0, 1, 2, and 3 mean that the strain rates are $3.68 \times 10^{-4}$, $3.68 \times 10^{-3}$, $3.68 \times 10^{-2}$, and $3.68 \times 10^{-1}$ $s^{-1}$, respectively), and *n* means the progressive test number.

Pull-out tests were carried out with MTS 322. For all specimens, a loading rate of 3 kN/min was initially applied for one minute. After that, dynamic loading was applied on specimens D10S0, D10S1, D10S2, and D10S3 up to the bond failure at a displacement rate of $7 \times 10^{-5}$, $7 \times 10^{-4}$, $7 \times 10^{-3}$, and $7 \times 10^{-2}$ m/s, respectively, as reported in [14]. The strain rate was calculated by dividing the displacement rate by the sum of the bond length (50 mm), the length of the PVC pipe (50 mm), and the length of BFRP bars on the loading end (90 mm). The static strain rate ranges from $0.5 \times 10^{-5}$ $s^{-1}$ to $5 \times 10^{-5}$ $s^{-1}$ [54]. The strain rate of $3.68 \times 10^{-4}$ $s^{-1}$ was close to the range of static strain rate. Therefore, the strain rate of $3.68 \times 10^{-4}$ $s^{-1}$ could be considered a fully-static strain rate. The load was measured with an electronic loading cell attached to the machine. Two LVDTs were installed on the surface of the concrete and the free end of BFRP bars. The data logger DH5922 was utilized to record the load and displacement for all specimens at a frequency of 50.0 kHz. Each test specimen was broken to visualize the nature of the bond failure mechanism at the interface between concrete and BFRP bars.

### 7.2.2 Failure Modes

The bond failure modes of concrete members reinforced with BFRP bars under different strain rates are shown in Fig. 7.16. Three failure modes: (1) crushing of concrete, (2) bond failure at the interface between resin and concrete, and (3) bond failure at the interface between fiber and resin were observed, as reported in [30]. Furthermore, the combined failure modes mostly appeared at the bond interface between concrete and BFRP bars. The interfacial failure modes of concrete members

**Table 7.6** Test results of bond strength and slip corresponding to bond strength

| Identification of specimens | Strain rate ($s^{-1}$) | Bond strength (MPa) | | | | Slip corresponding to bond strength (mm) | | | |
|---|---|---|---|---|---|---|---|---|---|
| | | Individual | Average | Results from Eq. (7.18) | Deviations (%) | Individual | Average | Results from Eq. (7.20) | Deviations (%) |
| D10S0-1 | $3.68 \times 10^{-4}$ | 13.87 | 12.60 | 12.60 | 0 | 3.10 | 3.15 | 3.15 | 0 |
| D10S0-2 | | 11.51 | | | | 3.06 | | | |
| D10S0-3 | | 12.74 | | | | 3.47 | | | |
| D10S0-4 | | 11.69 | | | | 3.16 | | | |
| D10S0-5 | | 13.17 | | | | 2.94 | | | |
| D10S1-1 | $3.68 \times 10^{-3}$ | 12.79 | 13.13 | 13.09 | − 0.3 | 3.05 | 2.89 | 2.93 | + 1.4 |
| D10S1-2 | | 13.71 | | | | 3.27 | | | |
| D10S1-3 | | 12.45 | | | | 2.45 | | | |
| D10S1-4 | | 12.24 | | | | 2.63 | | | |
| D10S1-5 | | 14.45 | | | | 3.03 | | | |
| D10S2-1 | $3.68 \times 10^{-2}$ | 15.12 | 14.01 | 14.05 | + 0.3 | 2.89 | 2.61 | 2.57 | − 1.5 |
| D10S2-2 | | 13.45 | | | | 2.81 | | | |
| D10S2-3 | | 13.84 | | | | 2.35 | | | |
| D10S2-4 | | 14.30 | | | | 2.53 | | | |
| D10S2-5 | | 13.36 | | | | 2.48 | | | |
| D10S3-1 | $3.68 \times 10^{-1}$ | 14.01 | 15.34 | 15.33 | − 0.1 | 2.24 | 2.11 | 2.12 | + 0.5 |
| D10S3-2 | | 16.05 | | | | 1.94 | | | |
| D10S3-3 | | 14.65 | | | | 2.21 | | | |
| D10S3-4 | | 15.88 | | | | 2.20 | | | |
| D10S3-5 | | 16.09 | | | | 1.94 | | | |

**Fig. 7.16** Failure modes for different specimens

reinforced with BFRP bars under different strain rates were similar. All specimens failed due to the crushing of concrete and partly accompanied by the stripping off of the fiber surface. For each specimen, brittle failure occurred in concrete members reinforced with BFRP bars during the pull-out test, and the brittleness was more evident as the strain rates increased. Therefore, the damages on the surface of BFRP bars were more severe for specimens D10S3 than those for specimens D10S2, D10S1, and D10S0, as shown in Fig. 7.16. Further examination of the failure zone indicated that damages on the BFRP bars' surface were local. In some areas, concrete was stuck on BFRP bars leaving some undamaged parts on the BFRP bars' surface, as shown in Fig. 7.16c. However, in some parts, the surface of BFRP bars was peeling off before the crushing of concrete, as shown in Fig. 7.16d.

### *7.2.3 Bond Stress-Slip Relationship of Concrete Members Reinforced with BFRP Bars*

Figure 7.17 shows the bond stress-slip relationships of concrete members reinforced with BFRP bars under the strain rates of $3.68 \times 10^{-4}$, $3.68 \times 10^{-3}$, $3.68 \times 10^{-2}$, and $3.68 \times 10^{-1}$ $s^{-1}$. The comparisons of bond stress-slip relationships for specimens under different strain rates are shown in Fig. 7.18. The characteristics of the curves were similar regardless of strain rate, and each phase of the curves was described as follows.

The micro-slip phase was featured by the almost linear portion of the curve. The bond stress between concrete and BFRP bars increased rapidly with a very small

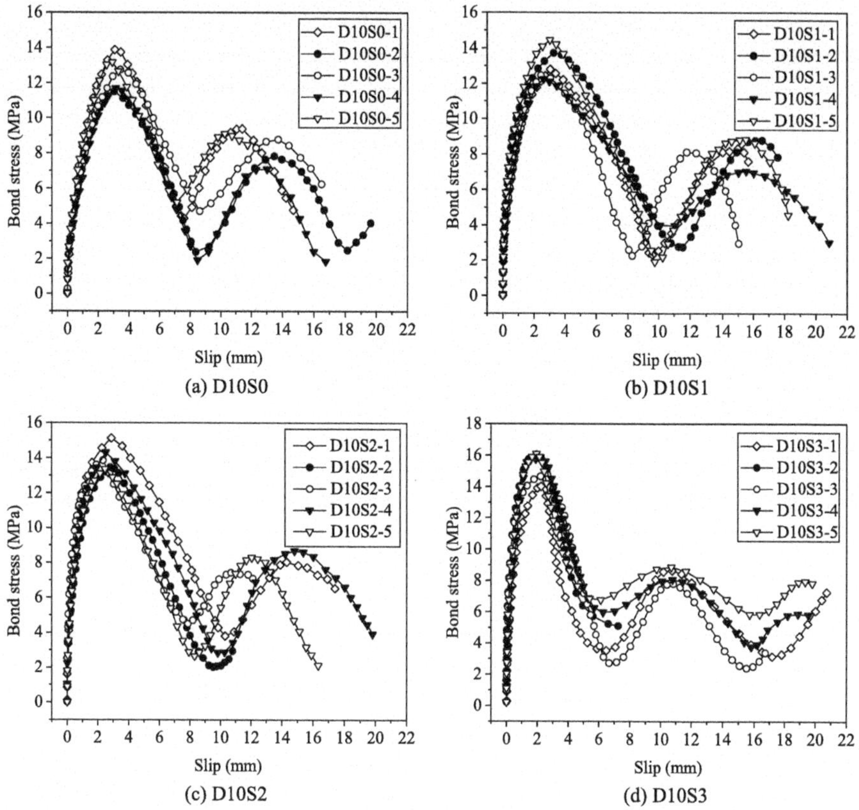


**Fig. 7.17** Bond stress-slip relationships for specimens under different strain rates

**Fig. 7.18** Comparisons on bond stress-slip relationships for specimens under different strain rates

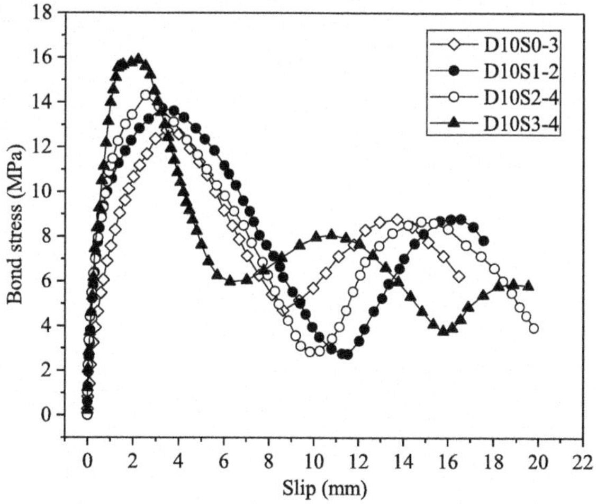

corresponding slip value. The slip in this portion is mainly caused by the elastic deformations of materials [55]. With the increase of pull-out load, the pull-out phase occurred due to the local crushing of concrete and cracking propagation, and the bond stress increased nonlinearly with slip. The result in [25] reports that after micro-slip phase, the bond stress-slip relationship curve of concrete members reinforced with GFRP ribbed bars shows a nonlinear relationship up to the bond strength because of the relative slip between GFRP ribbed bars and surrounding concrete. During the pull-out phase, the resistance to pull-out strength included two components: the interfacial frictional bond between the polymer matrix and fiber; and additional resisting force due to the mechanical deformations on the surface of fiber. Results in [37] show that the mechanical interlock between the bars and surrounding concrete is the most effective component in the bond strength during the pull-out phase. Also, as shown in Fig. 7.18, the stiffness of this portion increased as the strain rates increased. After reaching the bond strength, the residual phase occurred, and the bond stress decreased rapidly with the significant slip. Results in [56] show that the bond stress between the grout and FRP bars decreases almost linearly after reaching the peak value of bond stress. Results in (57] also show that the bond stress between high strength concrete and bars decreases with the abrupt increase in slip during the residual phase. The downtrend of bond stress stopped as it was restrained by another neck, which was formed due to uneven deformations on the surface of BFRP bars or crushed concrete along the bond length. Subsequently, the bond stress started increasing resembling the sinusoidal wave until the BFRP bars were completely pulled out from the specimen. The sinusoidal wave resemblance of the bond stress-slip relationship curves shown in Figs. 7.17 and 7.18 was related to the surface deformation of BFRP bars. Results in (56] show that the bond stress-slip relationship between the grout and FRP bars shows an oscillating behavior in the residual phase and the residual load depends not only on the friction resistance but also on the mechanical interlock between the grout and FRP bars. The reoccurrence of the bond strength and its interval largely depended on the pitch of the spirally wounded fiber. Generally, the bond strength at the second peak is assumed as 50% of the first peak [38]. The bond strength at the second peak was about 30% to 60% of the first peak. The residual curve indicated that the second peak had lower stiffness (more flat peak) than the previous peak, as reported in [6].

### *7.2.4 Bond Strength Under Cyclic Loading*

The bond strength obtained from the test was 12.60, 13.13, 14.01, and 15.34 MPa when the strain rate was $3.68 \times 10^{-4}$, $3.68 \times 10^{-3}$, $3.68 \times 10^{-2}$, and $3.68 \times 10^{-1}$ $s^{-1}$ for specimens D10S0, D10S1, D10S2, and D10S3, respectively, as shown in Table 7.6. The bond strength of concrete members reinforced with BFRP bars increased as the strain rates increased.

Based on the models of bond strength of concrete members reinforced with FRP bars proposed in [6, 23], the relationship between the bond strength $\tau_{max}$ and the cylinder compressive strength $f_c'$ could be expressed as Eq. (7.15).

$$\tau_{\max} = k\left(f_c'\right)^{0.5} \tag{7.15}$$

The bond strength is a power function of the cylinder compressive strength, and the value of the parameter $k$ varies for different types of FRP bars. According to CEB-FIP model code 2010 [58], for a given strain rate, a model for calculating the cylinder compressive strength of concrete under dynamic loading could be expressed as Eqs. (7.16) and (7.17).

$$\frac{f_c'^{d}}{f_c'^{s}} = \left(\frac{\dot{\varepsilon}_d}{\dot{\varepsilon}_s}\right)^{0.014}, \dot{\varepsilon}_d \leq 30\ \mathrm{s}^{-1} \tag{7.16}$$

$$\frac{f_c'^{d}}{f_c'^{s}} = 0.012\left(\frac{\dot{\varepsilon}_d}{\dot{\varepsilon}_s}\right)^{1/3}, \dot{\varepsilon}_d > 30\ \mathrm{s}^{-1} \tag{7.17}$$

where $f_c'^{d}$ and $f_c'^{s}$ are the dynamic and quasi-static cylinder compressive strength, respectively, in MPa; $\dot{\varepsilon}_d$ and $\dot{\varepsilon}_s$ are the dynamic and quasi-static strain rate, respectively, in $\mathrm{s}^{-1}$; and $\dot{\varepsilon}_s$ is calculated as $30 \times 10^{-6}\ \mathrm{s}^{-1}$.

To describe the effect of strain rate on the bond strength, the dynamic increase factor (DIF), which is the ratio of the bond strength under dynamic loading to that under quasi-static loading, is derived from Eqs. (7.15)–(7.17). The value of DIF could be utilized to illustrate the rate dependence of the bond behavior. As shown in Fig. 7.19, for the strain rate less than and more than $3.68 \times 10^{-3}\ \mathrm{s}^{-1}$, the result of the DIF obtained from the models developed in [6] overestimated the bond strength, and the result of the DIF obtained from the models developed in [6, 23] underestimated the bond strength. The deviations of the results of DIF obtained from the models in [6, 23] for FRP bars and test were large. Moreover, the previous models predicted that the DIF increased linearly as the logarithm of the ratio of $\dot{\varepsilon}_d$ to $\dot{\varepsilon}_s$ increased, which could not represent the actual bond behavior of concrete members reinforced with BFRP bars under dynamic loading. Therefore, the models for FRP bars could not be utilized to predict the bond strength of concrete members reinforced with BFRP bars under different strain rates.

Figure 7.19 shows that the relationship between DIF and the ratio of $\dot{\varepsilon}_d$ to $\dot{\varepsilon}_s$ is nonlinear and can be expressed as Eq. (7.18).

$$\mathrm{DIF} = \tau_{\max}^{d} / \tau_{\max}^{s} = 1 + \beta(\lg(\dot{\varepsilon}_d/\dot{\varepsilon}_s))^{\gamma} \tag{7.18}$$

where $\tau_{\max}^{d}$ and $\tau_{\max}^{s}$ are the bond strength under dynamic loading and quasi-static loading, respectively; and $\dot{\varepsilon}_s$ is calculated as $3.68 \times 10^{-4}\ \mathrm{s}^{-1}$.

Coefficients $\beta$ and $\gamma$ were determined from the regression analysis, and the results were 0.039 and 1.560, respectively. As shown in Table 7.6, the value of the bond strength obtained from Eq. (7.18) was 12.60, 13.09, 14.05, and 15.33 MPa when the strain rate was $3.68 \times 10^{-4}$, $3.68 \times 10^{-3}$, $3.68 \times 10^{-2}$, and $3.68 \times 10^{-1}\ \mathrm{s}^{-1}$

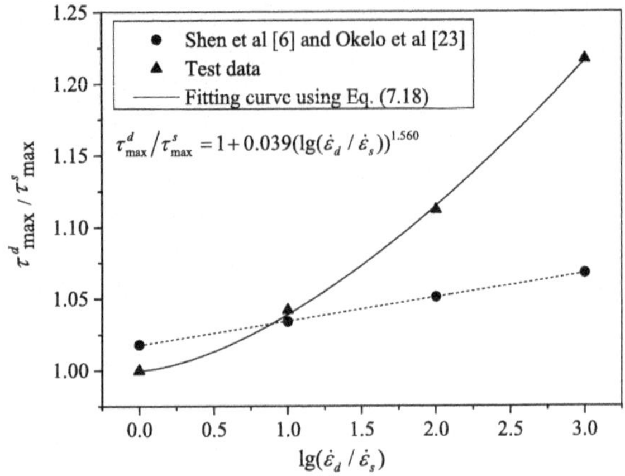


**Fig. 7.19** Relationship between bond strength and strain rates

for specimens D10S0, D10S1, D10S2, and D10S3, respectively. The results of the deviations from Eq. (7.18) and the test were 0, 0.3%, 0.3%, and 0.1%, respectively.

### 7.2.5 *Slip Corresponding to Bond Strength Under Cyclic Loading*

For reinforcing bars, a sophisticated slip corresponding to the bond strength model considering the effect of cylinder compressive strength is developed in [59] and could be expressed as Eq. (7.19).

$$s_0 = 0.011d\left(\frac{2}{(f_c'/33)+1}\right)\frac{d}{h}\left(\frac{l_a}{d}\right)^{1.6} \tag{7.19}$$

where $s_0$ is the slip corresponding to bond strength, in mm; and $h$ is the height of specimens, in mm.

The DIF of the slip ($DIF_s$), which is the ratio of the slip corresponding to the bond strength under dynamic loading to that under quasi-static loading, is derived from experimental and theoretical models using Eqs. (7.16), (7.17), and (7.19). The slip corresponding to the bond strength of concrete members reinforced with BFRP bars under strain rate of $3.68 \times 10^{-1}\ s^{-1}$ from the test was 2.11 mm, while the slip corresponding to the bond strength obtained from the model proposed in [59] was 2.93 mm. The deviations of the results obtained from the model in [59] for reinforcing bars and the test were 38.9%. The deviations were large. Furthermore, the value of $DIF_s$ obtained from the model proposed in [59] was overestimated, as shown in Fig. 7.20. Therefore, the models proposed in [59] for reinforcing bars could not be utilized to predict the slip corresponding to the bond strength of concrete members reinforced with BFRP bars under different strain rates.

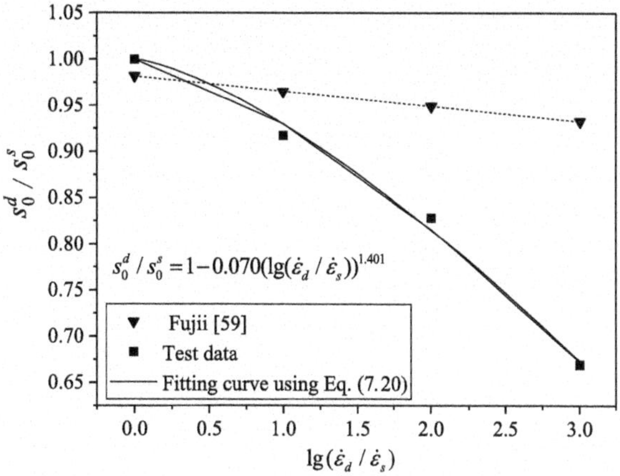


**Fig. 7.20** Relationship between slip corresponding to bond strength and strain rates

Figure 7.20 shows that the relationship between $DIF_s$ and the ratio of $\dot{\varepsilon}_d$ to $\dot{\varepsilon}_s$ is nonlinear and could be expressed as Eq. (7.20).

$$DIF_s = s_0^d / s_0^s = 1 - m(\lg(\dot{\varepsilon}_d / \dot{\varepsilon}_s))^n \tag{7.20}$$

where $s_0^d$ and $s_0^s$ are the slip corresponding to bond strength under dynamic loading and quasi-static loading, respectively.

Coefficients $m$ and $n$ were determined from the regression analysis, and the results were 0.070 and 1.401, respectively. The value of the slip corresponding to bond strength obtained from Eq. (7.20) was 3.15, 2.93, 2.57, and 2.12 mm when the strain rate was $3.68 \times 10^{-4}$, $3.68 \times 10^{-3}$, $3.68 \times 10^{-2}$, and $3.68 \times 10^{-1}\ s^{-1}$ for specimens D10S0, D10S1, D10S2, and D10S3, respectively, as shown in Table 7.6. The results of the deviations from Eq. (7.20) and the test were 0, 1.4%, 1.5%, and 0.5%, respectively.

### 7.2.6 *Prediction Model for Bond Stress-Slip Relationship Under Dynamic Loading*

The well-known BPE model was firstly developed for reinforcing bars and then utilized to represent the local bond stress-slip relationship of FRP bars. Therefore, the BPE model could be utilized to analyze the bond stress-slip relationship of concrete members reinforced with BFRP bars under dynamic loading. The parameters in the BPE model were calibrated according to the test results to describe the bond behavior of BFRP bars under different strain rates.

The BPE model proposed in [50] has been utilized for describing the bond stress-slip relationship of FRP bars effectively. The ascending branch of the BPE model could be expressed as Eq. (7.21).

$$\frac{\tau^d}{\tau^d_{\max}} = \left(\frac{s^d}{s^d_0}\right)^\alpha, 0 \le s^d \le s^d_0 \tag{7.21}$$

where $\tau^d$ and $s^d$ are bond stress and its corresponding slip value at any stage of loading, respectively; and $\alpha$ is a curve-fitting parameter, and the values of $\alpha$ must be from 0 to 1 to obtain physically meaningful results. The parameter $\alpha$ is evaluated by equating the areas $A_a$ underneath the ascending branch of the experimental curve [31, 60], which could be expressed as Eq. (7.22).

$$A_a = \int_0^{s^d_0} \tau^d_{\max}\left(\frac{s}{s^d_0}\right)^\alpha \mathrm{d}s = \frac{\tau^d_{\max} s^d_0}{1+\alpha} \tag{7.22}$$

Figure 7.21 shows the relationship between parameter $\alpha$ and strain rates. The value of the parameter $\alpha$ decreased as the strain rate increased. The value of the average parameter $\alpha$ was 0.411, 0.307, 0.269, and 0.248 when the strain rate was $3.68 \times 10^{-4}$, $3.68 \times 10^{-3}$, $3.68 \times 10^{-2}$, and $3.68 \times 10^{-1}$ $s^{-1}$ for specimens D10S0, D10S1, D10S2, and D10S3, respectively, as shown in Table 7.7. The relationship between the coefficient $\alpha$ and the logarithm of the ratio of $\dot{\varepsilon}_d$ to $\dot{\varepsilon}_s$ was linear and could be expressed as Eq. (7.23).

$$\alpha = \alpha_0 - \alpha_1 \lg(\dot{\varepsilon}_d/\dot{\varepsilon}_s) \tag{7.23}$$

Coefficients $\alpha_0$ and $\alpha_1$ were determined from the regression analysis, and the results were 0.388 and 0.053, respectively. As shown in Table 7.7, the value of the parameter $\alpha$ obtained from Eq. (7.23) was 0.388, 0.335, 0.282, and 0.229 when the

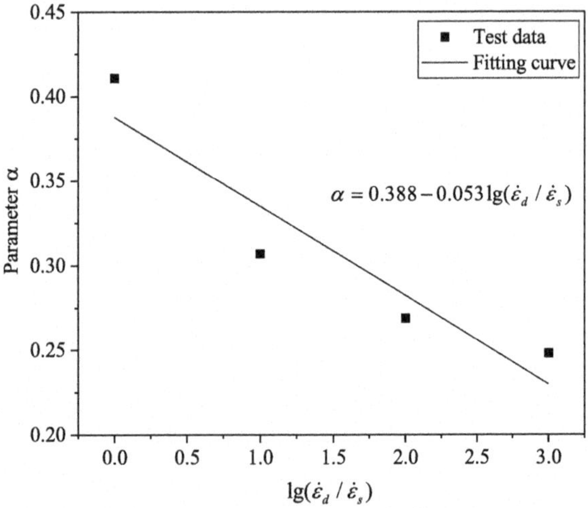


**Fig. 7.21** Relationship between parameter $\alpha$ and strain rates

**Table 7.7** Fitting results of parameters $\alpha$ and $P$

| Identification of specimen | Strain rate ($s^{-1}$) | Parameter $\alpha$ | | | Parameter $P$ | | |
|---|---|---|---|---|---|---|---|
| | | Average value | Results from Eq. (7.23) | Deviations (%) | Average value | Results from Eq. (7.24) | Deviations (%) |
| D10S0 | $3.68 \times 10^{-4}$ | 0.411 | 0.388 | – 5.6 | 0.488 | 0.371 | – 24.0 |
| D10S1 | $3.68 \times 10^{-3}$ | 0.307 | 0.335 | + 9.1 | 0.315 | | – 17.8 |
| D10S2 | $3.68 \times 10^{-2}$ | 0.269 | 0.282 | + 4.8 | 0.290 | | + 27.9 |
| D10S3 | $3.68 \times 10^{-1}$ | 0.248 | 0.229 | – 7.7 | 0.390 | | – 4.9 |

strain rate was $3.68 \times 10^{-4}$, $3.68 \times 10^{-3}$, $3.68 \times 10^{-2}$, and $3.68 \times 10^{-1}$ $s^{-1}$ for specimens D10S0, D10S1, D10S2, and D10S3, respectively. The results of the deviations from Eq. (7.23) and the test were 5.6%, 9.1%, 4.8%, and 7.7%, respectively.

Furthermore, the second branch proposed in the BPE model showed that the bond strength was constant as the slip increased. A linear descending branch occurred afterward. However, experimental bond stress-slip relationship curves of FRP bars showed a lack of the second branch in the BPE model. Therefore, the modified BPE model with the same ascending branch in the BPE model and a softening branch has been developed in [60]. The linear softening branch in the modified BPE model could be expressed as Eq. (7.24).

$$\frac{\tau^d}{\tau^d_{max}} = 1 - p\left(\frac{s^d}{s^d_0} - 1\right), s^d > s^d_0 \tag{7.24}$$

where $p$ is a parameter obtained from the regression analysis on the test results, which represents the slope of the softening branch.

The average parameter $p$ was 0.488, 0.315, 0.290, and 0.390 when the strain rate was $3.68 \times 10^{-4}$, $3.68 \times 10^{-3}$, $3.68 \times 10^{-2}$, and $3.68 \times 10^{-1}$ $s^{-1}$ for specimens D10S0, D10S1, D10S2, and D10S3, respectively, as shown in Table 7.7. Figure 7.22 shows the relationship between the parameter $p$ and strain rates. The average value of the parameter $p$ for specimens under different strain rates was 0.371. The parameter $p$ utilized in [6] ranges from 0.256 to 0.404. The result of the parameter $p$ was in the range of the result in [6].

Substituting Eq. (7.23) into Eq. (7.21) and combining Eqs. (7.18), (7.20), and (7.24), Eqs. (7.25)–(7.28) was obtained as follows:

$$\frac{\tau^d}{\tau^d_{max}} = \left(\frac{s^d}{s^d_0}\right)^{0.388-0.053\lg(\dot{\varepsilon}_d/\dot{\varepsilon}_s)}, 0 \le s^d \le s^d_0 \tag{7.25}$$

$$\frac{\tau^d}{\tau^d_{max}} = 1 - 0.371\left(\frac{s^d}{s^d_0} - 1\right), s^d > s^d_0 \tag{7.26}$$

$$\tau^d_{max}/\tau^s_{max} = 1 + 0.039(\lg(\dot{\varepsilon}_d/\dot{\varepsilon}_s))^{1.560} \tag{7.27}$$

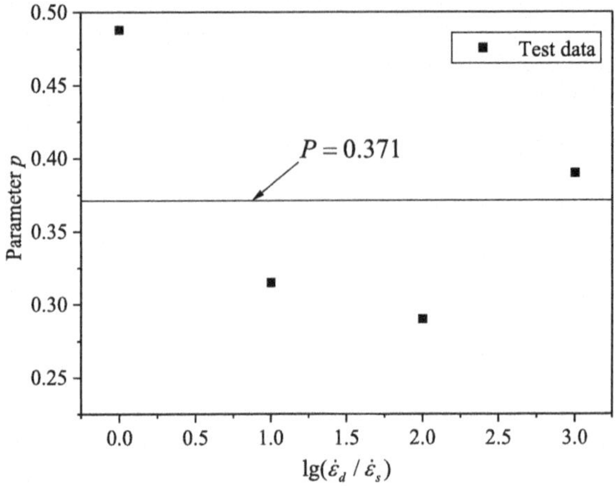


**Fig. 7.22** Relationship between parameter $p$ and strain rates

$$s_0^d / s_0^s = 1 - 0.070(\lg(\dot{\varepsilon}_d / \dot{\varepsilon}_s))^{1.401} \tag{7.28}$$

The bond stress-slip relationships for specimens D10S0-3, D10S1-1, D10S2-5, and D10S3-4 obtained from test results and Eqs. (7.25)–(7.28) are shown in Fig. 7.23. Generally, the predicted curves obtained from Eqs. (7.25)–(7.28) showed a good correlation with the experimental results. The model proposed could be utilized to predict the bond stress-slip relationship of concrete members reinforced with BFRP bars under dynamic loading when the diameter of BFRP bars was 10 mm, the strain rate was less than $3.68 \times 10^{-1}\ s^{-1}$, and the surface of BFRP bars had spiral indentations.

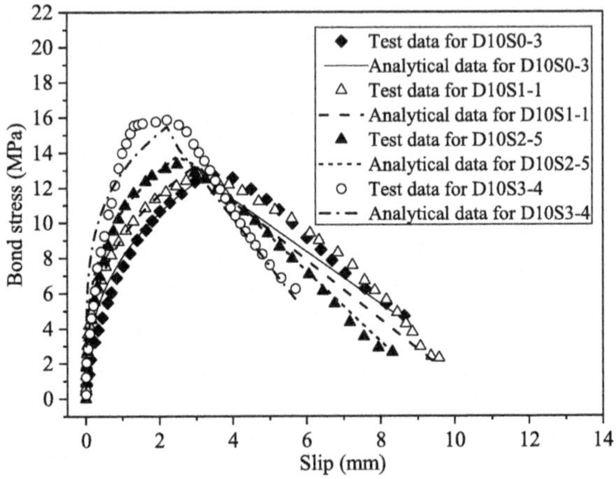


**Fig. 7.23** Comparisons on bond stress-slip relationships obtained from test data and analytical data for specimens under different strain rates

## 7.3 Bond Behavior Between BFRP Bars and Concrete Under Cyclic Loading

### *7.3.1 Experimental Program*

The configurations of the specimens are shown in Table 7.8. A label "FCX-*n*" is used to distinguish each test specimen. The alphabet FC in the specimen name means cyclic loading, X means the diameter of BFRP bars (i.e., 14 means that the diameter of the BFRP bars is 14 mm); and *n* is the succession specimen number. For example, FC14-2 indicates the second specimen of 14 mm BFRP bars diameter under cyclic loading.

As shown in the test setup scheme and schematic diagram (Fig. 7.24), the test was conducted on a servo-hydraulic testing machine (MTS 322). The special gripping system to avoid the possibility of premature failure in the gripping zone of the BFRP bars was used according to the size of the steel tube, as reported in [6]. A steel plate of 25 mm thick stiffened by a transverse steel plate was used at the top of the specimen. It was penetrated in the central portion, which would allow the BFRP bar to be fixed to the gripping system of the testing machine, as reported in [61]. The steel plate was tightened with the steel casting by four threaded bars and eight threaded nuts at four corners, pressing the steel plate to avoid the differential stress on the surface of the BFRP bars, which was in accordance with the method proposed in ACI Committee 408 [62]. Therefore, there was no vertical movement of the specimens during the cyclic loading. An electronic loading cell attached to the machine was used to measure the load. Two LVDTs were installed on the free end of BFRP bars and the surface of the steel plate, respectively, as shown in Fig. 7.24, and as reported in [55]. The displacement loading rate was 0.014 mm/s and was in accordance with

**Table 7.8** Test and fitting results of bond strength under cyclic loading

| Identification of specimens | Bond strength (MPa) | | | | | | | | | | Residual bond strength (MPa) |
|---|---|---|---|---|---|---|---|---|---|---|---|
| | 1st | 2nd | 3rd | 4th | 5th | 6th | 7th | 8th | 9th | 10th | |
| FC14-1 | 8.80 | 5.78 | 4.94 | 4.53 | 4.31 | 4.18 | 4.11 | 4.08 | 4.07 | 4.06 | 8.12 |
| FC14-2 | 8.77 | 5.92 | 5.22 | 4.94 | 4.83 | 4.76 | 4.74 | 4.72 | 4.71 | 4.71 | 8.59 |
| FC14-3 | 8.35 | 5.77 | 4.93 | 4.49 | 4.23 | 4.06 | 3.94 | 3.88 | 3.85 | 3.84 | 6.00 |
| FC14-4 | 9.86 | 6.50 | 4.97 | 4.30 | 3.91 | 3.66 | 3.52 | 3.44 | 3.40 | 3.38 | 8.82 |
| FC14-5 | 10.25 | 7.52 | 6.58 | 6.08 | 5.79 | 5.65 | 5.57 | 5.53 | 5.51 | 5.50 | 10.65 |
| Average value | 9.21 | 6.30 | 5.33 | 4.87 | 4.61 | 4.46 | 4.38 | 4.33 | 4.31 | 4.30 | 8.44 |
| $\tau_{max,N}/\tau_{max,1}$ | 1.00 | 0.68 | 0.58 | 0.53 | 0.50 | 0.48 | 0.48 | 0.47 | 0.47 | 0.47 | / |
| Result from Eq. (7.34) | 1.00 | 0.67 | 0.58 | 0.53 | 0.51 | 0.49 | 0.48 | 0.47 | 0.46 | 0.46 | |
| Deviation (%) | 0 | 1.5 | 0 | 0 | – 2.0 | – 2.1 | 0 | 0 | 2.1 | 2.1 | |

Fig. 7.24 Test setup layout of tested specimens

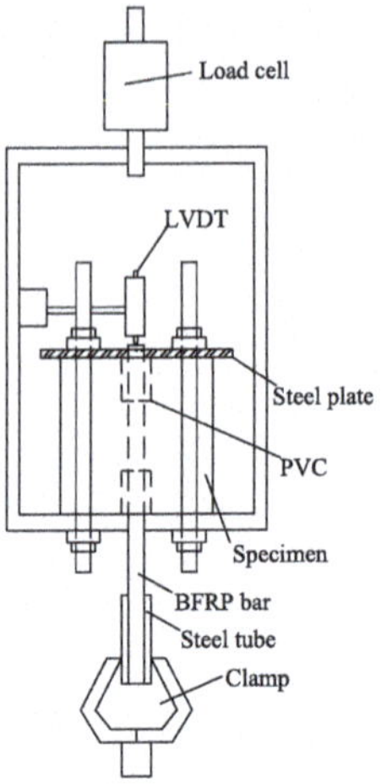


(a) Loading device

(b) Photography of loading device

the limit displacement loading rate of 1.27 mm/min. A DH5922 data logger with a frequency of 2 Hz was used to record the experimental data from the loading cell and LVDTs. The specimens were loaded up to pull-out failure by the displacement control, as reported in [63] and the loaded end slip displacement of each cycle was set to 3.50 mm for the test, which was smaller than the slip corresponding to bond strength under static loading (3.90 mm), as reported in [6].

### 7.3.2 Failure Modes

Failure modes of the specimens indicated the bond failure mechanism of BFRP bars and concrete and, could be utilized to predict bond behavior. Therefore, the failure modes of each specimen need to be investigated. The nature of the failure mechanism could be visualized after breaking the specimens, and examining the bar surface and the surrounding concrete. Specimens FC14-1, FC14-2, FC14-3, FC14-4, and FC14-5 showed identical failure modes on concrete and BFRP surface, and thus failure mode of specimen FC14-1 was chosen as a representative. The bond failure mode of specimen FC14-1 is shown in Fig. 7.25. A combined failure mode was observed during the pull-out test. In general, the failure of the specimens was the combination of two kinds of failure modes: (1) shear deformation of the bar surface and (2) concrete crushing. The shear deformation of the bar surface was due to the shear failure of successive layers, and it was evidenced by the delamination of the outer layer of BFRP bars. The concrete crushed around the BFRP bar mainly because the concrete strength was low. The buckling of the bars was not observed during the compression stage of the cyclic loading. All specimens failed due to concrete crushing, as well as the shear deformation of the bar surface.

**Fig. 7.25** Failure mode of the specimen

### 7.3.3 *Bond Stress-Slip Relationship Between BFRP Bars and Concrete Under Cyclic Loading*

Test results on the bond stress-slip relationship between BFRP bars and concrete under cyclic loading are shown in Fig. 7.26. Figure 7.27 shows the representative sketch of the bond stress-slip relationship under cyclic loading, which is considered to be able to represent the real bond stress–slip response. There are four phases of the bond stress-slip response. Considering the symmetrical response in loading and reverse loading shown in Fig. 7.27, only the first-half cycle was described as follows.

The first phase was the elastic phase (segment OA), which was characterized by the almost linear phase of the bond stress smaller than 40% of the bond strength (point A). At this phase, since the bond deterioration was negligible, the bond response was considered elastic. In the beginning (near point O), the major bond resistance was chemical adhesion. However, it was quite weak and would fail quickly once the slip at the free end occurred. Thereafter, the chemical adhesion was replaced by mechanical interlocking as the major resistance against the pull-out load. However, the mechanical interlocking was also quite weak and caused little damage in concrete, and the slip at this phase was primarily caused by the elastic deformations of BFRP bars.

The second phase was the cracking propagation phase (segment AD). Due to the local concrete crushing in front of the ribs of BFRP bars and multiple cracking propagations, the bond response became nonlinear with the increase of pull-out load. There were three parts in this phase: the initial cracking propagation part (segment AB), the stable cracking propagation part (segment BC), and the unstable cracking propagation part (segment CD). The first part was the initial cracking propagation part (segment AB). In this part (bond stress smaller than 80% of the bond strength), the bond deterioration was still not obvious. With the increase of pull-out load, some fine cracks started to originate from the top of the ribs and would be completely closed once unloaded. Therefore, the damage of the concrete was still not obvious. The second part was the stable cracking propagation part (segment BC). As the bond stress

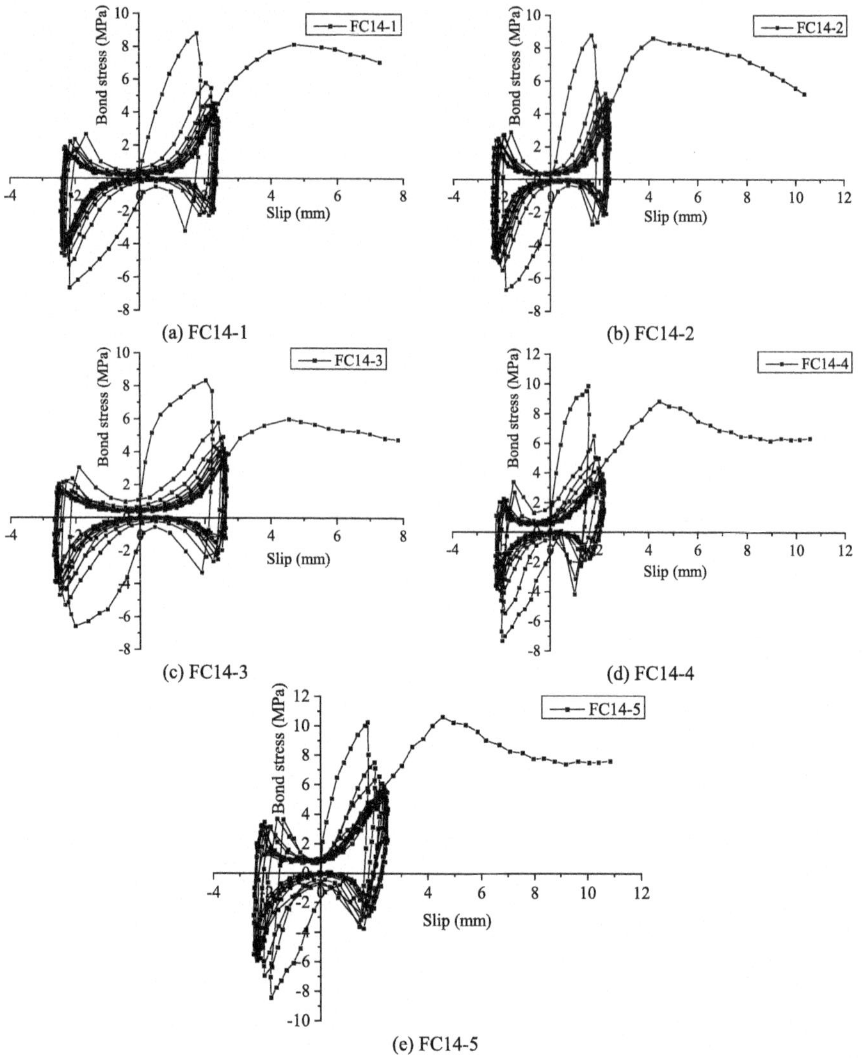


**Fig. 7.26** Relationships between bond stress and slip for different specimens

increased up to bond strength (point C), the originated fine cracks propagated forward and developed into multiple fine cracks. In this part, the bond deterioration was more obvious than in the previous part. Deterioration in bond strength occurred as the cyclic response around the bond strength was lower than the respective value under static loading. The third part was the unstable cracking propagation part (segment CD). After the bond strength (point C), a rapid decrease in bond stress was followed by a small increase in slip. Due to the unstable propagation of the multiple fine cracks, the constraint around the BFRP bars was weakened quickly. The microcracks began

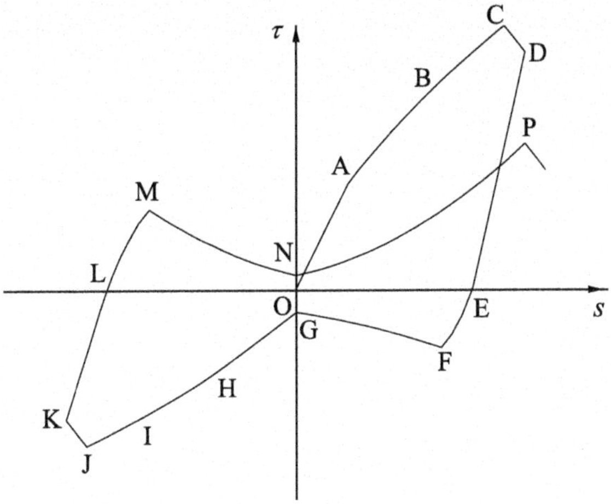


**Fig. 7.27** Sketch of the bond stress-slip relationships under cyclic loading

to combine into macro-cracks, and the concrete crushing in front of the ribs became severer.

The third phase was the cracking closure phase (segment DE). In this phase, although the ribs were still in contact with the surrounding concrete, the bond stress decreased quickly with the closing up of cracks. Once the ribs were fully in contact with the surrounding concrete in the opposite direction, these previously opened cracks would close totally. Therefore, as the bond stress decreased to zero (point E), residual slip caused by local damage of concrete or the inelastic deformation of BFRP bars was irrecoverable.

The fourth phase was the friction phase (segment EG). After point E, the pull-out load was applied reversely, and the friction branch was followed which could be divided into two parts: static friction part and kinetic friction part. The first part was the static friction part (segment EF). In this part, the bond stress increased quickly with a little slip. The maximum value of bond stress at point F was approximately 30% of the bond stress at point D. The second part was the kinetic friction part (segment FG). The bond stress of this part was quite small. Interfacial roughness and radial compression were the main factors of this part. As the slip decreased to zero (point G), the surrounding concrete was extruded by the ribs and was damaged because of the shearing and crushing. Therefore, the bond stress decreased due to the loose concrete particles.

Since the bond mechanism for the loading segment GN was similar to that of segment OG as described above, the response of the following second-half cycle was similar to that of the first-half cycle. However, the bond strength in the reverse direction (point J) was affected by the previous concrete damage in the first-half cycle as well as the Poisson's effect of the BFRP bars.

The positive and reverse ascending branches in the first cycle were protruding (segment OC). After the first cycle, the shape of the ascending branch became concave (segment NP). All the bond stress-slip relationships were presented as inverse sigmoid curves and were asymmetrical shapes, which may be due to the mixed nature of concrete, as reported in [64, 65].

### *7.3.4 Bond Strength Under Number of Cycles Loading*

RC structures are likely to be subjected to cyclic loading during their service life, either by design or by accident. Cyclic loading will lead to a successive deterioration of bond strength caused by the progress of micro-crushing of concrete in front of the irregularity of the bars surface and the propagation of microcracks [61]. Therefore, understanding the effect of the number of cycles on the bond strength between BFRP bars and concrete was necessary. The bond strength between BFRP bars and concrete was also necessary for evaluating the durability and integrity of RC structures. Such bond strength played a crucial role in the stress transferring from the surrounding concrete to the BFRP bars.

Figure 7.28 shows the relationship between the ratio of the bond strength of the $N$th cycle $\tau_{\max,N}$, to that of the first cycle $\tau_{\max,1}$, and the number of cycles. All specimens reached the maximum value of bond strength at the first cycle. Then, the deteriorative speed slowed down with the increase of the number of cycles.

The bond deterioration between bar and surrounding concrete is dependent on the damage of concrete, which has a close correlation to the energy dissipation $E$ during the cyclic loading. Therefore, a damage function $d$ is introduced in the microplane model proposed in [66], as given in Eq. (7.29).

$$d = 1 - \mathrm{e}^{-1.2(E/E_0)^{1.1}} \tag{7.29}$$

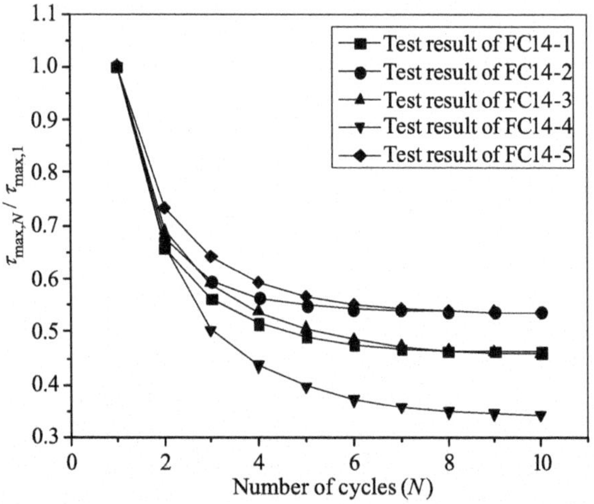


**Fig. 7.28** Relationship between the ratio of bond strength and the number of cycles for different specimens

where $d$ is a damage function; $E$ is the accumulated microplane shear energy dissipation; and $E_0$ is a constant representing the area under the monotonic (undamaged) microplane shear stress–strain curve. However, a simple formula with enough accuracy is better for practical designs. Therefore, Eq. (7.29) could not be used.

Investigations on the relationship between the bond strength and number of cycles are conducted in [55, 67]. Regardless of concrete types and bar types, models for predicting the relationships between the bond strength and number of cycles were listed below.

Based on the pull-out tests of bars embedded in concrete, a model for the bond strength deterioration proposed in [55] is shown in Eqs. (7.30) and (7.31):

$$u_n/u_0 = 0.97^n, \quad s \le s_{\text{peak}} \tag{7.30}$$

$$u_n/u_0 = 0.80^n, \quad s > s_{\text{peak}} \tag{7.31}$$

where $u_n$ is the bond strength of the $n$th cycle, in MPa; $u_0$ is the bond strength of the first cycle, in MPa; $n$ is the number of cycles; $s$ is the slip, in mm; and $s_{\text{peak}}$ is the slip corresponding to bond strength under monotonic loading, in mm.

Based on the pull-out tests of bars embedded in concrete, a model for the bond strength deterioration proposed in [67] is shown in Eqs. (7.32) and (7.33):

$$\frac{u_N}{u_0} = 1 - 0.1\ln(N+1), \quad s \le s_{sp} \tag{7.32}$$

$$\frac{u_N}{u_0} = 1 - 0.35\ln(N+1), \quad s > s_{sp} \tag{7.33}$$

where $u_N$ is the bond strength of the $N$th cycle, MPa; $u_0$ is the bond strength of the first cycle, MPa; $N$ is the number of cycles, $s$ is the slip, mm; and $s_{sp}$ is the slip at which the splitting bond strength is mobilized, mm.

The results of the bond strength obtained from the test and theoretical models proposed in [55, 67] are shown in Fig. 7.29. The test results of the ratio of the bond strength under the $N$th cycle to that under the first cycle were between the results obtained from the models proposed in [55, 67].

The relationship between the ratio of the bond strength under the $N$th cycle to that under the first cycle, and the number of cycles was nonlinear and could be expressed as Eq. (7.34).

$$\frac{\tau_{\max,N}}{\tau_{\max,1}} = \frac{A \cdot N + B}{C \cdot N + D} \tag{7.34}$$

where $\tau_{\max,N}$ is the bond strength of the $N$th cycle, in MPa; $\tau_{\max,1}$ is the bond strength of the first cycle, calculated as 9.21 MPa; and $N$ is the number of cycles.

Coefficients $A$, $B$, $C$, $D$ were determined from the regression analysis, and the results were 0.74, 0.75, 1.82, and – 0.33, respectively. The ratio of the bond strength

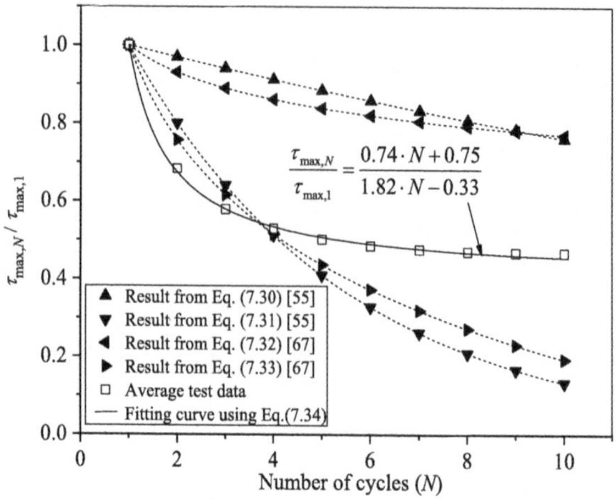


**Fig. 7.29** Comparisons on the ratio of the bond strength obtained from test results and theoretical models

under the $N$th cycle to that under the first cycle obtained from Eq. (7.34) was 1.00, 0.67, 0.58, 0.53, 0.51, 0.49, 0.48, 0.47, 0.46, and 0.46 when the number of cycles varied from 1 to 10, respectively, as shown in Table 7.8 and Fig. 7.29. The results of the deviations for the ratio of bond strength under the $N$th cycle to that under the first cycle obtained from the test and Eq. (7.34) were 0, 1.5%, 0, 0, – 2.0%, – 2.1%, 0, 0, 2.1%, and 2.1%, respectively, as shown in Table 7.8 and Fig. 7.29.

### 7.3.5 *Slip Corresponding to Bond Strength Under Number of Cycles Loading*

Figure 7.30 shows the relationship between the ratio of the slip corresponding to bond strength of the $N$th cycle $s_N$, to that of the first cycle $s_1$, and the number of cycles. The slip accumulated gradually with the increase of the number of cycles, but the rate of increase was changeable. When the specimen was cast, micro-voids between BFRP bars and concrete existed as BFRP bars were not in full contact with the surrounding concrete. Some of the micro-voids would gradually be closed when subjected to cyclic loading. Firstly, due to the existing voids, the slip was easier to develop. Then, most of the voids were closed up as the loading continued. As a result, the accumulation rate slowed down and the system was stabilized. Therefore, the increase in the slip was due to the first several cycles especially the first one. Then, with the increase of the number of cycles, the increase speed slowed down.

Investigations on the relationship between the slip corresponding to bond strength and number of cycles are conducted in [63, 68, 69]. Regardless of bar types, concrete types, and stress level, some models for predicting relationships between the slip corresponding to bond strength and number of cycles are listed below:

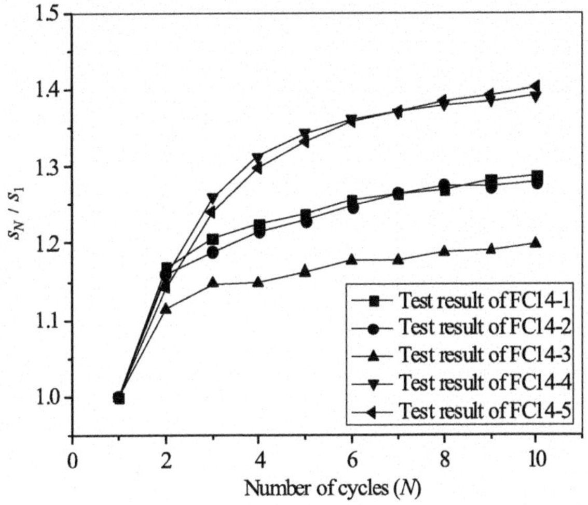


**Fig. 7.30** Relationship between the ratio of the slip corresponding to bond strength and the number of cycles for different specimens

Based on the test results of bars, the model for the slip corresponding to bond strength is proposed in [63], as shown in Eq. (7.35).

$$s_N = s_1 \cdot N^{\alpha},\ \alpha = 0.098 \tag{7.35}$$

Based on the pull-out test of ribbed bars under cyclic loading, the model for the slip corresponding to bond strength is proposed in [68], as shown in Eq. (7.36).

$$s_N = s_1 \cdot N^{\beta},\ \beta = 0.107 \tag{7.36}$$

Based on the constant stress level $\tau_{rep}/\tau_{\max} = 0.5$ (i.e., the ratio of repeated bond stress to bond strength is 0.5), the model for the slip corresponding to bond strength is proposed in [69], as shown in Eq. (7.37).

$$s_N = s_1 \cdot N^{\gamma},\ \gamma = 0.119 \tag{7.37}$$

where $s_N$ is the slip corresponding to the bond strength of the $N$th cycle, in mm; $s_1$ is the initial slip corresponding to the bond strength of the first cycle, in mm; and $N$ is the number of cycles.

The results of the ratio of the slip corresponding to bond strength under the $N$th cycle to that under the first cycle obtained from the test and theoretical models proposed in [63, 68, 69] are shown in Fig. 7.31. The predicted values calculated with the models proposed in [63, 68, 69] all underestimated the ratio of the slip corresponding to bond strength under the $N$th cycle to that under the first cycle. Therefore, based on the models proposed in [63, 68, 69], the relationship between the ratio of the slip corresponding to bond strength under the $N$th cycle to that under the first cycle and the number of cycles could be expressed with Eq. (7.38).

$$s_N = s_1 \cdot N^r \tag{7.38}$$

Coefficient $r$ was determined from the regression analysis, and the result was 0.131. The value of the ratio of the slip corresponding to bond strength under the $N$th cycle to that under the first cycle obtained from Eq. (7.38) was 1.00, 1.10, 1.15, 1.20, 1.23, 1.26, 1.29, 1.31, 1.33, and 1.35 when the number of cycles varied from 1 to 10, respectively, as shown in Table 7.9 and Fig. 7.31. The results of the deviations for the ratio of slip corresponding to bond strength under the $N$th cycle to that under the first cycle obtained from the test and Eq. (7.38) were 0, 4.3%, 5.0%, 3.2%, 2.4%, 1.6%, 0, − 0.8%, − 2.3%, and − 3.1%, respectively, as shown in Table 7.9 and Fig. 7.31.

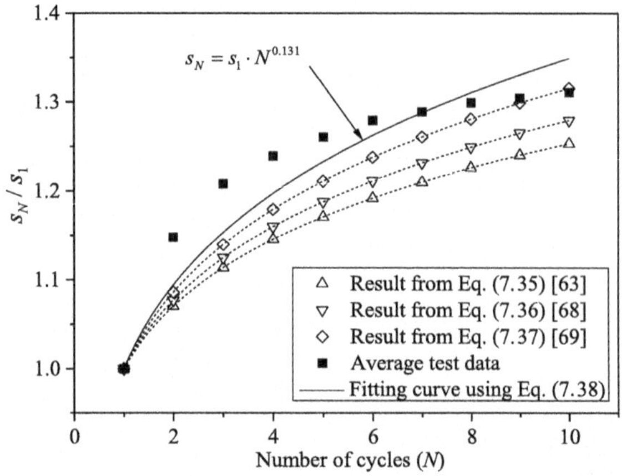


**Fig. 7.31** Comparisons on the ratio of the slip corresponding to bond strength obtained from test results and theoretical models

**Table 7.9** Test and fitting results of the slip corresponding to bond strength under cyclic loading

| Identification of specimens | Slip corresponding to bond strength (mm) | | | | | | | | | | Residual slip corresponding to bond strength (mm) |
|---|---|---|---|---|---|---|---|---|---|---|---|
| | 1st | 2nd | 3rd | 4th | 5th | 6th | 7th | 8th | 9th | 10th | |
| FC14-1 | 1.88 | 2.20 | 2.27 | 2.30 | 2.33 | 2.36 | 2.38 | 2.39 | 2.41 | 2.42 | 4.70 |
| FC14-2 | 1.89 | 2.19 | 2.24 | 2.29 | 2.32 | 2.35 | 2.39 | 2.41 | 2.40 | 2.41 | 4.17 |
| FC14-3 | 2.23 | 2.49 | 2.56 | 2.57 | 2.60 | 2.63 | 2.63 | 2.65 | 2.66 | 2.68 | 4.54 |
| FC14-4 | 1.58 | 1.83 | 1.99 | 2.08 | 2.13 | 2.16 | 2.17 | 2.19 | 2.19 | 2.20 | 4.43 |
| FC14-5 | 1.78 | 2.03 | 2.20 | 2.30 | 2.37 | 2.41 | 2.43 | 2.46 | 2.47 | 2.49 | 4.55 |
| Average value | 1.87 | 2.15 | 2.25 | 2.31 | 2.35 | 2.38 | 2.40 | 2.42 | 2.43 | 2.44 | 4.48 |
| $s_N/s_1$ | 1.00 | 1.15 | 1.21 | 1.24 | 1.26 | 1.28 | 1.29 | 1.30 | 1.30 | 1.31 | / |
| Result from Eq. (7.38) | 1.00 | 1.10 | 1.15 | 1.20 | 1.23 | 1.26 | 1.29 | 1.31 | 1.33 | 1.35 | |
| Deviation (%) | 0 | 4.3 | 5.0 | 3.2 | 2.4 | 1.6 | 0 | -0.8 | -2.3 | -3.1 | |

### 7.3.6 Hysteretic Curve Area Under Number of Cycles Loading

The energy dissipation in one cycle could be derived by calculating the area of hysteretic loop of bond stress-slip relationship curve [55]. Energy dissipation capacity is taken as a crucial index to evaluate the seismic and fatigue performance of specimens under cyclic loading [66].

Figure 7.32 shows the relationship between the ratio of the hysteretic curve area of the $N$th cycle $A_N$, to that of the first cycle $A_1$, and the number of cycles. All specimens reached the maximum value of hysteretic curve area at the first cycle. The deterioration of the hysteretic curve area was due to the first several cycles especially the first one. Then the deteriorative speed slowed down with the increase of the number of cycles, which was similar with the results of bond strength.

Based on the model for the relationship between the ratio of the bond strength under the $N$th cycle to that under the first cycle, and the number of cycles, the relationship between the ratio of the hysteretic curve area under the $N$th cycle to that under the first cycle and the number of cycles could be expressed with Eq. (7.39).

$$\frac{A_N}{A_1} = \frac{a \cdot N + b}{c \cdot N + d} \tag{7.39}$$

where $A_N$ is the hysteretic curve area of the $N$th cycle, in N/mm; $A_1$ is the hysteretic curve area of the first cycle, calculated as 27.62 N/mm; and $N$ is the number of cycles.

Coefficients $a$, $b$, $c$, $d$ were determined from the regression analysis, and the results were 0.52, 0.97, 2.01, and − 0.52, respectively. The value of the ratio of the hysteretic curve area under the $N$th cycle to that under the first cycle obtained from Eq. (7.39) was 1.00, 0.57, 0.46, 0.41, 0.38, 0.36, 0.34, 0.33, 0.32, and 0.31 when the number of cycles varied from 1 to 10, respectively, as shown in Table 7.10 and Fig. 7.32. The results of the deviations of the ratio of the hysteretic curve area under the $N$th

**Fig. 7.32** Relationship between the ratio of the hysteretic curve area and the number of cycles for different specimens

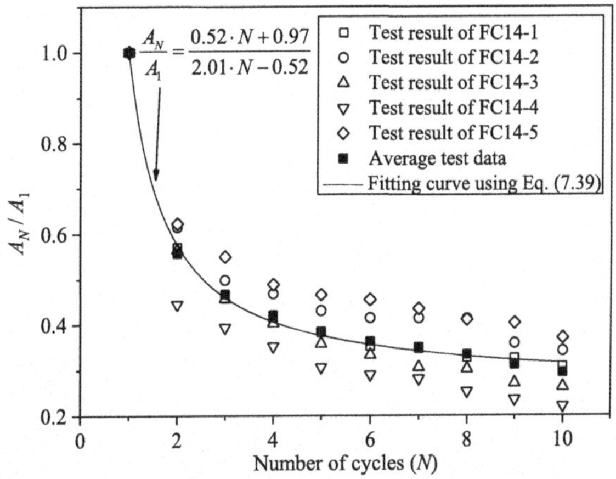

**Table 7.10** Test and fitting results of hysteretic curve area under cyclic loading

| Identification of specimens | Hysteretic curve area (N/mm) | | | | | | | | | |
|---|---|---|---|---|---|---|---|---|---|---|
| | 1st | 2nd | 3rd | 4th | 5th | 6th | 7th | 8th | 9th | 10th |
| FC14-1 | 24.31 | 13.87 | 11.15 | 10.09 | 9.35 | 8.57 | 8.44 | 7.95 | 7.94 | 7.48 |
| FC14-2 | 23.91 | 14.71 | 11.92 | 11.20 | 10.31 | 9.93 | 9.91 | 9.89 | 8.60 | 8.20 |
| FC14-3 | 32.12 | 18.16 | 14.72 | 12.96 | 11.50 | 10.71 | 9.83 | 9.71 | 8.70 | 8.43 |
| FC14-4 | 32.12 | 14.31 | 12.66 | 11.32 | 9.84 | 9.31 | 9.01 | 8.14 | 7.55 | 7.09 |
| FC14-5 | 25.62 | 15.96 | 14.09 | 12.52 | 11.93 | 11.66 | 11.16 | 10.55 | 10.36 | 9.51 |
| Average value | 27.62 | 15.40 | 12.91 | 11.62 | 10.59 | 10.04 | 9.67 | 9.25 | 8.63 | 8.14 |
| $A_N/A_1$ | 1.00 | 0.56 | 0.47 | 0.42 | 0.38 | 0.36 | 0.35 | 0.33 | 0.31 | 0.29 |
| Result from Eq. (7.39) | 1.00 | 0.57 | 0.46 | 0.41 | 0.38 | 0.36 | 0.34 | 0.33 | 0.32 | 0.31 |
| Deviation (%) | 0 | − 1.8 | 2.1 | 2.4 | 0 | 0 | 2.9 | 0 | − 3.2 | − 6.9 |

cycle to that under the first cycle obtained from the test and Eq. (7.39) were 0, − 1.8%, 2.1%, 2.4%, 0, 0, 2.9%, 0, − 3.2%, and − 6.9%, respectively, as shown in Table 7.10 and Fig. 7.32. Equation (7.39) could be used for predicting the hysteretic curve area considering the number of cycles. It could also be used for evaluating the seismic and fatigue performance of concrete structures reinforced with BFRP bars under cyclic loading.

### 7.3.7 Comparisons on Bond Strength and Corresponding Slip Under Static and Cyclic Loading

The average bond strength between BFRP bars with the diameter of 14 mm and concrete under static loading was 11.58 MPa, as reported in [6]. The bond strength was 9.21 and 8.44 MPa when the specimens were subjected to cyclic loading, and after cyclic loading, respectively. Figure 7.33 shows the bond strength of BFRP bars with the diameter of 14 mm under static and cyclic loading, and the residual bond strength. The bond strength decreased by 20.5%, and 27.1% when the specimens were subjected to cyclic loading, and after cyclic loading when compared to that under static loading, respectively. Similar results in [70] show that the ratio of the residual bond strength to the bond strength under monotonic loading decreased with the increase of the controlling displacement. The ratio decreased to 0.08 as the controlling displacement was 4.50 mm. The results in [71] also show that the bond strength under cyclic loading is less than that under monotonic loading regardless of number of cycles, concrete types, and bar diameter.

The average slip corresponding to the bond strength between BFRP bars with a diameter of 14 mm and concrete under static loading is 3.90 mm [6]. The slip

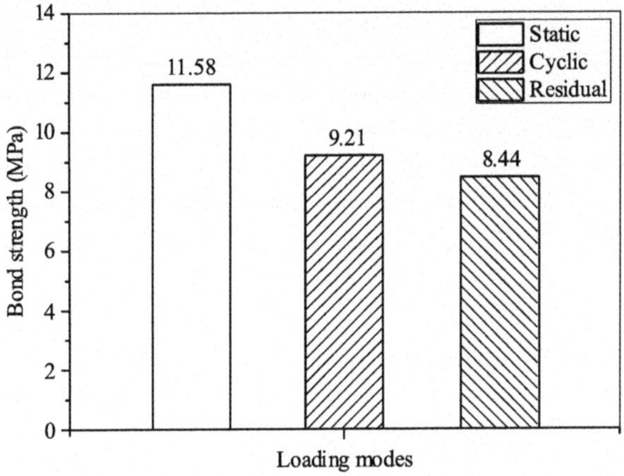


**Fig. 7.33** Comparisons on the bond strength for specimens under different loading modes

corresponding to bond strength was 2.44 and 4.48 mm when the specimens were subjected to cyclic loading, and after cyclic loading, respectively. Figure 7.34 shows the slip corresponding to the bond strength of BFRP bars with the diameter of 14 mm under static and cyclic loading, and the residual slip. The slip corresponding to bond strength decreased by 37.4% and increased by 14.9% when the specimens were subjected to cyclic loading, and after cyclic loading when compared to that under static loading, respectively. The results in [63] also show that the slip corresponding to bond strength under monotonic loading is larger than that under cyclic loading regardless of the stress level.

The bond stress-slip model under cyclic loading is also investigated in [72], which shows that the bond stress-slip envelope curve under cyclic loading coincides with the bond stress-slip curve under static loading when the number of cycles and the loaded end slip displacement are both small. A modified model is proposed in [73] based on

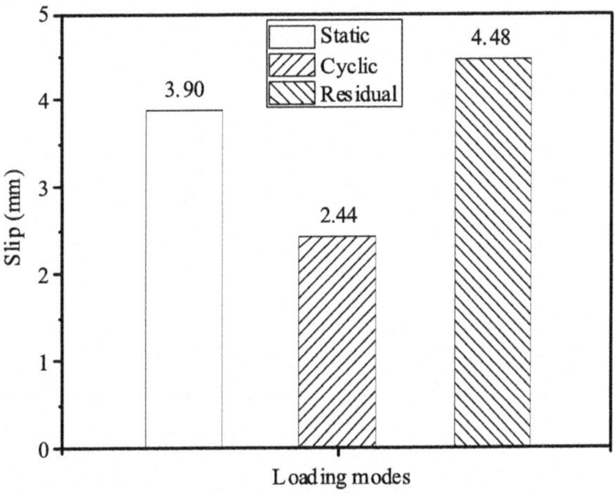


**Fig. 7.34** Comparisons on the slip corresponding to bond strength for specimens under different loading modes

the results in [72]. The results in [73] show that the bond stress-slip envelope curve under cyclic loading coincides with the bond stress-slip curve under static loading only when the loaded end slip displacement is smaller than the slip corresponding to bond strength under static loading. The loaded end slip displacement was smaller than the slip corresponding to bond strength under static loading, and both the bond strength and slip corresponding to bond strength under cyclic loading were smaller than that under static loading.

## 7.4 Summary

Investigations on the bond behavior between BFRP bars and concrete under static, dynamic, and cyclic loadings were conducted. All conclusions of the result summarized were listed below:

(1) The model for the bond stress-slip relationship between BFRP bars and concrete under static loading was proposed in consideration of the ratio of concrete cover depth to diameter of BFRP bars and concrete cubic compressive strength. The bond strength or the slip corresponding to the bond strength decreased or increased by 8.1% and 30.1%, or by 23.8% and 38.7% when the BFRP bar diameter increased from 10 mm to 14 and 20 mm, respectively. The bond strength increased by 18.8% when the concrete compressive strength increased by 64.7%. The mechanisms of the bond failure between BFRP bars and concrete were due to the crushing of concrete and the failure of BFRP surface. The failure mode was equally contributed from crushing of concrete or failure of BFRP surface and mainly from delaminated resin-rich surface for specimens.

(2) The bond behavior between BFRP bars and concrete was sensitive to strain rate. The bond strength increased log-nonlinearly as the strain rates increased, and the slip corresponding to bond strength decreased considerably as the strain rates increased. The models for the bond stress-slip relationship, bond strength, and the slip corresponding to the bond strength were proposed in consideration of the strain rate.

(3) The cyclic loading affected the bond behavior between BFRP bars and concrete. Both the bond strength and hysteretic curve area or the slip corresponding to bond strength at free end decreased or increased with the increase of the number of cycles. The models for the bond strength, slip corresponding to bond strength, and hysteretic curve area were proposed considering the number of cycles. The bond strength or the slip corresponding to bond strength after cyclic loading decreased or increased compared with the corresponding values for the first cycle and static loading, respectively.

## References


1. Elgabbas F, Vincent P, Ahmed EA et al (2016) Experimental testing of basalt-fiber-reinforced polymer bars in concrete beams. Compos B Eng 91:205–218
2. Antonopoulos CP, Triantafillou TC (2002) Analysis of FRP-strengthened RC beam-column joints. J Compos Constr 6(1):41–51
3. Shen DJ, Liu XZ, Zhou BZ et al (2021) Influence of initial cracks on the frequency of a 60-year-old reinforced-concrete box beam. Mag Concr Res 73(3):121–134
4. Shen DJ, Yang Q, Jiao Y et al (2017) Experimental investigations on reinforced concrete shear walls strengthened with basalt fiber-reinforced polymers under cyclic load. Constr Build Mater 136:217–229
5. Mahroug MEM, Ashour AF, Lam D (2014) Experimental response and code modelling of continuous concrete slabs reinforced with BFRP bars. Compos Struct 107:664–674
6. Shen DJ, Ojha B, Shi X et al (2016) Bond stress–slip relationship between basalt fiber-reinforced polymer bars and concrete using a pull-out test. J Reinf Plast Compos 35(9):747–763
7. Shen DJ, Zeng X, Zhang JY et al (2019) Behavior of RC box beam strengthened with basalt FRP using end anchorage with grooving. J Compos Mater 53(23):3307–3324
8. Shen DJ, Yang Q, Huang CB et al (2019) Tests on seismic performance of corroded reinforced concrete shear walls repaired with basalt fiber-reinforced polymers. Constr Build Mater 209:508–521
9. Sim J, Park C (2005) Characteristics of basalt fiber as a strengthening material for concrete structures. Compos B Eng 36(6–7):504–512
10. Shen DJ, Deng SC, Zhang JY et al (2015) Behavior of reinforced concrete box beam with initial cracks repaired with basalt fiber-reinforced polymer sheet. J Reinf Plast Compos 34(18):1540–1554
11. Artemenko SE (2003) Polymer composite materials made from carbon, basalt, and glass fibres. Structure and properties. Fibre Chem 35(3):226–229
12. Shen DJ, Shi X, Ji Y et al (2015) Strain rate effect on bond stress–slip relationship between basalt fiber-reinforced polymer sheet and concrete. J Reinf Plast Compos 34(7):547–563
13. Dong ZQ, Wu G, Xu B et al (2016) Bond durability of BFRP bars embedded in concrete under seawater conditions and the long-term bond strength prediction. Mater Des 92:552–562
14. Shen DJ, Shi HF, Ji Y et al (2015) Strain rate effect on effective bond length of basalt FRP sheet bonded to concrete. Constr Build Mater 82:206–218
15. Shen DJ, Ji Y, Yin FF et al (2015) Dynamic bond stress-slip relationship between basalt FRP sheet and concrete under initial static loading. J Compos Constr 19(6):04015012
16. Lü NC, Cheng J, Cheng YH (2005) A dynamic model of bridging fiber pull-out of composite materials. Mech Res Commun 32(1):1–14
17. Michal M, Keuser M (2014) Bond of steel and concrete under high loading rates. In: Proceedings of the 9th international conference on structural dynamics, Eurodyn, pp 3491–3495
18. Filippou FC (1986) A simple model for reinforcing bar anchorages under cyclic excitations. J Struct Eng 112(7):1639–1659
19. Ojha B (2015) Experimental study on bond behavior between basalt FRP bar and concrete. Hohai University, Nanjing
20. Shen DJ, Li CC, Feng ZZ et al (2019) Influence of strain rate on bond behavior of concrete members reinforced with basalt fiber-reinforced polymer rebars. Constr Build Mater 228:116755
21. Shen DJ, Wen CY, Zhu PF et al (2020) Bond behavior between basalt fiber-reinforced polymer bars and concrete under cyclic loading. Constr Build Mater 258:119518
22. Ministry of Housing and Urban Rural Development of the People's Republic of China (2019) Standard for test methods of concrete physical and mechanical properties: GB/T 50081-2019. China Architecture and Building Press, Beijing (in Chinese)
23. Okelo R, Yuan RL (2005) Bond strength of fiber reinforced polymer bars in normal strength concrete. J Compos Constr 9(3):203–213

24. ACI (2004) Guide test methods for fiber-reinforced polymers (FRPs) for reinforcing or strengthening concrete structures: ACI 440-3R-04. American Concrete Institute, Farmington Hills
25. Hao QD, Wang YL, He Z et al (2009) Bond strength of glass fiber reinforced polymer ribbed bars in normal strength concrete. Constr Build Mater 23(2):865–871
26. Achillides Z, Pilakoutas K (2004) Bond behavior of fiber reinforced polymer bars under direct pullout conditions. J Compos Constr 8(2):173–181
27. Chitsazan I, Kobraei M, Jumaat MZ et al (2010) An experimental study on the flexural behaviour of FRP RC bemas and comparison of the ultimate moment capacity with ACI. J Civil Eng Constr Technol 1(2):27–42
28. Lorenzis LD, Nanni A (2001) Characterization of FRP rods as near-surface mounted reinforcement. J Compos Constr 5(2):114–121
29. Kocaoz S, Samaranayake VA, Nanni A (2005) Tensile characterization of glass FRP bars. Compos B Eng 36(2):127–134
30. Lee JY, Kim TY, Kim TJ et al (2008) Interfacial bond strength of glass fiber reinforced polymer bars in high-strength concrete. Compos B Eng 39(2):258–270
31. El Refai A, Ammar MA, Masmoudi R (2015) Bond performance of basalt fiber-reinforced polymer bars to concrete. J Compos Constr 19(3):04014050
32. El Refai A, Abed F, Altalmas A (2015) Bond durability of basalt fiber–reinforced polymer bars embedded in concrete under direct pullout conditions. J Compos Constr 19(5):04014078
33. Katz A (1999) Bond mechanism of FRP bars to concrete. Mater Struct 32:761–768
34. Lees JM, Burgoyne CJ (1999) Transfer bond stresses generated between FRP tendons and concrete. Mag Concr Res 51(4):229–239
35. De Lorenzis L, Miller B, Nanni A (2001) Bond of FRP laminates to concrete. ACI Mater J 98(3):256–264
36. Xiao JZ, Falkner H (2007) Bond behaviour between recycled aggregate concrete and steel bars. Constr Build Mater 21(2):395–401
37. Albarwary IHM, Haido JH (2013) Bond strength of concrete with the reinforcement bars polluted with oil. Eur Sci J 9(6):255–272
38. Won JP, Park CG, Kim HH et al (2008) Effect of fibers on the bonds between FRP reinforcing bars and high-strength concrete. Compos B Eng 39(5):747–755
39. Arias JPM, Vazquez A, Escobar MM (2012) Use of sand coating to improve bonding between GFRP bars and concrete. J Compos Mater 46(18):2271–2278
40. Ozbakkaloglu T, Saatcioglu M (2009) Tensile behavior of FRP anchors in concrete. J Compos Constr 13(2):82–92
41. Tighiouart B, Benmokrane B, Gao D (1998) Investigation of bond in concrete member with fibre reinforced polymer (FRP) bars. Constr Build Mater 12(8):453–462
42. Focacci F, Nanni A, Bakis CE (2000) Local bond-slip relationship for FRP reinforcement in concrete. J Compos Constr 4(1):24–31
43. Ahmad FS, Foret G, le Roy R (2011) Bond between carbon fibre-reinforced polymer (CFRP) bars and ultra high performance fibre reinforced concrete (UHPFRC): experimental study. Constr Build Mater 25(2):479–485
44. Baena M, Torres LI, Turon A et al (2009) Experimental study of bond behaviour between concrete and FRP bars using a pull-out test. Compos B Eng 40(8):784–797
45. Quayyum S (2010) Bond behaviour of fibre reinforced polymer (FRP) bars in concrete. University of British Columbia, Vancouver
46. ACI (2006) Guide for the design and construction of concrete reinforced with FRP bars: ACI 440.1R-06. American Concrete Institute, Farmington Hills
47. Faza SS, GangaRao HVS (1991) Bending and bond behavior of concrete beams reinforced with plastic bars. Transp Res Rec 1290:185–193
48. Diab AM, Elyamany HE, Hussein MA et al (2014) Bond behavior and assessment of design ultimate bond stress of normal and high strength concrete. Alex Eng J 53(2):355–371
49. Cosenza E, Manfredi G, Realfonzo R (2002) Development length of FRP straight bars. Compos B Eng 33(7):493–504

50. Eligehausen R, Popov EP, Bertero VV (1982) Local bond stress-slip relationships of deformed bars under generalized excitations. In: Proceedings of the 7th European conference on earthquake engineering. Chicago, Illinois, United States, pp 69–80
51. Aiello MA, Leone M, Pecce M (2007) Bond performances of FRP bars-reinforced concrete. J Mater Civ Eng 19(3):205–213
52. De Lorenzis L, Rizzo A, La Tegola A (2002) A modified pull-out test for bond of near-surface mounted FRP rods in concrete. Compos B Eng 33(8):589–603
53. ACI (1992) State-of-the-art report on high-strength concrete: ACI 363R-92. American Concrete Institute, Farmington Hills
54. Bischoff PH, Perry SH (1991) Compressive behaviour of concrete at high strain rates. Mater Struct 24:425–450
55. Huang L, Chi Y, Xu LH et al (2016) Local bond performance of rebar embedded in steel-polypropylene hybrid fiber reinforced concrete under monotonic and cyclic loading. Constr Build Mater 103:77–92
56. Zhang BR, Benmokrane B, Chennouf A (2000) Prediction of tensile capacity of bond anchorages for FRP tendons. J Compos Constr 4(2):39–47
57. Huang Z, Engström B, Magnusson J (1996) Experimental investigation of the bond and anchorage behaviour of deformed bars in high strength concrete. In: Proceedings, fourth international symposium on utilization of high strength/high performance concrete
58. Fib (2013) Fib model code for concrete structures 2010: fib model code 2010. Ernst and Sohn, Berlin
59. Fujii S (1992) Bond and anchorage characteristics of reinforced concrete structures. Kyoto University, Kyoto
60. Cosenza E, Manfredi G, Realfonzo R (1997) Behavior and modeling of bond of FRP bars to concrete. J Compos Constr 1(2):40–51
61. Campione G, Cucchiara C, la Mendola L et al (2005) Steel–concrete bond in lightweight fiber reinforced concrete under monotonic and cyclic actions. Eng Struct 27(6):881–890
62. ACI (1992) Bond under cyclic loading: ACI 408.2R-92. American Concrete Institute, Farmington Hills
63. Oh BH, Kim SH (2007) Realistic models for local bond stress-slip of reinforced concrete under repeated loading. J Struct Eng 133(2):216–224
64. Lee JY, Kim KH, Kim SW et al (2014) Strength degradation of glass fiber reinforced polymer bars subjected to reversed cyclic load. Strength Mater 46:235–240
65. Zhou H, Lu J, Xv X et al (2015) Experimental study of bond-slip performance of corroded reinforced concrete under cyclic loading. Adv Mech Eng 7(3):1–10
66. Huang L, Xu L, Chi Y et al (2015) Experimental investigation on the seismic performance of steel–polypropylene hybrid fiber reinforced concrete columns. Constr Build Mater 87:16–27
67. Harajli MH (2009) Bond stress–slip model for steel bars in unconfined or steel, FRC, or FRP confined concrete under cyclic loading. J Struct Eng 135(5):509–518
68. Rehm G, Eligehausen R (1979) Bond of ribbed bars under high cycle repeated loads. ACI J 76(2):297–310
69. Balázs GL, Koch R, Harre W (1994) Verbund von stahl in beton unter betriebsbeanspruchung. Deutscher ausschuss fr stahlbeton-30, Forschungskolloquium. Sttutgart 1567–1590
70. Li J, Gao X, Zhang P (2007) Experimental investigation on the bond of reinforcing bars in high performance concrete under cyclic loading. Mater Struct 40:1027–1044
71. Pul S (2010) Loss of concrete-steel bond strength under monotonic and cyclic loading of lightweight and ordinary concretes. Iran J Sci Technol Trans B: Eng 34(B4):397–406
72. Morita S (1973) Local bond stress-slip relationship under repeated loading. In: Proceedings, IABSE symposium on the resistance and ultimate deformability of structures (Lisbon 1973). International Association for Bridge and Structural Engineering, pp 221–227
73. Tassios TP (1979) Properties of bond between concrete and steel under load cycles idealizing seismic action. In: Proceedings of AICAP-CEB symposium: Structural concrete under seismic action, Comité Euro-International du Beton (CEB), Switzerland, pp 67–122

SPRINGER NATURE

# GPSR Compliance

*The European Union's (EU) General Product Safety Regulation (GPSR) is a set of rules that requires consumer products to be safe and our obligations to ensure this.*

*If you have any concerns about our products, you can contact us on ProductSafety@springernature.com*

In case Publisher is established outside the EU, the EU authorized representative is:

Springer Nature Customer Service Center GmbH
Europaplatz 3
69115 Heidelberg, Germany

The manufacturer's authorised representative in the EU is Springer Nature Customer Service Centre GmbH, Europaplatz 3, 69115 Heidelberg, Germany. If you have any concerns regarding our products, please contact ProductSafety@springernature.com

Printed and bound by CPI Group (UK) Ltd, Croydon, CR0 4YY
01/10/2026
02225962-0014